Study Guide and Student Solutions Manual to accompany

Physics

FOR SCIENTISTS AND ENGINEERS

Volume 2 Fourth Edition

by Serway

John R. Gordon
James Madison University

Ralph McGrew
Broome Community College

Steven Van Wyk
Olympic College

Raymond Serway
James Madison University

Saunders College Publishing
Harcourt Brace College Publishers

Fort Worth Philadelphia San Diego New York Orlando Austin
San Antonio Toronto Montreal London Sydney Tokyo

Printed in the United States of America.

Gordon, McGrew, Van Wyk, Serway; Student Solutions Manual and Study Guide, Volume II, to accompany Physics for Scientists and Engineers, 4E. Serway.

ISBN 0-03-016487-7

890123 018 9876543

Preface

This is the second volume (Chapters 23-47) of the Student Solution Manual and Study Guide, and has been written to accompany the textbook **Physics for Scientists and Engineers**, Fourth Edition, by Raymond A. Serway. The purpose of this Student Solution Manual and Study Guide is to provide the students with a convenient review of the basic concepts and applications presented in the textbook, together with solutions to selected end-of-chapter problems from the textbook. This is not an attempt to rewrite the textbook in a condensed fashion. Rather, emphasis is placed upon clarifying typical troublesome points, and providing further drill for methods of problem solving.

Each chapter is divided into several parts, and every textbook chapter has a matching chapter in this book. Very often, reference is made to specific equations or figures in the textbook. Every feature of this Study Guide has been included to insure that it serves as a useful supplement to the textbook. Most chapters contain the following components:

- **Notes From Selected Chapter Sections:** This is a summary of important concepts, newly defined physical quantities, and rules governing their behavior.

- **Equations and Concepts:** This represents a review of the chapter, with emphasis on highlighting important concepts, and describing important equations and formalisms.

- **Suggestions, Skills, and Strategies:** This offers hints and strategies for solving typical problems that the student will often encounter in the course. In some sections, suggestions are made concerning mathematical skills that are necessary in the analysis of problems.

- **Review Checklist:** This is a list of topics and techniques the student should master after reading the chapter and working the assigned problems.

- **Solutions to Selected End-of-Chapter Problems:** Solutions are shown for approximately half of the odd-numbered problems from the text which illustrate the important concepts of the chapter.

An important note concerning significant figures: The answers to all end-of-chapter problems are stated to the smallest number of significant figures that were given in the problem. The sole exception to this is when integers are used, and are assumed to be of infinite precision (such as 0 or 1, in some cases). In this case, answers were given to three significant figures.

We sincerely hope that this Student Solution Manual and Study Guide will be useful to you in reviewing the material presented in the text, and in improving your ability to solve problems and score well on exams. We welcome any comments or suggestions which could help improve the content of this study guide in future editions; and we wish you success in your study.

John R. Gordon
Raymond A. Serway
James Madison University
Harrisonburg, VA 22807

Ralph McGrew
Broome Community College
Binghamton, NY 13902-1017

Steve Van Wyk
Olympic College
Bremerton, WA 98310

Acknowledgments

It is a pleasure to acknowledge the excellent work of Mrs. Linda Miller who typed the manuscript for this Fourth Edition. Her attention to detail has made our work easier. We thank Michael Rudmin for revising and creating original art work for this fourth edition. His graphics skills and technical expertise have combined to produce illustrations which complement the text. The overall appearance of the Study Guide reflects his work on the final camera-ready copy.

We also thank the professional staff at Saunders, especially Susan Pashos for managing all phases of the project. Finally, we express our appreciation to our families for their inspiration, patience, and encouragement.

Suggestions for Study

Very often we are asked "How should I study this subject, and prepare for examinations?" There is no simple answer to this question, however, we would like to offer some suggestions which may be useful to you.

1. It is essential that you understand the basic concepts and principles before attempting to solve assigned problems. This is best accomplished through a careful reading of the textbook before attending your lecture on that material, jotting down certain points which are not clear to you, taking careful notes in class, and asking questions. You should reduce memorization of material to a minimum. Memorizing sections of a text, equations, and derivations does not necessarily mean you understand the material. Perhaps the best test of your understanding of the material will be your ability to solve the problems in the text, or those given on exams.

2. Try to solve as many problems at the end of the chapter as possible. You will be able to check the accuracy of your calculations to the odd-numbered problems, since the answers to these are given at the back of the text. Furthermore, detailed solutions to approximately half of the odd-numbered problems are provided in this study guide. Many of the worked examples in the text will serve as a basis for your study.

3. The method of solving problems should be carefully planned. First, read the problem several times until you are confident you understand what is being asked. Look for key words which will help simplify the problem, and perhaps allow you to make certain assumptions. You should also pay special attention to the information provided in the problem. In many cases a simple diagram is a good starting point; and it is always a good idea to write down the given information before proceeding with a solution. After you have decided on the method you feel is appropriate for the problem, proceed with your solution. If you are having difficulty in working problems, we suggest that you again read the text and your lecture notes. It may take several readings before you are ready to solve certain problems, though the solved problems in this Study Guide should be of value to you in this regard. However, your solution to a problem does not have to look just like the one presented here. A problem can sometimes be solved in different ways, starting from different principles. If you wonder about the validity of an alternative approach, ask your instructor.

4. After reading a chapter, you should be able to define any new quantities that were introduced, and discuss the first principles that were used to derive fundamental formulas. A review is provided in each chapter of the Study Guide for this purpose, and the marginal notes in the textbook (or the index) will help you locate these topics. You should be able to correctly associate with each physical quantity the symbol used to represent that quantity (including vector notation if appropriate) and the SI unit in which the quantity is specified. Furthermore, you should be able to express each important formula or equation in a concise and accurate prose statement.

5. We suggest that you use this Study Guide to review the material covered in the text, and as a guide in preparing for exams. You should also use the Chapter Review, Notes From Selected Chapter Sections, and Equations and Concepts to focus in on any points which require further study. Remember that the main purpose of this Study Guide is to improve upon the efficiency and effectiveness of your study hours and your overall understanding of physical concepts. However, it should not be regarded as a substitute for your textbook or individual study and practice in problem solving.

Table of Contents

Chapter 23

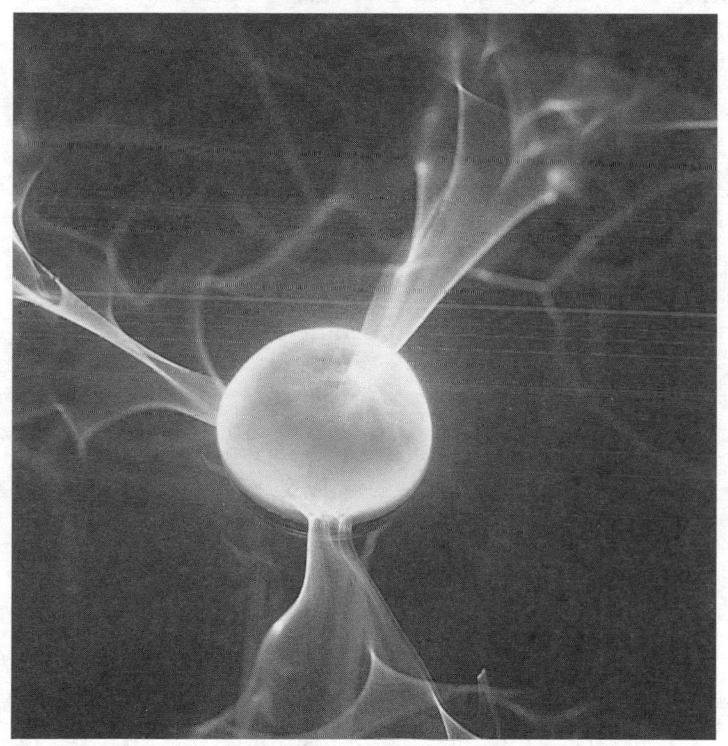

Electric Fields

Chapter 23

ELECTRIC FIELDS

INTRODUCTION

The electromagnetic force between charged particles is one of the fundamental forces of nature. In this chapter, we begin by describing some of the basic properties of electric forces. We then discuss Coulomb's law, which is the fundamental law of force between any two charged particles. The concept of an electric field associated with a charge distribution is then introduced, and its effect on other charged particles is described. The method of using Coulomb's law to calculate electric fields of a given charge distribution is discussed, and several examples are given. Then the motion of a charged particle in a uniform electric field is discussed. We conclude the chapter with a brief discussion of the oscilloscope.

NOTES FROM SELECTED CHAPTER SECTIONS

23.1 Properties of Electric Charges

Electric charge has the following important properties:

- There are two kinds of charges in nature, positive and negative, with the property that unlike charges attract one another and like charges repel one another.
- The force between charges varies as the inverse square of their separation.
- Charge is conserved.
- Charge is quantized.

23.2 Insulators and Conductors

Conductors are materials in which electric charges move freely under the influence of an electric field; *insulators* are materials that do not readily transport charge.

23.3 Coulomb's Law

Experiments show that an *electric force* has the following properties:

- It is inversely proportional to the square of the separation, r, between the two particles and is along the line joining them.

- It is proportional to the product of the magnitudes of the charges, $|q_1|$ and $|q_2|$, on the two particles.
- It is attractive if the charges are of opposite sign and repulsive if the charges have the same sign.

23.4 The Electric Field

An electric field exists at some point if a test charge at rest placed at that point experiences an electrical force.

The electric field vector **E** at some point in space is defined as the electric force **F** acting on a positive test charge placed at that point divided by the magnitude of the test charge q_0.

The total electric field due to a group of charges equals the *vector sum* of the electric fields of all the charges at some point.

23.6 Electric Field Lines

A convenient aid for visualizing electric field patterns is to draw lines pointing in the same direction as the electric field vector at any point. These lines, called electric field lines, are related to the electric field in any region of space in the following manner:

- The electric field vector **E** is *tangent* to the electric field line at each point.
- The number of lines per unit area through a surface perpendicular to the lines is proportional to the strength of the electric field in that region. Thus, **E** is large when the field lines are close together and small when they are far apart.

The rules for drawing electric field lines for any charge distribution are as follows:

- The lines must begin on positive charges and terminate on negative charges, or at infinity in the case of an excess of charge.
- The number of lines drawn leaving a positive charge or approaching a negative charge is proportional to the magnitude of the charge.
- No two field lines can cross.

EQUATIONS AND CONCEPTS

The *magnitude* of the *electrostatic force* between two stationary point charges, q_1 and q_2, separated by a distance r is given by Coulomb's law.

$$F = k_e \frac{|q_1||q_2|}{r^2} \tag{23.1}$$

The *direction* of the force on each charge is determined from the experimental observation that like charges repel each other and charges of unlike sign attract each other.

$$k_e = \frac{1}{4\pi \epsilon_0} \quad \text{where}$$

$$\epsilon_0 = 8.8542 \times 10^{-12}\ \text{C}^2/\text{N}\cdot\text{m}^2$$

In calculations, an approximate value for k_e may be used.

$$k_e = 8.99 \times 10^9 \frac{\text{N}\cdot\text{m}^2}{\text{C}^2}$$

The direction of the electrostatic force on each charge is determined from the experimental observation that like sign charges experience forces of mutual repulsion and unlike sign charges attract each other. By virtue of Newton's third law, the magnitude of the force on each of the two charges is the same regardless of the relative magnitude of the values of q_1 and q_2.

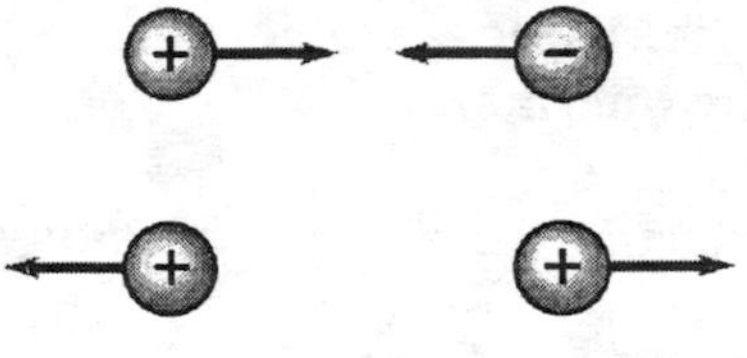

In cases where there are more than two charges present, the resultant force on any one charge is the vector sum of the forces exerted on that charge by the remaining individual charges present.

When more than two point charges are present, the *total electrostatic force* exerted on the ith charge is the vector sum of the forces exerted on that charge by the others individually. The principle of superposition applies.

$$\mathbf{F}_i = \sum_{\substack{i,j=1 \\ j\neq i}}^{N} \mathbf{F}_{ij}$$

The electric force between two charges can be expressed in vector form. $\mathbf{F}_{21}$ is the force *on* q_2 *due to* q_1 and $\hat{\mathbf{r}}$ is a unit vector directed from q_1 to q_2. Coulomb's law applies exactly only to point charges or particles.

$$\mathbf{F}_{21} = k_e \frac{q_1 q_2}{r^2} \hat{\mathbf{r}} \qquad (23.2)$$

The *electric field* at any point in space is defined as the ratio of electric force per unit charge exerted on a small positive test charge placed at the point where the field is to be determined.

$$\mathbf{E} \equiv \frac{\mathbf{F}}{q_0} \qquad (23.3)$$

This definition together with Coulomb's law leads to an expression for calculating the *electric field a distance r from a point charge, q*. In this case the unit vector $\hat{\mathbf{r}}$ is directed away from q and toward the point P where the field is to be calculated.

$$\mathbf{E} = k_e \frac{q}{r^2} \hat{\mathbf{r}} \qquad (23.4)$$

The definition of the electric field combined with Coulomb's law leads to an expression for calculating the electric field a distance, r, from a point charge, q_0. The direction of the electric field is radially outward from a positive point charge and radially inward toward a negative point charge. The superposition principle holds when the electric field at a point is due to a number of point charges.

$$\mathbf{E} = k_e \sum_i \frac{q_i}{r_i^2} \hat{\mathbf{r}}_i \qquad (23.5)$$

(vector sum)

Electric field lines are a convenient graphical representation of electric field patterns. These lines are drawn so that the electric field vector, **E**, is tangent to the electric field lines at each point. Also, the number of lines per unit area through a surface perpendicular to the lines is proportional to the strength or magnitude of the electric field over the region. In every case, electric field lines must begin on positive charges and terminate on negative charges or at infinity; the number of lines leaving or approaching a charge is proportional to the magnitude of the charge; and no two field lines can cross.

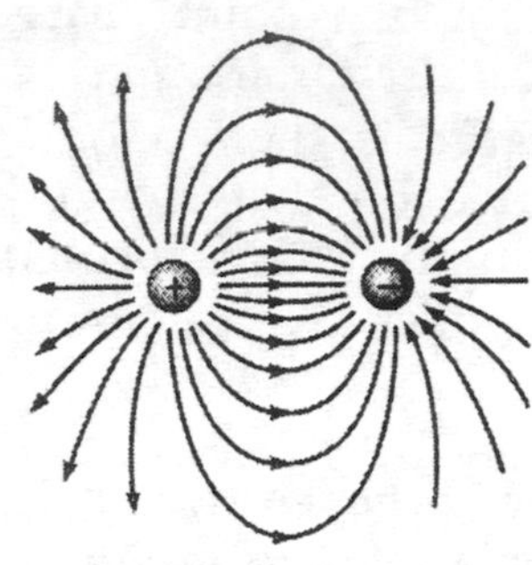

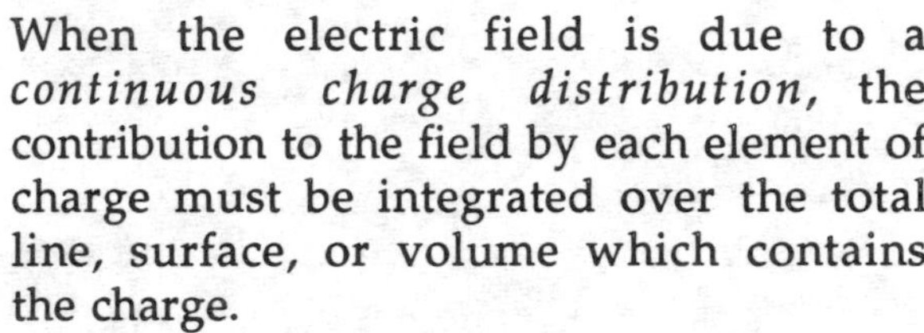
When the electric field is due to a *continuous charge distribution*, the contribution to the field by each element of charge must be integrated over the total line, surface, or volume which contains the charge.

$$\mathbf{E} = k_e \int \frac{dq}{r^2}\hat{\mathbf{r}} \qquad (23.6)$$

In order to perform the integration described above, it is convenient to represent a charge increment dq as the product of an element of length, area, or volume and the *charge density over that region*. *Note*: For those cases in which the charge is not uniformly distributed, the densities λ, σ, and ρ must be stated as functions of position.

For an element of length dx

$$dq = \lambda\, dx$$

For an element of area dA

$$dq = \sigma\, dA$$

For an element of volume dV

$$dq = \rho\, dV$$

For uniform charge distributions the volume charge density (ρ), the surface charge density (σ), and the linear charge density (λ) can be calculated from mass properties.

$$\rho \equiv \frac{Q}{V}$$

$$\sigma \equiv \frac{Q}{A}$$

$$\lambda \equiv \frac{Q}{L}$$

SUGGESTIONS, SKILLS, AND STRATEGIES

Here is a problem-solving strategy for electric forces and fields:

- Units: When performing calculations that involve the use of the Coulomb constant k_e that appears in Coulomb's law, charges must be in coulombs and distances in meters. If they are given in other units, you must convert them to SI.

- Applying Coulomb's law to point charges: It is important to use the superposition principle properly when dealing with a collection of interacting point charges. When several charges are present, the resultant force on any one of them is found by finding the individual force that every other charge exerts on it and then finding the vector sum of all these forces. The magnitude of the force that any charged object exerts on another is given by Coulomb's law, and the direction of the force is found by noting that the forces are repulsive between like charges and attractive between unlike charges.

- Calculating the electric field of point charges: Remember that the superposition principle can also be applied to electric fields, which are also vector quantities. To find the total electric field at a given point, first calculate the electric field at the point due to each individual charge. The resultant field at the point is the vector sum of the fields due to the individual charges.

- To evaluate the electric field of a continuous charge distribution, it is convenient to employ the concept of charge density. Charge density can be written in different ways: charge per unit volume, ρ; charge per unit area, σ; or charge per unit length, λ. The total charge distribution is then subdivided into a small element of volume dV, area dA, or length dx. Each element contains an increment of charge dq (equal to ρdV, σdA, or λdx). If the charge is *nonuniformly* distributed over the region, then the charge densities must be written as functions of position. For example, if the charge density along a line or long bar of length b is proportional to the distance from one end of the bar, then the linear charge density could be written as $\lambda = bx$ and the charge increment dq becomes $dq = bx\ dx$.

- Symmetry: Whenever dealing with either a distribution of point charges or a continuous charge distribution, take advantage of any symmetry in the system to simplify your calculations.

REVIEW CHECKLIST

- ▷ Describe the fundamental properties of electric charge and the nature of electrostatic forces between charged bodies.
- ▷ Use Coulomb's law to determine the net electrostatic force on a point electric charge due to a known distribution of a finite number of point charges.
- ▷ Calculate the electric field **E** (magnitude and direction) at a specified location in the vicinity of a group of point charges.
- ▷ Calculate the electric field due to a continuous charge distribution. The charge may be distributed uniformly or nonuniformly along a line, over a surface, or throughout a volume.
- ▷ Describe quantitatively the motion of a charged particle in a uniform electric field.

SOLUTIONS TO SELECTED END-OF-CHAPTER PROBLEMS

7. Three point charges are located at the corners of an equilateral triangle as in Figure P23.7. Calculate the net electric force on the 7.0-μC charge.

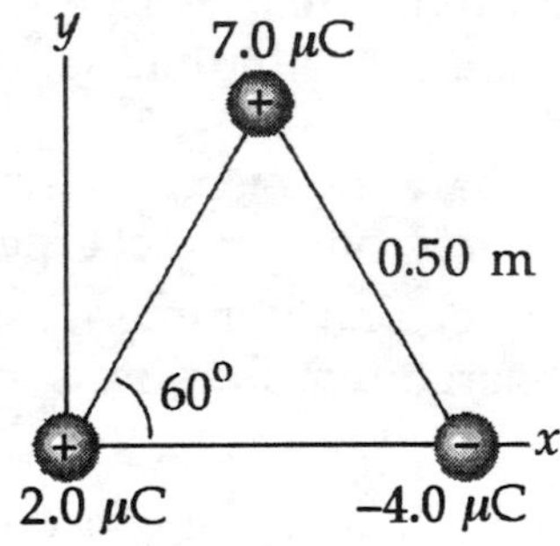

Figure P23.7

Solution These physical steps account for the force on the 7.0-μC charge: The 2.0-μC and −4.0-μC charges create electric fields at the upper corner of the triangle. These two fields add as vectors. The total **E**-field exerts the force on the third charge. We follow these steps in the solution.

The 2.0-μC charge creates a field

$$\mathbf{E} = k_e \frac{q}{r^2}\hat{\mathbf{r}} = \frac{\left(8.99\times10^9\ \mathrm{N\cdot m^2/C^2}\right)\left(2.0\times10^{-6}\ \mathrm{C}\right)}{(0.50\ \mathrm{m})^2} \text{ at } 60° \text{ above the } x \text{ direction}$$

$$= 7.19\times10^4\ \frac{\mathrm{N}}{\mathrm{C}}\ (\cos 60°\ \mathbf{i} + \sin 60°\ \mathbf{j})$$

$$= \left(3.60\times10^4\ \mathbf{i} + 6.23\times10^4\mathbf{j}\right)\frac{\mathrm{N}}{\mathrm{C}} \text{ (away from the 2.0-}\mu\text{C charge)}$$

The -4.0-μC charge creates a field

$$\mathbf{E} = \frac{\left(8.99\times10^{9}\ \mathrm{N\cdot m^2/C^2}\right)\left(-4.0\times10^{-6}\ \mathrm{C}\right)}{(0.50\ \mathrm{m})^2}\ \text{at } 120^\circ \text{ counterclockwise from the } x \text{ axis}$$

$$= 1.44\times10^{5}\ \frac{\mathrm{N}}{\mathrm{C}}\ \text{ at } 60^\circ \text{ below the } x \text{ direction}$$

$$= 1.44\times10^{5}\ \frac{\mathrm{N}}{\mathrm{C}}\ (\cos 60^\circ\ \mathbf{i} - \sin 60^\circ\ \mathbf{j})$$

$$= \left(7.19\times10^{4}\mathbf{i} - 1.25\times10^{5}\mathbf{j}\right)\ \frac{\mathrm{N}}{\mathrm{C}}\ \text{ (towards the negative charge)}$$

The total field at the location of the 7.0-μC charge is the vector sum of these two fields:

$$\mathbf{E} = \left(3.60\times10^{4} + 7.19\times10^{4}\right)\mathbf{i}\ \frac{\mathrm{N}}{\mathrm{C}} + \left(6.23\times10^{4} - 1.25\times10^{5}\right)\mathbf{j}\ \frac{\mathrm{N}}{\mathrm{C}}$$

$$= \left(1.08\times10^{5}\mathbf{i} - 6.23\times10^{4}\mathbf{j}\right)\ \frac{\mathrm{N}}{\mathrm{C}}$$

The 7.0-μC charge exerts no force on itself, but experiences a force

$$\mathbf{F} = q\mathbf{E} = \left(7\times10^{-6}\ \mathrm{C}\right)\left(1.08\times10^{5}\ \frac{\mathrm{N}}{\mathrm{C}}\mathbf{i} - 6.23\times10^{4}\ \frac{\mathrm{N}}{\mathrm{C}}\mathbf{j}\right)$$

$$= (0.755\mathbf{i} - 0.436\mathbf{j})\ \mathrm{N} \quad \Diamond$$

This is the answer, expressed as a set of components. We can also write it as:

$$\mathbf{F} = \sqrt{(0.755\ \mathrm{N})^2 + (0.436\ \mathrm{N})^2}\ \text{ at } \tan^{-1}\left(\frac{0.436\ \mathrm{N}}{0.755\ \mathrm{N}}\right) \text{ below the } x \text{ axis direction}$$

$$= 0.872\ \mathrm{N} \text{ at } 30^\circ \text{ below the } x \text{ direction}$$

---

12. Richard Feynman once said that if two persons stood at arm's length from each other and each person had 1% more electrons than protons, the force of repulsion between them would be enough to lift a "weight" equal to that of the entire Earth. Carry out an order-of-magnitude calculation to substantiate this assertion.

Solution Suppose each person has mass 70 kg. In terms of elementary particles, each consists of precisely equal numbers of protons and electrons and a nearly equal number of neutrons. The electrons comprise very little of the mass, so we find the number of protons-and-neutrons in each person:

$$70\text{ kg}\left(\frac{1\text{ u}}{1.66\times10^{-27}\text{ kg}}\right)=4.0\times10^{28}\text{ u}$$

Of these, nearly one half, 2.0×10^{28}, are protons, and 1% of this is 2.0×10^{26}, constituting a charge of $(2\times10^{26})(1.6\times10^{-19}\text{ C})=3\times10^{7}\text{ C}$. Thus, Feynman's force is

$$F=\frac{k_e q_1 q_2}{r^2}=\frac{(8.99\times10^{9}\text{ N}\cdot\text{m}^2/\text{C}^2)(3.0\times10^{7}\text{ C})^2}{(0.50\text{ m})^2}=4.0\times10^{25}\text{ N}$$

where we have used a half-meter arm's length.

The mass of the Earth in a gravitational field of magnitude 9.8 m/s^2 would weigh

$$w=mg=(6.0\times10^{24}\text{ kg})(10\text{ m}/\text{s}^2)=6.0\times10^{25}\text{ N}$$

Thus, the forces are of the same order of magnitude. ◊

---

15. What are the magnitude and direction of the electric field that will balance the weight of (a) an electron and (b) a proton? (Use the data in Table 23.1.)

Solution Require $m\mathbf{g}+q\mathbf{E}=0$ or $\mathbf{E}=-\dfrac{m\mathbf{g}}{q}$

(a) For an electron: $$\mathbf{E}=\frac{-(9.11\times10^{-31}\text{ kg})(9.8\text{ m/s}^2)(-\mathbf{j})}{-1.6\times10^{-19}\text{ C}}=(-5.57\times10^{-11}\ \mathbf{j})\text{ N/C}\quad ◊$$

(b) For a proton: $$\mathbf{E}=\frac{-(1.67\times10^{-27}\text{ kg})(9.8\text{ m/s}^2)(-\mathbf{j})}{+1.6\times10^{-19}\text{ C}}=(1.02\times10^{-7}\ \mathbf{j})\text{ N/C}\quad ◊$$

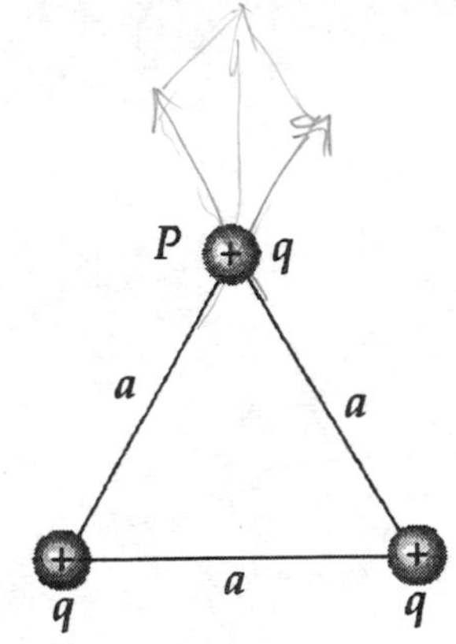

Figure P23.23

23. Three equal positive charges q are at the corners of an equilateral triangle of sides a as in Figure P23.23. (a) At what point in the plane of the charges (other than ∞) is the electric field zero? (b) What are the magnitude and direction of the electric field at P due to the two charges at the base?

Solution

(a) The electric field has the general appearance shown. It is zero at the center, where the three charges individually produce fields that cancel out. (There are also three other points, but they are hard to find.)

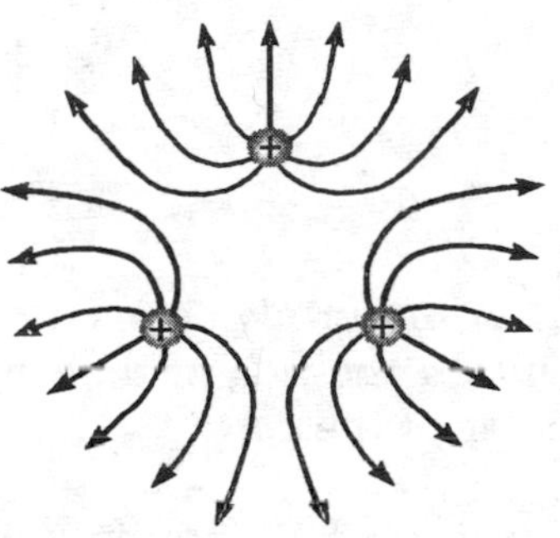

(b) You will likely need to review vector addition in Chapter Three.

$$\mathbf{E} = k_e \sum_i \frac{q_i}{r_i^2} \hat{\mathbf{r}}_i$$

$$\mathbf{E} = k_e \frac{q}{a^2} \text{ to the right and upward at } 60° + k_e \frac{q}{a^2} \text{ to the left and upward at } 60°$$

The x components, positive and negative $k_e q \cos \frac{60°}{a^2}$, add to zero, leaving

$$\mathbf{E} = k_e q \sin 60° \,\mathbf{j}/a^2 + k_e q \sin 60° \,\mathbf{j}/a^2 = 2k_e q \sin 60° \,\mathbf{j}/a^2 \quad \Diamond$$

29. A continuous line of charge lies along the x axis, extending from $x = +x_0$ to positive infinity. The line carries a uniform linear charge density λ_0. What are the magnitude and direction of the electric field at the origin?

Solution A segment of the line between x and $x + dx$ has charge $\lambda_0\, dx$ and creates an electric field at the origin of

$$d\mathbf{E} = \left(k_e \frac{dq}{r^2}\right)\hat{\mathbf{r}} = \left(k_e \lambda_0 \frac{dx}{x^2}\right)(-\mathbf{i})$$

The total field at the origin is

$$\mathbf{E} = \int_{\text{All charge}} d\mathbf{E} = \int_{x=x_0}^{\infty} k_e \lambda_0 (-\mathbf{i}) x^{-2}\, dx$$

$$\mathbf{E} = (-k_e \lambda_0 \mathbf{i}) \frac{x^{-1}}{-1}\Bigg]_{x_0}^{\infty} = (k_e \lambda_0 \mathbf{i})\left(\frac{1}{\infty} - \frac{1}{x_0}\right)$$

$$\mathbf{E} = \left(k_e \frac{\lambda_0}{x_0}\right)(-\mathbf{i}) \quad \lozenge$$

---

31. A uniformly charged ring of radius 10 cm has a total charge of 75 μC. Find the electric field on the axis of the ring at (a) 1.0 cm, (b) 5.0 cm, (c) 30 cm, and (d) 100 cm from the center of the ring.

Solution Using the result of Example 23.11:

$$E = \frac{k_e x Q}{(x^2 + a^2)^{3/2}} = \frac{(8.99 \times 10^9)(75 \times 10^{-6})x}{(x^2 + 0.10^2)^{3/2}} = \frac{6.75 \times 10^5\ x}{(x^2 + 0.010)^{3/2}}$$

Now, using your calculator,

(a) At $x = 0.010$ m $\quad \mathbf{E} = 6.7 \times 10^6 \mathbf{i}$ N/C $\lozenge$

(b) At $x = 0.050$ m $\quad \mathbf{E} = 2.4 \times 10^7 \mathbf{i}$ N/C $\lozenge$

(c) At $x = 0.30$ m $\quad \mathbf{E} = 6.4 \times 10^6 \mathbf{i}$ N/C $\lozenge$

(d) At $x = 1.0$ m $\quad \mathbf{E} = 6.7 \times 10^5 \mathbf{i}$ N/C $\lozenge$

---

39. A uniformly charged insulating rod of length 14 cm is bent into the shape of a semicircle as in Figure P23.39. If the rod has a total charge of –7.5 μC, find the magnitude and direction of the electric field at O, the center of the semicircle.

Solution Let λ be the charge per unit length.

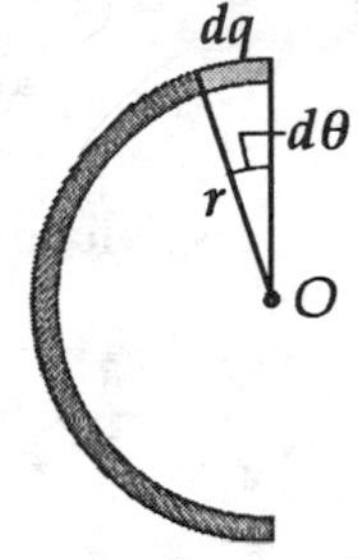

Figure P23.39 (modified)

Then, $dq = \lambda ds = \lambda r d\theta$

and $dE = \dfrac{kdq}{r^2}$

In component form, $E_y = 0$ (from symmetry)

$$dE_x = dE\cos\theta$$

Integrating, $E_x = \int dE_x = \int \dfrac{k\lambda r\cos\theta\, d\theta}{r^2}$

and $E_x = \dfrac{k\lambda}{r}\int_{-\pi/2}^{\pi/2}\cos\theta\, d\theta = \dfrac{2k\lambda}{r}$

But $Q_{total} = \lambda\ell$, where $\ell = 0.14$ m, and $r = \ell/\pi$.

Thus, $E_x = \dfrac{2\pi kQ}{\ell^2} = \dfrac{(2\pi)(8.99\times10^9\ \text{N}\cdot\text{m}^2/\text{C}^2)(-7.5\times10^{-6}\ \text{C})}{(0.14\ \text{m})^2}$

$\mathbf{E} = (-2.2\times10^7\ \text{N/C})\mathbf{i}$ ◊

---

41. A negatively charged rod of finite length has a uniform charge per unit length. Sketch the electric field lines in a plane containing the rod.

Solution

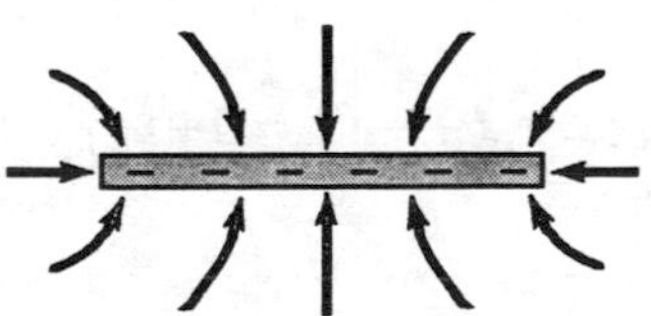

Since the rod has negative charge, field lines point inwards. Any field line points nearly toward the center of the rod at large distances, where the rod would look like just a point charge. The lines curve to reach the rod perpendicular to its surface, where they end at equally-spaced points.

---

43. Figure P23.43 shows the electric field lines for two point charges separated by a small distance. (a) Determine the ratio q_1/q_2. (b) What are the signs of q_1 and q_2?

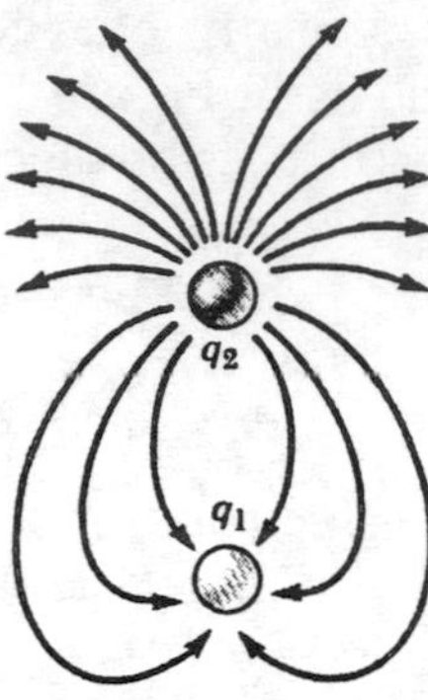

Figure P23.43

Solution 6 field lines terminate on q_1

18 field lines originate from q_2

(a) $\dfrac{q_1}{q_2} = \dfrac{-N_1}{N_2} = -\dfrac{1}{3}$ ◊

(b) q_1 is negative; q_2 is positive ◊

---

45. A proton accelerates from rest in a uniform electric field of 640 N/C. At some later time, its speed is 1.20×10^6 m/s (nonrelativistic since v is much less than the speed of light). (a) Find the acceleration of the proton. (b) How long does it take the proton to reach this speed? (c) How far has it moved in this time? (d) What is its kinetic energy at this time?

Solution

(a) $a = \dfrac{F}{m} = \dfrac{qE}{m} = \dfrac{(1.6 \times 10^{-19}\ \text{C})(640\ \text{N/C})}{1.67 \times 10^{-27}\ \text{kg}}$

$a = 6.13 \times 10^{10}\ \text{m/s}^2$ ◊

(b) $t = \dfrac{\Delta v}{a} = \dfrac{1.2 \times 10^6\ \text{m/s}}{6.13 \times 10^{10}\ \text{m/s}^2} = 19.5\ \mu\text{s}$ ◊

(c) $x = v_0 t + \frac{1}{2}at^2 = 0 + \left(\frac{1}{2}\right)(6.13 \times 10^{10}\ \text{m/s}^2)(19.5 \times 10^{-5}\ \text{s})^2$

$x = 11.7\ \text{m}$ ◊

(d) $K = \frac{1}{2}mv^2 = \frac{1}{2}(1.67 \times 10^{-27}\ \text{kg})(1.2 \times 10^6\ \text{m/s})^2$

$K = 1.20 \times 10^{-15}\ \text{J}$ ◊

47A. The electrons in a particle beam each have a kinetic energy K. What are the magnitude and direction of the electric field that will stop these electrons in a distance d?

Solution

$$K_i + W = K_f$$

To stop electrons of kinetic energy K requires a force $\mathbf{F} = q\mathbf{E}$ applied opposite $\mathbf{v}$ for a distance d. If we take the x axis in the direction of motion,

$$K + (-e)\mathbf{E} \cdot d\mathbf{i} = 0$$

$$e\mathbf{E} \cdot d\mathbf{i} = K$$

$$\mathbf{E} = (K/ed)\mathbf{i} \quad \Diamond$$

Additional Calculation: Repeat the above calculation, with $K = 1.6 \times 10^{-17}$ J, and $d = 10.0$ cm.

$$E = \frac{\frac{1}{2}mv_0^2}{ed} = \frac{1.6 \times 10^{-17}\text{ J}}{(1.6 \times 10^{-19}\text{ C})(0.1\text{ m})} = 1.00\text{ kN/C} \quad \Diamond$$

From our original solution, the field again must be in the direction of the electron's motion $\Diamond$

---

49. A proton is projected in the positive x direction into a region of a uniform electric field $\mathbf{E} = -6.00 \times 10^5\mathbf{i}$ N/C. The proton travels 7.00 cm before coming to rest. Determine (a) the acceleration of the proton, (b) its initial speed, and (c) the time it takes the proton to come to rest.

Solution

The proton has charge of $+e$. It experiences a force

$$\mathbf{F} = q\mathbf{E} = \left(1.6 \times 10^{-19}\text{ C}\right)\left(-6.00 \times 10^5\mathbf{i}\text{ N/C}\right)$$

$$= -9.60 \times 10^{-14}\mathbf{i}\text{ N}$$

(a) Its acceleration, from $\Sigma\mathbf{F} = m\mathbf{a}$, is

$$\mathbf{a} = \frac{\left(-9.60\times10^{-14}\mathbf{i}\text{ N}\right)}{1.67\times10^{-27}\text{ kg}} = -5.75\times10^{13}\text{ m/s}^2\,\mathbf{i} \quad \Diamond$$

(b) From $v^2 - v_0{}^2 = 2a(x - x_0)$,

$$0 - v_0{}^2 = 2(-5.75\times10^{13}\text{ m/s}^2)(7\times10^{-2}\text{ m})$$

$$v_0 = 2.84\times10^{6}\text{ m/s} \quad \Diamond$$

(c) The initial speed is not the average speed, so we know that the time t is not given by the constant velocity equation $t = \Delta x/v$. The velocity varies with both time and position, so

$$t = \int\frac{dx}{v(x)} \neq \frac{1}{v(x)}\int dx \neq \frac{\Delta x}{v}$$

Instead, the time is given by the constant acceleration equation $v = v_0 + at$:

$$t = \frac{(v - v_0)}{a} = \frac{(0 - 2.84\times10^{6}\text{ m/s})}{(-5.75\times10^{13}\text{ m/s}^2)} = 4.93\times10^{-8}\text{ s} \quad \Diamond$$

Alternatively, we can calculate the time from $x = x_0 + \frac{1}{2}(v_0 + v)t$, as follows:

$$t = \frac{x - x_0}{\frac{1}{2}(v_0 + v)} = \frac{7\times10^{-2}\text{ m}}{\frac{1}{2}(2.84\times10^{6}\text{ m/s})} = 4.93\times10^{-8}\text{ s}$$

---

51. A proton moves at 4.50×10^5 m/s in the horizontal direction. It enters a uniform electric field of 9.60×10^3 N/C directed vertically. Ignore any gravitational effects and find (a) the time it takes the proton to travel 5.00 cm horizontally, (b) its vertical displacement after it has traveled 5.00 cm horizontally, and (c) the horizontal and vertical components of its velocity after it has traveled 5.00 cm horizontally.

Solution E is directed along the y direction, therefore, $a_x = 0$ and $x = v_{ox}t$

(a) $t = \dfrac{x}{v_{ox}} = \dfrac{0.0500\text{ m}}{4.50\times10^{5}\text{ m/s}} = 1.11\times10^{-7}\text{ s} \quad \Diamond$

(b) $a_y = \frac{qE_y}{m} = \frac{(1.6\times10^{-19}\text{ C})(9.6\times10^{3}\text{ N/C})}{1.67\times10^{-27}\text{ kg}} = 9.20\times10^{11}\text{ m/s}^2$

$y = v_{oy}t + \frac{1}{2}a_y t^2$

$y = \left(\frac{1}{2}\right)(9.2\times10^{11}\text{ m/s}^2)(1.11\times10^{-7}\text{ s})^2 = 5.68\text{ mm}$ ◊

(c) $v_x = v_{ox} = 4.5\times10^{5}\text{ m/s}$ ◊

$v_y = v_{oy} + a_y t = 0 + (9.20\times10^{11}\text{ m/s}^2)(1.11\times10^{-7}\text{ s}) = 1.02\times10^{5}\text{ m/s}$ ◊

---

55. A charged cork ball of mass 1.00 g is suspended on a light string in the presence of a uniform electric field as in Figure P23.55. When **E** = (3.00**i** + 5.00**j**) × 10⁵ N/C, the ball is in equilibrium at $\theta = 37.0°$. Find (a) the charge on the ball and (b) the tension in the string.

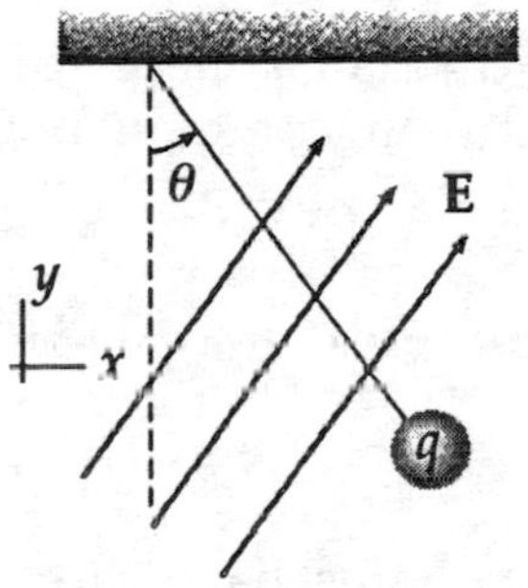

Figure P23.55

Solution

(a) $E_x = 3\times10^{5}\text{ N/C}$ and $E_y = 5\times10^{5}\text{ N/C}$

(1) $\Sigma F_x = qE_x - T\sin 37.0° = 0$

(2) $\Sigma F_y = qE_y + T\cos 37.0° - mg = 0$

Substitute T from Equation (1) into Equation (2):

$$q = \frac{mg}{\left(E_y + \frac{E_x}{\tan 37.0°}\right)} = \frac{(1.00\times10^{-3}\text{ kg})(9.80\text{ m/s}^2)}{\left(5.00 + \frac{3.00}{\tan 37.0°}\right)\times10^{5}\text{ N/C}} = 1.09\times10^{-8}\text{ C}$$ ◊

(b) From Equation (1), $T = \frac{qE_x}{\sin 37°} = 5.43\times10^{-3}\text{ N}$ ◊

Additional Calculation: Repeat the above problem, using a cork ball of mass m, an angle θ, and an electric field of $\mathbf{E} = (E_x\mathbf{i} + E_y\mathbf{j})$ N/C.

Again, we have: $-T\sin\theta + qE_x = 0$ and $+T\cos\theta + qE_y - mg = 0$

(a) Substituting $T = \dfrac{qE_x}{\sin\theta}$,

$$\frac{qE_x\cos\theta}{\sin\theta} + qE_y = mg$$

$$q = mg(E_x\cot\theta + E_y)^{-1} \quad \lozenge$$

(b)

$$T = \frac{mg\,E_x}{E_x\cos\theta + E_y\sin\theta} \quad \lozenge$$

---

57. Two small spheres of mass m are suspended from strings of length L that are connected at a common point. One sphere has charge Q; the other has charge $2Q$. Assume the angles, θ_1 and θ_2, that the strings make with the vertical are small. (a) How are θ_1 and θ_2 related? (b) Show that the distance r between the spheres is $r \cong \left(\dfrac{4k_eQ^2L}{mg}\right)^{1/3}$.

Solution

(a) The spheres have different charges, but each exerts an equal force on the other, given by $F_e = k_e(Q)(2Q)/r^2$, where r is the distance between them. Since their masses are equal, $\theta_1 = \theta_2$. $\lozenge$

(b) For equilibrium, $\Sigma F_y = 0 \quad T\cos\theta - mg = 0$

$\Sigma F_x = 0 \quad F_e - T\sin\theta = 0$

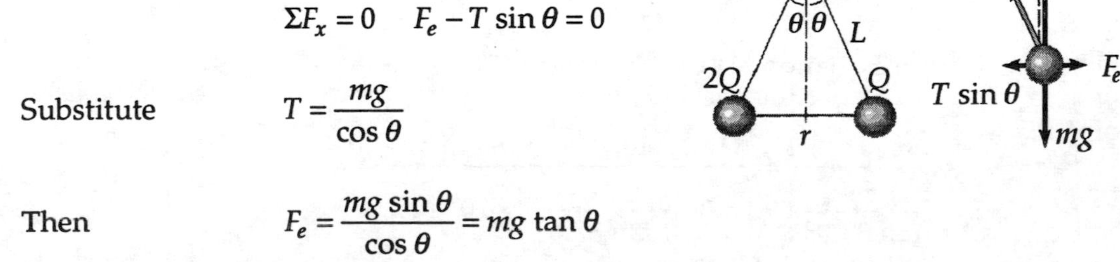

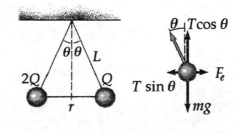

Substitute $T = \dfrac{mg}{\cos\theta}$

Then $F_e = \dfrac{mg\sin\theta}{\cos\theta} = mg\tan\theta$

For small angles, $\tan\theta \cong \sin\theta = \dfrac{r}{2L}$. Putting it all together by substitution,

$$F_e \cong mg \sin\theta \qquad \frac{k_e Q(2Q)}{r^2} \cong mg\frac{r}{2L} \qquad 4k_e Q^2 L \cong mg\, r^3$$

$$r \cong \left(\frac{4k_e Q^2 L}{mg}\right)^{1/3} \quad \Diamond$$

---

63. Three charges of equal magnitude q reside at the corners of an equilateral triangle of side length a (Fig. P23.63). (a) Find the magnitude and direction of the electric field at point P, midway between the negative charges, in terms of k_e, q, and a. (b) Where must a $-4q$ charge be placed so that any charge located at P will experience no net electric force? In part (b), let the distance between the $+q$ charge and P be 1.00 m.

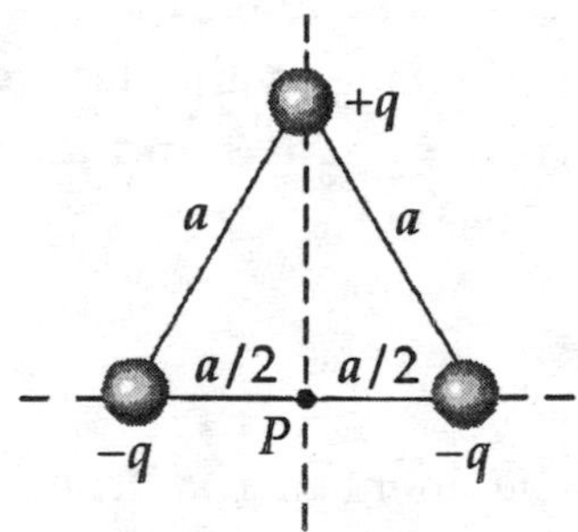

Figure P23.63

Solution

Label the charges and field contributions as shown in the figure. By symmetry $E_1 = E_2$.

(a) $E_x = -E_1 + E_2 + 0; \quad E_x = 0$

$$E_y = 0 - E_3 + 0 = \frac{k_e q}{\left(3a^2/4\right)} = -\frac{4k_e q}{3a^2}$$

$$E = \frac{4k_e q}{3a^2}(-\mathbf{j}) \quad \Diamond$$

(b) Let $q_4 = -4q$ and require that $E_4 = |\mathbf{E}|$, and let y equal location of q_4 on the axis above point P.

Then, $\quad \dfrac{k_e q_4}{y^2} = \dfrac{4k_e q}{3a^2} \quad$ or $\quad \dfrac{k_e(4q)}{y^2} = \dfrac{4k_e q}{3a^2} \quad$ and $\quad y = \sqrt{3}a.$

But given that the height of the triangle is 1 m, $\quad a = \dfrac{2}{\sqrt{3}} \quad$ and $\quad y = 2\text{ m} \quad \Diamond$

---

69. A thin rod of length L and uniform charge per unit length λ lies along the x axis as shown in Figure P23.69. (a) Show that the electric field at P, a distance y from the rod, along the perpendicular bisector has no x component and is given by $E = 2k_e\lambda \sin\theta_0/y$. (b) Using your result to part (a), show that the field of a rod of *infinite* length is $E = 2k_e\lambda/y$. (*Hint:* First calculate the field at P due to an element of length dx, which has a charge $\lambda\,dx$. Then change variables from x to θ using the facts that $x = y\tan\theta$ and $dx = y\sec^2\theta\,d\theta$ and integrate over θ.)

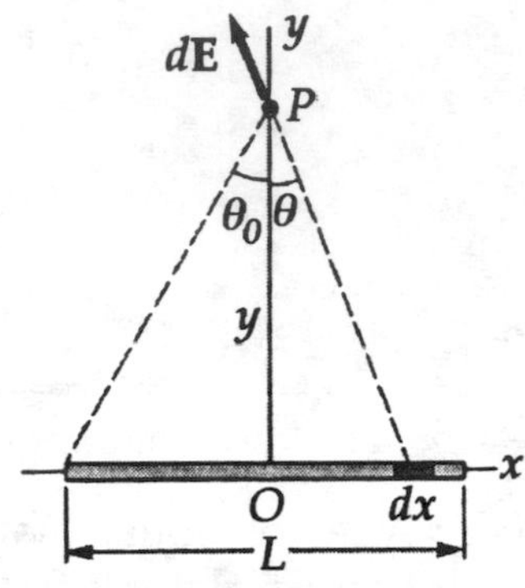

Figure P23.69 (modified)

Solution

(a) The segment of rod from x to $(x + dx)$ has a charge *of* $\lambda\,dx$, and creates an electric field

$$d\mathbf{E} = \frac{k_e dq}{r^2}\hat{\mathbf{r}} = \frac{k_e\lambda\,dx}{y^2+x^2} \quad \text{upward along the line from } dq \text{ to } P.$$

This bit of field has an x component of $$dE_x = \frac{k_e\lambda\,dx}{y^2+x^2}(-\sin\theta) = \frac{-k_e\lambda x\,dx}{(y^2+x^2)^{3/2}}$$

and a y component, $$dE_y = \frac{k_e\lambda\,dx}{y^2+x^2}(\cos\theta) = \frac{k_e\lambda y\,dx}{(y^2+x^2)^{3/2}}$$

The total field has an x component $$E_x = \int_{\text{All } q} dE_x = \int_{-L/2}^{L/2} -\frac{k_e\lambda x\,dx}{(y^2+x^2)^{3/2}}$$

To integrate change variables to θ, such that $x = y\tan\theta$. When $x = -L/2$, $\theta = -\theta_0$; when $x = L/2$, $\theta = \theta_0$. Further,

$$(y^2+x^2)^{3/2} = (y^2+y^2\tan^2\theta)^{3/2} = y^3\sec^3\theta \quad \text{and} \quad dx = y\sec^2\theta\,d\theta$$

Thus
$$E_x = \int_{-\theta_0}^{\theta_0} -\frac{k_e\lambda y(\tan\theta)y(\sec^2\theta)\,d\theta}{y^3\sec^3\theta} = -\frac{k_e\lambda}{y}\int_{-\theta_0}^{\theta_0}\sin\theta\,d\theta$$

$$= +\frac{k_e\lambda}{y}\cos\theta\Big|_{-\theta_0}^{\theta_0} = \frac{k_e\lambda}{y}(\cos\theta_0 - \cos(-\theta_0))$$

$$= \frac{k_e\lambda}{y}(\cos\theta_0 - \cos\theta_0) = 0$$

This answer has to be zero because each segment of rod on the left produces a field whose contribution cancels out that of the corresponding segment of rod on the right. But every incremental bit of charge produces at P a contribution to the field with upward y component:

$$E_y = \int_{\text{All } q} dE_y = \int_{-L/2}^{L/2} \frac{k_e\lambda y\,dx}{(y^2+x^2)^{3/2}}$$

Think of k_e, λ, and y as known constants. Now E_y is the unknown and x is the variable of integration, which we again change to θ, with $x = y\tan\theta$:

$$E_y = \int_{-\theta_0}^{\theta_0} \frac{k_e\lambda y\,(y\sec^2\theta)\,d\theta}{y^3\sec^3\theta} = \frac{k_e\lambda}{y}\int_{-\theta_0}^{\theta_0}\cos\theta\,d\theta$$

$$= \frac{k_e\lambda}{y}\sin\theta\Big]_{-\theta_0}^{\theta_0} = \frac{k_e\lambda}{y}(\sin\theta_0 - \sin(-\theta_0))$$

$$= \frac{k_e\lambda}{y}(\sin\theta_0 + \sin\theta_0) = \frac{2k_e\lambda\sin\theta_0}{y} \quad \Diamond$$

(b) As L goes to infinity, θ_0 goes to 90° and $\sin\theta_0$ becomes 1. Then the infinite amount of charge produces a finite field at P:

$$\mathbf{E} = 0\mathbf{i} + \frac{2k_e\lambda}{y}\mathbf{j} \quad \Diamond$$

---

75. A *negatively* charged particle $-q$ is placed at the center of a uniformly charged ring, where the ring has a total positive charge Q as in Example 23.11. The particle, confined to move along the x axis, is displaced a *small* distance x along the axis (where $x \ll a$) and released. Show that the particle oscillates with simple harmonic motion along the x axis with a frequency given by

$$f = \frac{1}{2\pi}\left(\frac{k_e qQ}{ma^3}\right)^{1/2}$$

Solution From Example 23.11, the electric field at points along the x axis is

$$\mathbf{E} = \frac{k_e xQ\mathbf{i}}{(x^2 + a^2)^{3/2}}$$

The field is zero at $x = 0$, so the negative charge is in equilibrium at this point. When it is displaced by an amount x that is small compared to a,

$$\Sigma\mathbf{F} = m\frac{d^2\mathbf{x}}{dt^2}$$

$$(-q)\frac{k_e xQ\mathbf{i}}{a^3} = m\frac{d^2\mathbf{x}}{dt^2}$$

The particle's acceleration is proportional to its distance (x) from the equilibrium position and is oppositely directed, so it moves in simple harmonic motion. ◊

Continuing, since $\frac{d^2\mathbf{x}}{dt^2} = -\omega^2 x$,

$$\omega = \left(\frac{k_e qQ}{ma^3}\right)^{1/2} = 2\pi f$$

and

$$f = \frac{1}{2\pi}\left(\frac{k_e qQ}{ma^3}\right)^{1/2} \quad ◊$$

Chapter 24

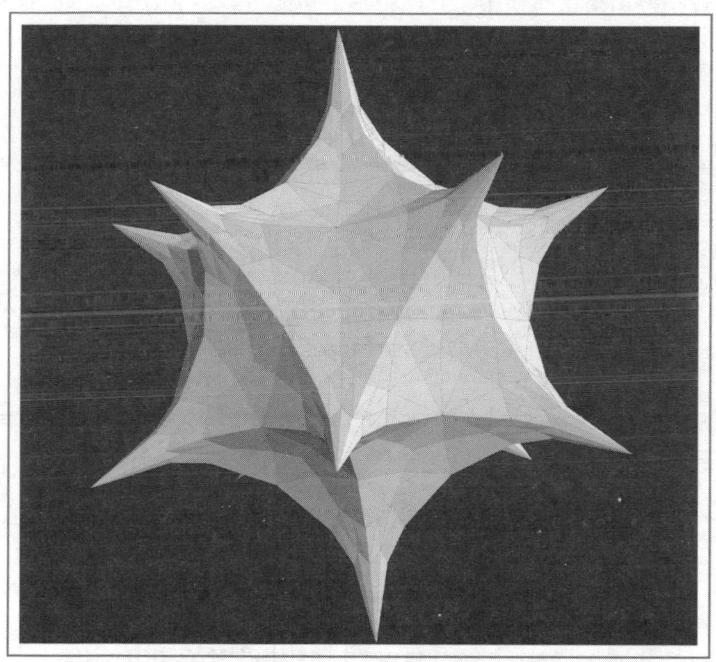

Gauss's Law

GAUSS'S LAW

INTRODUCTION

In the preceding chapter we showed how to calculate the electric field generated by a given charge distribution from Coulomb's law. This chapter describes an alternative procedure for calculating electric fields known as *Gauss's law.* This formulation is based on the fact that the fundamental electrostatic force between point charges is an inverse-square law. Although Gauss's law is a consequence of Coulomb's law, Gauss's law is much more convenient for calculating the electric field of highly symmetric charge distributions. Furthermore, Gauss's law serves as a guide for understanding more complicated problems.

NOTES FROM SELECTED CHAPTER SECTIONS

24.2 Gauss's Law

Gauss's law states that the net electric flux through a closed gaussian surface is equal to the net charge inside the surface divided by ϵ_0.

The gaussian surface should be chosen so that it has the same symmetry as the charge distribution.

24.4 Conductors in Electrostatic Equilibrium

A conductor in electrostatic equilibrium has the following properties:

- The electric field is zero everywhere inside the conductor.
- Any excess charge on an isolated conductor resides entirely on its surface.
- The electric field just outside a charged conductor is perpendicular to the conductor's surface and has a magnitude σ/ϵ_0, where σ is the charge per unit area at that point.
- On an irregularly shaped conductor, charge tends to accumulate at locations where the radius of curvature of the surface is the smallest, that is, at sharp points.

Chapter 24

EQUATIONS AND CONCEPTS

The *electric flux* is a measure of the number of electric field lines that penetrate some surface. For a *plane surface* in a *uniform field*, the flux depends on the angle between the normal to the surface and the direction of the field.

$$\Phi = EA\cos\theta \qquad (24.2)$$

In the case of a *general surface* in the region of a *nonuniform* field, the flux is calculated by integrating the normal component of the field over the surface in question.

$$\Phi = \int_{\text{surface}} \mathbf{E}\cdot d\mathbf{A} \qquad (24.3)$$

Gauss's law states that when Φ (Eq. 24.3) is evaluated over a *closed* surface (gaussian surface), the result equals *the net charge enclosed by the surface* divided by the constant ϵ_0. In this equation, the symbol $\oint$ indicates that the integral must be evaluated over a *closed* surface.

$$\oint \mathbf{E}\cdot d\mathbf{A} = \frac{q_{in}}{\epsilon_0} \qquad (24.6)$$

The field exterior to a charged sphere of radius a is equivalent to that of a point charge located at the center of the sphere. At an *interior point* of a uniformly charged sphere, the electric field is proportional to the distance from the center of the sphere.

$$E = k_e\frac{Q}{r^2} \quad \text{for} \quad r > a$$

$$E = k_e\left(\frac{Q}{a^3}\right)r \quad \text{for} \quad r < a$$

The electric field due to a uniformly charged, *nonconducting, infinite plate* is uniform everywhere.

$$E = \frac{\sigma}{2\,\epsilon_0} \qquad (24.8)$$

The electric field just outside the surface of a charged conductor in equilibrium can be expressed in terms of the surface charge density on the conductor. Just outside the conductor, the field is normal to the surface.

$$E_n = \frac{\sigma}{\epsilon_0} \qquad (24.9)$$

SUGGESTIONS, SKILLS, AND STRATEGIES

Gauss's law is a very powerful theorem which relates any charge distribution to the resulting electric field at any point in the vicinity of the charge. In this chapter you should learn how to apply Gauss's law to those cases in which the charge distribution has a sufficiently high degree of symmetry. As you review the examples presented in Section 24.3 of the text, observe how each of the following steps have been included in the application of the equation $\oint \mathbf{E} \cdot d\mathbf{A} = \frac{q}{\epsilon_0}$ to that particular situation.

- The gaussian surface should be chosen to have the *same symmetry as the charge distribution.*

- The dimensions of the surface must be such that the surface includes the point where the electric field is to be calculated.

- From the symmetry of the charge distribution, you should be able to correctly describe the direction of the electric field vector, **E**, relative to the direction of an element of surface area vector, $d\mathbf{A}$, over each region of the gaussian surface.

- From the symmetry of the charge distribution, you should also be able to identify one or more portions of the closed surface (and in some cases the entire surface) over which the magnitude of **E** remains constant.

- Write $\mathbf{E} \cdot d\mathbf{A}$ as $E\, dA \cos\theta$ and use the results of (c) and (d) to divide the surface into separate regions such that over each region:

  $E\, dA \cos\theta$ will equal 0 when $\mathbf{E} \perp d\mathbf{A}$ (as is the case over each end of the cylindrical gaussian surface in Figure 24.1) or when $\mathbf{E} = 0$ as is the case over a surface inside of a conductor.

  $E\, dA \cos\theta$ will equal $E\, dA$ when $\mathbf{E} \parallel d\mathbf{A}$ (as is the case over the curved portion of the cylindrical gaussian surface in Figure 24.1).

  $E\, dA \cos\theta = -E\, dA$ when **E** and $d\mathbf{A}$ are oppositely directed.

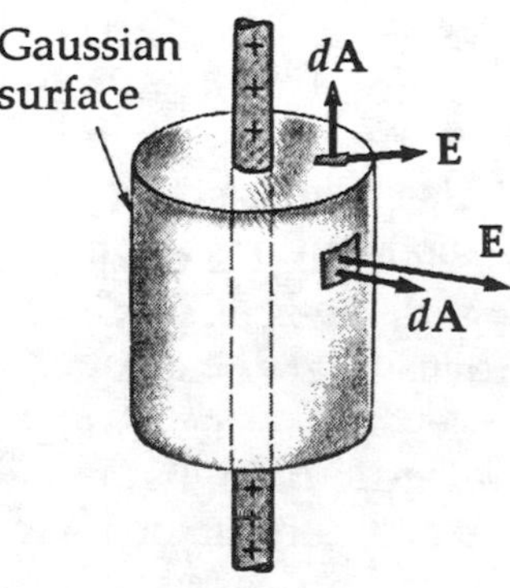

Figure 24.1

- If the gaussian surface has been chosen and subdivided so that the magnitude of **E** is *constant* over those regions where $\mathbf{E}\cdot d\mathbf{A} = E\,dA$, then over each of those regions

$$\int \mathbf{E}\cdot d\mathbf{A} = E\int dA = E\text{ (area of region)}$$

- The total charge enclosed by the gaussian surface is $q = \int dq$. It is often convenient to represent the charge distribution in terms of the charge density ($dq = \lambda dx$ for a line of charge, $dq = \sigma dA$ for a surface of charge, or $dq = \rho dV$ for a volume of charge). The integral of dq is then evaluated only over that length, area, or volume which includes that portion of the charge *inside* the gaussian surface.

- Once the left and right sides of Gauss's law have been evaluated, you can calculate the electric field on the gaussian surface, assuming the charge distribution is given in the problem. Conversely, if the electric field is known, you can calculate the charge distribution that produces the field.

REVIEW CHECKLIST

▷ Calculate the *electric* flux through a surface; in particular, find the net electric flux through a *closed* surface.

▷ Understand that a gaussian surface must be a real or imaginary *closed* surface within a conductor, a dielectric, or in space. Also remember that the net electric flux through a closed gaussian surface is equal to the net charge enclosed by the surface divided by the constant ϵ_0.

▷ Use Gauss's law to evaluate the electric field at points in the vicinity of charge distributions which exhibit spherical, cylindrical, or planar symmetry.

▷ Describe the properties which characterize an electrical conductor in electrostatic equilibrium.

SOLUTIONS TO SELECTED END-OF-CHAPTER PROBLEMS

5A. A loop of diameter d is rotated in a uniform electric field until the position of maximum electric flux is found. The flux in this position is measured to be Φ. What is the electric field strength?

Solution $\quad Flux = \mathbf{E}\cdot\mathbf{A} = \mathbf{E}\cdot\hat{\mathbf{n}}\ \pi\ (d/2)^2$

The maximum value occurs when the field is along the axis of the loop:

$$\Phi = E\,\pi\, d^2/4 \qquad\qquad E = \frac{4\Phi}{\pi d^2} \quad \Diamond$$

Additional Calculation: Solve for the electric field strength when the flux is measured to be 5.2×10^5 N·m²/C, and the loop is 40 cm in diameter.

$$\Phi = \mathbf{E}\cdot\mathbf{A} = E\pi r^2 \cos 0° = 5.2\times10^5\ \mathrm{N\cdot m^2/C} = E\pi(0.20\ \mathrm{m})^2$$

$$E = 4.1\times10^6\ \mathrm{N/C} \quad \Diamond$$

---

13. A point charge of 12 μC is placed at the center of a *spherical* shell of radius 22 cm. What is the total electric flux through (a) the surface of the shell and (b) any hemispherical surface of the shell? (c) Do the results depend on the radius? Explain.

Solution

(a) $\Phi = 4\pi k_e q = 4\pi\left(8.99\times10^9\ \frac{\mathrm{N\cdot m^2}}{\mathrm{C^2}}\right)(12\times10^{-6}\ \mathrm{C}) = 1.36\times10^6\ \frac{\mathrm{N\cdot m^2}}{\mathrm{C}} \quad \Diamond$

(b) For half shell, $\quad \Phi = \frac{1}{2}\left(1.36\times10^6\ \frac{\mathrm{N\cdot m^2}}{\mathrm{C^2}}\right) = 6.79\times10^5\ \frac{\mathrm{N\cdot m^2}}{\mathrm{C}} \quad \Diamond$

(c) $\Phi \propto EA$, and $EA = \left(\frac{k_e Q}{r^2}\right)(4\pi r^2) = 4\pi k_e Q$, so Φ is independent of r $\quad \Diamond$

---

15. The following charges are located inside a submarine: 5.0 μC, –9.0 μC, 27 μC, and –84 μC. Calculate the net electric flux through the submarine. Compare the number of electric field lines leaving the submarine with the number entering it.

Solution The total charge within the closed surface is

$$5.0\ \mu\text{C} - 9.0\ \mu\text{C} + 27\ \mu\text{C} - 84\ \mu\text{C} = -61\ \mu\text{C}$$

so the total electric flux is

$$\Phi = \frac{q}{\epsilon_0} = \frac{-61\times10^{-6}\ \text{C}}{(8.85\times10^{-12}\ \text{C}^2/\text{N}\cdot\text{m}^2)} \quad \text{or} \quad \Phi = -6.9\times10^{6}\ \text{N}\cdot\text{m}^2/\text{C} \quad \lozenge$$

The minus sign means that more lines enter the surface than leave it.

---

23. A point charge Q is located just above the center of the flat face of a hemisphere of radius R as in Figure P24.23. (a) What is the electric flux (a) through the curved surface and (b) through the flat face?

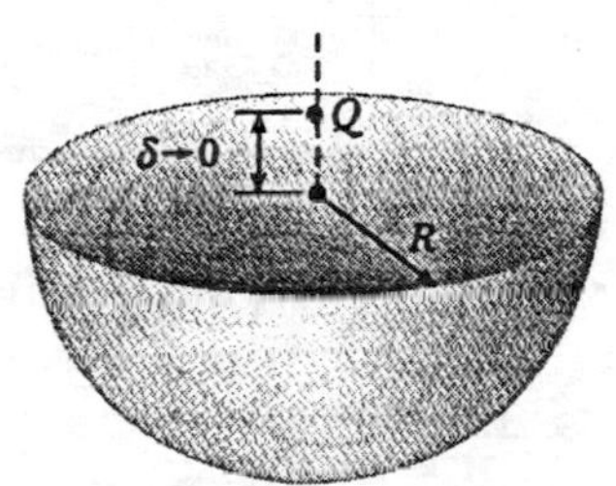

Figure P24.23

Solution

(a) With δ very small, all points on the hemisphere are nearly at distance R from the charge, so the field everywhere on the curved surface is k_eQ/R^2 radially outward. Then the flux is this field strength times the area of half a sphere:

$$\Phi_{\text{curved}} = \left(k_e\frac{Q}{R^2}\right)\left(\tfrac{1}{2}\right)\left(4\pi R^2\right) = \frac{1}{4\pi\epsilon_0}Q(2\pi) = \frac{Q}{2\epsilon_0} \quad \lozenge$$

(b) The closed surface encloses zero charge so Gauss's law gives

$$\Phi_{\text{curved}} + \Phi_{\text{flat}} = 0$$

$$\Phi_{\text{flat}} = -\Phi_{\text{curved}} = \frac{-Q}{2\epsilon_0} \quad \lozenge$$

---

25. Consider a thin spherical shell of radius 14.0 cm with a total charge of 32.0 μC distributed uniformly on its surface. Find the electric field (a) 10.0 cm and (b) 20.0 cm from the center of the charge distribution.

Solution

(a) A sphere of radius 10.0 cm encloses zero charge, so $\mathbf{E} = 0$ ◊

(b) For a gaussian sphere of radius 20.0 cm, $\oint \mathbf{E} \cdot d\,\mathbf{A} = \frac{q_{\text{in}}}{\epsilon_0}$

The field is radially outward, and $E\,4\pi r^2 = \frac{q}{\epsilon_0}$

$$E = \frac{k_e q}{r^2} = \frac{8.99 \times 10^9 \text{ N} \cdot \text{m}^2\,(32.0 \times 10^{-6}\text{ C})}{\text{C}^2\,(0.200\text{ m})^2} = 7.19 \times 10^6 \text{ N/C}$$

So $\mathbf{E} = (7.19 \times 10^6 \text{N/C})\hat{\mathbf{r}}$ ◊

---

33. A uniformly charged, straight filament 7.00 m in length has a total positive charge of 2.00 μC. An uncharged cardboard cylinder 2.00 cm in length and 10.0 cm in radius surrounds the filament at its center, with the filament as the axis of the cylinder. Using any reasonable approximations, find (a) the electric field at the surface of the cylinder and (b) the total electric flux through the cylinder.

Solution The approximation in this case is that the filament length is so large when compared to the cylinder length that the "infinite line" of charge can be assumed.

(a) $E = \frac{2k_e\lambda}{r}$ where $\lambda = \frac{2.00 \times 10^{-6}\text{ C}}{7.00\text{ m}} = 2.86 \times 10^{-7}\text{ C/m}$

so $$E = \frac{(2)(8.99 \times 10^9 \text{ N} \cdot \text{m}^2/\text{C})(2.86 \times 10^{-7}\text{ C/m})}{0.100\text{ m}} = 5.15 \times 10^4 \text{ N/C}$$ ◊

(b) $\Phi = 2\pi r L E = 2\pi r L\left(\frac{2k_e\lambda}{r}\right) = 4\pi k_e \lambda L$

so $$\Phi = 4\pi\left(8.99 \times 10^9 \frac{\text{N} \cdot \text{m}^2}{\text{C}^2}\right)(2.86 \times 10^{-7}\text{ C/m})(0.0200\text{ m}) = 6.47 \times 10^2 \frac{\text{N} \cdot \text{m}^2}{\text{C}}$$ ◊

---

35. Consider a long cylindrical charge distribution of radius R with a uniform charge density ρ. Find the electric field at distance r from the axis where $r < R$.

Solution If ρ is positive, the field must everywhere be radially outward. Choose as the gaussian surface a cylinder of length L and radius r, contained inside the charged rod. Its volume is $\pi r^2 L$ and it encloses charge $\rho \pi r^2 L$. The circular end caps have no electric flux through them; there $\mathbf{E} \cdot d\mathbf{A} = 0$. The curved surface has $\mathbf{E} \cdot d\mathbf{A} = E dA \cos 0°$, and E must be the same strength everywhere over the curved surface.

Then $\oint \mathbf{E} \cdot d\mathbf{A} = \dfrac{q}{\epsilon_0}$ becomes $E \displaystyle\int_{\text{Curved Surface}} dA = \dfrac{\rho \pi r^2 L}{\epsilon_0}$

Noting that $2\pi r L$ is the lateral surface area of the cylinder,

$$E(2\pi r)L = \frac{\rho \pi r^2 L}{\epsilon_0}$$

Thus, $\mathbf{E} = \dfrac{\rho r}{2 \epsilon_0}$ radially away from the axis ◊

---

37. A large flat sheet of charge has a charge per unit area of 9.0 $\mu C/m^2$. Find the electric field intensity just above the surface of the sheet, measured from its midpoint.

Solution For a large insulating sheet, $E = \dfrac{\sigma}{2 \epsilon_0} = 2\pi k_e \sigma$

so $$E = (2\pi)\left(8.99 \times 10^9 \ \frac{\text{N} \cdot \text{m}^2}{\text{C}^2}\right)(8.99 \times 10^{-6} \ \text{C/m}^2) = 5.09 \times 10^5 \ \text{N/C} \quad ◊$$

E will be perpendicular to the sheet.

---

39. A long, straight metal rod has a radius of 5.00 cm and a charge per unit length of 30.0 nC/m. Find the electric field (a) 3.00 cm, (b) 10.0 cm, and (c) 100 cm from the axis of the rod, where distances are measured perpendicular to the rod.

Solution (a) Inside the conductor, $E = 0$ ◊

(b) Outside the conductor, $E = 2\dfrac{k_e\lambda}{r}$

At $r = 0.10$ m,

$$E = 2\frac{(8.99\times10^{9}\ \text{N}\cdot\text{m}^2/\text{C}^2)(30.0\times10^{-9}\ \text{C/m})}{0.100\ \text{m}} = 5.40\times10^{3}\ \text{N/C} \quad ◊$$

(c) At $r = 1.0$ m,

$$E = 2\frac{(8.99\times10^{9}\ \text{N}\cdot\text{m}^2/\text{C}^2)(30.0\times10^{-9}\ \text{C/m})}{1.00\ \text{m}} = 540\ \text{N/C} \quad ◊$$

---

41. A thin conducting plate 50.0 cm on a side lies in the xy plane. If a total charge of 4.00×10^{-8} C is placed on the plate, find (a) the charge density on the plate, (b) the electric field just above the plate, and (c) the electric field just below the plate.

Solution In this problem ignore "edge" effects and assume that the total charge distributes uniformly over each side of the plate (one half the total charge on each side).

(a) $\sigma = \dfrac{q}{A} = \left(\dfrac{1}{2}\right)\dfrac{4.00\times10^{-8}\ \text{C}}{(0.500\ \text{m})^2} = 8.00\times10^{-8}\ \text{C/m}^2$ ◊

(b) Just above the plate,

$$E = \frac{\sigma}{\epsilon_0} = \frac{8.00\times10^{-8}\ \text{C/m}^2}{(8.85\times10^{-12}\ \text{C}^2/\text{N}\cdot\text{m}^2)} = 9.04\times10^{3}\ \text{N/C upward} \quad ◊$$

(c) Just below the plate, $E = \dfrac{\sigma}{\epsilon_0} = 9.04\times10^{3}$ N/C downward ◊

---

47. A long, straight wire is surrounded by a hollow metal cylinder whose axis coincides with that of the wire. The wire has a charge per unit length of λ, and the cylinder has a net charge per unit length of 2λ. From this information, use Gauss's law to find (a) the charge per unit length on the inner and outer surfaces of the cylinder and (b) the electric field outside the cylinder, a distance r from the axis.

Solution (a) Use a cylindrical gaussian surface S_1 within the conducting cylinder.

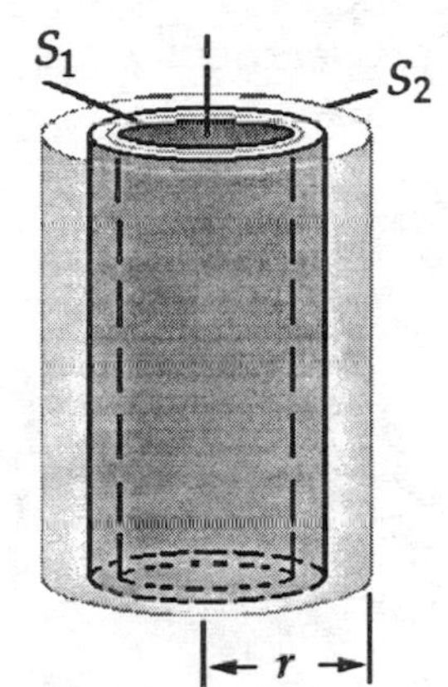

$E = 0$; Thus, $\oint E_n dA = \left(\frac{1}{\varepsilon_0}\right) q_{in} = 0$

and $\lambda_{inner} = -\lambda$ ◊

Also, $\lambda_{inner} + \lambda_{outer} = 2\lambda$

Thus, $\lambda_{outer} = 3\lambda$ ◊

(b) For a gaussian surface S_2 outside the conducting cylinder,

$$\oint E_n dA = \left(\frac{1}{\epsilon_0}\right) q_{in}, \text{ or } E(2\pi r L) = \frac{1}{\epsilon_0}(\lambda - \lambda + 3\lambda)L, \text{ and } E = \frac{3\lambda}{2\pi \epsilon_0 r} \quad ◊$$

---

49. A sphere of radius R surrounds a point charge Q, located at its center. (a) Show that the electric flux through a circular cap of half-angle θ (Fig. P24.49) is

$$\Phi = \frac{Q}{2\epsilon_0}(1 - \cos\theta)$$

What is the flux for (b) $\theta = 90°$ and (c) $\theta = 180°$?

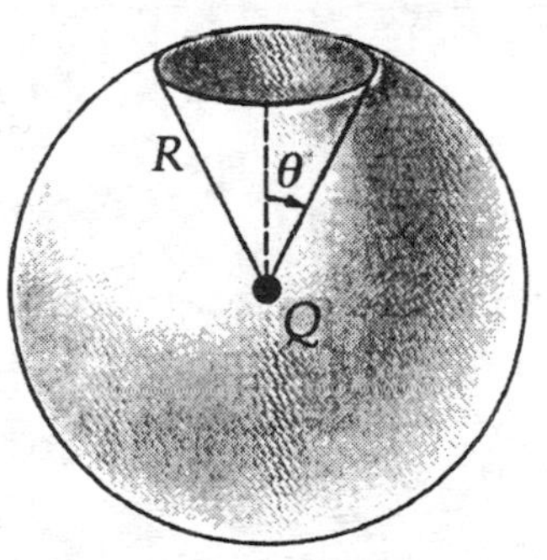

Figure P24.49

Solution

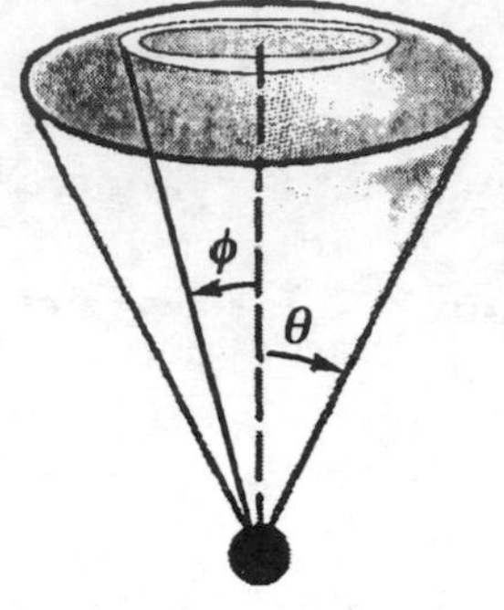

(a) The electric field of the point charge has constant strength $k_e Q/R^2$ over the cap and points radially outward. To find the area of the curved cap, we think of it as formed of rings, each of radius $r = R \sin \phi$, where ϕ ranges from 0 to θ. The width of each ring is $ds = R\, d\phi$, so its area is the product of its two perpendicular dimensions,

$$dA = (2\pi r)ds = 2\pi (R \sin \phi)(R\, d\phi)$$

The whole cap has an area of:

$$A = \int dA = \int_0^\theta 2\pi R^2 \sin \phi\, d\phi = 2\pi R^2(-\cos \phi)\Big]_0^\theta = 2\pi R^2(-\cos \theta + 1)$$

The flux through it is

$$\Phi = \int \mathbf{E} \cdot d\mathbf{A} = \int E\, dA \cos 0° = E \int dA = EA$$

$$= \frac{k_e Q}{R^2} 2\pi R^2 (1 - \cos \theta) = \left(\frac{1}{4\pi \epsilon_0}\right)(2\pi Q)(1 - \cos \theta) = \frac{Q}{2\,\epsilon_0}(1 - \cos \theta) \quad \Diamond$$

(b) For $\theta = 90°$, the cap is a hemisphere and intercepts half the flux from the charge:

$$\Phi = \frac{Q}{2\,\epsilon_0}(1 - \cos 90°) = \frac{Q}{2\,\epsilon_0} \quad \Diamond$$

(c) For $\theta = 180°$, the cap is a full sphere and all the field lines go through it:

$$\Phi = \frac{Q}{2\,\epsilon_0}(1 - \cos 180°) = \frac{Q}{\epsilon_0} \quad \Diamond$$

51. A solid insulating sphere of radius a has a uniform charge density ρ and a total charge Q. Concentric with this sphere is an uncharged, conducting hollow sphere whose inner and outer radii are b and c, as in Figure P24.50. (a) Find the magnitude of the electric field in the regions $r < a$, $a < r < b$, $b < r < c$, and $r > c$. (b) Determine the induced charge per unit area on the inner and outer surfaces of the hollow sphere.

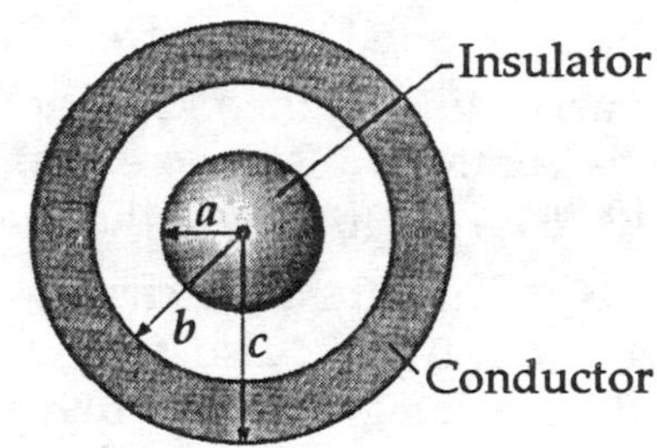

Figure P24.50

Solution

(a) Choose as the gaussian surface a concentric sphere of radius r. The electric field will be perpendicular to its surface, and will be uniform in strength over its surface.

The sphere of radius $r < a$ encloses charge $\rho\frac{4}{3}\pi r^3$,

so $\Phi = \dfrac{q}{\varepsilon_0}$ becomes $E \cdot 4\pi r^2 = \dfrac{\rho\frac{4}{3}\pi r^3}{\epsilon_0}$, and $E = \dfrac{\rho r}{3\,\epsilon_0}$ ◊

For $a < r < b$, we have $E\left(4\pi r^2\right) = \dfrac{\rho 4}{3}\pi a^3 / \epsilon_0 = \dfrac{Q}{\epsilon_0}$ and $E = \dfrac{\rho a^3}{3\,\epsilon_0\, r^2} = \dfrac{Q}{4\pi\,\epsilon_0\, r^2}$ ◊

For $b < r < c$, we must have $E = 0$ ◊ because any nonzero field would be moving charges in the metal. Free charges did move in the metal to deposit charge $-Q_b$ on its inner surface, at radius b, leaving charge $+Q_c$ on its outer surface, at radius c. Since the shell as a whole is neutral, $Q_c - Q_b = 0$.

For $r > c$, $\Phi = \dfrac{q}{\epsilon_0}$ reads $E 4\pi r^2 = \dfrac{Q + Q_c - Q_b}{\epsilon_0}$, and $E = \dfrac{Q}{4\pi\,\epsilon_0\, r^2}$ ◊

(b) For a gaussian surface of radius $b < r < c$, we have $0 = \dfrac{Q - Q_b}{\epsilon_0}$

so $Q_b = Q$ and the charge density on the inner surface is $\dfrac{-Q_b}{A} = \dfrac{-Q}{4\pi b^2}$ ◊

Then $Q_c = Q_b = Q$, and the charge density on the outer surface is $+\dfrac{Q}{4\pi c^2}$ ◊

55. A solid insulating sphere of radius R has a nonuniform charge density that varies with r according to the expression $\rho = Ar^2$, where A is a constant and $r < R$ is measured from the center of the sphere. (a) Show that the electric field outside ($r > R$) the sphere is $E = AR^5/5\epsilon_0 r^2$. (b) Show that the electric field inside ($r < R$) the sphere is $E = Ar^3/5\epsilon_0$. (*Hint:* Note that the total charge Q on the sphere is equal to the integral of $\rho\, dV$, where r extends from 0 to R; also note that the charge q within a radius $r < R$ is *less* than Q. To evaluate the integrals, note that the volume element dV for a spherical shell of radius r and thickness dr is equal to $4\pi r^2\, dr$.)

Solution

(a) We call the constant A', reserving A to denote area. The whole charge of the ball is

$$Q = \int_{\text{ball}} dQ = \int_{\text{ball}} \rho dV = \int_{r=0}^{R} A'r^2 4\pi r^2 dr = 4\pi A' \frac{r^5}{5}\bigg]_0^R = \frac{4\pi A'R^5}{5}$$

To find the electric field, consider as gaussian surface a concentric sphere of radius r outside the ball of charge: $\int \mathbf{E}\cdot d\mathbf{A} = \frac{Q}{\epsilon_0}$

Solving, $$E A \cos 0 = \frac{Q}{\epsilon_0}$$

$$E\, 4\pi r^2 = \frac{4\pi A'R^5}{5\,\epsilon_0}$$

$$E = \frac{A'R^5}{5\,\epsilon_0\, r^2} \quad \Diamond$$

(b) Let the gaussian sphere lie inside the ball of charge:

$$\int_{\text{sphere, radius } r} \mathbf{E}\cdot d\mathbf{A} = \int_{\text{sphere, radius } r} dQ/\epsilon_0$$

Solving, $E(\cos 0)\int dA = \int \frac{\rho dV}{\epsilon_0}$ becomes $E\,A = \int_0^r \frac{A'r^2\,4\pi r^2\,dr}{\epsilon_0}$,

and $E\,4\pi r^2 = \left(\frac{A'4\pi}{\epsilon_0}\right)\left(\frac{r^5}{5}\right)\Big]_0^r = \frac{A'4\pi\,r^5}{5\,\epsilon_0}$

or $E = \frac{A'r^3}{5\epsilon_0}$ ◊

---

59. Repeat the calculations for Problem 58 when both sheets have *positive* uniform charge densities σ.

Solution For each sheet, the magnitude of the field at any point is $|\mathbf{E}| = \frac{\sigma}{2\,\epsilon_0}$

(a) At point to the left of the two parallel sheets

$$\mathbf{E} = E_1(-\mathbf{i}) + E_2(-\mathbf{i}) = 2E(-\mathbf{i})$$

$$\mathbf{E} = -\frac{\sigma}{\epsilon_0}\mathbf{i} \quad ◊$$

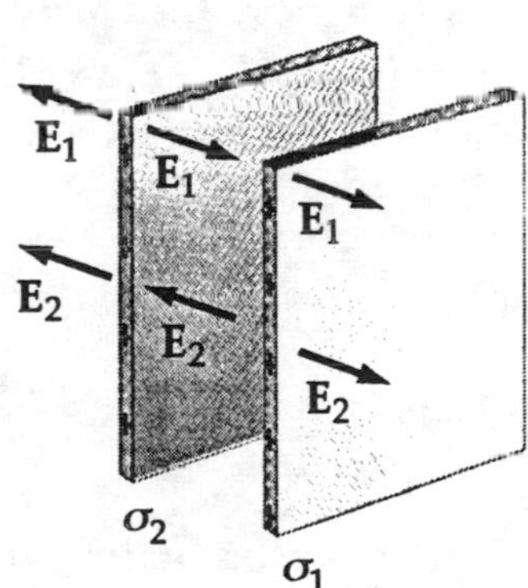

Figure P24.58
(modified)

(b) At point between the two sheets

$$\mathbf{E} = E_1\mathbf{i} + E_2(-\mathbf{i}) = 0$$

$$\mathbf{E} = 0 \quad ◊$$

(c) At point to the right of the two parallel sheets

$$\mathbf{E} = E_1\mathbf{i} + E_2\mathbf{i} = 2E\mathbf{i}$$

$$\mathbf{E} = \frac{\sigma}{\epsilon_0}\mathbf{i} \quad ◊$$

---

61. A slab of insulating material (infinite in two of its three dimensions) has a uniform positive charge density ρ. An edge view of the slab is shown in Figure P24.61. (a) Show that the electric field a distance x from its center and inside the slab is $E = \rho x / \epsilon_0$. (b) Suppose an electron of charge $-e$ and mass m is placed inside the slab. If it is released from rest at a distance x from the center, show that the electron exhibits simple harmonic motion with a frequency

$$f = \frac{1}{2\pi}\sqrt{\frac{\rho e}{m\,\epsilon_0}}$$

Figure P24.61

Solution

(a) The slab has left-to-right symmetry, so its field must be equal in strength at x and at $-x$. It points everywhere away from the central plane. Take as gaussian surface a rectangular box of thickness $2x$ and height and width L, centered on the $x = 0$ plane. The charge it contains is $\rho V = \rho 2xL^2$. The total flux leaving it is EL^2 through the right face, EL^2 through the left face, and zero through each of the other four sides.

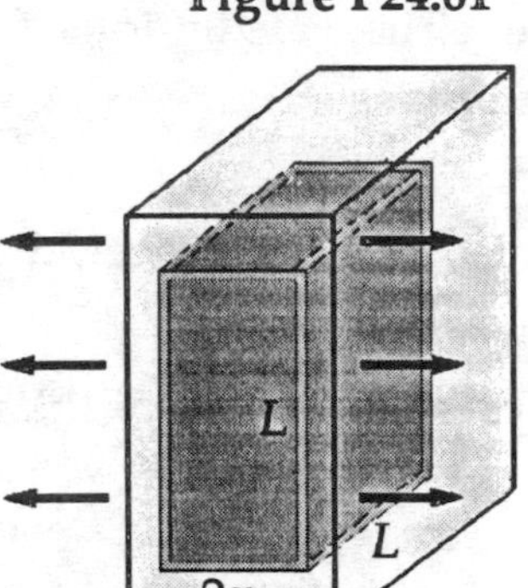

Thus Gauss's law $\int \mathbf{E}\cdot d\mathbf{A} = \dfrac{q}{\epsilon_0}$ becomes $2EL^2 = \dfrac{\rho 2xL^2}{\epsilon_0}$, so $E = \dfrac{\rho x}{\epsilon_0}$ ◊

(b) The electron experiences a force opposite to $\mathbf{E}$. When displaced to $x > 0$, it experiences a restoring force to the left. For it, $\Sigma\mathbf{F} = m\mathbf{a}$ reads $q\mathbf{E} = m\mathbf{a}$:

$$\frac{-e\rho x\mathbf{i}}{\epsilon_0} = m\mathbf{a} \qquad \mathbf{a} = -\left(\frac{e\rho}{m\,\epsilon_0}\right)x\mathbf{i} \qquad \text{or} \qquad \mathbf{a} = -\omega^2\mathbf{x}.$$

That is, its acceleration is proportional to its displacement and oppositely directed, as is required for simple harmonic motion.

Solving for the frequency, $\omega^2 = \dfrac{e\rho}{m\,\epsilon_0}$, and $f = \dfrac{\omega}{2\pi} = \dfrac{1}{2\pi}\sqrt{\dfrac{e\rho}{m\,\epsilon_0}}$ ◊

Chapter 25

Electric Potential

Chapter 25

ELECTRIC POTENTIAL

INTRODUCTION

The concept of potential energy was first introduced in Chapter 8 in connection with such conservative forces as the force of gravity and the elastic force of a spring. By using the law of energy conservation, we were often able to avoid working directly with forces when solving various mechanical problems. In this chapter we see that the energy concept is also of great value in the study of electricity. Since the electrostatic force given by Coulomb's law is conservative, electrostatic phenomena can be described conveniently in terms of an electrical potential energy. This idea enables us to define a scalar quantity called *electric potential.* Because the potential is a scalar function of position, it offers a simpler way of describing electrostatic phenomena than does the electric field.

NOTES FROM SELECTED CHAPTER SECTIONS

25.1 Potential Difference and Electric Potential

The potential difference between two points $V_B - V_A$ equals the work per unit charge that an *external agent* must perform in order to move a test charge, q_0, from point A to point B *without* a change in kinetic energy.

The potential at an arbitrary point is the work required per unit charge to bring a *positive test charge from infinity to that point.*

25.2 Potential Differences in a Uniform Electric Field

Electric field lines always point in the direction of decreasing electric potential. A positive electric charge loses electric potential energy when it moves in the direction of the electric field. An equipotential surface is any surface consisting of a continuous distribution of points having the same electric potential. Equipotential surfaces are perpendicular to electric field lines.

25.3 Electric Potential and Potential Energy Due to Point Charges

Equipotential surfaces for a point charge are a family of spheres concentric with the charge.

25.4 Obtaining E From the Electric Potential

If the electric potential (which is a scalar) is known as a function of coordinates (x, y, z), the components of the electric field (a vector quantity) can be obtained by taking the negative derivative of the potential with respect to the coordinates.

25.6 Potential of a Charged Conductor

The surface of any charged conductor in equilibrium is an equipotential surface. Also (*since the electric field is zero inside the conductor*), the potential is constant everywhere inside the conductor and equal to its value at the surface.

EQUATIONS AND CONCEPTS

The *potential difference* between two points a and b in an electric field, $\Delta V = V_b - V_a$, can be found by integrating $\mathbf{E} \cdot d\mathbf{s}$ along *any path* from a to b.

$$\Delta V = -\int_A^B \mathbf{E} \cdot d\mathbf{s} \tag{25.3}$$

If the field is *uniform*, the potential difference depends only on the displacement d in the direction parallel to **E**.

$$\Delta V = -E \int_A^B ds = -Ed \tag{25.6}$$

The *change in potential energy*, ΔU, of a charge in moving from point a to point b in an electric field depends on the sign and magnitude of the charge as well as on the change in potential, ΔV.

$$\Delta U = q_0 \, \Delta V = -q_0 Ed \tag{25.7}$$

In the special case where the electric field is uniform, the change in potential energy is proportional to the distance the charge moves along a direction parallel to the electric field. Note that a positive charge loses electric potential energy when it moves in the direction of the electric field.

The *electric potential* at a point in the vicinity of several point charges is calculated in a manner which assumes that the potential is zero at infinity.

$$V = k_e \sum_i \frac{q_i}{r_i} \qquad (25.12)$$

The *potential energy of a pair of charges* separated by a distance r represents the work required to assemble the charges from an infinite separation. Hence, the negative of the potential energy equals the minimum work required to separate them by an infinite distance. The electric potential energy associated with a system of two charged particles is positive if the two charges have the same sign, and negative if they are of opposite sign.

$$U = k_e \frac{q_1 q_2}{r_{12}} \qquad (25.13)$$

If there are more than two charged particles in the system, the *total potential energy* is found by calculating U for each pair of charges and summing the terms algebraically.

$$U = \frac{k_e}{2} \sum_{i=1}^{N} \sum_{j=1}^{N} \frac{q_i q_j}{r_{ij}}$$

when $i \neq j$

If the scalar electric potential function throughout a region of space is known, then the vector electric field can be calculated from the potential function. The *components of the electric field* in rectangular coordinates are given in terms of partial derivatives of the potential.

$$E_x = -\frac{\partial V}{\partial x}$$

$$E_y = -\frac{\partial V}{\partial y}$$

$$E_z = -\frac{\partial V}{\partial z}$$

The vector expression for the electric field can be evaluated at any point $P\ (x, y, z)$ within the region.

$$\mathbf{E} = -\mathbf{i}\frac{\partial V}{\partial x} - \mathbf{j}\frac{\partial V}{\partial y} - \mathbf{k}\frac{\partial V}{\partial z}$$

In other words, the electric field is equal to the negative gradient of the potential.

$$\mathbf{E} = -\nabla V$$

The *potential* (relative to zero at infinity) *for a continuous charge distribution* can be calculated by integrating the contribution due to a charge element dq over the line, surface, or volume which contains all the charge. Here, as in the case of a continuous charge distribution, it is convenient to represent dq in terms of the appropriate charge density.

$$V = k_e \int \frac{dq}{r} \qquad (25.19)$$

SUGGESTIONS, SKILLS, AND STRATEGIES

The vector expressions giving the electric field **E** over a region can be obtained from the scalar function which describes the electric potential, V, over the region by using a vector differential operator called the gradient operator, ∇:

$$\mathbf{E} = -\nabla V$$

This is equivalent to

$$\mathbf{E} = -\mathbf{i}\frac{\partial V}{\partial x} - \mathbf{j}\frac{\partial V}{\partial y} - \mathbf{k}\frac{\partial V}{\partial z}$$

The derivatives in the above expression are called *partial derivatives*. This means that when the derivative is taken with respect to any one coordinate, any other coordinates which appear in the expression for the potential function are treated as constants.

Since the electrostatic force is a conservative force, the work done by the electrostatic force in moving a charge q from an initial point a to a final point b depends only on the location of the two points and is independent of the path taken between a and b. When calculating potential differences using the equation

$$V_b - V_a = -\int_a^b \mathbf{E} \cdot d\mathbf{s} \qquad (25.3)$$

any path between a and b may be chosen to evaluate the integral; therefore you should select a path for which the evaluation of the "line integral" in Equation 25.3 will be as convenient as possible. For example; where A, B, and C are constants, if **E** is in the form

$$\mathbf{E} = Ax\mathbf{i} + By\mathbf{j} + Cz\mathbf{k}$$

The potential is integrated as $\int_a^b \mathbf{E}\cdot d\mathbf{s} = \int_a^b (Ax\,dx + By\,dy + Cz\,dz)$

and Equation 25.3 becomes

$$V_b - V_a = -\left[A\int_{x_a}^{x_b} x\,dx + B\int_{y_a}^{y_b} y\,dy + C\int_{z_a}^{z_b} z\,dz\right]$$

$$V_b - V_a = \frac{A}{2}\left(x_a^2 - x_b^2\right) + \frac{B}{2}\left(y_a^2 - y_b^2\right) + \frac{C}{2}\left(z_a^2 - z_b^2\right)$$

Problem-Solving Strategy

- When working problems involving electric potential, remember that potential is a *scalar quantity* (rather than a vector quantity like the electric field), so there are no components to worry about. Therefore, when using the superposition principle to evaluate the electric potential at a point due to a system of point charges, you simply take the algebraic sum of the potentials due to each charge. However, you must keep track of signs. The potential for each positive charge ($V = k_e q/r$) is positive, while the potential for each negative charge is negative.

- Just as in mechanics, only *changes* in electric potential are significant, hence the point where you choose the potential to be zero is arbitrary. When dealing with point charges or a finite-sized charge distribution, we usually define $V = 0$ to be at a point infinitely far from the charges. However, if the charge distribution itself extends to infinity, some other nearby point must be selected as the reference point.

- The electric potential at some point P due to a continuous distribution of charge can be evaluated by dividing the charge distribution into infinitesimal elements of charge dq located at a distance r from the point P. You then treat this element as a point charge, so that the potential at P due to the element is $dV = k_e\,dq/r$. The total potential at P is obtained by integrating dV over the entire charge distribution. In performing the integration for most problems, it is necessary to express dq and r in terms of a single variable. In order to simplify the integration, it is important to give careful consideration of the geometry involved in the problem.

- Another method that can be used to obtain the potential due to a finite continuous charge distribution is to start with the definition of the potential difference given by Equation 25.3. If $\mathbf{E}$ is known or can be obtained easily (say from Gauss's law), then the line integral of $\mathbf{E}\cdot d\mathbf{s}$ can be evaluated. An example of this method is given in Example 25.11.

- Once you know the electric potential at a point, it is possible to obtain the electric field at that point by remembering that *the electric field is equal to the negative of the derivative of the potential with respect to some coordinate.*

- Until now, we have been using the symbols V to represent the electric potential at some point and ΔV to represent the potential difference between two points. In descriptions of electrical devices, however, it is common practice to use the symbol V to represent the potential difference across the device. Hence, in this book both symbols will be used to denote potential differences, depending on the circumstances.

  In practice, a variety of phrases are used to describe the potential difference between two points, the most common being "voltage." A voltage *applied* to a device or *across* a device has the same meaning as the potential difference across the device. For example, if we say that the voltage across a certain capacitor is 12 volts, we mean that the potential difference between the capacitor's plates is 12 volts.

REVIEW CHECKLIST

▷ Understand that each point in the vicinity of a charge distribution can be characterized by a scalar quantity called the electric potential, V. The values of this potential function over the region (a scalar field) are related to the values of the electrostatic field over the region (a vector field).

▷ Calculate the electric potential difference between any two points in a uniform *electric field,* and the electric potential difference between any two points in the vicinity of a *group of point charges.*

▷ Calculate the electric *potential energy* associated with a group of point charges.

▷ Calculate the electric potential due to *continuous charge distributions* of reasonable symmetry—such as a charged ring, sphere, line, or disk.

▷ Obtain an expression for the electric field (a *vector* quantity) over a region of space if the scalar electric potential function for the region is known.

▷ Calculate the work done by an external force in moving a charge q between any two points in an electric field when (a) an expression giving the field as a function of position is known, or when (b) the charge distribution (either point charges or a continuous distribution of charge) giving rise to the field is known.

SOLUTIONS TO SELECTED END-OF-CHAPTER PROBLEMS

3. (a) Calculate the speed of a proton that is accelerated from rest through a potential difference of 120 V. (b) Calculate the speed of an electron that is accelerated through the same potential difference.

Solution

(a) Energy is conserved as the proton moves from high to low potential; we take it as from 120 V to ground:

$$K_i + U_i + \Delta K_{nc} = K_f + U_f$$

(Review this work-energy theory of motion from Chapter 8 to the full extent necessary for you.)

$$0 + qV + 0 = \tfrac{1}{2}mv^2 + 0$$

$$(1.6\times10^{-19}\text{ C})(120\text{ V})\left(\frac{1\text{ J}}{1\text{ V}\cdot\text{C}}\right) = \tfrac{1}{2}(1.67\times10^{-27}\text{ kg})v^2$$

$$v = 1.52\times10^{5}\text{ m/s} \quad \Diamond$$

(b) The electron will gain speed in moving the other way, from $V_i = 0$ to $V_f = 120$ V:

$$K_i + U_i + \Delta K_{nc} = K_f + U_f$$

$$0 + 0 + 0 = \tfrac{1}{2}mv^2 + qV$$

$$0 = \tfrac{1}{2}(9.11\times10^{-31}\text{ kg})v^2 - (1.6\times10^{-19}\text{ C})(120\text{ J/C})$$

Note how the negative charge of the electron means it can have positive kinetic energy with zero total energy.

$$v = 6.49\times10^{6}\text{ m/s} \quad \Diamond$$

This is less than one tenth the speed of light, so we need not use the relativistic kinetic energy formula.

9. A positron has the same mass as an electron. When a positron is accelerated from rest between two points at a fixed potential difference, it acquires a speed that is 30% of the speed of light. What speed is achieved by a proton accelerated from rest between the same two points?

Solution $\Delta K = qV$ and $q_{e+} = q_H$

Thus, $\Delta K_{e^+} = \Delta K_H$ and $\left(\frac{1}{2}mv^2\right)_{e^+} = \left(\frac{1}{2}mv^2\right)_H$

Therefore, $v_H = v_{e^+}\sqrt{\frac{m_{e^+}}{m_H}} = (9.00\times10^7 \text{ m/s})\sqrt{\frac{9.11\times10^{-31}\text{ kg}}{1.67\times10^{-27}\text{ kg}}} = 2.10\times10^6 \text{ m/s}$ ◊

---

15. An electron moving parallel to the x axis has an initial speed of 3.7×10^6 m/s at the origin. Its speed is reduced to 1.4×10^5 m/s at the point $x = 2.0$ cm. Calculate the potential difference between the origin and this point. Which point is at the higher potential?

Solution Use the work-energy theorem to equate the energy of the electron at $x = 0$ and at $x = 2.0$ cm. The unknown will be the difference in potential $V_2 - V_0$.

$$K_i + U_i + \Delta K_{nc} = K_f + U_f$$

$$\tfrac{1}{2}mv_i^2 + qV_0 + 0 = \tfrac{1}{2}mv_f^2 + qV_2$$

Solving, $\frac{1}{2}m\left(v_i^2 - v_f^2\right) = q(V_2 - V_0)$, or $V_2 - V_0 = \frac{m(v_i^2 - v_f^2)}{2q}$

Noting that the electron's charge is negative, and evaluating the potential,

$$V_2 - V_0 = \frac{(9.11\times10^{-31}\text{ kg})\left((3.7\times10^6\text{ m/s})^2 - (1.4\times10^5\text{ m/s})^2\right)}{2(-1.6\times10^{-19}\text{ C})} = -39\text{ V} \quad ◊$$

The negative sign means that the 2.0-cm location is lower in potential than the origin. A positive charge would slow in free flight toward higher voltage, but the negative electron slows as it moves into lower potential. The 2.0-cm distance was unnecessary information for this problem. *If* the field were uniform, we could find it from $\Delta V = -Ed$.

---

23. At a distance r away from a point charge q, the electric potential is $V = 400$ V and the magnitude of the electric field is $E = 150$ N/C. Determine the values of q and r.

Solution We have $V = \frac{k_e q}{r}$ and $|\mathbf{E}| = \left|\frac{k_e q}{r^2}\hat{\mathbf{r}}\right| = \frac{k_e q}{r^2}$

We can substitute $q = \frac{Vr}{k_e}$ into our second equation, to get $E = \frac{k_e}{r^2}\frac{Vr}{k_e} = \frac{V}{r}$

Solving for r and q,
$$r = \frac{V}{E} = \left(\frac{400\text{ V}}{150\text{ N/C}}\right)\left(\frac{1\text{ J}}{\text{V}\cdot\text{C}}\right)\left(\frac{1\text{ N}\cdot\text{m}}{\text{J}}\right) = 2.67\text{ m} \quad \lozenge$$

and
$$q = \frac{Vr}{k_e} = \frac{(400\text{ V})(2.67\text{ m})}{8.99\times 10^9\text{ N}\cdot\text{m}^2/\text{C}^2}\left(\frac{1\text{ J}}{\text{V}\cdot\text{C}}\right)\left(\frac{1\text{ N}\cdot\text{m}}{\text{J}}\right) = 1.19\times 10^{-7}\text{ C} \quad \lozenge$$

You should check the answers by showing that $k_e q/r^2$ works out to be 150 N/C.

---

27. The three charges in Figure P25.27 are at the vertices of an isosceles triangle. Calculate the electric potential at the midpoint of the base, taking $q = 7.00\ \mu$C.

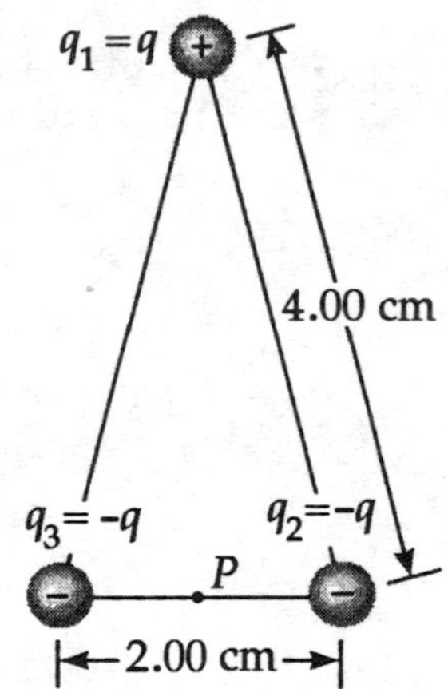

Figure P25.27

Solution Let $q_1 = q$; and $q_2 = q_3 = -q$

The charges are at distances

$$r_1 = \sqrt{(0.0400\text{ m})^2 - (0.0100\text{ m})^2} = 3.87\times 10^{-2}\text{ m}$$

and $r_2 = r_3 = 0.0100$ m

The voltage at point p is $V_p = \frac{k_e q_1}{r_1} + \frac{k_e q_2}{r_2} + \frac{k_e q_3}{r_3}$,

so
$$V_p = \left(7\times 10^{-6}\text{ C}\right)\left(8.99\times 10^9\ \frac{\text{N}\cdot\text{m}^2}{\text{C}^2}\right)\left(\frac{1}{0.0387} - \frac{1}{0.0100} - \frac{1}{0.0100}\right)\text{m}^{-1}$$

and $V_p = -11.0\times 10^6\text{ V}$ $\lozenge$

Additional Calculation: Calculate the electric field vector at the same point due to the three charges.

The separate fields of the two negative charges are in opposite directions and add to zero:

$$\mathbf{E}_p = \frac{k_e q_1}{r_1^2}\hat{\mathbf{r}}_1 = \frac{(8.99\times10^9\ \mathrm{N\cdot m^2/C^2})(7\times10^{-6}\ \mathrm{C})}{((0.0400\ \mathrm{m})^2-(0.0100\ \mathrm{m})^2)}\ \text{down} = (42.0\times10^6\ \mathrm{N/C})(-\mathbf{j})$$

---

31. The Bohr model of the hydrogen atom states that the electron can exist only in certain allowed orbits. The radius of each Bohr orbit is $r = n^2(0.0529\ \mathrm{nm})$ where $n = 1, 2, 3, \ldots$. Calculate the electric potential energy of a hydrogen atom when the electron is in the (a) first allowed orbit, $n = 1$, (b) second allowed orbit, $n = 2$, and (c) when the electron has escaped from the atom, $r = \infty$. Express your answers in electron volts.

Solution The electric potential energy is given by $U = k_e \frac{q_1 q_2}{r}$

(a) For the first allowed Bohr orbit,

$$U = (8.99\times10^9\ \mathrm{N\cdot m^2/C^2})\frac{(-1.6\times10^{-19}\ \mathrm{C})(1.6\times10^{-19}\ \mathrm{C})}{(0.0529\times10^{-9}\ \mathrm{m})}$$

$$U = -4.37\times10^{-18}\ \mathrm{J} = \frac{-4.37\times10^{-18}\ \mathrm{J}}{1.6\times10^{-19}\ \mathrm{J/eV}} = -27.3\ \mathrm{eV}\quad\Diamond$$

(b) For the second allowed orbit,

$$U = (8.99\times10^9\ \mathrm{N\cdot m^2/C^2})\frac{(-1.6\times10^{-19}\ \mathrm{C})(1.6\times10^{-19}\ \mathrm{C})}{2^2(0.0529\times10^{-9}\ \mathrm{m})}$$

$$U = -1.092\times10^{-18}\ \mathrm{J} = -6.81\ \mathrm{eV}\quad\Diamond$$

(c) When the electron is at $r = \infty$,

$$U = (8.99\times10^9\ \mathrm{N\cdot m^2/C^2})\frac{(-1.6\times10^{-19}\ \mathrm{C})(1.6\times10^{-19}\ \mathrm{C})}{\infty\ \mathrm{m}} = 0\ \mathrm{J}\quad\Diamond$$

---

32. Calculate the energy required to assemble the array of charges shown in Figure P25.32, where $a = 0.20$ m, $b = 0.40$ m, and $q = 6.0\ \mu$C.

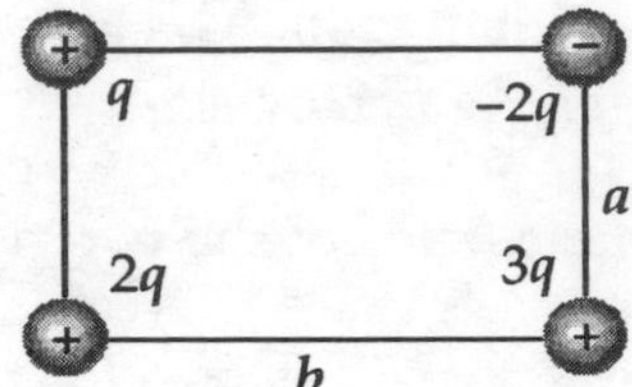

Figure P25.32

Solution Imagine starting with all charges very far away, so that they exert negligible forces on each other. Zero work is required to bring up the charge q to its final location. Then to bring up the $-2q$ charge requires work to change the potential energy to

$$U = \frac{k_e q_1 q_2}{r} = \frac{k_e q(-2q)}{b}$$

To bring up the $2q$ charge close to the other two requires extra work input

$$\frac{k_e q(2q)}{a} + \frac{k_e(-2q)(2q)}{\sqrt{a^2+b^2}}$$

Then to bring up the $3q$ charge requires still more work in amount

$$\frac{k_e q(3q)}{\sqrt{a^2+b^2}} + \frac{k_e(-2q)(3q)}{a} + \frac{k_e(2q)(3q)}{b}$$

The total energy required to assemble the array is

$$U = k_e q^2\left(-\frac{2}{b} + \frac{2}{a} - \frac{4}{\sqrt{a^2+b^2}} + \frac{3}{\sqrt{a^2+b^2}} - \frac{6}{a} + \frac{6}{b}\right)$$

$$= k_e q^2\left(\frac{4}{b} - \frac{4}{a} - \frac{1}{\sqrt{a^2+b^2}}\right)$$

$$= (8.99\times10^9\ \mathrm{N\cdot m^2/C^2})(6.0\times10^{-6}\ \mathrm{C})^2\left(\frac{4}{0.40\ \mathrm{m}} - \frac{4}{0.20\ \mathrm{m}} - \frac{1}{0.447\ \mathrm{m}}\right)$$

Evaluating, $U = -3.96$ J ◊

33. Show that the amount of work required to assemble four identical point charges of magnitude Q at the corners of a square of side s is $5.41k_eQ^2/s$.

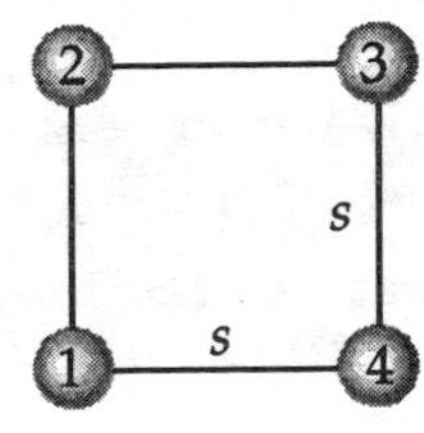

Solution Each charge creates potential on its own, and so injects energy into each other charge. We must add up $U = q\,V$ contributions for all pairs:

$$U = q_1V_2 + q_1V_3 + q_1V_4 + q_2V_3 + q_2V_4 + q_3V_4$$

$$U = \frac{q_1k_eq_2}{r_{12}} + \frac{q_1k_eq_3}{r_{13}} + \frac{q_1k_eq_4}{r_{14}} + \frac{q_2k_eq_3}{r_{23}} + \frac{q_2k_eq_4}{r_{24}} + \frac{q_3k_eq_4}{r_{34}}$$

$$U = \frac{Qk_eQ}{s} + \frac{Qk_eQ}{\sqrt{2}s} + \frac{Qk_eQ}{s} + \frac{Qk_eQ}{s} + \frac{Qk_eQ}{\sqrt{2}s} + \frac{Qk_eQ}{s}$$

$$U = \frac{k_eQ^2}{s}\left(4 + \frac{2}{\sqrt{2}}\right) = 5.414k_e\frac{Q^2}{s} \quad \Diamond$$

---

39. Over a certain region of space, the electric potential is $V = 5x - 3x^2y + 2yz^2$. Find the expressions for the x, y, and z components of the electric field over this region. What is the magnitude of the field at the point P, which has coordinates (1, 0, −2) m?

Solution $V = 5x - 3x^2y + 2yz^2$ Evaluate E at (1, 0, −2) m.

$$E_x = -\frac{\partial V}{\partial x} = -5 + 6xy = -5 + 6(1)(0) = -5$$

$$E_y = \frac{-\partial V}{\partial y} = 3x^2 - 2z^2 = 3(1)^2 - 2(-2)^2 = -5$$

$$E_z = \frac{-\partial V}{\partial z} = -4yz = -4(0)(-2) = 0$$

$$E = \sqrt{E_x{}^2 + E_y{}^2 + E_z{}^2} = \sqrt{(-5)^2 + (-5)^2 + 0^2} = 7.08\ \text{N/C} \quad \Diamond$$

---

47. A rod of length L (Fig. P25.47) lies along the x axis with its left end at the origin and has a nonuniform charge density $\lambda = \alpha x$ (where α is a positive constant). (a) What are the units of the α? (b) Calculate the electric potential at A.

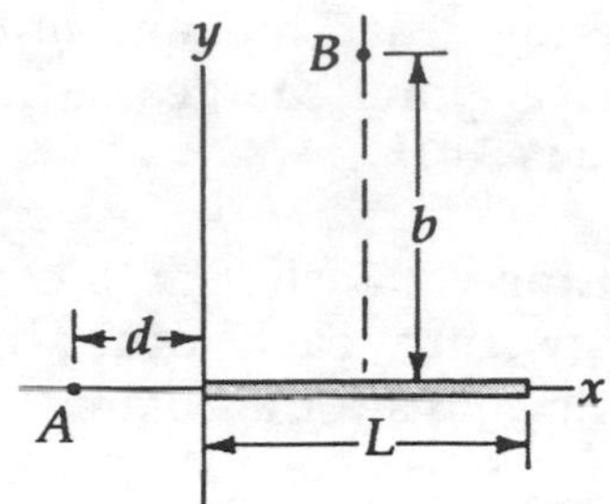

Figure P25.47

Solution

(a) As a linear charge density, λ has units of C/m.
So $\alpha = \lambda/x$ must have units of C/m^2. ◊

b) Consider one bit of the rod at location x and of length dx. The amount of charge on it is $\lambda dx = (\alpha x)dx$. Its distance from A is $d + x$, so the bit of electric potential it creates at A is

$$dV = k_e \frac{dq}{r} = k_e \alpha x \frac{dx}{(d+x)}$$

We must integrate all these contributions for the whole rod, from $x = 0$ to $x = L$:

$$V = \int_{\text{all } q} dV = \int_0^L \frac{k_e \alpha x\, dx}{d+x}$$

To perform the integral, make a change of variables to

$$u = d + x, \quad du = dx, \quad u \text{ (at } x = 0) = d, \quad \text{and} \quad u \text{ (at } x = L) = d + L$$

$$V = \int_{u=d}^{d+L} \frac{k_e \alpha (u-d)du}{u} = k_e \alpha \int_d^{d+L} du - k_e \alpha d \int_d^{d+L} \left(\frac{1}{u}\right) du$$

[Keep track of symbols: the <u>unknown</u> is V. The k_e, α, d, and L are <u>known</u> and constant. And x and u are variables, and will not appear in the answer.]

$$V = k_e \alpha u \Big|_d^{d+L} - k_e \alpha d \ln u \Big|_d^{d+L} = k_e \alpha (d + L - d) - k_e \alpha d (\ln(d+L) - \ln d)$$

$$V = k_e \alpha L - k_e \alpha d \ln\left(\frac{d+L}{d}\right) \quad ◊$$

We have the answer when the unknown is expressed in terms of the d, L, and α mentioned in the problem and the universal constant k_e.

51. How many electrons should be removed from an initially uncharged spherical conductor of radius 0.300 m to produce a potential of 7.50 kV at the surface?

Solution $V = \frac{k_e Q}{R}$, but $Q = Ne$

Thus, $N = \frac{VR}{k_e e}$ where N = number of electrons removed.

$$N = \frac{(7.50 \times 10^3 \text{ V})(0.300 \text{ m})}{\left(8.99 \times 10^9 \frac{\text{N} \cdot \text{m}^2}{\text{C}^2}\right)(1.60 \times 10^{-19} \text{ C})} = 1.56 \times 10^{12} \text{ electrons} \quad \lozenge$$

---

53. A spherical conductor has a radius of 14.0 cm and charge of 26.0 μC. Calculate the electric field and the electric potential (a) r = 10.0 cm, (b) r = 20.0 cm, and (c) r = 14.0 cm from the center.

Solution

(a) Inside a conductor when charges are not moving, the electric field is zero and the potential is uniform, the same as on the surface.

$$\mathbf{E} = 0 \quad \lozenge$$

$$V = \frac{k_e q}{R} = \frac{8.99 \times 10^9 \text{ N} \cdot \text{m}^2 (26 \times 10^{-6} \text{ C})}{\text{C}^2 (0.14 \text{ m})} = 1.67 \text{ MV} \quad \lozenge$$

(b) The sphere behaves like a point charge at its center when you stand outside.

$$\mathbf{E} = \frac{k_e q}{r^2}\hat{\mathbf{r}} = \frac{8.99 \times 10^9 \text{ N} \cdot \text{m}^2 (26 \times 10^{-6} \text{ C})}{\text{C}^2 (0.20 \text{ m})^2}\hat{\mathbf{r}} = (5.84 \text{ MN/C})\hat{\mathbf{r}} \quad \lozenge$$

$$V = \frac{k_e q}{r} = 1.17 \text{ MV} \quad \lozenge$$

(c) $$\mathbf{E} = \frac{k_e q}{r^2}\hat{\mathbf{r}} = (11.9 \text{ MN/C})\hat{\mathbf{r}} \quad \lozenge$$

$$V = 1.67 \text{ MV as in part (a)} \quad \lozenge$$

---

55. Two charged spherical conductors are connected by a long conducting wire, and a charge of 20.0 μC is placed on the combination. (a) If one sphere has a radius of 4.00 cm and the other has a radius of 6.00 cm, what is the electric field near the surface of each sphere? (b) What is the electric potential of each sphere?

Solution The two balls have different charges and different-size electric fields outside their surfaces, but they are at the same potential.

Our two equations are $V_4 = V_6 = \dfrac{k_e Q_4}{r_4} = \dfrac{k_e Q_6}{r_6}$ and $Q_4 + Q_6 = 20.0\ \mu\text{C}$.

Substituting $Q_6 = \dfrac{6.00\text{ cm}}{4.00\text{ cm}} Q_4$, we have $Q_4 + 1.50 Q_4 = 20.0\ \mu\text{C}$,

Solving, $Q_4 = 20.0\ \mu\text{C} / 2.50 = 8.00\ \mu\text{C}$ and $Q_6 = (1.50)(8.00\ \mu\text{C}) = 12.0\ \mu\text{C}$

Then

(a) $$\mathbf{E}_4 = \frac{k_e Q_4}{r_4^2}\hat{\mathbf{r}} = \frac{(8.99\times10^9\ \text{N}\cdot\text{m}^2/\text{C}^2)(8.00\times10^{-6}\ \text{C})}{(4.00\times10^{-2}\ \text{m})^2}\hat{\mathbf{r}} = 45.0\times10^6\ \text{N/C outward}\ \lozenge$$

$$\mathbf{E}_6 = \frac{k_e Q_6}{r_6^2}\hat{\mathbf{r}} = \frac{(8.99\times10^9\ \text{N}\cdot\text{m}^2/\text{C}^2)(12.0\times10^{-6}\ \text{C})}{(6.00\times10^{-2}\ \text{m})^2}\hat{\mathbf{r}} = 30.0\times10^6\ \text{N/C outward}\ \lozenge$$

(The fact that the wire is long guarantees that charge will be distributed with uniform density all over the surface of each single sphere, so the point-charge equations for **E** and V apply.)

(b) $$V_4 = \frac{k_e Q_4}{r} = \frac{(8.99\times10^9\ \text{N}\cdot\text{m}^2/\text{C}^2)(8.00\times10^{-6}\ \text{C})}{(4.00\times10^{-2}\ \text{m})} = 1.80\times10^6\ \text{V}\ \lozenge$$

You should check that *V* comes out the same for the larger ball.

57. Consider a Van de Graaff generator with a 30.0-cm-diameter dome operating in dry air. (a) What is the maximum potential of the dome? (b) What is the maximum charge on the dome?

Solution You must look up the dielectric strength of air. The maximum-size electric field that can exist in air without ionizing molecules to produce a spark is $E_{max} = 3 \times 10^6$ V/m. Now it is easier to do part (b) first.

(b) $|E| = \frac{k_e Q}{r^2}$

$$Q = \frac{Er^2}{k_e} = \frac{(3.00 \times 10^6 \text{ V/m})(0.15 \text{ m})^2}{8.99 \times 10^9 \text{ N} \cdot \text{m}^2/\text{C}^2}(1 \text{ N} \cdot \text{m}/\text{V} \cdot \text{C}) = 7.51 \ \mu\text{C} \ \Diamond$$

(a) $V = \frac{k_e Q}{r} = \frac{(8.99 \times 10^9 \text{ N} \cdot \text{m}^2)(7.51 \times 10^{-6} \text{ C})}{\text{C}^2 \ (0.15 \text{ m})} = 450 \text{ kV} \ \Diamond$

---

61. At a certain distance from a point charge, the magnitude of the electric field is 500 V/m and the electric potential is –3.00 kV. (a) What is the distance to the charge? (b) What is the magnitude of the charge?

Solution At a distance r from a point charge $V_r = \frac{k_e q}{r}$ and $E_r = \frac{k_e q}{r^2}$

Thus, $E_r = \frac{rV_r}{r^2} = \frac{V_r}{r}$

(a) $r = \frac{V}{E_r} = \frac{3000 \text{ V}}{500 \text{ N/C}} = 6.00 \frac{\text{N} \cdot \text{m/C}}{\text{N/C}} = 6.00 \text{ m} \ \Diamond$

(b) $q = \frac{rV_r}{k_e} = \frac{(6.00 \text{ m})(-3000 \text{ V})}{8.99 \times 10^9 \text{ N} \cdot \text{m}^2/\text{C}^2}$

$q = -2.00 \ \mu\text{C} \ \Diamond$

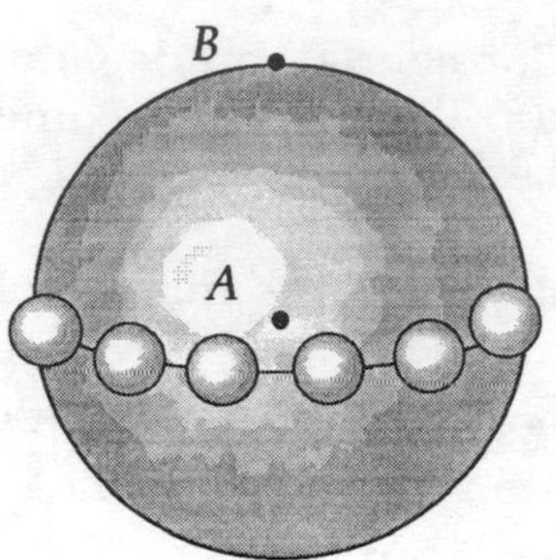

65. Equal charges ($q = 2.0\ \mu C$) are placed at 30° intervals around the equator of a sphere that has a radius of 1.2 m. What is the electric potential (a) at the center of the sphere and (b) at its north pole?

Solution

(a) The number of charges is $\frac{360}{30} = 12$. Each is equidistant from point A, the center, and so produces an equal contribution to the voltage there.

So $$V_A = 12\frac{k_e Q}{r} = (12)\frac{(8.99\times10^9\ \mathrm{N\cdot m^2/C^2})(2.0\times10^{-6}\ \mathrm{C})}{(1.2\ \mathrm{m})}\left(\frac{\mathrm{V\cdot C}}{\mathrm{J}}\right)\left(\frac{\mathrm{J}}{\mathrm{Nm}}\right) = 180\ \mathrm{kV}\ \Diamond$$

(b) The distance from each charge to the pole B is

$$\sqrt{(1.2\ \mathrm{m})^2 + (1.2\ \mathrm{m})^2} = 1.7\ \mathrm{m}$$

and $$V_B = 12\frac{k_e Q}{r} = 12\ \frac{(8.99\times10^9\ \mathrm{N\cdot m^2/C^2})(2.0\times10^{-6}\ \mathrm{C})}{(1.7\ \mathrm{m})} = 127\ \mathrm{kV}\ \ \Diamond$$

Additional Calculation: Work out the electric field (a) at the center, and (b) at the poles.

In both cases, the horizontal components of the field cancel out, because the charges have radial symmetry about the pole-to-pole axis of the sphere.

(a) At the center, there is no vertical component of the field, and $\mathbf{E} = 0$.

(b) The poles are a distance of $R\sqrt{2}$ away from the equator, where R is the radius of the sphere. The field due to one charge is then

$$E = \frac{k_e Q}{r^2} = \frac{(8.99\times10^9\ \mathrm{N\cdot m^2/C^2})(2.0\times10^{-6}\ \mathrm{C})}{(1.70\ \mathrm{m})^2} = 6250\ \mathrm{N/C}$$

and its vertical component is (6250 N/C) sin(45°) = 4420 N/C.

Twelve charges will produce a field of (12)(4420 N/C) = 53.0 kN/C outward. $\Diamond$

70. The liquid-drop model of the nucleus suggests that high-energy oscillations of certain nuclei can split the nucleus into two unequal fragments plus a few neutrons. The fragments acquire kinetic energy from their mutual Coulomb repulsion. Calculate the electric potential energy (in electron volts) of two spherical fragments from a uranium nucleus having the following charges and radii: $38e$ and 5.5×10^{-15} m; $54e$ and 6.2×10^{-15} m. Assume that the charge is distributed uniformly throughout the volume of each spherical fragment and that their surfaces are initially in contact at rest. (The electrons surrounding the nucleus can be neglected.)

Solution The problem is equivalent to finding the potential energy of a point charge $38e$ at distance 11.7×10^{-15} m from a point charge $54e$.

$$U = qV = k_e \frac{q_1 q_2}{r_{12}}$$

$$U = (8.99 \times 10^9)\frac{(38)(54)(1.6 \times 10^{-19})^2}{(5.5 + 6.2) \times 10^{-15}}$$

$$U = 4.04 \times 10^{-11} \text{ J}$$

$$U = 253 \text{ MeV} \quad \Diamond$$

---

73. Calculate the work that must be done to charge a spherical shell of radius R to a total charge Q.

Solution When the potential of the shell is V due to a charge q, the work required to add an additional increment of charge dq is

$$dW = V dq \quad \text{where} \quad V = \frac{k_e q}{R}$$

$$dW = \left(\frac{k_e q}{R}\right) dq \quad \text{and} \quad W = \frac{k_e}{R}\int_0^Q q\,dq$$

Therefore,

$$W = \left(\frac{k_e}{R}\right)\left(\frac{Q^2}{2}\right) \quad \Diamond$$

---

75. From Gauss's law, the electric field set up by a uniform line of charge is

$$\mathbf{E} = \left(\frac{\lambda}{2\pi \epsilon_0 r}\right)\hat{\mathbf{r}}$$

where $\hat{\mathbf{r}}$ is a unit vector pointing radially away from the line and λ is the charge per meter along the line. Derive an expression for the potential difference between $r = r_1$ and $r = r_2$.

Solution In Equation 25.3, $V_2 - V_1 = \Delta V = -\int_1^2 \mathbf{E} \cdot d\mathbf{s}$; think about stepping from distance r_1 out to the larger distance r_2 away from the charged line. Then $d\mathbf{s} = dr\,\hat{\mathbf{r}}$, and we can make r the variable of integration:

$$V_2 - V_1 = -\int_{r_1}^{r_2} \frac{\lambda}{2\pi \epsilon_0 r}\hat{\mathbf{r}} \cdot dr\,\hat{\mathbf{r}} \quad \text{with} \quad \hat{\mathbf{r}} \cdot \hat{\mathbf{r}} = 1 \cdot 1 \cdot \cos 0° = 1$$

The potential difference is $$V_2 - V_1 = -\frac{\lambda}{2\pi \epsilon_0}\int_{r_1}^{r_2} \frac{dr}{r} = -\frac{\lambda}{2\pi \epsilon_0} \ln r \Big]_{r_1}^{r_2}$$

and $$V_2 - V_1 = -\frac{\lambda}{2\pi \epsilon_0}(\ln r_2 - \ln r_1) = -\frac{\lambda}{2\pi \epsilon_0}\ln\frac{r_2}{r_1} \quad \Diamond$$

If $r_2 > r_1$, then $V_2 - V_1$ is negative. This means the potential decreases as we move away from a positively-charged filament.

---

79. It is shown in Example 25.10 that the potential at a point P a distance d above one end of a uniformly charged rod of length L lying along the x axis is

$$V = \frac{k_e Q}{L}\ln\left(\frac{L + \sqrt{L^2 + d^2}}{d}\right)$$

Use this result to derive an expression for the y component of the electric field at P. (*Hint:* Replace d with y.)

Solution Replacing d with y as directed, we can differentiate the potential as shown:

$$E_y = -\frac{\partial V}{\partial y} = -\frac{k_e Q}{L}\frac{d}{dy}\left[\ln(L + \sqrt{L^2 + y^2}) - \ln y\right] = \frac{-k_e Q}{L}\left[\frac{2y/\left(2\sqrt{L^2 + y^2}\right)}{L + \sqrt{L^2 + y^2}} - \frac{1}{y}\right] = \frac{k_e Q}{y\sqrt{L^2 + y^2}} \quad \Diamond$$

---

81. A dipole is located along the y axis as in Figure P25.81. (a) At a point P, which is far from the dipole ($r \gg a$), the electric potential is

$$V = k_e \frac{p\cos\theta}{r^2}$$

where $p = 2qa$. Calculate the radial component of the associated electric field, E_r, and the perpendicular component, E_θ. Note that

$$E_\theta = \frac{1}{r}\left(\frac{\partial V}{\partial \theta}\right).$$

Do these results seem reasonable for $\theta = 90°$ and $0°$? for $r = 0$? (b) For the dipole arrangement shown, express V in terms of rectangular coordinates using $r = (x^2 + y^2)^{1/2}$ and

$$\cos\theta = \frac{y}{(x^2 + y^2)^{1/2}}$$

Using these results and taking $r \gg a$, calculate the field components E_x and E_y.

Solution

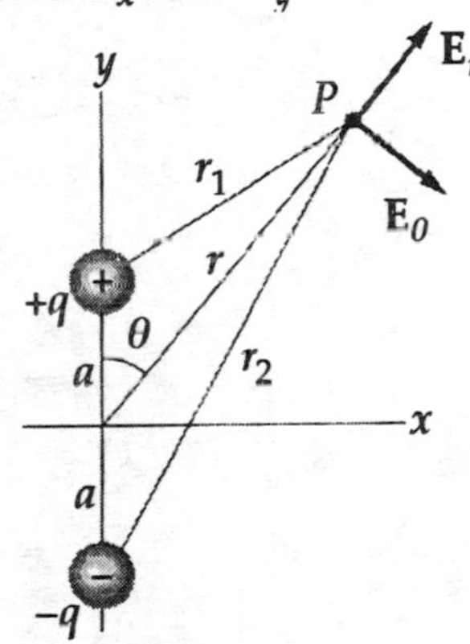

Figure P25.81

(a) $E_r = -\frac{\partial V}{\partial r} = -\frac{\partial}{\partial r}\left(\frac{k_e p\cos\theta}{r^2}\right) = \frac{2k_e p\cos\theta}{r^3} = E_r$ ◊

In spherical coordinates,

$$E_\theta = \frac{1}{r}\left(\frac{\partial V}{\partial \theta}\right) = \frac{1}{r}\frac{\partial}{\partial \theta}\left(\frac{k_e p\cos\theta}{r^2}\right) = \frac{k_e p\sin\theta}{r^3} = E_\theta \quad ◊$$

(b) For $r = \sqrt{x^2 + y^2}$, $\quad \cos\theta = \frac{y}{\sqrt{x^2 + y^2}}$ $\quad$ and $\quad V = \frac{k_e p y}{\left(x^2 + y^2\right)^{3/2}}$

$$E_x = -\frac{\partial V}{\partial x} = -\frac{\partial}{\partial x}\left(\frac{k_e p y}{\left(x^2 + y^2\right)^{3/2}}\right) = \frac{3k_e p x y}{\left(x^2 + y^2\right)^{5/2}} = E_x \quad ◊$$

$$E_y = -\frac{\partial V}{\partial y} = -\frac{\partial}{\partial y}\left(\frac{k_e p y}{\left(x^2 + y^2\right)^{3/2}}\right) = \frac{k_e p\left(2y^2 - x^2\right)}{\left(x^2 + y^2\right)^{5/2}} \quad ◊$$

34691 Serway Ancillary SAUNC Electronic Artwork Oper: kvk 10-6-95

Chapter 26

Capacitance and Dielectrics

CAPACITANCE AND DIELECTRICS

INTRODUCTION

This chapter is concerned with the properties of capacitors, devices that store charge. Capacitors are commonly used in a variety of electrical circuits. For instance, they are used (1) to tune the frequency of radio receivers, (2) as filters in power supplies, (3) to eliminate sparking in automobile ignition systems, and (4) as energy-storing devices in electronic flash units.

A capacitor basically consists of two conductors separated by an insulator. We shall see that the capacitance of a given device depends on its geometry and on the material separating the charged conductors, called a *dielectric*.

NOTES FROM SELECTED CHAPTER SECTIONS

26.1 Definition of Capacitance

The capacitance of a capacitor depends on the physical characteristics of the device (size, shape, and separation of plates and the nature of the dielectric medium filling the space between the plates). Since the potential difference between the plates is proportional to the quantity of charge on each plate, the value of the capacitance is independent of the charge on the capacitor.

26.3 Combinations of Capacitors

When two or more unequal capacitors are connected in *series*, they carry the same charge, but the potential differences are not the same. Their capacitances add as reciprocals, and the equivalent capacitance of the combination is always *less* than the smallest individual capacitor.

When two or more capacitors are connected in *parallel*, the potential difference across each is the same. The charge on each capacitor is proportional to its capacitance, hence the capacitances add directly to give the equivalent capacitance of the parallel combination.

26.4 Energy Stored in a Charged Capacitor

The electrostatic potential energy stored in a charged capacitor equals the work done in the charging process—moving charges from one conductor at a lower potential to another conductor at a higher potential.

26.5 Capacitors with Dielectrics

A dielectric is a nonconducting material characterized by a dimensionless parameter—the dielectric constant, κ. In general, the use of a dielectric has the following effects:

- increases the capacitance
- increases the maximum operating voltage
- provides mechanical support for the two conductors

EQUATIONS AND CONCEPTS

The *capacitance* of a capacitor is defined as the ratio of the charge on either conductor (or plate) to the magnitude of the potential difference between the conductors.

$$C \equiv \frac{Q}{V} \qquad (26.1)$$

The *capacitance of an air-filled parallel-plate capacitor* is proportional to the area of the plates and inversely proportional to the separation of the plates.

$$C = \frac{\epsilon_0 A}{d} \qquad (26.3)$$

When the region between the plates is completely filled by a material of dielectric constant κ, the capacitance increases by the factor κ.

$$C = \kappa \frac{\epsilon_0 A}{d} \qquad (26.14)$$

The *equivalent capacitance* of a *parallel combination* of capacitors is larger than any individual capacitor in the group.

$$C_{eq} = C_1 + C_2 + C_3 + \ldots \qquad (26.7)$$

The *equivalent capacitance* of a *series combination* of capacitors is smaller than the smallest capacitor in the group.

$$\frac{1}{C_{eq}} = \frac{1}{C_1} + \frac{1}{C_2} + \frac{1}{C_3} + \ldots \qquad (26.9)$$

In the special case of only *two capacitors in series*, the equivalent capacitance is equal to the ratio of the product to the sum of their capacitance.

$$C_{eq} = \frac{C_1 C_2}{C_1 + C_2}$$

The *electrostatic energy* stored in the electrostatic field of a charged capacitor equals the work done (by a battery or other source) in charging the capacitor from $q = 0$ to $q = Q$.

$$U = \frac{Q^2}{2C} = \tfrac{1}{2}QV = \tfrac{1}{2}CV^2 \qquad (26.10)$$

The *energy density* at any point in the electrostatic field of a charged capacitor is proportional to the square of the electric field intensity at that point.

$$u_E = \tfrac{1}{2}\,\epsilon_0\,E^2 \qquad (26.12)$$

Two equal and opposite charges of magnitude q separated by a distance $2a$ constitute an electric dipole. This configuration is characterized by an *electric dipole moment*, $\mathbf{p}$. The direction of the vector $\mathbf{p}$ is from the negative to the positive charge.

$$p \equiv 2aq \qquad (26.15)$$

An external uniform electric field will exert a *net torque* on an electric dipole when the dipole moment makes an angle θ with the direction of the field.

$$\boldsymbol{\tau} = \mathbf{p} \times \mathbf{E} \qquad (26.17)$$

There is *potential energy* associated with the dipole-electric field system.

$$U = -\mathbf{p} \cdot \mathbf{E} \qquad (26.19)$$

SUGGESTIONS, SKILLS, AND STRATEGIES

Problem-Solving Hints for Capacitance:

- When analyzing a series-parallel combination of capacitors to determine the equivalent capacitance, you should make a sequence of circuit diagrams which show the successive steps in the simplification of the circuit; combine at each step those capacitors which are in simple-parallel or simple-series relationship to each other and use appropriate equations for series or parallel capacitors at each step of the simplification. At each step, you know two of the three quantities: Q, V, and C. You will be able to determine the remaining quantity using the relation $Q = CV$.

- When calculating capacitance, be careful with your choice of units. To calculate capacitance in farads, make sure that distances are in meters and use the SI value of ϵ_0. When checking consistency of units, remember that the units for electric fields are newtons per coulomb (N/C) or the equivalent volts per meter (V/m).

- When two or more unequal capacitors are connected in series, they carry the same charge, but their potential differences are not the same. The capacitances add as reciprocals, and the equivalent capacitance of the combination is always less than the smallest individual capacitor.

- When two or more capacitors are connected in parallel, the potential differences across them are the same. The charge on each capacitor is proportional to its capacitance; hence, the capacitances add directly to give the equivalent capacitance of the parallel combination.

- A dielectric increases capacitance by the factor κ (the dielectric constant) because induced surface charges on the dielectric reduce the electric field inside the material from E to E/κ.

- Be careful about problems in which you may be connecting or disconnecting a battery to a capacitor. It is important to note whether modifications to the capacitor are being made while the capacitor is connected to the battery or after it is disconnected. If the capacitor remains connected to the battery, the voltage across the capacitor necessarily remains the same (equal to the battery voltage), and the charge is proportional to the capacitance, *however it may be modified* (say, by insertion of a dielectric). On the other hand, if you disconnect the capacitor from the battery before making any modifications to the capacitor, then its charge remains the same. In this case, as you vary the capacitance, the voltage across the plates changes in inverse proportion to capacitance, according to $V = Q/C$.

REVIEW CHECKLIST

▷ Use the basic definition of capacitance and the equation for finding the potential difference between two points in an electric field in order to calculate the capacitance of a capacitor for cases of relatively simple geometry—parallel plates, cylindrical, spherical.

▷ Determine the equivalent capacitance of a network of capacitors in series-parallel combination and calculate the final charge on each capacitor and the potential difference across each when a known potential is applied across the combination.

▷ Make calculations involving the relationships among potential, charge, capacitance, stored energy, and energy density for capacitors, and apply these results to the particular case of a parallel plate capacitor.

▷ Calculate the capacitance, potential difference, and stored energy of a capacitor which is partially or completely filled with a *dielectric*.

SOLUTIONS TO SELECTED END-OF-CHAPTER PROBLEMS

7. An isolated charged conducting sphere of radius 12.0 cm creates an electric field of 4.90×10^4 N/C at a distance of 21.0 cm from its center. (a) What is its surface charge density? (b) What is its capacitance?

Solution

(a) The electric field outside a spherical charge distribution of radius R is $E = \dfrac{k_e q}{r^2}$.

Therefore, $q = \dfrac{Er^2}{k_e}$. Since the surface charge density is $\sigma = \dfrac{q}{A}$,

$$\sigma = \frac{Er^2}{k_e 4\pi R^2} = \frac{(4.90\times 10^4 \text{ N/C})(0.21 \text{ m})^2}{\left(8.99\times 10^9 \dfrac{\text{N}\cdot\text{m}^2}{\text{C}^2}\right)(4\pi)(0.12 \text{ m})^2} = 1.33\ \mu\text{C/m}^2 \quad \Diamond$$

(b) For an isolated charged sphere of radius R,

$$C = 4\pi \epsilon_0 R = (4\pi)\left(8.85\times 10^{-12} \frac{\text{C}^2}{\text{N}\cdot\text{m}^2}\right)(0.12 \text{ m}) = 13.3 \text{ pF} \quad \Diamond$$

9A. Two spherical conductors with radii R_1 and R_2 are separated by a distance large enough to make induction effects negligible. The spheres are connected by a thin conducting wire and are brought to the same potential V relative to $V = 0$ at $r = \infty$. (a) Determine the capacitance C of the system, where $C = (Q_1 + Q_2)/V$. (b) What is the charge ratio Q_1/Q_2?

Solution

(a) "Induction effects are negligible" means that the charge on each ball is distributed equally over its surface. Then each charge creates potential at its surface according to $V = k_e Q_1/R_1$ and $V = k_e Q_2/R_2$. Thus,

$$Q_1 = \frac{VR_1}{k_e} = 4\pi \epsilon_0 R_1 V \quad \text{and} \quad Q_2 = \frac{VR_2}{k_e} = 4\pi \epsilon_0 R_2 V$$

So

$$Q_1 + Q_2 = (4\pi \epsilon_0 R_1 + 4\pi \epsilon_0 R_2)V$$

The proportionality of $Q_1 + Q_2$ to V shows that the system constitutes a capacitor, and gives for its capacitance

$$\frac{(Q_1 + Q_2)}{V} = 4\pi \epsilon_0 (R_1 + R_2) \quad \Diamond$$

The two balls are the positive plates of two capacitor in parallel, with their negative plates being the ground at infinite distance.

(b) From above,

$$\frac{Q_1}{Q_2} = \frac{4\pi \epsilon_0 R_1 V}{4\pi \epsilon_0 R_2 V} = \frac{R_1}{R_2} \quad \Diamond$$

Related Calculation: **9.** Evaluate your answers with conductors of radii $R_1 = 0.15$ cm and $R_2 = 0.23$ cm. Use a potential of $V = 775$ V relative to $V = 0$ at $r = \infty$.

(a) $$C = \frac{(Q_1 + Q_2)}{V} = 4\pi \epsilon_0 (R_1 + R_2) = 4\pi\left(8.85 \times 10^{-12} \frac{\text{C}^2}{\text{N} \cdot \text{m}^2}\right)(0.38 \times 10^{-2} \text{ m}) = 4.23 \times 10^{-13} \text{ F} \quad \Diamond$$

(b) $$\frac{Q_1}{Q_2} = \frac{4\pi \epsilon_0 R_1 V}{4\pi \epsilon_0 R_2 V} = \frac{R_1}{R_2} = \frac{0.15}{0.23} = 0.652 \quad \Diamond$$

13. When a potential difference of 150 V is applied to the plates of a parallel-plate capacitor, the plates carry a surface charge density of 30 nC/cm². What is the spacing between the plates?

Solution We have $Q = CV$ with $C = \frac{\epsilon_0 A}{d}$ Thus, $Q = \frac{\epsilon_0 AV}{d}$

The surface charge density on each plate is the same in magnitude, $\sigma = \frac{Q}{A} = \frac{\epsilon_0 V}{d}$

Thus, $d = \frac{\epsilon_0 V}{Q/A}$

$$d = \frac{(8.85\times10^{-12}\ \mathrm{C^2})\ 150\ \mathrm{V\ cm^2}}{\mathrm{N\cdot m^2}\ (30\times10^{-9}\ \mathrm{C})}\left(\frac{1\ \mathrm{m^2}}{10^4\mathrm{cm^2}}\right)\left(\frac{\mathrm{J}}{\mathrm{VC}}\right)\left(\frac{\mathrm{N\cdot m}}{\mathrm{J}}\right) = 4.42\ \mu\mathrm{m} \quad \Diamond$$

Another starting point: Recall from Example 24.8 that the electric field between the plates is $E = \frac{\sigma}{\epsilon_0}$. And note that the difference in potential from the negative to the positive plate is $V = Ed = \frac{\sigma d}{\epsilon_0}$. This gives again $d = \frac{V\epsilon_0}{\sigma}$

---

15. An air-filled capacitor consists of two parallel plates, each with an area of 7.60 cm², separated by a distance of 1.80 mm. If a 20.0-V potential difference is applied to these plates, calculate (a) the electric field between the plates, (b) the surface charge density, (c) the capacitance, and (d) the charge on each plate.

Solution

(a) The potential difference between two points in a uniform electric field is $V = Ed$, so

$$E = \frac{V}{d} = \frac{20.0\ \mathrm{V}}{1.80\times10^{-3}\ \mathrm{m}} = 1.11\times10^4\ \mathrm{V/m} \quad \Diamond$$

(b) The electric field between capacitor plates is $E = \frac{\sigma}{\epsilon_0}$, therefore

$$\sigma = \epsilon_0 E = \left(8.85\times10^{-12}\ \frac{\mathrm{C^2}}{\mathrm{N\cdot m^2}}\right)(1.11\times10^4\ \mathrm{V/m}) = 9.83\times10^{-8}\ \mathrm{C/m^2} = 98.3\ \mathrm{nC/m^2} \quad \Diamond$$

(c) For a parallel-plate capacitor,

$$C = \frac{\epsilon_0 A}{d} = \frac{\left(8.85 \times 10^{-12} \frac{\text{C}^2}{\text{N}\cdot\text{m}^2}\right)(7.6 \times 10^{-4} \text{ m}^2)}{1.80 \times 10^{-3} \text{ m}} = 3.74 \times 10^{-12} \text{ F} = 3.74 \text{ pF} \quad \Diamond$$

(d) $Q = CV = (3.74 \times 10^{-12} \text{ F})(20.0 \text{ V}) = 7.48 \times 10^{-11} \text{ C} = 74.8 \text{ pC} \quad \Diamond$

---

21. A 50.0-m length of coaxial cable has an inner conductor that has a diameter of 2.58 mm and carries a charge of 8.10 μC. The surrounding conductor has an inner diameter of 7.27 mm and a charge of –8.10 μC. (a) What is the capacitance of this cable? (b) What is the potential difference between the two conductors? Assume the region between the conductors is air.

(a) $$C = \frac{\ell}{2k_e \ln(2b/2a)} = \frac{50 \text{ m}}{2(8.99 \times 10^9 \text{ N}\cdot\text{m}^2/\text{C}^2)\ln(7.27/2.58)} = 2.68 \times 10^{-9} \text{ F} \quad \Diamond$$

(b) $$V = \frac{Q}{C} = \frac{8.10 \times 10^{-6} \text{ C}}{2.68 \times 10^{-9} \text{ F}} = 3.02 \text{ kV} \quad \Diamond$$

---

23. An air-filled spherical capacitor is constructed with inner and outer shell radii of 7.00 and 14.0 cm, respectively. (a) Calculate the capacitance of the device. (b) What potential difference between the spheres results in a charge of 4.00 μC on the capacitor?

Solution

(a) For a spherical capacitor, $C = \dfrac{ab}{k(b-a)}$

$$C = \frac{(0.0700 \text{ m})(0.140 \text{ m})}{\left(8.99 \times 10^9 \frac{\text{N}\cdot\text{m}^2}{\text{C}^2}\right)(0.140 - 0.0700) \text{ m}} = 1.56 \times 10^{-11} \text{ F} = 15.6 \text{ pF} \quad \Diamond$$

(b) $$V = \frac{Q}{C} = \frac{(4.00 \times 10^{-6} \text{ C})}{1.56 \times 10^{-11} \text{ F}} = 2.56 \times 10^5 \text{ V} = 256 \text{ kV} \quad \Diamond$$

---

31. Four capacitors are connected as shown in Figure P26.31. (a) Find the equivalent capacitance between points a and b. (b) Calculate the charge on each capacitor if $V_{ab} = 15$ V.

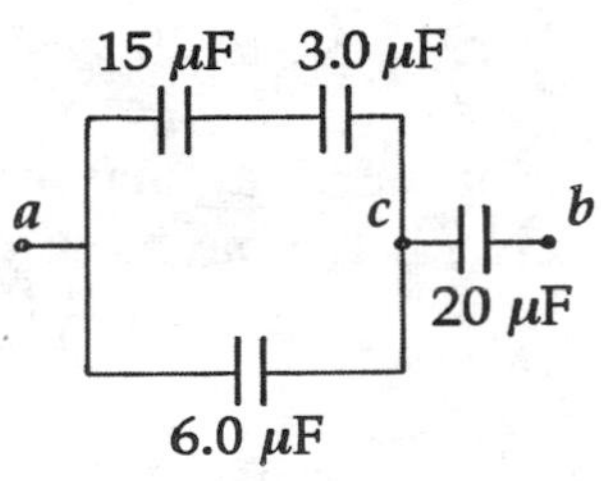

Figure P26.31

(a) We successively simplify the circuit, proceeding from the given diagram through solution figures (a) - (c).

First, the 15 μF and 3.0 μF in series are equivalent to

$$\frac{1}{\left(\frac{1}{15\ \mu\text{F}}+\frac{1}{3.0\ \mu\text{F}}\right)} = 2.5\ \mu\text{F}$$

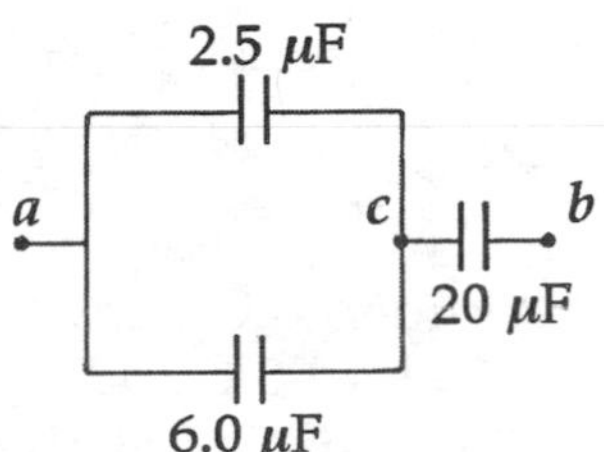

(a)

Next, 2.5 μF parallels 6.0 μF, and is equivalent to 8.5 μF.

At last, 8.5 μF and 20 μF are in series, equivalent to

$$\frac{1}{\left(\frac{1}{8.5}+\frac{1}{20}\right)} = 5.96\ \mu F \quad \Diamond$$

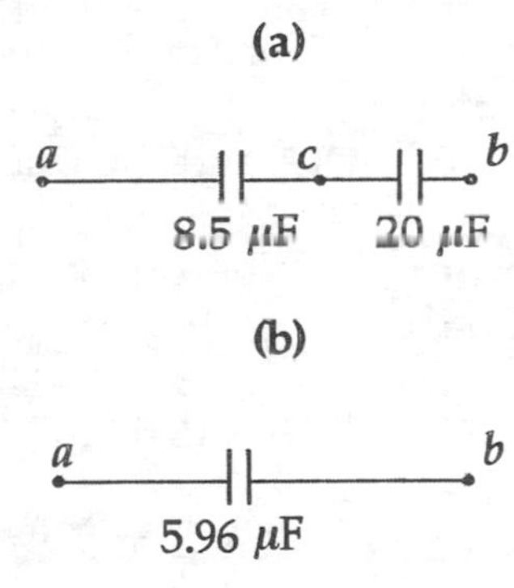

(c)

(b) We find the charge on and the voltage across each capacitor by working backwards through solution figures (c) - (a), alternately applying $Q = CV$ and $V = Q/C$ to every capacitor, real or equivalent. For the 5.96 μF capacitor, we have $Q = CV = (5.96\ \mu\text{F})(15\ \text{V}) = 89.5\ \mu\text{C}$.

Thus, if a is higher in potential than b, just 89.5 μC flows to the right past a and past b to charge the capacitors in each picture. In (b) we have, for the 8.5 μF capacitor,

$$V_{ac} = \frac{Q}{C} = \frac{89.5\ \mu\text{C}}{8.5\ \mu\text{F}} = 10.5\ \text{V}$$

and for the 20 μF in (b), (a), and the original circuit, we have $Q_{20} = 89.5\ \mu\text{C}$ ◊

$$V_{cb} = \frac{Q}{C} = \frac{89.5\ \mu\text{C}}{20\ \mu\text{F}} = 4.47\ \text{V}$$

Next, (a) is equivalent to (b), so $V_{cb} = 4.47$ V and $V_{ac} = 10.5$ V.

For the 2.5 μF, $V = 10.5$ V and $Q = CV = (2.5\ \mu\text{F})(10.5\ \text{V}) = 26.3\ \mu\text{C}$

For the 6.0 μF, $V = 10.5$ V and $Q_6 = CV = (6.0\ \mu\text{F})(10.5\ \text{V}) = 63.2\ \mu\text{C} = Q_6$ ◊

Now, 26.3 μC having flowed in the upper parallel branch in (a), back in the original circuit we have

$$Q_{15} = 26.3\ \mu\text{C} \ \lozenge \qquad \text{and} \qquad Q_3 = 26.3\ \mu\text{C} \ \lozenge$$

Related Calculation: An exam problem would also ask for the voltage across each:

$$V_{15} = \frac{Q}{C} = \frac{26.3\ \mu\text{C}}{15\ \mu\text{F}} = 1.75\ \text{V} \qquad \text{and} \qquad V_3 = \frac{Q}{C} = \frac{26.3\ \mu\text{C}}{3.00\ \mu\text{F}} = 8.77\ \text{V} \ \lozenge$$

---

33. Consider the circuit shown in Figure P26.33, where $C_1 = 6.00\ \mu$F, $C_2 = 3.00\ \mu$F, and $V = 20.0$ V. Capacitor C_1 is first charged by the closing of switch S_1. Switch S_1 is then opened, and the charged capacitor is connected to the uncharged capacitor by the closing of S_2. Calculate the initial charge acquired by C_1 and the final charge on each.

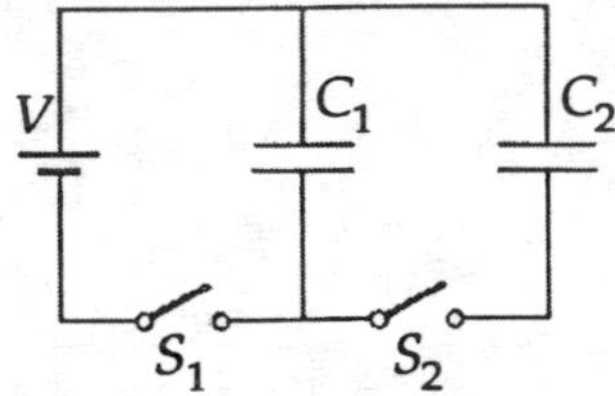

Figure P26.33

Solution When S_1 is closed, the charge on C_1 will be

$$Q_1 = C_1V_1 = (6.00\ \mu\text{F})(20.0\ \text{V}) = 120\ \mu\text{C} \ \lozenge$$

When S_1 is opened and S_2 is closed, the total charge will remain constant and be shared by the two capacitors:

$$Q_1' = 120\ \mu\text{C} - Q_2'$$

The potential across the two capacitors will be equal.

$$V' = \frac{Q_1'}{C_1} = \frac{Q_2'}{C_2} \qquad \text{or} \qquad \frac{120\ \mu\text{C} - Q_2'}{6.00\ \mu\text{F}} = \frac{Q_2'}{3.00\ \mu\text{F}}$$

and

$$Q_2' = 40\ \mu\text{C} \ \lozenge$$

$$Q_1' = 120\ \mu\text{C} - 40\ \mu\text{C} = 80\ \mu\text{C} \ \lozenge$$

---

37. A group of identical capacitors is connected first in series and then in parallel. The combined capacitance in parallel is 100 times larger than for the series connection. How many capacitors are in the group?

Solution Name your ignorance. Call C the capacitance of one capacitor and n the number of capacitors.

$$C_p = C_1 + C_2 + \ldots + C_n = nC, \quad \text{so} \quad \frac{1}{C_s} = \frac{1}{C_1} + \frac{1}{C_2} + \ldots + \frac{1}{C_n} = \frac{n}{C}$$

So $$C_s = \frac{C}{n}.$$

From this, $C_p = 100\ C_s$ requires $nC = 100\ C/n$

$$n^2 = 100, \text{ and } \quad n = 10 \quad \Diamond$$

---

43. A conducting slab of a thickness d and area A is inserted into the space between the plates of a parallel-plate capacitor with spacing s and surface area A, as in Figure P26.43. What is the capacitance of the system?

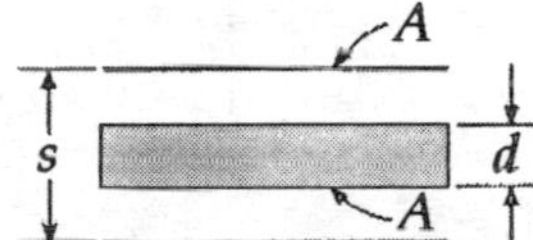

Figure P26.43

Solution If the capacitor is charged with charge Q, free charges will move across the slab to neutralize the electric field inside it, with the top and bottom faces of the slab then carrying charges $+Q$ and $-Q$. Then the capacitor with slab is electrically equivalent to two capacitors in series. Call x the upper gap, so $s - d - x$ is the distance between the lower two surfaces.

The upper capacitor has $C_1 = \dfrac{\epsilon_0 A}{x}$ and the lower has $C_2 = \dfrac{\epsilon_0 A}{s - d - x}$

So the combination has $$C = \frac{1}{1/C_1 + 1/C_2} = \frac{1}{\dfrac{x}{\epsilon_0 A} + \dfrac{s-d-x}{\epsilon_0 A}} = \frac{\epsilon_0 A}{s-d} \quad \Diamond$$

Example 26.10 performed this same analysis in the special case when the slab is just halfway between the plates.

---

45. Calculate the energy stored in a 18.0-μF capacitor when it is charged to a potential of 100 V.

Solution The energy stored in a charged capacitor is

$$U = \tfrac{1}{2}CV^2 = \left(\tfrac{1}{2}\right)(18 \times 10^{-6}\ \text{F})(100\ \text{V})^2 = 0.0900\ \text{J} = 90.0\ \text{mJ} \quad \Diamond$$

---

49. A 16.0-pF parallel-plate capacitor is charged by a 10.0-V battery. If each plate of the capacitor has an area of 5.00 cm^2, what is the energy stored in the capacitor? What is the energy density (energy per unit volume) in the electric field of the capacitor if the plates are separated by air?

Solution $U = \tfrac{1}{2}CV^2 = \tfrac{1}{2}\left(16.0 \times 10^{-12}\ \text{F}\right)(10\ \text{V})^2\left(\dfrac{1\ \text{C}}{\text{F}\cdot\text{V}}\right)\left(\dfrac{1\ \text{J}}{\text{V}\cdot\text{C}}\right) = 8.00 \times 10^{-10}\ \text{J} \quad \Diamond$

Here are two methods for finding the energy density:

Method one: We can find the distance between plates from $C = \dfrac{\epsilon_0 A}{d}$

$$d = \frac{\epsilon_0 A}{C} = \frac{(8.85 \times 10^{-12}\ \text{C}^2/\text{N}\cdot\text{m}^2)(5.00 \times 10^{-4}\ \text{m}^2)}{(16.0 \times 10^{-12}\ \text{F})}\left(\frac{1\ \text{F}\cdot\text{V}}{\text{C}}\right)\left(\frac{1\ \text{N}\cdot\text{m}}{\text{V}\cdot\text{C}}\right) = 2.77 \times 10^{-4}\ \text{m}$$

Now the volume of the space between the plates is

$$Ad = (5.00 \times 10^{-4}\ \text{m}^2)(2.77 \times 10^{-4}\ \text{m}) = 1.38 \times 10^{-7}\ \text{m}^3$$

It is in this volume that the 800 pJ is stored, since the energy belongs to the separation of the charges on the plates, to the electric field in this volume. Thus, its density is

$$u = \frac{U}{V} = \frac{8.00 \times 10^{-10}\ \text{J}}{1.38 \times 10^{-7}\ \text{m}^3} = 5.79 \times 10^{-3}\ \text{J/m}^3 \quad \Diamond$$

Method two: The charge on each plate is:

$$Q = CV = 16 \times 10^{-12}\ \text{F}\ (10\ \text{V}) = 16 \times 10^{-11}\ \text{C}$$

so its surface density is $\sigma = \frac{Q}{A} = \frac{16.0 \times 10^{-11}\ \text{C}}{5.00 \times 10^{-4}\ \text{m}^2} = 3.20 \times 10^{-7}\ \text{C/m}^2$

and the strength of the electric field between the plates is

$$E = \frac{\sigma}{\epsilon_0} = \frac{(3.20 \times 10^{-7}\ \text{C/m}^2)}{(8.85 \times 10^{-12}\ \text{C}^2/\text{N}\cdot\text{m}^2)} = 3.62 \times 10^4\ \text{N/C}$$

Then the energy density is

$$u = \tfrac{1}{2}\epsilon_0 E^2 = \tfrac{1}{2}(8.85 \times 10^{-12}\ \text{C}^2/\text{N}\cdot\text{m}^2)\left(3.62 \times 10^4\ \frac{\text{N}}{\text{C}}\right)^2 = 5.79 \times 10^{-3}\ \text{J/m}^3 \quad \lozenge$$

---

51. A parallel-plate capacitor has a charge Q and plates of area A. Show that the force exerted on each plate by the other is $F = Q^2/2\epsilon_0 A$. (*Hint:* Let $C = \epsilon_0 A/x$ for an arbitrary plate separation x; then, require that the work done in separating the two charged plates be $W = \int F\,dx$.)

Solution The electric field in the space between the plates is $E = \frac{\sigma}{\epsilon_0} = \frac{Q}{A\epsilon_0}$.

You might think that the force on one plate is $F = QE = \frac{Q^2}{A\epsilon_0}$, but this is two times too large, because neither plate exerts a force on itself. The force *on* one plate is exerted *by* the other, through its electric field $\mathbf{E} = \sigma/2\epsilon_0 = Q/2A\epsilon_0$. The force on each plate is

$$F = (Q_{\text{self}})(E_{\text{other}}) = Q^2/2A\epsilon_0.$$

To prove this, we follow the hint, and calculate that the work done in separating the plates is the potential energy stored in the charged capacitor:

$$U = \frac{1}{2}\frac{Q^2}{C} = \int F\,dx$$

From the fundamental theorem of calculus, $dU = F\,dx$, and

$$F = \frac{d}{dx}U = \frac{d}{dx}\left(\frac{Q^2}{2C}\right) = \frac{1}{2}\frac{d}{dx}\left(\frac{Q^2}{\epsilon_0 A/x}\right) = \frac{1}{2}\frac{d}{dx}\left(\frac{Q^2 x}{\epsilon_0 A}\right) = \frac{1}{2}\left(\frac{Q^2}{\epsilon_0 A}\right) \quad \lozenge$$

---

57. A parallel-plate capacitor has a plate area of 0.64 cm^2. When the plates are in a vacuum, the capacitance of the device is 4.9 pF. (a) Calculate the value of the capacitance if the space between the plates is filled with nylon. (b) What is the maximum potential difference that can be applied to the plates without causing dielectric breakdown?

Solution For a capacitor filled with a dielectric, $\kappa = 3.4$ (nylon):

(a) $C = \kappa C_0 = (3.4)(4.9 \text{ pF}) = 16.7 \text{ pF}$ ◊

From $C_0 = \dfrac{\epsilon_0 A}{d}$, we get $$d = \frac{\epsilon_0 A}{C_0} = \frac{\left(8.85\times 10^{-12}\ \dfrac{\text{C}^2}{\text{N}\cdot\text{m}^2}\right)(6.4\times 10^{-5}\text{ m}^2)}{4.9\times 10^{-12}\text{ F}} = 1.16\times 10^{-4}\text{ m}$$

and $$V_{max} = E_{max} d$$

(b) For nylon, $E_{max} = 14 \times 10^6$ V/m,

so $$V_{max} = (14\times 10^6 \text{ V/m})(1.16\times 10^{-4}\text{ m}) = 1.62\times 10^3 \text{ V} = 1.6 \text{ kV} \quad ◊$$

---

59. A commercial capacitor is constructed as in Figure 26.12a. This particular capacitor is "rolled" from two strips of aluminum separated by two strips of paraffin-coated paper. Each strip of foil and paper is 7.0 cm wide. The foil is 0.0040 mm thick, and the paper is 0.025 mm thick and has a dielectric constant of 3.7. What length should the strips be if a capacitance of 9.5 × 10^{-8} F is desired? (Use the parallel-plate formula.)

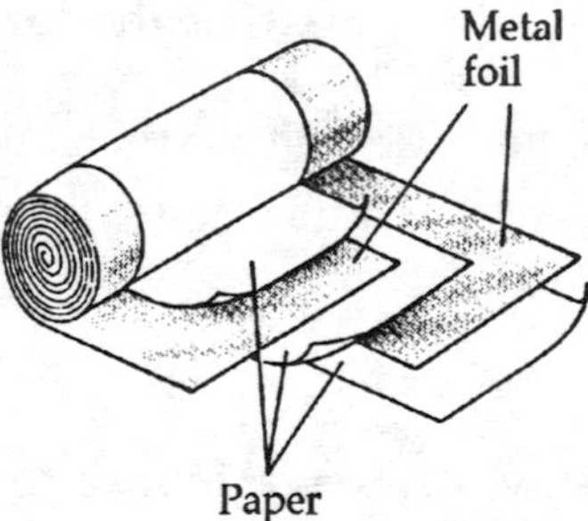

Figure 26.12a

Solution

$$C = \frac{\kappa \epsilon_0 A}{d} = \frac{3.7(8.85\times 10^{-12}\text{ C}^2/\text{N}\cdot\text{m}^2)(0.07\text{ m})L}{2.5\times 10^{-5}\text{ m}} = 9.5\times 10^{-8}\text{ F}$$

Solving, we get $L = 1.04$ m ◊

The distance between positive and negative charges is the thickness of one sheet of paper because these charges sit on adjacent surfaces of the metal foils.

---

67. When two capacitors are connected in parallel, the equivalent capacitance is 4.00 μF. If the same capacitors are reconnected in series, the equivalent capacitance is one-fourth the capacitance of one of the two capacitors. Determine the two capacitances.

Solution For the parallel connection, $C_1 + C_2 = 4\ \mu\text{F}$. For the series,

$$\frac{1}{C_1}+\frac{1}{C_2}=\frac{1}{C_1/4}=\frac{4}{C_1} \quad \text{so} \quad \frac{1}{C_2}=\frac{3}{C_1}$$

$$C_1 = 3C_2$$

We solve by substitution:

$$3C_2 + C_2 = 4\ \mu\text{F}$$

So $\qquad C_2 = 1.00\ \mu\text{F}$ ◊ $\qquad$ and $\qquad C_1 = 3C_2 = 3.00\ \mu\text{F}$ ◊

---

73. Three capacitors—8.0 μF, 10 μF, and 14 μF—are connected to the terminals of a 12-V battery. How much energy does the battery supply if the capacitors are connected (a) in series and (b) in parallel?

Solution

(a) For a series combination of capacitors,

$$\frac{1}{C}=\frac{1}{C_1}+\frac{1}{C_2}+\frac{1}{C_3}=\frac{1}{8.0\ \mu\text{F}}+\frac{1}{10\ \mu\text{F}}+\frac{1}{14\ \mu\text{F}}$$

This gives $\qquad C = 3.37\ \mu\text{F}$

and $\qquad U=\frac{1}{2}CV^2=\frac{1}{2}(3.37\times10^{-6}\ \text{F})(12\ \text{V})^2=2.43\times10^{-4}\ \text{J}=243\ \mu\text{J}$ ◊

(b) For a parallel combination of capacitors,

$$C = C_1 + C_2 + C_3 = 8\ \mu\text{F} + 10\mu\text{F} + 14\ \mu\text{F} = 32\ \mu\text{F}$$

and $\qquad U=\frac{1}{2}CV^2=\frac{1}{2}(32\times10^{-6}\ \text{F})(12\ \text{V})^2=2.30\times10^{-3}\ \text{J}=2.30\ \text{mJ}$ ◊

---

75. An isolated capacitor of unknown capacitance has been charged to a potential difference of 100 V. When the charged capacitor is then connected in parallel to an uncharged 10-μF capacitor, the voltage across the combination is 30 V. Calculate the unknown capacitance.

Solution

Call the unknown capacitance C_u. The charge originally deposited one *each* plate, + on one, – on the other, is

$$Q = C_u V = C_u \, 100 \text{ V}$$

Now in the new connection this same conserved charge redistributes itself between the two capacitors according to $Q = Q_1 + Q_2$.

$$Q_1 = C_u \, (30 \text{ V})$$

$$Q_2 = (10 \, \mu\text{F})(30 \text{ V}) = 300 \, \mu \text{C}$$

We can eliminate Q and Q_1 by substitution:

$$C_u(100 \text{ V}) = C_u(30 \text{ V}) + 300 \, \mu\text{C}$$

$$C_u = \frac{300 \, \mu\text{C}}{70 \text{ V}} = 4.29 \, \mu\text{F} \quad \lozenge$$

Related Calculation Solve this problem for the unknown capacitance, using an initial potential difference of V_0, a capacitor of known capacitance C, and a voltage across the two combined capacitors, $V < V_0$.

Solution

Call the unknown C_u. The total charge is:

$$Q = C_u V_0 = (C_u + C) \, V.$$

So

$$C_u = \frac{CV}{V_0 - V} \quad \lozenge$$

---

79. A parallel-plate capacitor is constructed using a dielectric material whose dielectric constant is 3.0 and whose dielectric strength is 2.0×10^8 V/m. The desired capacitance is 0.25 μF, and the capacitor must withstand a maximum potential difference of 4000 V. Find the minimum area of the capacitor plates.

Solution $\kappa = 3,\qquad E_{max} = 2.0 \times 10^8 \text{ V/m} = \dfrac{V_{max}}{d}\qquad$ so $\qquad d = \dfrac{V_{max}}{E_{max}}$

For $\quad C = \dfrac{\kappa \epsilon_0 A}{d} = 0.25 \times 10^{-6}$ F,

$$A = \frac{Cd}{\kappa \epsilon_0} = \frac{CV_{max}}{\kappa \epsilon_0 E_{max}}$$

$$A = \frac{(0.25 \times 10^{-6})(4000)}{3(8.85 \times 10^{-12})(2.0 \times 10^8)} = 0.19 \text{ m}^2 \quad \Diamond$$

---

83. A parallel-plate capacitor of plate separation d is charged to a potential difference V_0. A dielectric slab of thickness d and dielectric constant κ is introduced between the plates *while the battery remains connected to the plates.* (a) Show that the ratio of energy stored after the dielectric is introduced to the energy stored in the empty capacitor is $U/U_0 = \kappa$. Give a physical explanation for this increase in stored energy. (b) What happens to the charge on the capacitor? (Note that this situation is not the same as Example 26.7, in which the battery was removed from the circuit before the dielectric was introduced.)

Solution

(a) The capacitance changes from, say, C_0 to κC_0. The battery will maintain constant voltage across it by pumping out extra charge. The original energy is $U_0 = \frac{1}{2}C_0V^2$ and the final energy is $U = \frac{1}{2}\kappa C_0 V^2$, so $U/U_0 = \kappa$. The extra energy comes from (part of the) electrical work done by the battery in separating extra charge.

(b) The original charge is $Q_0 = C_0 V$ and the final value is $Q = \kappa C_0 V$, so the charge increases by the factor κ. $\Diamond$

---

89. The inner conductor of a coaxial cable has a radius of 0.80 mm and the outer conductor's inside radius is 3.0 mm. The space between the conductors is filled with polyethylene, which has a dielectric constant of 2.3 and a dielectric strength of 18×10^6 V/m. What is the maximum potential difference that this cable can withstand?

Solution We can increase the potential difference between core and sheath until the electric field in the polyethylene starts to punch a hole through it, passing a spark. This will first happen at the surface of the inner conductor, where the electric field is strongest, according to Example 24.7. For application here, Gauss's law must be modified to

$$\oint \mathbf{E} \cdot d\mathbf{A} = \frac{q}{\epsilon} = \frac{q}{\kappa \epsilon_0}$$

to give, in place of Equation 24.7, $$E = \frac{\lambda}{2\pi \epsilon_0 r} = \frac{\lambda}{2\pi \kappa \epsilon_0 r}$$

Now from Example 26.2, if there were vacuum between the conductors, the voltage between them would be

$$|V_b - V_a| = 2k_e \lambda \ln\left(\frac{b}{a}\right) = \frac{\lambda}{2\pi \epsilon_0} \ln\left(\frac{b}{a}\right)$$

With a dielectric, it is $$V = \frac{\lambda}{2\pi \epsilon_0} \ln\left(\frac{b}{a}\right) = \frac{\lambda}{2\pi \kappa \epsilon_0} \ln\left(\frac{b}{a}\right)$$

So when $E = E_{\text{max}}$ at $r = a$, we have $$\frac{\lambda_{\text{max}}}{2\pi \kappa \epsilon_0} = E_{\text{max}} a$$

and $$V_{\text{max}} = \frac{\lambda_{\text{max}}}{2\pi \kappa \epsilon_0} \ln\left(\frac{b}{a}\right) = E_{\text{max}} a \ln\left(\frac{b}{a}\right)$$

Thus, $$V_{\text{max}} = \left(18 \times 10^6 \ \frac{\text{V}}{\text{m}}\right)\left(0.80 \times 10^{-3} \text{ m}\right) \ln\left(\frac{3.0 \text{ mm}}{0.80 \text{ mm}}\right) = 19 \text{ kV} \quad \Diamond$$

Chapter 27

Current and Resistance

CURRENT AND RESISTANCE

INTRODUCTION

Thus far our discussion of electrical phenomena has been confined to charges at rest, or electrostatics. We now consider situations involving electric charges in motion. The term *electric current,* or simply *current,* is used to describe the rate of flow of charge through some region of space.

In this chapter we first define current and current density. A microscopic description of current is given, and some of the factors that contribute to the resistance to the flow of charge in conductors are discussed. Mechanisms responsible for the electrical resistance of various materials depend on the composition of the material and on temperature. A classical model is used to describe electrical conduction in metals, and some of the limitations of this model are pointed out.

NOTES FROM SELECTED CHAPTER SECTIONS

27.1 Electric Current

The direction of conventional current is designated as the direction of motion of positive charge. In an ordinary metal conductor, the direction of current will be *opposite* the *direction of flow of electrons* (which are the charge carriers in this case).

27.2 Resistance and Ohm's Law

For ohmic materials, the ratio of the current density to the electric field (that gives rise to the current) is equal to a constant σ, the conductivity of the material. The reciprocal of the conductivity is called the resistivity, ρ. Each ohmic material has a characteristic resistivity which depends only on the properties of the specific material and is a function of temperature.

27.5 A Model for Electrical Conduction

In the classical model of electronic conduction in a metal, electrons are treated like molecules in a gas and, in the absence of an electric field, have a *zero average velocity*.

Under the influence of an electric field, the electrons move along a direction opposite the direction of the applied field with a *drift velocity* which is proportional to the average time between collisions with atoms of the metal and inversely proportional to the number of free electrons per unit volume.

EQUATIONS AND CONCEPTS

Under the action of an electric field, electric charges will move through gases, liquids, and solid conductors. *Electric current, I,* is defined as the rate at which charge moves through a cross section of the conductor.

$$I \equiv \frac{dQ}{dt} \qquad (27.2)$$

The direction of the current is in the direction of the flow of positive charges. The SI unit of current is the *ampere* (A).

$$1\,\text{A} = 1\,\text{C/s} \qquad (27.3)$$

The current in a conductor can be related to the number of mobile charge carriers per unit volume, n; the quantity of charge associated with each carrier, q; and the *drift velocity*, v_d, of the carriers.

$$I = nqv_dA \qquad (27.4)$$

The *current density, J,* in a conductor is a vector quantity which is proportional to the electric field in the conductor.

$$J \equiv \frac{I}{A} \qquad (27.5)$$

or

$$J = nqv_d \qquad (27.6)$$

For many practical applications, a more useful form of Ohm's law relates the potential difference across a conductor and the current in the conductor to a composite of several physical characteristics of the conductor called the *resistance, R.*

$$R \equiv \frac{V}{I} \qquad (27.8)$$

The *resistance* of a given conductor of uniform cross section depends on the length, cross-sectional area, and a characteristic property of the material of which the conductor is made. The parameter, ρ, is the *resistivity* of the material of which the conductor is made. The resistivity is the inverse of the *conductivity* and has units of ohm-meters. The unit of resistance is the ohm (Ω).

$$R = \rho \frac{L}{A} \tag{27.11}$$

$$1\ \Omega = 1\ \text{V/A}$$

The *resistivity* and therefore the resistance of a conductor vary with temperature in an approximately linear manner. In these expressions, α is the *temperature coefficient of resistance* and T_0 is a stated reference temperature (usually 20°C).

$$\rho = \rho_0[1 + \alpha(T - T_0)] \tag{27.12}$$

$$R = R_0[1 + \alpha(T - T_0)] \tag{27.14}$$

Power will be supplied to a resistor or other current-carrying devices when a potential difference is maintained between the terminals of the circuit element. The quantities can be related in an equation called Joule's law and the SI unit of power is the watt (W). When the device obeys Ohm's law, the power dissipated can be expressed in alternative forms.

$$P = IV \tag{27.22}$$

$$P = I^2 R = \frac{V^2}{R} \tag{27.23}$$

The average time between collisions with atoms of a metal is an important parameter in the description of the classical model of electronic conduction in metals. This characteristic time is denoted by τ and can be related to the drift velocity (see Eq. 27.4) or the resistivity (see Eq. 27.11) associated with the conductor. In these equations, m and q represent the mass and charge of the electron, E is the magnitude of the applied electric field, and n is the number of free electrons per unit volume.

$$v_d = \frac{qE}{m}\tau \tag{27.17}$$

$$\rho = \frac{m}{nq^2\tau} \tag{27.20}$$

SUGGESTIONS, SKILLS, AND STRATEGIES

Equation 27.11, $R = \rho \frac{L}{A}$, can be used directly to calculate the resistance of a conductor of uniform cross-sectional area and constant resistivity. For those cases in which the area, resistivity, or both vary along the length of the conductor, the resistance must be determined as an integral of dR.

The conductor is subdivided into elements of length dx over which ρ and A may be considered constant in value and the total resistance is

$$R = \frac{\rho}{A} \int dx$$

Consider, for example, the case of a truncated cone of constant resistivity, radii a and b and height h. The conductor should be subdivided into disks of thickness dx, radius r, area = πr^2 and oriented parallel to the faces of the cone as shown in Figure 27.1. Note from the geometry that

$$x = \left(\frac{r-a}{b-a}\right)h \quad \text{so that} \quad r = \frac{x}{h}(b-a) + a$$

and

$$R = \int_0^h \frac{\rho}{\pi r^2} dx$$

The remainder of this calculation is left as a problem for you to work out (see Problem 27.68 of the text).

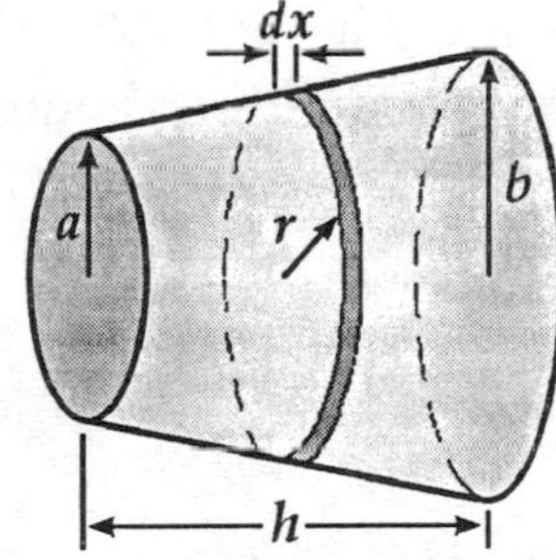

Figure 27.1

REVIEW CHECKLIST

▷ Define the term, electric current, in terms of rate of charge flow, and its corresponding unit of measure, the ampere. Calculate electron drift velocity, and quantity of charge passing a point in a given time interval in a specified current-carrying conductor.

- ▷ Determine the resistance of a conductor using Ohm's law. Also, calculate the resistance based on the physical characteristics of a conductor. Distinguish between ohmic and nonohmic conductors.

- ▷ Make calculations of the variation of resistance with temperature, which involves the concept of the temperature coefficient of resistivity.

- ▷ Use Joule's law to calculate the power dissipated in a resistor.

SOLUTIONS TO SELECTED END-OF-CHAPTER PROBLEMS

6. Suppose that the current through a conductor decreases exponentially with time according to

$$I(t) = I_0 e^{-t/\tau}$$

where I_0 is the initial current (at $t = 0$), and τ is a constant having dimensions of time. Consider a fixed observation point within the conductor. (a) How much charge passes this point between $t = 0$ and $t = \tau$? (b) How much charge passes this point between $t = 0$ and $t = 10\tau$? (c) How much charge passes this point between $t = 0$ and $t = \infty$?

Solution From $I = \dfrac{dQ}{dt}$ $\quad dQ = I\,dt \quad Q = \int dQ = \int I\,dt$

(a) $Q = \int_0^\tau I_0\, e^{-t/\tau}\, dt$

Let $u = -\dfrac{t}{\tau}$; $\quad du = -\dfrac{dt}{\tau}$; $\quad$ at $t = 0,\ u = 0$; $\quad$ at $t = \tau,\ u = -1$:

$$Q = \int_{u=0}^{-1} I_0\, e^u(-\tau)\, du = -I_0 \tau\, e^u\Big]_{u=0}^{u=-1}$$

$$= -I_0\tau(e^{-1} - e^0) = I_0\tau(e^0 - e^{-1}) = 0.6321\, I_0\tau \quad \Diamond$$

In both case (b) and case (c),

$$Q = \int I_0 e^{-t/\tau} dt$$

$$= \int (-I_0 \tau) e^{-t/\tau} \left(-\frac{dt}{\tau}\right)$$

$$= (-I_0 \tau) e^{-t/\tau}$$

(b) $Q = (-I_0\tau)\, e^{-t/\tau}\Big]_{t=0}^{10\tau} = (-I_0\tau)\left(e^{-10} - e^0\right) = 0.99995\, I_0\tau$ ◊

(c) $Q = (-I_0\tau)\, e^{-t/\tau}\Big]_0^{\infty} = (-I_0\tau)\left(e^{-\infty} - e^0\right) = I_0\tau$ ◊

---

11. A coaxial conductor with a length of 20 m consists of an inner cylinder with a radius of 3.0 mm and a concentric outer cylindrical tube with an inside radius of 9.0 mm. A uniformly distributed leakage current of 10 μA flows between the two conductors. Determine the leakage current density (in A/m^2) through a cylindrical surface (concentric with the conductors) that has a radius of 6.0 mm.

Solution The current density is $J = I/A$, where A is the area through which the current passes perpendicularly. Here A is the lateral surface area of the cylinder of radius 6.0 mm and length 20 m. To see its area, make a straight cut down the side, and flatten out the surface to obtain a rectangle of length 20 m and width 2π (6.0 mm). Then,

$$J = \frac{I}{A} = \frac{10 \times 10^{-6}\ \text{A}}{2\pi(6 \times 10^{-3}\ \text{m})(20\ \text{m})} = 1.3 \times 10^{-5}\ \text{A}/\text{m}^2 \quad \Diamond$$

---

21. A wire with a resistance R is lengthened to 1.25 times its original length by pulling it through a small hole. Find the resistance of the wire after it is stretched.

Solution We assume its density and mass are unchanged, so its volume is constant. Call its original cross-section area A_1 and the area of the hole A_2.

Then $L_1A_1 = L_2A_2$, $L_2 = 1.25\,L_1$, and $A_1 = 1.25\,A_2$.

$R = \dfrac{\rho L_1}{A_1}$ changes to $$R_2 = \frac{\rho L_2}{A_2} = \rho\frac{L_1(1.25)}{(A_1/1.25)} = (1.25)^2\left(\frac{\rho L_1}{A_1}\right)$$

$$R_2 = (1.25)^2 R = 1.56R \quad \Diamond$$

---

23. Suppose that you wish to fabricate a uniform wire out of 1.0 g of copper. If the wire is to have a resistance of $R = 0.50\ \Omega$, and all of the copper is to be used, what will be (a) the length and (b) the diameter of this wire?

Solution Don't mix up symbols! Call the density ρ_d and the resistivity ρ_r. Then from $\rho_d = m/V$, the volume is $V = A\,L = \dfrac{m}{\rho_d}$. The resistance is $R = \dfrac{\rho_r L}{A}$

(a) We can solve for L by eliminating A:

$$A = \frac{m}{L\rho_d} \qquad R = \frac{\rho_r L}{(m/L\rho_d)} \qquad R = \frac{\rho_r \rho_d L^2}{m}$$

$$L = \sqrt{\frac{mR}{\rho_r\rho_d}} = \sqrt{\frac{(1.0\times 10^{-3}\ \text{kg})(0.50\ \Omega)}{(1.7\times 10^{-8}\ \Omega\cdot\text{m})(8.93\times 10^{3}\ \text{kg/m}^3)}} = 1.8\ \text{m} \quad \Diamond$$

(b) To have a single diameter, the wire has a circular cross section:

$$A = \pi r^2 = \pi\left(\frac{d}{2}\right)^2 = \frac{m}{L\rho_d}$$

$$d = \sqrt{\frac{4\,m}{\pi L\rho_d}} = \sqrt{\frac{4(1.0\times 10^{-3}\ \text{kg})}{\pi(1.8\ \text{m})(8.93\times 10^{3}\ \text{kg/m}^3)}}$$

$$d = 0.28\ \text{mm} \quad \Diamond$$

---

25. A 0.90-V potential difference is maintained across a 1.5-m length of tungsten wire that has a cross-sectional area of 0.60 mm². What is the current in the wire?

Solution From Ohm's law, $I = \frac{V}{R}$ where $R = \frac{\rho L}{A}$. Therefore,

$$I = \frac{VA}{\rho L} = \frac{(0.90 \text{ V})(6.0 \times 10^{-7} \text{ m}^2)}{(5.6 \times 10^{-8} \ \Omega \cdot \text{m})(1.5 \text{ m})} = 6.4 \text{ A} \quad \lozenge$$

---

26. The electron beam emerging from a certain high-energy electron accelerator has a circular cross section of radius 1.00 mm. (a) If the beam current is 8.00 μA, find the current density in the beam, assuming that it is uniform throughout. (b) The speed of the electrons is so close to the speed of light that their speed can be taken as $c = 3.00 \times 10^8$ m/s with negligible error. Find the electron density in the beam. (c) How long does it take for an Avogadro's number of electrons to emerge from the accelerator?

Solution

(a) $J = \frac{I}{A} = \frac{8 \times 10^{-6} \text{ A}}{\pi(1 \times 10^{-3} \text{ m})^2} = 2.55 \text{ A/m}^2 \quad \lozenge$

(b) From $J = nev_d$, we have

$$n = \frac{J}{ev_d} = \frac{2.55 \text{ A/m}^2}{(1.60 \times 10^{-19} \text{ C})(3.0 \times 10^8 \text{ m/s})}$$

$$n = 5.31 \times 10^{10} \text{ m}^{-3} \quad \lozenge$$

(c) From $I = \frac{\Delta Q}{\Delta t}$, we have

$$\Delta t = \frac{\Delta Q}{I} = \frac{N_A e}{I} = \frac{(6.02 \times 10^{23})(1.60 \times 10^{-19} \text{ C})}{8.00 \times 10^{-6} \text{ A}}$$

$$\Delta t = 1.20 \times 10^{10} \text{ s} \quad \lozenge \quad \text{(or about 381 years!)}$$

---

29. An aluminum wire with a diameter of 0.10 mm has a uniform electric field of 0.20 V/m imposed along its entire length. The temperature of the wire is 50°C. Assume one free electron per atom. (a) Use the information in Table 27.1 and determine the resistivity. (b) What is the current density in the wire? (c) What is the total current in the wire? (d) What is the drift speed of the conduction electrons? (e) What potential difference must exist between the ends of a 2.0-m length of the wire to produce the stated electric field strength?

Solution

(a) $\rho = \rho_0(1+\alpha(T-T_0)) = (2.82\times10^{-8}\ \Omega\cdot\text{m})\left(1+(3.9\times10^{-3}\ {}^\circ\text{C}^{-1})\ (30\ {}^\circ\text{C})\right) = 3.15\times10^{-8}\ \Omega\cdot\text{m}$ ◊

(b) $J = \sigma E = E/\rho = \dfrac{(0.20\ \text{V}/\text{m})}{(3.15\times10^{-8}\ \Omega\cdot\text{m})}\left(1\dfrac{\Omega\cdot\text{A}}{\text{V}}\right) = 6.35\times10^{6}\ \text{A}/\text{m}^2$ ◊

(c) $J = \dfrac{I}{A}$ $\quad I = JA = J\pi r^2$

$$I = \pi(0.050\times10^{-3}\ \text{m})^2\left(6.35\times10^{6}\ \frac{\text{A}}{\text{m}^2}\right) = 50\ \text{mA}$$ ◊

(d) The number-density of free electrons is given by the mass density:

$$n = 2.70\times10^{3}\ \frac{\text{kg}}{\text{m}^3}\left(\frac{1\ \text{mole}}{26.98\ \text{g}}\right)\left(\frac{10^3\ \text{g}}{\text{kg}}\right)\left(\frac{6.02\times10^{23}\ \text{atoms}}{1\ \text{mole}}\right)\left(\frac{1\ \text{free e}^-}{\text{atom}}\right)$$

$n = 6.02\times10^{28}\ \text{electron}/\text{m}^3$

Now $J = nqv_d$ gives

$$v_d = \frac{J}{nq} = \frac{(6.35\times10^{6}\ \text{A}/\text{m}^2)}{(6.02\times10^{28}\ \text{e}^-/\text{m}^3)(-1.6\times10^{-19}\ \text{C}/\text{e}^-)} = -6.6\times10^{-4}\ \text{m}/\text{s}$$ ◊

The sign indicates that the electrons drift opposite to the field and current.

(e) $V = EL = (0.20\ \text{V/m})(2.0\ \text{m}) = 0.40\ \text{V}$ ◊

31. What is the fractional change in the resistance of an iron filament when its temperature changes from 25°C to 50°C?

Solution $R = R_0[1 + \alpha\,\Delta T]$ or $R - R_0 = R_0\alpha\,\Delta T$

The fractional change in resistance $= f = \dfrac{R - R_0}{R_0}$, therefore

$$f = \frac{R_0\,\alpha\,\Delta T}{R_0} = \alpha\,\Delta T = (5.0\times10^{-3}\ (°\mathrm{C})^{-1})(50.0°\mathrm{C} - 25.0°\mathrm{C}) = 0.125 \quad \Diamond$$

---

40. If the drift velocity of free electrons in a copper wire is 7.84 × 10⁻⁴ m/s, calculate the electric field in the conductor.

Solution The electron density in copper, from Example 27.1, is $8.48\times10^{28}\,/\mathrm{m}^3$. The current density in this wire is

$$J = nqv_d = (8.48\times10^{28}\,/\,\mathrm{m}^3)(1.6\times10^{-19}\ \mathrm{C})(7.84\times10^{-4}\ \mathrm{m/s})$$

$$J = 1.06\times10^{7}\ \mathrm{A/m^2}$$

Now the microscopic form of Ohm's law is

$$J = \sigma E = E/\rho$$

$$E = \rho J = (1.7\times10^{-8}\ \Omega\cdot\mathrm{m})(1.06\times10^{7}\ \mathrm{A/m^2})$$

$$E = 0.180\ \mathrm{V/m} \quad \Diamond$$

---

41. Use data from Example 27.1 to calculate the collision mean free path of electrons in copper if the average thermal speed of conduction electrons is 8.6×10^{5} m/s.

Solution The resistivity can be expressed as $\rho = \dfrac{m}{nq^2\tau}$ where τ is the average time between collisions.

$$\tau = \frac{m}{np q^2} = \frac{9.11\times10^{-31}\text{ kg}}{(8.48\times10^{28}/\text{m}^3)(1.7\times10^{-8}\ \Omega\cdot\text{m})(1.60\times10^{-19}\text{ C})^2}$$

$$\tau = 2.46\times10^{-14}\text{ s}$$

The mean free path $L = \bar{v}\tau$ where $\bar{v}$ is the average thermal speed. Therefore,

$$L = (8.6\times10^5\text{ m/s})(2.46\times10^{-14}\text{ s}) = 2.16\times10^{-8}\text{ m} = 21.6\text{ nm} \quad \lozenge$$

---

45. Suppose that a voltage surge produces 140 V for a moment. By what percentage will the output of a 120-V, 100-W light bulb increase, assuming its resistance does not change?

Solution We find the resistance:

$$P_1 = V_1 I_1$$

$$I_1 = \frac{P_1}{V_1} = \frac{100\text{ W}}{120\text{ V}} = 0.833\text{ A}$$

$$R = \frac{V_1}{I_1} = \frac{120\text{ V}}{0.833\text{ A}} = 144\ \Omega$$

Now the current is larger, $I_2 = \frac{V_2}{R} = \frac{140\text{ V}}{144\ \Omega} = 0.972\text{ A}$

and the power is much larger: $P_2 = I_2 V_2 = (0.972\text{ A})(140\text{ V}) = 136\text{ W}$

The percent increase is $\frac{136\text{ W} - 100\text{ W}}{100\text{ W}} = 0.361 = 36.1\% \quad \lozenge$

---

51. What is the required resistance of an immersion heater that will increase the temperature of 1.5 kg of water from 10 °C to 50 °C in 10 min while operating at 110 V?

Solution Assume $E_{(\text{thermal})} = E_{(\text{electrical})}$

$$E_{(\text{thermal})} = mc\,\Delta T \quad \text{and} \quad E_{(\text{electrical})} = \left(\frac{V^2}{R}\right)t$$

Therefore, since $c = 4186\ \text{J/kg}\cdot{}^\circ\text{C}$

$$R = \frac{V^2 t}{cm\,\Delta T} = \frac{(110\ \text{V})^2(600\ \text{s})}{(4186\ \text{J/kg}\cdot{}^\circ\text{C})(1.50\ \text{kg})(40\ {}^\circ\text{C})} = 29\ \Omega \quad \lozenge$$

---

53. Compute the cost per day of operating a lamp that draws 1.7 A from a 110-V line if the cost of electrical energy is \$0.06/kWh.

Solution The power of the lamp is $P = VI = U/t$, where U is the energy transformed. Then the energy you buy is

$$U = VIt = (110\ \text{V})(1.7\ \text{A})(1\ \text{day})\left(\frac{24\ \text{h}}{1\ \text{day}}\right)\left(\frac{3600\ \text{s}}{\text{h}}\right)\left(\frac{1\ \text{J}}{\text{V}\cdot\text{C}}\right)\left(\frac{1\ \text{C}}{\text{A}\cdot\text{s}}\right) = 16.2\ \text{MJ in standard units;}$$

In kilowatt hours,

$$U = VIt = (110\ \text{V})(1.7\ \text{A})(1\ \text{day})\left(\frac{24\ \text{h}}{1\ \text{day}}\right)\left(\frac{\text{J}}{\text{V}\cdot\text{C}}\right)\left(\frac{\text{C}}{\text{A}\cdot\text{s}}\right)\left(\frac{\text{W}\cdot\text{s}}{\text{J}}\right) = 4.49\ \text{kW}\cdot\text{h}$$

So the lamp costs $(4.49\ \text{kWh})\left(\frac{\$0.06}{\text{kWh}}\right) = 26.9\ \text{cents} \quad \lozenge$

---

55. A certain toaster has a heating element made of Nichrome resistance wire. When first connected to a 120-V voltage source (and the wire is at a temperature of 20.0°C), the initial current is 1.80 A but begins to decrease as the resistive element heats up. When the toaster has reached its final operating temperature, the current has dropped to 1.53 A. (a) Find the power the toaster consumes when it is at its operating temperature. (b) What is the final temperature of the heating element?

Solution

(a) $P = VI = (120\text{ V})(1.53\text{ A}) = 184\text{ W}$ ◊

(b) The resistance at 20°C is $R_0 = \frac{V}{I} = \frac{120\text{ V}}{1.8\text{ A}} = 66.7\ \Omega$

At operating temperature, $R = \frac{120\text{ V}}{1.53\text{ A}} = 78.4\ \Omega$

Neglecting thermal expansion, we have

$$R = \frac{\rho L}{A} = \frac{\rho_0(1+\alpha(T-T_0))L}{A} = R_0(1+\alpha(T-T_0))$$

$$T = T_0 + \frac{R/R_0 - 1}{\alpha} = 20°\text{C} + \frac{78.4\ \Omega/66.7\ \Omega - 1}{0.4\times 10^{-3}/\text{C}°} = 461°\text{C}$$ ◊

---

59. An electric car is designed to run off a bank of 12-V batteries with total energy storage of 2.0×10^7 J. (a) If the electric motor draws 8.0 kW, what is the current delivered to the motor? (b) If the electric motor draws 8.0 kW as the car moves at a steady speed of 20 m/s, how far will the car travel before it is "out of juice"?

Solution

(a) Since $P = VI$ $\quad I = \frac{P}{V} = \frac{8.0\times 10^3\text{ W}}{12\text{ V}} = 667\text{ A}$ ◊

(b) We find the time the car runs from $P = \frac{U}{t}$

$$t = \frac{U}{P} = \left(\frac{2.0\times 10^7\text{ J}}{8.0\times 10^3\text{ W}}\right)\left(\frac{1\text{ W}\cdot\text{s}}{\text{J}}\right) = 2.5\times 10^3\text{ s}$$

So it moves a distance of $\quad x = vt = (20\text{ m/s})(2.5\times 10^3\text{ s}) = 50\text{ km}$ ◊

---

61. A Wheatstone bridge can be used to measure the strain ($\Delta L/L_0$) of a wire (see Section 12.4), where L_0 is the length before stretching, L is the length after stretching, and $\Delta L = L - L_0$. Let $\alpha = \Delta L/L_0$. Show that the resistance is $R = R_0\,(1 + 2\alpha + \alpha^2)$ for any length where $R_0 = \rho L_0 / A_0$. Assume that the resistivity and volume of the wire stay constant.

Solution On the construction of a Wheatstone bridge, see Section 28.6.

The original resistance is $R_0 = \dfrac{\rho L_0}{A_0}$. Assuming constant volume gives $L_0 A_0 = LA$, where A is the final cross-sectional area.

With α representing the strain, $\alpha = \dfrac{\Delta L}{L_0} = \dfrac{L - L_0}{L_0}$, so $L = L_0 + \alpha L_0$,

We then have $L_0 A_0 = L_0(1+\alpha)A$, and $A = \dfrac{A_0}{1+\alpha}$

The new resistance is
$$R = \frac{\rho L}{A} = \frac{\rho L_0(1+\alpha)}{A_0/(1+\alpha)} = \left(\frac{\rho_0 L_0}{A_0}\right)(1+\alpha)^2$$

$$R = R_0(1+2\alpha+\alpha^2) \quad \Diamond$$

---

63. A resistor is constructed by forming a material of resistivity ρ into the shape of a hollow cylinder of length L and inner and outer radii r_a and r_b, respectively (Fig. P27.63). In use, a potential difference is applied between the ends of the cylinder, producing a current parallel to the axis. (a) Find a general expression for the resistance of such a device in terms of L, ρ, r_a, and r_b.

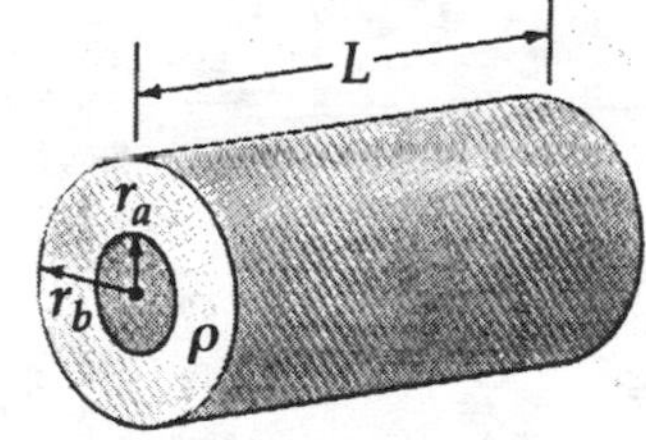

Figure P27.63

(b) Obtain a numerical value for R when L = 4.0 cm, r_a = 0.50 cm, r_b = 1.2 cm, and the resistivity $\rho = 3.5 \times 10^5\ \Omega\cdot\text{m}$. (c) Suppose now that the potential difference is applied between the inner and outer surfaces so that the resulting current flows radially outward. Find a general expression for the resistance of the device in terms of L, ρ, r_a, and r_b. (d) Calculate the value of R using the parameter values given in part (b).

Solution

(a) $R = \dfrac{\rho L}{A} = \dfrac{\rho L}{\pi(r_b^2 - r_a^2)}$ ◊

where A is the cross-sectional area of the conductor in the shape of a *hollow* cylinder.

(b) $R = \dfrac{\rho L}{\pi(r_b^2 - r_a^2)} = \dfrac{(3.5\times10^5\ \Omega\cdot\text{m})(0.040\ \text{m})}{\pi\left[(0.012\ \text{m})^2 - (0.0050\ \text{m})^2\right]} = 3.74\times10^7\ \Omega = 37\ \text{M}\Omega$ ◊

(c) A cylindrical shell of radius r, length L, and thickness dr will have resistance

$$dR = \frac{\rho dr}{2\pi rL} \quad \text{for current flowing radially.}$$

The whole resistance of the hollow cylinder is

$$R = \int dR = \int_{r_a}^{r_b} \frac{\rho dr}{2\pi rL} = \frac{\rho}{2\pi L}\ln r\Big]_{r_a}^{r_b} = \frac{\rho}{2\pi L}\ln\left(\frac{r_b}{r_a}\right) \quad ◊$$

(d) $$R = \frac{3.5\times10^5\ \Omega\cdot\text{m}}{2\pi(4.0\times10^{-2}\ \text{m})}\ln\left(\frac{1.2}{0.50}\right) = 1.2\ \text{M}\Omega \quad ◊$$

---

65. A more general definition of the temperature coefficient of resistivity is

$$\alpha = \frac{1}{\rho}\frac{d\rho}{dT}$$

where ρ is the resistivity at temperature T. (a) Assuming that α is constant, show that

$$\rho = \rho_0 e^{\alpha(T-T_0)}$$

where ρ_0 is the resistivity at temperature T_0. (b) Using the series expansion ($e^x \approx 1 + x$; $x \ll 1$), show that the resistivity is given approximately by the expression $\rho = \rho_0[1 + \alpha(T - T_0)]$ for $\alpha(T - T_0) \ll 1$.

Solution $\alpha = \frac{1}{\rho}\frac{d\rho}{dT}$

(a) Separating variables, $\int_{\rho_0}^{\rho} \frac{d\rho}{\rho} = \int_{T_0}^{T} \alpha\, dT$

$$\ln\left(\frac{\rho}{\rho_0}\right) = \alpha(T - T_0)$$

$$\rho = \rho_0 e^{\alpha(T - T_0)} \quad \Diamond$$

(b) From the series expansion $e^x \approx 1 + x$, with x much less than 1,

$$\rho = \rho_0[1 + \alpha(T - T_0)] \quad \Diamond$$

---

67. Material with uniform resistivity ρ is formed into a wedge as shown in Figure P27.67. Show that the resistance between face A and face B of this wedge is

$$R = \rho \frac{L}{w(y_2 - y_1)} \ln\left(\frac{y_2}{y_1}\right)$$

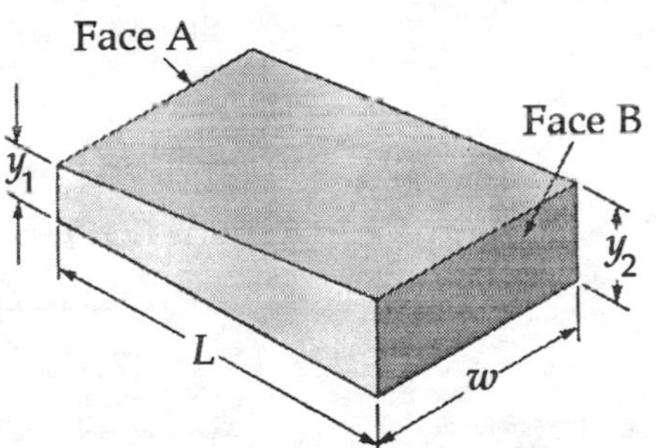

Figure P27.67

Solution The current flows generally parallel to L. Consider a slice of the material perpendicular to this current, of thickness dx, and at distance x from face A. Then the other dimensions of the slice are w and y, where

$$\frac{y - y_1}{x} = \frac{y_2 - y_1}{L} \text{ by proportion, so } y = y_1 + (y_2 - y_1)\frac{x}{L}.$$

The bit of resistance which this slice contributes is

$$dR = \frac{\rho dx}{A} = \frac{\rho dx}{wy} = \frac{\rho dx}{w(y_1 + (y_2 - y_1)x/L)}$$

and the whole resistance is that of all the slices:

$$R = \int_{x=0}^{L} dR = \int_0^L \frac{\rho dx}{w(y_1 + (y_2 - y_1)x/L)}$$

$$= \frac{\rho}{w} \frac{L}{y_2 - y_1} \int_{x=0}^{L} \frac{((y_2 - y_1)/L)dx}{y_1 + (y_2 - y_1)x/L}$$

With $u = y_1 + (y_2 - y_1)x/L$, this is of the form $\int du/u$, so

$$R = \frac{\rho L}{w(y_2 - y_1)} \ln \left(y_1 + (y_2 - y_1)x / L\right)\Big]_{x=0}^{L}$$

$$= \frac{\rho L}{w(y_2 - y_1)} (\ln y_2 - \ln y_1)$$

$$= \frac{\rho L}{w(y_2 - y_1)} \ln \frac{y_2}{y_1} \quad \Diamond$$

---

69. An experiment is conducted to measure the electrical resistivity of Nichrome in the form of wires with different lengths and cross-sectional areas. For one set of measurements, a student uses #30 gauge wire, which has a cross-sectional area of 7.3×10^{-8} m^2. The voltage across the wire and the current in the wire are measured with a voltmeter and ammeter, respectively. For each of the measurements given in the table below taken on wires of three different lengths, calculate the resistance of the wires and the corresponding values of the resistivity. What is the average value of the resistivity, and how does it compare with the value given in Table 27.1?

L (m)	V (V)	I (A)	R (Ω)	ρ (Ω·m)
0.540	5.22	0.500		
1.028	5.82	0.276		
1.543	5.94	0.187		

Solution Find each resistance from $R = \frac{V}{I}$, as in $\frac{5.22\ \text{V}}{0.500\ \text{A}} = 10.4\ \Omega$.

Find each resistivity from $R = \frac{V}{I} = \frac{\rho L}{A}$

$$\rho = \frac{VA}{IL} = \frac{(5.22\ \text{V})(7.3\times10^{-8}\ \text{m}^2)}{(0.500\ \text{A})(0.540\ \text{m})}$$

$$\rho = 1.41\times10^{-6}\ \Omega\cdot\text{m}$$

To obtain

L (m)	R (Ω)	ρ (Ω·m)
0.540	10.4	1.41×10^{-6}
1.028	21.1	1.50×10^{-6}
1.543	31.8	1.50×10^{-6}

Thus the average resistivity is $\rho = 1.47\times10^{-6}\ \Omega\cdot\text{m}$ ◊

This differs from the tabulated $1.50\times10^{-6}\ \Omega\cdot\text{m}$ by 2%. The difference is accounted for by the experimental uncertainty, which we may estimate as

$$\frac{1.47-1.41}{1.47} = 4\%$$

Chapter 28

Direct Current Circuits

DIRECT CURRENT CIRCUITS

INTRODUCTION

This chapter analyzes some simple circuits whose elements include batteries, resistors, and capacitors in various combinations. Such analysis is simplified by the use of two rules known as *Kirchhoff's rules*, which follow from the laws of conservation of energy and conservation of charge. Most of the circuits are assumed to be in *steady state*, which means that the currents are constant in magnitude and direction. We close the chapter with a discussion of circuits containing resistors and capacitors, in which current varies with time.

NOTES FROM SELECTED CHAPTER SECTIONS

28.2 Resistors in Series and in Parallel

The *current* must be the same for each of a group of resistors connected in *series*.

The *potential difference* must be the same across each of a group of resistors in *parallel*.

28.3 Kirchhoff's Rules

- The sum of the currents entering any junction must equal the sum of the currents leaving that junction. (A junction is any point in the circuit where the current can split.)

- The algebraic sum of the changes in potential around any closed circuit loop must be *zero*.

The first rule is a statement of *conservation of charge*; the second rule follows from the *conservation of energy*.

28.4 *RC* Circuits

Consider an uncharged capacitor in series with a resistor, a battery, and a switch. In the charging process, charges are transferred from one plate of the capacitor to the other moving along a path *through the resistor, battery, and switch.* The charges *do not move across the gap between the plates of the capacitor.*

The battery does work on the charges to increase their electrostatic potential energy as they move from one plate to the other.

EQUATIONS AND CONCEPTS

When a battery is providing a current to an external circuit, the *terminal voltage* of the battery will be less than the emf due to *internal resistance* of the battery.

$$V = \mathcal{E} - Ir \qquad (28.1)$$

The *current, I,* delivered by a battery in a simple *dc* circuit, depends on the value of the *emf* of the source, $\mathcal{E}$; the total *load resistance* in the circuit, R; and the *internal resistance* of the source, r.

$$I = \frac{\mathcal{E}}{R + r} \qquad (28.3)$$

The total or equivalent resistance of a series combination of resistors is equal to the sum of the resistances of the individual resistors.

$$R_{eq} = R_1 + R_2 + R_3 + \ldots \qquad (28.6)$$

(Series combination)

A group of resistors connected in parallel has an equivalent resistance which is less than the smallest individual value of resistance in the group.

$$\frac{1}{R_{eq}} = \frac{1}{R_1} + \frac{1}{R_2} + \frac{1}{R_3} + \ldots \qquad (28.8)$$

(Parallel combination)

Resistors in series are connected so that they have only one common circuit point per pair; there is a common current through each resistor in the group.

Resistors in parallel are connected so that each resistor in the group has two circuit points in common with each of the other resistors; there is a common potential difference across each resistor in the group.

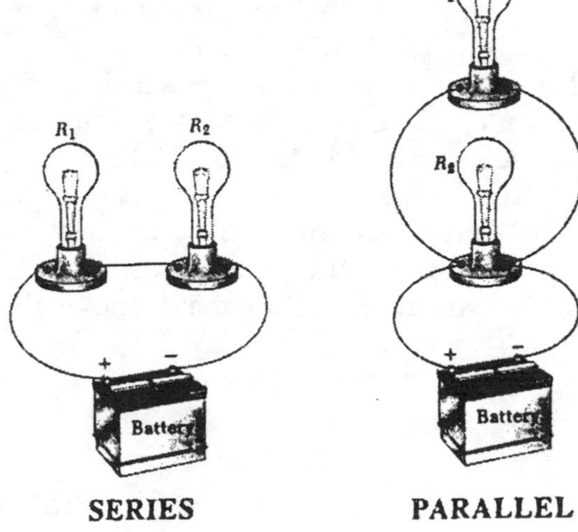

Many circuits which contain several resistors can be reduced to an equivalent single-loop circuit by successive step-by-step combinations of groups of resistors in series and parallel.

In the most general case, however, successive reduction is not possible and you must solve a true multiloop circuit by use of Kirchhoff's rules. Review the procedure suggested in the next section to apply Kirchhoff's rules.

When a potential difference is suddenly applied across an uncharged capacitor, the *current* in the circuit and the *charge* on the capacitor are functions of time. The *instantaneous values* of I and q depend on the capacitance and on the resistance in the circuit.

$$I(t) = \frac{\mathcal{E}}{R} e^{-t/RC} \qquad (28.13)$$

$$q(t) = C\mathcal{E}\left[1 - e^{-t/RC}\right] \qquad (28.12)$$

When a battery is used to charge a capacitor in series with a resistor, a quantity τ, called the time constant of the circuit, is used to describe the manner in which the charge on the capacitor varies with time. The charge on the capacitor increases from zero to 63% of its maximum value in a time interval equal to one time constant. Also, during one time constant, the charging current decreases from its initial maximum value of $I_0 = \frac{\mathcal{E}}{R}$ to 37% of I_0.

$$\tau = RC$$

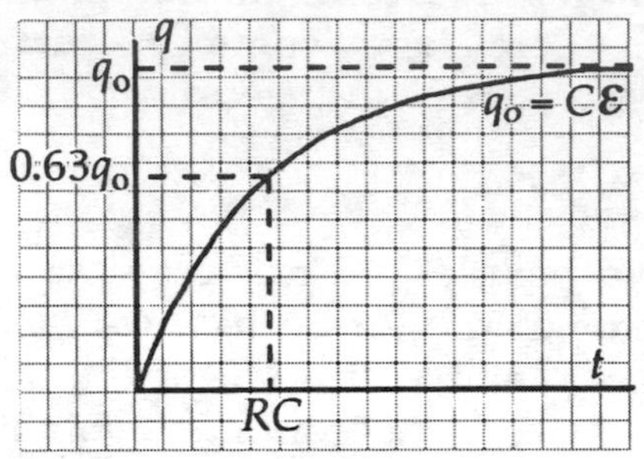

When a capacitor with an initial charge Q is discharged through a resistor, the *charge* and *current* decrease exponentially in time.

$$q(t) = Qe^{-t/RC} \tag{28.15}$$

$$I(t) = I_0 e^{-t/RC} \tag{28.16}$$

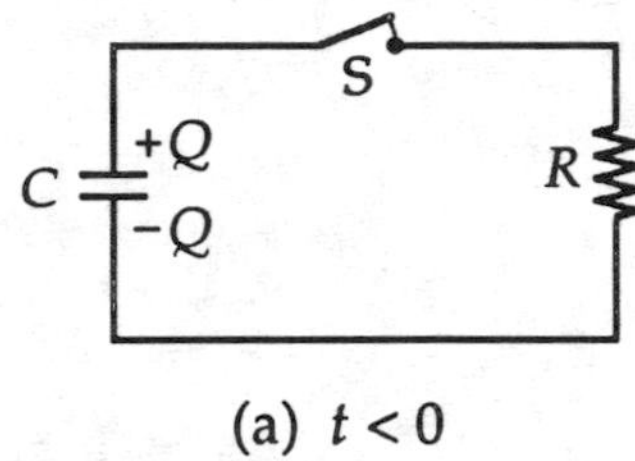

(a) $t < 0$

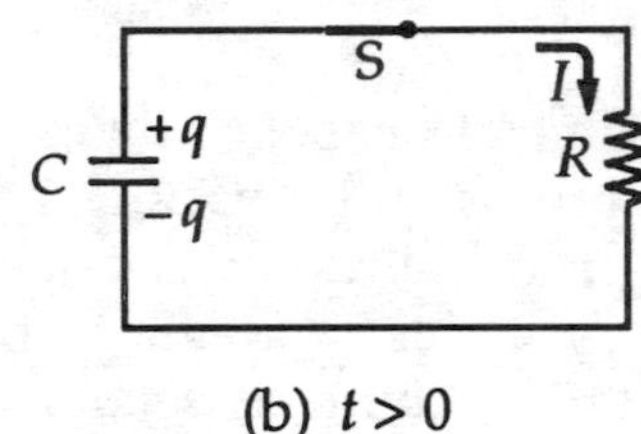

(b) $t > 0$

SUGGESTIONS, SKILLS, AND STRATEGIES

A problem-solving strategy for resistors:

- Be careful with your choice of units. To calculate the resistance of a device in ohms, make sure that distances are in meters and use the SI value of ρ.

- When two or more unequal resistors are connected in *series*, they carry the same current, but the potential differences across them are not the same. The resistors add directly to give the equivalent resistance of the series combination.

- When two or more unequal resistors are connected in *parallel*, the potential differences across them are the same. Since the current is inversely proportional to the resistance, the currents through them are not the same. The equivalent resistance of a parallel combination of resistors is found through reciprocal addition, and the equivalent resistor is always *less* than the smallest individual resistor.

- A complicated circuit consisting of resistors can often be reduced to a simple circuit containing only one resistor. To do so, repeatedly examine the circuit and replace any resistors that are in series or in parallel, using the procedures outlined above. Sketch the new circuit after each set of changes has been made. Continue this process until a single equivalent resistance is found.

- If the current through, or the potential difference across, a resistor in the complicated circuit is to be identified, start with the final equivalent circuit found in the last step, and gradually work your way back through the circuits, using $V = IR$ to find the voltage drop across each equivalent resistor.

A strategy for using Kirchhoff's rules:

- First, draw the circuit diagram and assign labels and symbols to all the known and unknown quantities. You must assign *directions* to the currents in each part of the circuit. Do not be alarmed if you guess the direction of a current incorrectly; the resulting value will be negative, but *its magnitude will be correct*. Although the assignment of current directions is arbitrary, you must stick with your guess throughout as you apply Kirchhoff's rules.

- Apply the junction rule to any junction in the circuit. The junction rule may be applied as many times as a new current (one not used in a previous application) appears in the resulting equation. In general, the number of times the junction rule can be used is one fewer than the number of junction points in the circuit.

- Now apply Kirchhoff's loop rule to as many loops in the circuit as are needed to solve for the unknowns. Remember you must have as many equations as there are unknowns (I's, R's, and $\mathcal{E}$'s). In order to apply this rule, you must correctly identify the change in potential as you cross each element in traversing the closed loop. Watch out for signs!

  Convenient "rules of thumb" which you may use to determine the increase or decrease in potential as you cross a resistor or seat of emf in traversing a circuit loop are illustrated in the following figure. Notice that the potential *decreases* (changes by

−*IR*) when the resistor is traversed *in the direction of the current*. There is an *increase* in potential of $+IR$ if the direction of travel is *opposite* the direction of current. If a seat of emf is traversed *in* the direction of the emf (from − to + on the battery), the potential *increases* by $\mathcal{E}$. If the direction of travel is from + to −, the potential *decreases* by $\mathcal{E}$ (changes by $-\mathcal{E}$).

$\Delta V = V_b - V_a = -IR$

$\Delta V = V_b - V_a = IR$

$\Delta V = V_b - V_a = \mathcal{E}$

$\Delta V = V_b - V_a = -\mathcal{E}$

- Finally, you must solve the equations simultaneously for the unknown quantities. Be careful in your algebraic steps, and check your numerical answers for consistency.

As an illustration of the use of Kirchhoff's rules, consider a three-loop circuit which has the *general form* shown in the following figure on the left. In this illustration, the actual circuit elements, *R*'s and $\mathcal{E}$'s are not shown but assumed known. There are six possible different values of *I* in the circuit; therefore you will need six independent equations to solve for the six values of *I*. There are four junction points in the circuit (at points *a*, *d*, *f*, and *h*). The first rule applied at *any three* of these points will yield three equations. The circuit can be thought of as a group of three "blocks" or meshes as shown in the following figure on the right. Kirchhoff's second law, when applied to each of these loops (*abcda*, *ahfga*, and *defhd*), will yield three additional equations. You can then solve the total of six equations simultaneously for the six values of I_1, I_2, I_3, I_4, I_5, and I_6. You can, of course, expect that the sum of the changes in potential difference around *any other closed loop* in the circuit will be zero (for example, *abcdefga* or *ahfedcba*); however the equations found by applying Kirchhoff's second rule to these additional loops *will not be independent* of the six equations found previously.

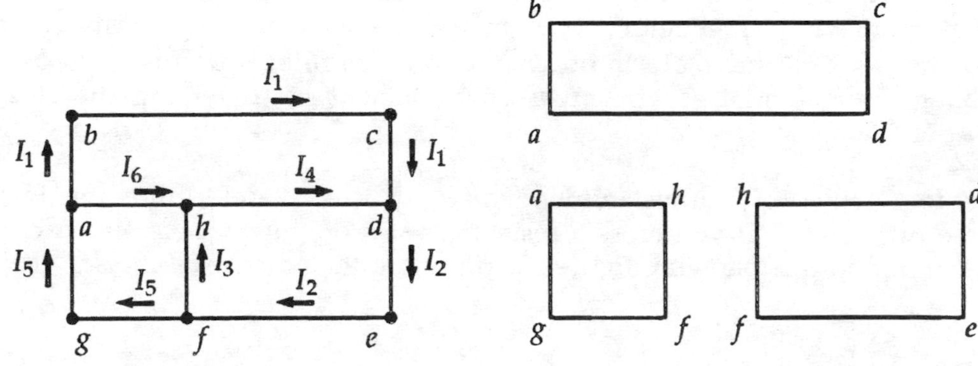

REVIEW CHECKLIST

- ▷ Calculate the equivalent resistance of a group of resistors in parallel, series, or series-parallel combination.
- ▷ Calculate the current in a single-loop circuit and the potential difference between any two points in the circuit. Use Ohm's law to calculate the current in a circuit and the potential difference between any two points in a circuit which can be reduced to an equivalent single-loop circuit.
- ▷ Apply Kirchhoff's rules to solve multiloop circuits; that is, find the currents and the potential difference between any two points.
- ▷ Calculate the charging (discharge) current $i(t)$ and the accumulated (residual) charge $q(t)$ during charging (and discharge) of a capacitor in an RC circuit.
- ▷ Understand the circuitry and make calculations for an unknown resistance, R_x, using the ammeter-voltmeter method and the Wheatstone bridge method. Determine the value of an unknown emf, $\mathcal{E}_x$, using a potentiometer circuit.

SOLUTIONS TO SELECTED END-OF-CHAPTER PROBLEMS

7. A battery has an emf of 15.0 V. The terminal voltage of the battery is 11.6 V when it is delivering 20.0 W of power to an external load resistor R. (a) What is the value of R? (b) What is the internal resistance of the battery?

Solution

(a) Combining Joule's law, $P = VI$, and Ohm's law, $V = IR$, gives

$$R = \frac{V^2}{P} = \frac{(11.6\text{ V})^2}{20.0\text{ W}} = 6.73\ \Omega \quad \Diamond$$

(b) $\mathcal{E} = IR + Ir$

$$r = \frac{\mathcal{E} - IR}{I} \quad \text{where} \quad I = \frac{V}{R}$$

Therefore, $$r = \frac{(\mathcal{E} - V)R}{V} = \frac{(15.0\text{ V} - 11.6\text{ V})(6.73\ \Omega)}{11.6\text{ V}} = 1.97\ \Omega \quad \Diamond$$

9. Two 1.50-V batteries—with their positive terminals in the same direction—are inserted in series into the barrel of a flashlight. One battery has an internal resistance of 0.255 Ω, the other an internal resistance of 0.153 Ω. When the switch is closed, a current of 600 mA occurs in the lamp. (a) What is the lamp's resistance? (b) What fraction of the power dissipated is dissipated in the batteries?

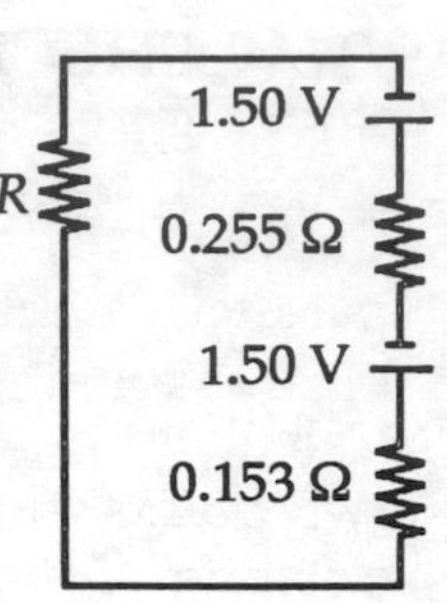

Solution Kirchhoff's loop theorem says

(a) $1.50\text{ V} - (0.255\ \Omega)(0.600\text{ A}) + 1.50\text{ V} - (0.153\ \Omega)(0.600\text{ A}) - R(0.600\text{ A}) = 0$

Solving, we get $R = \dfrac{2.76\text{ V}}{0.6\text{ A}} = 4.59\ \Omega$ ◊

(b) The total power converted in the circuit is the power output of the emfs:

$$(1.50\text{ V})(0.600\text{ A}) + (1.50\text{ V})(0.600\text{ A}) = 1.800\text{ W}$$

The power dissipated into heat in the batteries is

$$I^2R_1 + I^2R_2 = (0.600\text{ A})^2(0.255\ \Omega + 0.153\ \Omega) = 0.147\text{ W}$$

So the fractional inefficiency is $\dfrac{0.147\text{ W}}{1.800\text{ W}} = 0.0816$ ◊

---

13. The current in a circuit is tripled by connecting a 500-Ω resistor in parallel with the resistance of the circuit. Determine the resistance of the circuit in the absence of the 500-Ω resistor.

Solution In both pictures, the resistor R has the same voltage $\mathcal{E}$ across it, and the same current I through it.

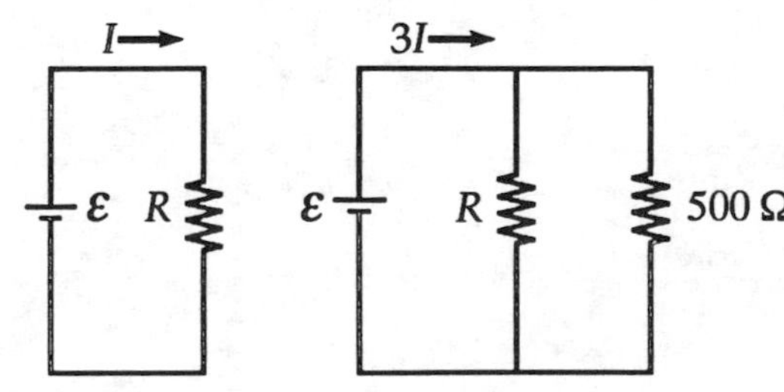

Kirchhoff's junction rule says that the current in the 500 Ω resistor is $3I - I = 2I$. So, $\mathcal{E} = 2I(500\ \Omega)$ and $\mathcal{E} = IR$

Thus, $IR = 2I(500\ \Omega)$ and $R = 1000\ \Omega$ ◊

---

19. Calculate the power dissipated in each resistor in the circuit of Figure P28.19.

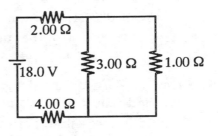

2.00 Ω
18.0 V
3.00 Ω
1.00 Ω
4.00 Ω

Figure P28.19

Solution

To find the power we must find the current in each resistor, so we find the resistance seen by the battery.

The given circuit reduces as shown in the figures, since

$$\frac{1}{\dfrac{1}{1.00\ \Omega}+\dfrac{1}{3.00\ \Omega}} = 0.750\ \Omega.$$

In (b), $I = \dfrac{18\ \text{V}}{6.75\ \Omega} = 2.67\ \text{A}$

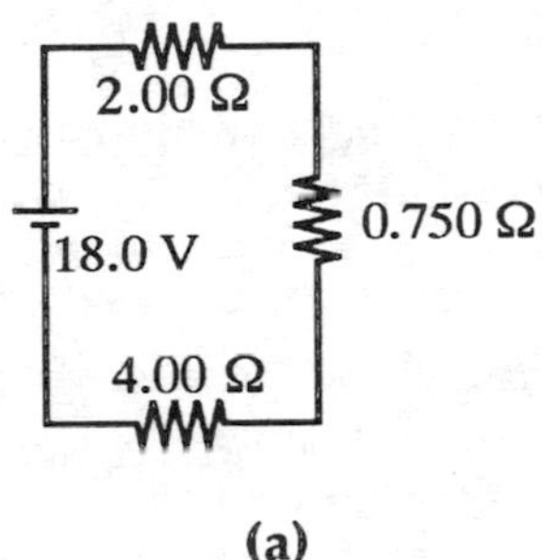

(a)

This is also the current in (a), so the power dissipation of the 2.00-Ω and 4.00-Ω resistors are

$$P_2 = IV = I^2R = (2.67\ \text{A})^2(2.00\ \Omega) = 14.2\ \text{W} \quad \lozenge$$

and $P_4 = I^2R = (2.67\ \text{A})^2(4.00\ \Omega) = 28.4\ \text{W} \quad \lozenge$

The voltage across the 0.750 Ω in (a), and across both the 3.00-Ω and the 1.00-Ω resistor in Figure P28.19, is

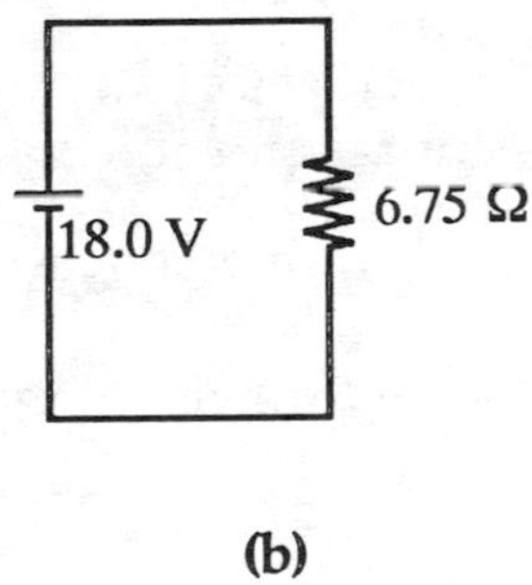

(b)

$$V = IR = (2.67\ \text{A})(0.750\ \Omega) = 2.00\ \text{V}$$

Then for the 3.00-Ω resistor, $I = \dfrac{V}{R} = \dfrac{2.00\ \text{V}}{3.00\ \Omega}$ and

$$P = IV = \left(\frac{2.00\ \text{V}}{3.00\ \Omega}\right)(2.00\ \text{V}) = 1.33\ \text{W} \quad \lozenge$$

For the 1.00-Ω resistor, $I = \dfrac{2.00\ \text{V}}{1.00\ \Omega}$ and $P = \left(\dfrac{2.00\ \text{V}}{1.00\ \Omega}\right)(2.00\ \text{V}) = 4.00\ \text{W} \quad \lozenge$

21. In Figure P28.21, find (a) the current in the 20-Ω resistor and (b) the potential difference between points *a* and *b*.

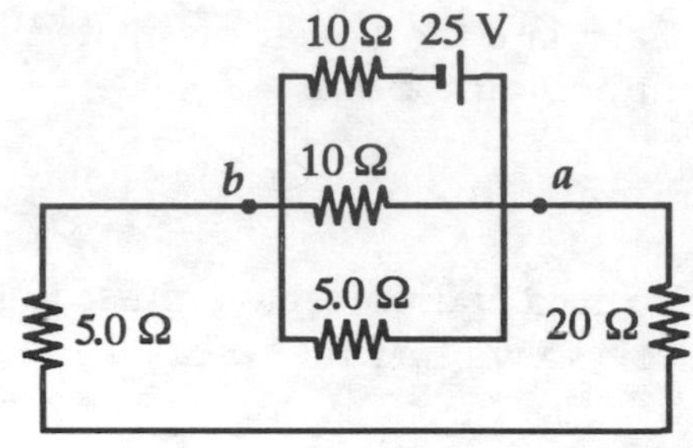

Figure P28.21

Solution If we turn the given diagram on its side, we find that it is the same as figure (a). The 20-Ω and 5.0-Ω resistors are in series, so the first reduction is as shown in (b). In addition, since the 10-Ω, 5.0-Ω, and 25-Ω resistors are then in parallel, we can solve for their equivalent resistance as:

$$R_{eq} = \frac{1}{\left(\frac{1}{10\ \Omega} + \frac{1}{5.0\ \Omega} + \frac{1}{25\ \Omega}\right)} = 2.94\ \Omega.$$

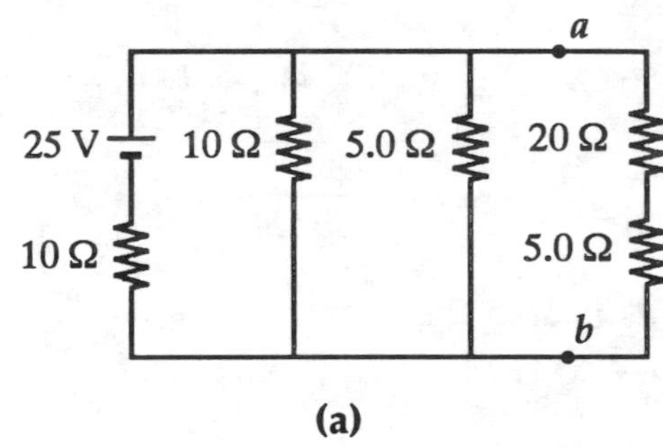

(a)

This is shown in figure (c), which in turn reduces to the circuit shown in (d).

Next, we work backwards through the pictures, applying $I = V/R$ and $V = IR$. The 12.94-Ω resistor is connected across 25-V, so the current through the voltage source in every diagram is

$$I = \frac{V}{R} = \frac{25\ \text{V}}{12.94\ \Omega} = 1.93\ \text{A}$$

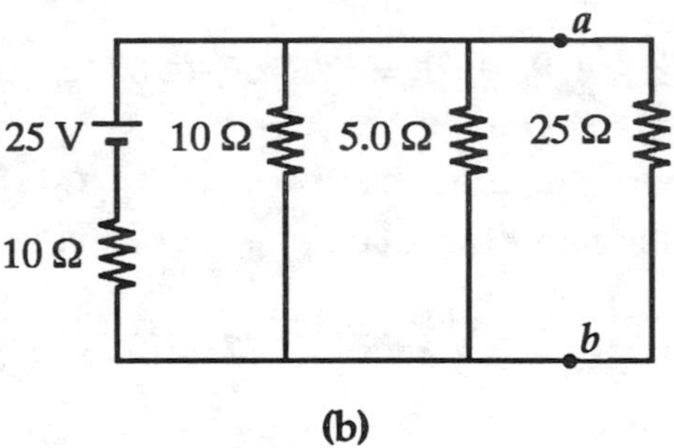

(b)

In figure (c), this 1.93 A goes through the 2.94-Ω equivalent resistor to give a voltage drop of:

$$V = IR = (1.93\ \text{A})(2.94\ \Omega) = 5.68\ \text{V}$$

From figure (b), we see that this voltage drop is the same across V_{ab}, the 10-Ω resistor, and the 5.0-Ω resistor.

(b) Therefore, $V_{ab} = 5.7$ V ◊

Since the current through the 20-Ω resistor is also the current through the 25-Ω line *ab*,

(a) $I = \frac{V_{ab}}{R_{ab}} = \frac{5.68\ \text{V}}{25\ \Omega} = 0.28\ \text{A}$ ◊

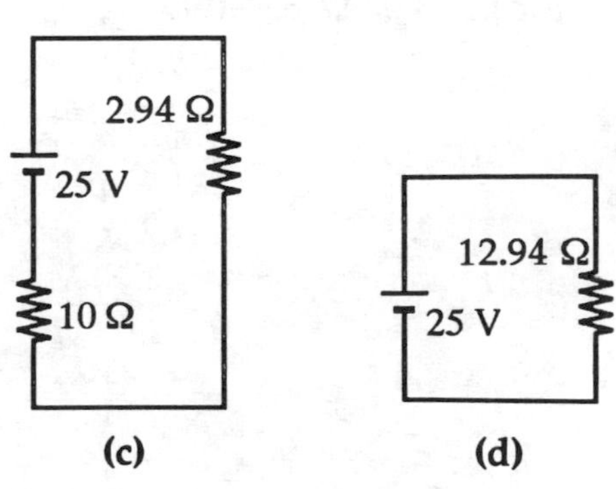

(c) **(d)**

27. Determine the current in each branch in Figure P28.27.

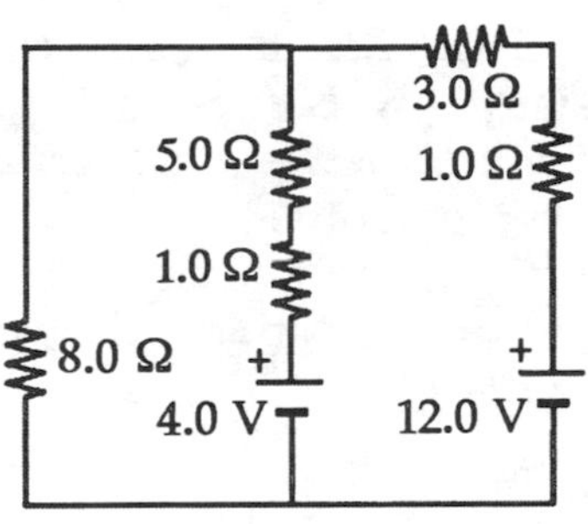

Figure P28.27

Solution First, we arbitrarily define the initial current directions and names, as shown in the figure below.

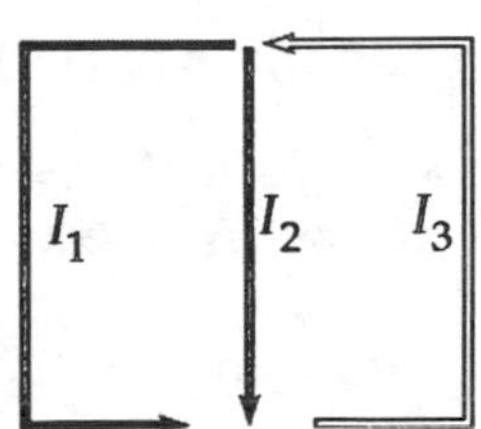

The current rule then says that $I_3 = I_1 + I_2$. (1)

The voltage rule says that in a clockwise trip around the left-hand loop,

$$+I_1(8.0\ \Omega) - I_2(5.0\ \Omega) - I_2(1.0\ \Omega) - 4.0\ \text{V} = 0 \quad (2)$$

and in a clockwise trip around the right-hand mesh,

$$4.0\ \text{V} + I_2(1.0\ \Omega + 5.0\ \Omega) + I_3(3.0\ \Omega + 1.0\ \Omega) - 12.0\ \text{V} = 0 \quad (3)$$

Solving by substitution rather than by determinants has the advantage that (just as when a cat has kittens) the answers, after the first, come out much more easily. Thus we substitute $(I_1 + I_2)$ for I_3, and reduce our three-equations to:

$$\left\{\begin{array}{l}(8.0\ \Omega)I_1 - (6.0\ \Omega)I_2 - 4.0\ \text{V} = 0 \\ \\ 4.0\ \text{V} + (6.0\ \Omega)I_2 + (4.0\ \Omega)(I_1 + I_2) - 12.0\ \text{V} = 0\end{array}\right\} \quad \text{or} \quad \left\{\begin{array}{l}I_2 = \dfrac{(8.0\ \Omega)I_1 - 4.0\ \text{V}}{6.0\ \Omega} \\ \\ -8.0\ \text{V} + (4.0\ \Omega)I_1 + (10.0\ \Omega)I_2 = 0\end{array}\right\}$$

Again substituting for I_2, we end up with: $-8.0\ \text{V} + (4.0\ \Omega)I_1 + \frac{10.0}{6.0}\left((8.0\ \Omega)I_1 - 4.0\ \text{V}\right) = 0$

$$(17.3\ \Omega)I_1 - (14.7\ \text{V}) = 0,$$

and I_1 0.85 A down in the 8 Ω resistor ◊

Thus, $I_2 = \dfrac{(8.0\ \Omega)(0.846\ \text{A}) - 4.0\ \text{V}}{6.0\ \Omega} = 0.46\ \text{A}$ down in middle branch ◊

$I_3 = 0.85\ \text{A} + 0.46\ \text{A} = 1.3\ \text{A}$ up in the right-hand branch ◊

It is a good idea to check the math by substituting back into the voltage-rule equations. If you are new at this, it is worth your time to solve the whole problem again, by taking counterclockwise trips around the loops.

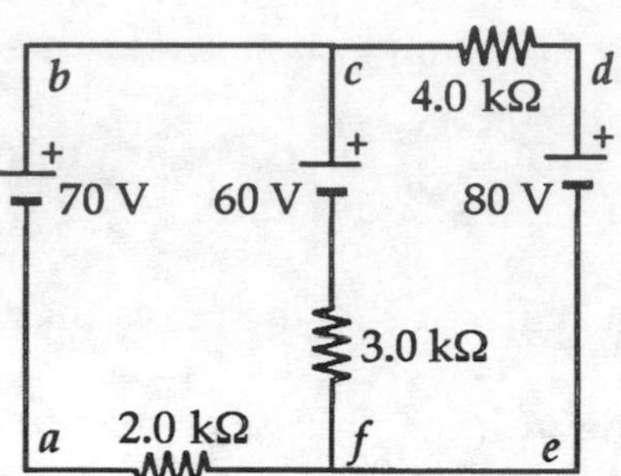

Figure P28.33

33. Using Kirchhoff's rules, (a) find the current in each resistor in Figure P28.33. (b) Find the potential difference between points *c* and *f*. Which point is at the higher potential?

Solution Define currents as shown in the figure, such that $I_1 + I_3 = I_2$:

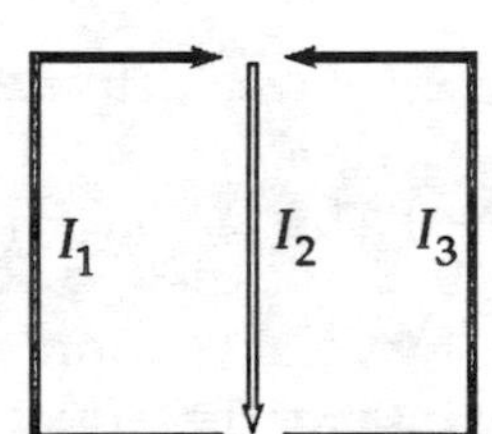

(a) For loop *abcfa*,

$$+70\text{ V} - (2.0\text{ k}\Omega)I_1 - (3.0\text{ k}\Omega)I_2 - 60\text{ V} = 0$$

And for *fedcf*,

$$+80\text{ V} - (3.0\text{ k}\Omega)I_2 - 60\text{ V} - (4.0\text{ k}\Omega)I_3 = 0$$

Now substitute $I_2 = (I_1 + I_3)$, to reduce this to two equations in two unknowns:

$$10\text{ V} - (2.0\text{ k}\Omega)I_1 - (3.0\text{ k}\Omega)I_1 - (3.0\text{ k}\Omega)I_3 = 0$$
$$20\text{ V} - (3.0\text{ k}\Omega)I_1 - (3.0\text{ k}\Omega)I_3 - (4.0\text{ k}\Omega)I_3 = 0$$

Solving the first equation for I_1, $\quad I_1 = \dfrac{10\text{ V} - (3.0\text{ k}\Omega)I_3}{5.0\text{ k}\Omega}$

Substituting this into the second yields an equation that can be solved for I_3:

$$20\text{ V} - (3.0\text{ k}\Omega)\left(\frac{10\text{ V} - (3.0\text{ k}\Omega)I_3}{5.0\text{ k}\Omega}\right) - (7.0\text{ k}\Omega)I_3 = 0$$

Thus, in the 4-$k\Omega$ resistor, $\quad I_3 = 2.7\text{ mA}$ ◊

In the 2-$k\Omega$ resistor, $\quad I_1 = 0.39\text{ mA}$ ◊

In the 3-$k\Omega$ resistor, $\quad I_2 = 3.1\text{ mA}$ ◊

(b) Going from *c* to *f*, the potential change is $-60\text{ V} - (3.0\text{ k}\Omega)(3.1\text{ mA}) = -69.1\text{ V}$.

Therefore, *c* is higher in potential than *f* by 69 V ◊

37. For the circuit shown in Figure P28.37, calculate (a) the current in the 2.0-Ω resistor and (b) the potential difference between points a and b.

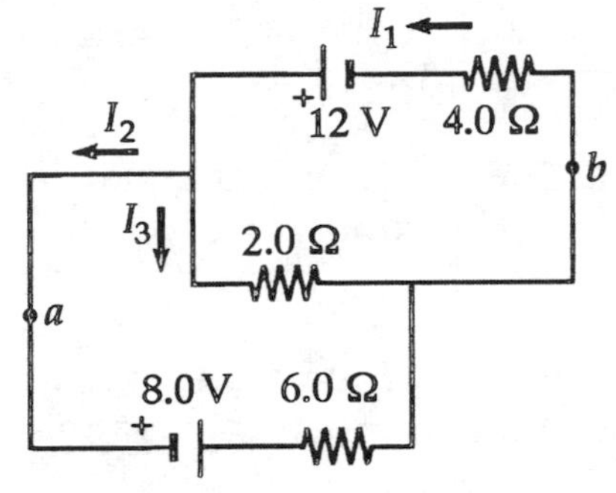

Figure P28.37 (modified)

Solution Arbitrarily choose current directions as labeled in the figure to the right.

(a) From the junction point rule, we have

$$I_1 = I_2 + I_3 \qquad (1)$$

Traversing the top loop counterclockwise gives

$$(12\ \text{V}) - (2.0\ \Omega)I_3 - (4.0\ \Omega)I_1 = 0 \qquad (2)$$

Traversing the bottom loop counterclockwise,

$$8.0\ \text{V} - (6.0\ \Omega)I_2 + (2.0\ \Omega)(-I_3) = 0 \qquad (3)$$

From Equation (2), $I_1 = (3.0\ \text{A}) - \dfrac{I_3}{2.0}$

From Equation (3), $I_2 = \dfrac{(4.0\ \text{A}) + I_3}{3.0}$

Substituting these values into Equation (1), we find that $I_3 = 0.909$ A.

Therefore, the current in the 2.0-Ω resistor is 0.91 A ◊

(b) $V_a - (0.909\ \text{A})(2.0\ \Omega) = V_b$,

Therefore, $V_a - V_b = 1.8\ \text{V}$, with $V_a > V_b$ ◊

39. Calculate the power dissipated in each resistor in Figure P28.39.

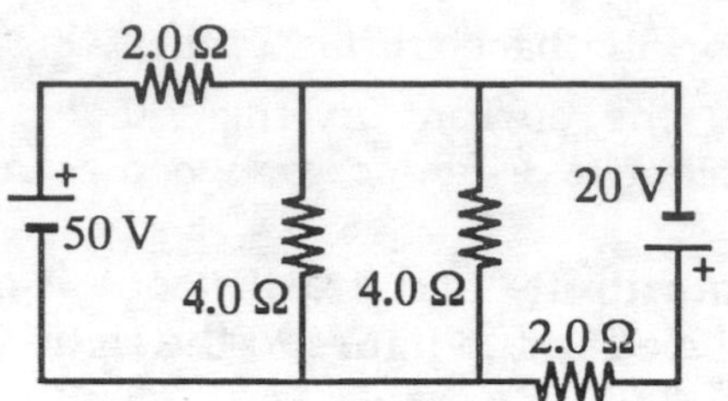

Figure P28.39

Solution There are six branches. Arbitrarily, choose directions and names for the six currents, as shown in the figure at the lower right:

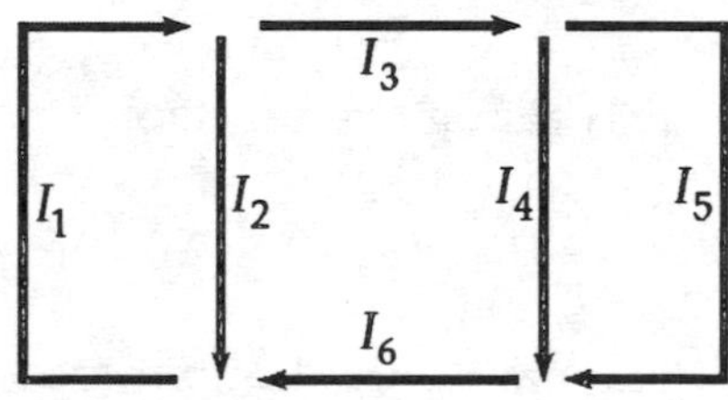

There are four junctions, so we write three independent junction equations:

$$I_1 = I_2 + I_3 \qquad I_3 = I_4 + I_5 \qquad I_4 + I_5 = I_6$$

(If we write the fourth, $I_2 + I_6 = I_1$, it is just the sum of the first three; it is not an independent equation, and makes no progress toward the solution.) There are three meshes, so we write a voltage equation for each, arbitrarily going clockwise:

In the left-hand loop, $\quad 50\text{ V} - (2.0\ \Omega)I_1 - (4.0\ \Omega)I_2 = 0$

In the center loop, $\quad (4.0\ \Omega)I_2 - (4.0\ \Omega)I_4 = 0$

and in the right-hand loop $\quad (4.0\ \Omega)I_4 + 20\text{ V} - (2.0\ \Omega)I_5 = 0$

At this point we have six equations in six unknowns. Note, however, that I_6 appears in only one equation. In addition, when we eliminate I_3 between the first two equations, only four equations in four unknowns remain:

$$I_1 = I_2 + I_4 + I_5 \qquad 50 - 2.0I_1 - 4.0I_2 = 0$$

$$4.0I_2 = 4.0I_4 \qquad 4.0I_4 + 20\text{ V} - 2.0I_5 = 0$$

Since $I_2 = I_4$,

$$I_1 = 2.0I_2 + I_5 \qquad 50 = 2.0I_1 + 4.0I_2 \qquad 20 = 2.0I_5 - 4.0I_2$$

Substituting the first equation into the other two,

$$50 = 8.0I_2 + 2.0I_5 \qquad I_5 = 10 + 2.0I_2$$

Finally,

$$50 = 8.0I_2 + 20 + 4.0I_2 \qquad I_2 = 2.5\text{ A}$$

Working back through the equations we used for substitution,

$I_5 = 10 + 2(2.5\text{ A}) = 15\text{ A}$ $\quad$ $I_1 = 2(2.5\text{ A}) + 15\text{ A} = 20\text{ A}$ $\quad$ $I_4 = I_2 = 2.5\text{ A}$

For the left-hand 2.0-Ω resistor, $\quad P = I^2R = (20\text{ A})^2(2.0\ \Omega) = 800\text{ W}$ ◊

For each of the two 4.0-Ω resistors, $\quad P = (2.5\text{ A})^2(4.0\ \Omega) = 25\text{ W}$ ◊

For the bottom 2.0-Ω resistor, $\quad P = (15\text{ A})^2(2.0\ \Omega) = 450\text{ W}$ ◊

As a profound check, we compare the total thermal energy output of the resistors, $800 + 25 + 25 + 450 = 1300\text{ W}$, with the total rate of conversion of chemical into electric energy by the batteries: $VI + VI = (50\text{ V})(20\text{ A}) + (20\text{ V})(15\text{ A}) = 1300\text{ W}$. They are equal, as they must be.

---

42. Consider a series RC circuit (Figure 28.15) for which $R = 1.00\text{ M}\Omega$, $C = 5.00\ \mu\text{F}$, and $\mathcal{E} = 30.0\text{ V}$. Find (a) the time constant of the circuit and (b) the maximum charge on the capacitor after the switch is closed. (c) If the switch is closed at $t = 0$, find the current in the resistor 10.0 s later.

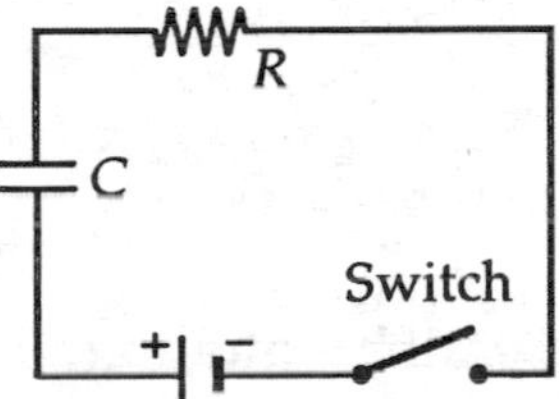

Figure 28.15 (b)

Solution

(a) $\tau = RC = (1.00\times 10^6\ \Omega)(5.00\times 10^{-6}\text{ F}) = 5.00\ \Omega\cdot\text{F} = 5.00\text{ s}$ ◊

(b) After a long time, the capacitor is "charged to thirty volts," separating charges

$$Q = CV = (5.00\times 10^{-6}\text{ F})(30\text{ V}) = 150\ \mu\text{C} \quad ◊$$

(c) $I = I_0 e^{-t/\tau}$ $\quad$ where $\quad I_0 = \dfrac{\mathcal{E}}{R}$ $\quad$ and $\quad \tau = RC$

$$I = \frac{\mathcal{E}}{R}e^{-t/RC} = \left(\frac{30\text{ V}}{1.00\times 10^6\ \Omega}\right)e^{-10.0\text{ s}/((1.00\times 10^6\ \Omega)(5.00\times 10^{-6}\text{ F}))}$$

$$I = 4.06\times 10^{-6}\text{ A} = 4.06\ \mu\text{A} \quad ◊$$

---

45. A 4.00-MΩ resistor and a 3.00-μF capacitor are connected in series with a 12.0-V power supply. (a) What is the time constant for the circuit? (b) Express the current in the circuit and the charge on the capacitor as functions of time.

Solution We suppose the switch is closed at $t = 0$.

R
12 V
3 μF
Switch

(a) $\tau = RC$

$$\tau = (4.00\times10^{6}\ \Omega)(3.00\times10^{-6}\ \text{F})\left(\frac{1\ \text{V}}{\text{A}\cdot\Omega}\right)\left(\frac{1\ \text{C}}{\text{F}\cdot\text{V}}\right)\left(\frac{1\ \text{A}\cdot\text{s}}{\text{C}}\right)$$

$$\tau = 12.0\ \text{s} \quad \Diamond$$

(b) $q = \mathcal{E}C\left(1 - e^{-t/RC}\right)$

$$q = (12.0\ \text{V})(3.00\ \mu\text{F})\left[1 - e^{-t/12\ \text{s}}\right]$$

$$q = 36.0\ \mu\text{C}\left[1 - e^{-t/12\ \text{s}}\right] \quad \Diamond$$

Differentiating,
$$I = \frac{dq}{dt} = (36.0\ \mu\text{C})\left[0 - e^{-t/12\ \text{s}}\right]\left(-\frac{1}{12\ \text{s}}\right)$$

$$I = 3.00e^{-t/12\ \text{s}}\ \mu\text{A} \quad \Diamond$$

---

47. The circuit in Figure P28.47 has been connected for a long time. (a) What is the voltage across the capacitor? (b) If the battery is disconnected, how long does it take the capacitor to discharge to 1/10 of its initial voltage?

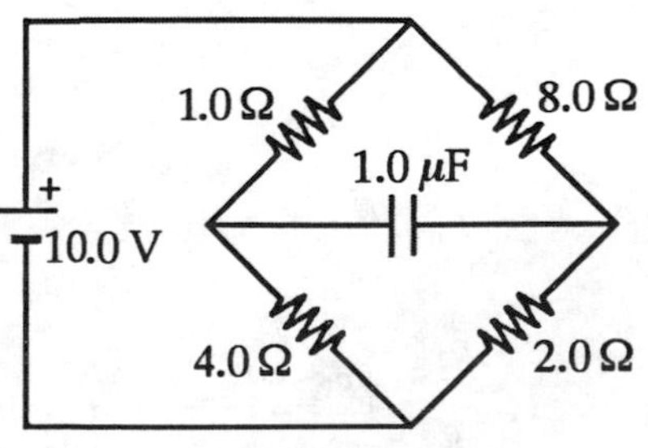

Figure P28.47

Solution

(a) After a long time the capacitor branch will carry negligible current. The current flow is as shown in Figure (a).

To find the voltage at point a, we first find the current, using the voltage rule:

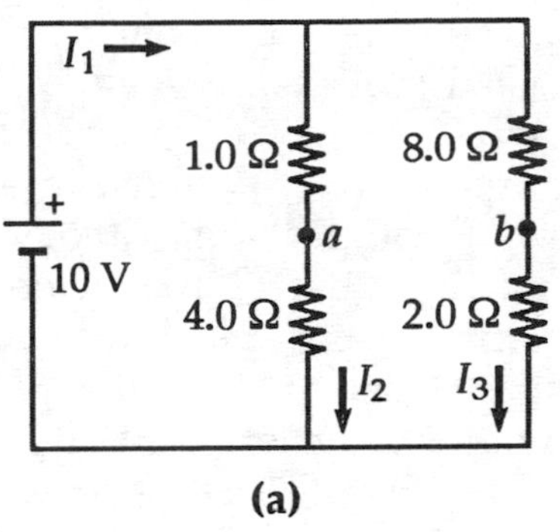

(a)

$$10\text{ V} - (1.0\ \Omega)I_2 - (4.0\ \Omega)I_2 = 0$$

$$I_2 = 2.0\text{ A}$$

$$V_a = (4.0\ \Omega)I_2 = 8.0\text{ V}$$

Similarly, $\quad 10\text{ V} - (8.0\ \Omega)I_3 - (2.0\ \Omega)I_3 = 0$

$$I_3 = 1.0\text{ A}$$

At point b, $\quad V_b = (2.0\ \Omega)I_3 = 2.0\text{ V}$

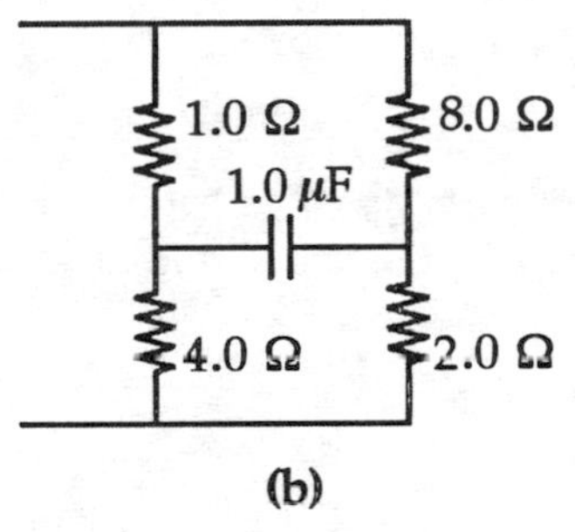

(b)

Thus, the voltage across the capacitor is

$$V_a - V_b = 8.0\text{ V} - 2.0\text{ V} = 6.0\text{ V} \quad \lozenge$$

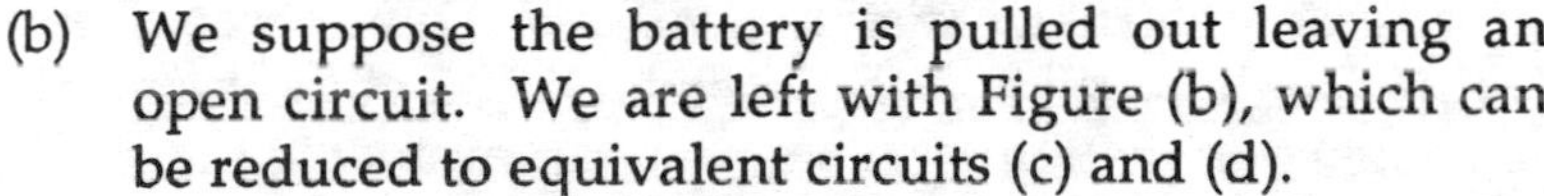
(b) We suppose the battery is pulled out leaving an open circuit. We are left with Figure (b), which can be reduced to equivalent circuits (c) and (d).

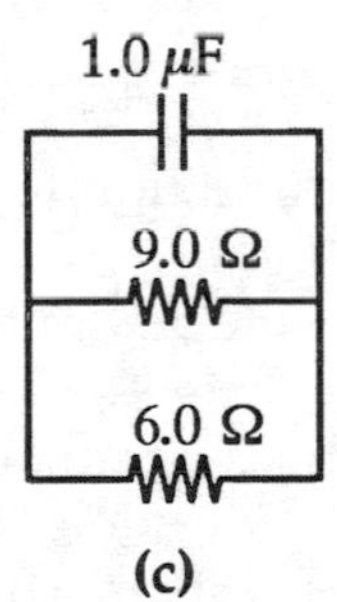

(c)

From (d), we see that the capacitor sees 3.60 Ω in its discharge. According to $q = Qe^{-t/RC}$, we calculate that

$$qC = QCe^{-t/RC} \quad \text{and} \quad V = V_0e^{-t/RC}$$

Solving, $\quad \frac{1}{10}V_0 = V_0e^{-t/(3.60\ \Omega)(1.00\ \mu\text{F})}$

$$e^{-t/3.60\ \mu\text{s}} = 0.100$$

$$(-t / 3.60\ \mu\text{s}) = \ln 0.100 = -2.30$$

$$\frac{t}{3.60\ \mu\text{s}} = 2.30$$

$$t = (2.30)(3.60\ \mu\text{s}) = 8.3\ \mu\text{s} \quad \lozenge$$

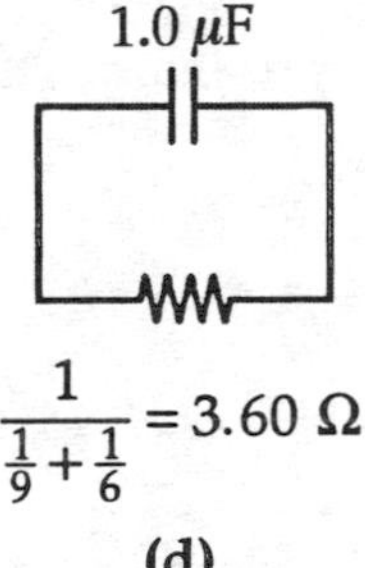

$$\frac{1}{\frac{1}{9}+\frac{1}{6}} = 3.60\ \Omega$$

(d)

53. The same galvanometer described in Problem 52 may be used to measure voltages. In this case a large resistor is wired in series with the galvanometer similar to Figure P28.23 (b), which in effect limits the current that flows through the galvanometer when large voltages are applied. Most of the potential drop occurs across the resistor placed in series. Calculate the value of the resistor that enables the galvanometer to measure an applied voltage of 25.0 V at full-scale deflection.

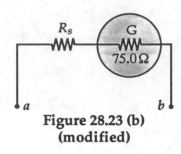

Figure 28.23 (b) (modified)

Solution $V_{ab} = 25.0$ V; From problem 52, $I = 1.50 \times 10^{-3}$ A and $R_g = 75.0\ \Omega$

$V_{ab} = I(R_s + R_g)$

$$R_g = \frac{V_{ab}}{I} - R_g = \frac{25.0\ \text{V}}{1.50 \times 10^{-3}\ \text{A}} - 75.0\ \Omega = 16.6\ \text{k}\Omega \quad \Diamond$$

---

55. Assume that a galvanometer has an internal resistance of 60.0 Ω and requires a current of 0.500 mA to produce full-scale deflection. What resistance must be connected in parallel with the galvanometer if the combination is to serve as an ammeter that has a full-scale deflection for a current of 0.100 A?

Solution Consider making the constructed ammeter read full scale. Then the current through the shunt (parallel) resistor must be 0.1 A – 0.5 mA = 99.5 mA.

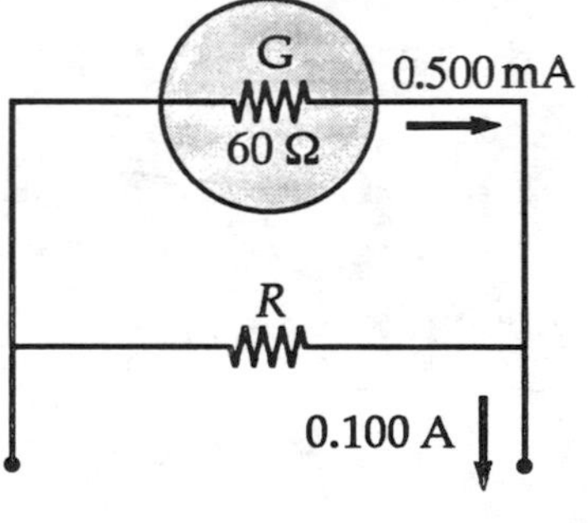

The voltage rule applied around the loop says

$$(-0.5\ \text{mA})(60\ \Omega) + (99.5\ \text{mA})R = 0$$

$$R = \frac{30\ \text{mV}}{99.5\ \text{mA}} = 0.302\ \Omega$$

---

57. A galvanometer having a full-scale sensitivity of 1.00 mA requires a 900-Ω series resistor to make a voltmeter reading full scale when 1.00 V is measured across the terminals. What series resistor is required to make the same galvanometer into a 50.0-V (full-scale) voltmeter?

Solution We find the resistance of the galvanometer:

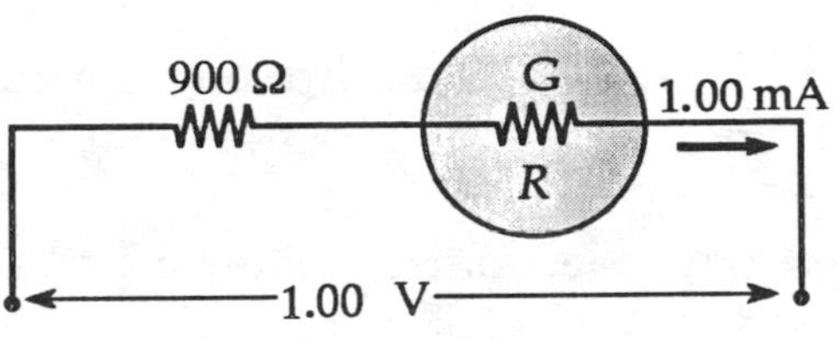

$$1.00\text{ V} = (900\ \Omega)(1.00\text{ mA}) + R\,(1.00\text{ mA})$$

$$1000\ \Omega = 900\ \Omega + R$$

$$R = 100\ \Omega$$

To establish full scale at 50 V, we have

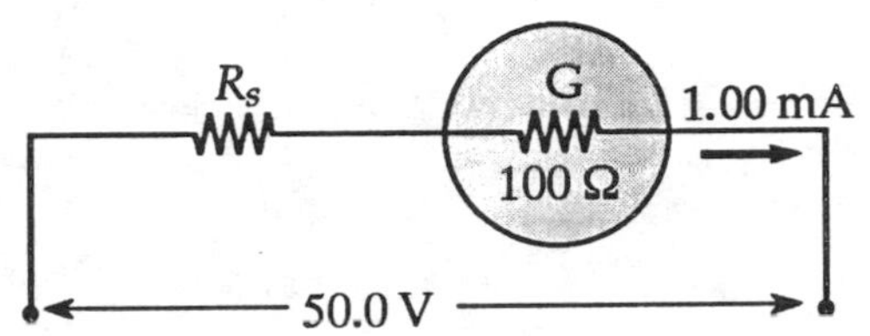

$$50.0\text{ V} = (1.00\text{ mA})R_s + (1.00\text{ mA})(100\ \Omega)$$

$$R_s = \frac{49.9\text{ V}}{1.00\text{ mA}} = 49.9\text{ k}\Omega \quad \lozenge$$

---

61. Consider the case when the Wheatstone bridge shown in Figure 28.24 of the text is unbalanced. Calculate the current through the galvanometer when $R_x = R_3 = 7.00\ \Omega$, $R_2 = 21.0\ \Omega$, and $R_1 = 14.0\ \Omega$. Assume the voltage across the bridge is 70.0 V, and neglect the galvanometer's resistance.

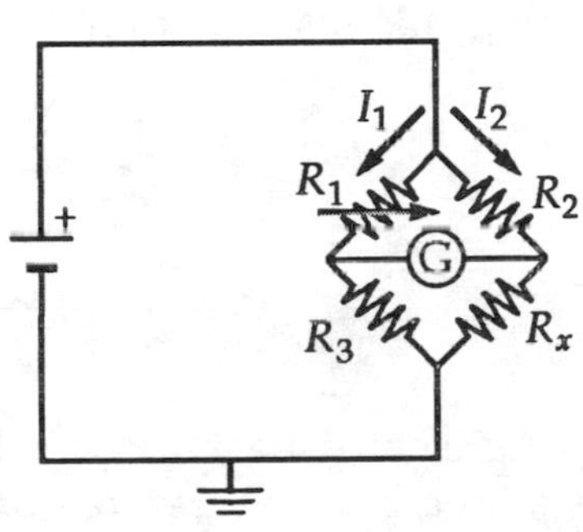

Figure 28.24

Solution There are six branches, so we write six independent equations.

Three for junctions :

$$I_0 = I_1 + I_2 \qquad I_1 = I_G + I_3 \qquad I_3 + I_4 = I_0$$

and three for loops:

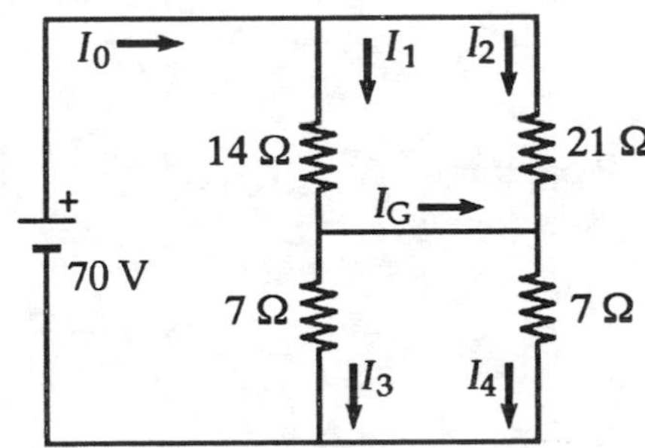

$70.0 - 14.0I_1 - 7.00I_3 = 0$ (left-hand loop)

$14.0I_1 - 21.0I_2 = 0$ (top-right loop)

$7.00I_3 - 7.00I_4 = 0$ (lower-right loop)

From our second and third loop equations, we have $I_3 = I_4$ and $I_1 = \left(\frac{21}{14}\right) I_2 = \frac{3I_2}{2}$

Thus, our other four equations become $I_0 = \frac{3I_2}{2} + I_2$ or $I_0 = \frac{5I_2}{2}$

$$2I_4 = I_0 \quad \text{or} \quad I_4 = \frac{5I_2}{4}$$

$$\frac{3I_2}{2} = I_G + I_4 \quad \text{or} \quad I_G = \frac{I_2}{4}$$

$$70 - 21I_2 - 7I_4 = 0 \quad \text{or} \quad 70 - 21I_2 - \frac{35I_2}{4} = 0$$

Solving this last equation, $I_2 = \frac{70\ \text{V}}{29.75\ \Omega} = 2.35\ \text{A}$

and $I_G = \frac{2.35\ \text{A}}{4} = 0.588\ \text{A}$ ◊

---

65. An electric heater is rated at 1500 W, a toaster at 750 W, and an electric grill at 1000 W. The three appliances are connected to a common 120-V circuit. (a) How much current does each draw? (b) Is a 25-A circuit sufficient in this situation? Explain.

Solution

(a) Heater: $I = \frac{P}{V} = \frac{1500\ \text{W}}{120\ \text{V}}\left(\frac{1\ \text{J/s}}{1\ \text{W}}\right)\left(\frac{1\ \text{V}}{1\ \text{J/C}}\right)\left(\frac{1\ \text{A}}{1\ \text{C/s}}\right) = 12.5\ \text{A}$ ◊

Toaster: $I = \frac{750\ \text{W}}{120\ \text{V}} = 6.25\ \text{A}$ ◊

Grill: $I = \frac{1000\ \text{W}}{120\ \text{V}} = 8.33\ \text{A}$ ◊

(b) Together in parallel they pass current 12.5 + 6.25 + 8.33 A = 27.1 A, so 25-A wiring cannot feed all of them energy at once. ◊

---

75. Three 60-W, 120-V lightbulbs are connected across a 120-V power source, as shown in Figure P28.75. Find (a) the total power dissipated in the three bulbs and (b) the voltage across each. Assume that the resistance of each bulb conforms to Ohm's law (even though in reality the resistance increases markedly with current).

Solution

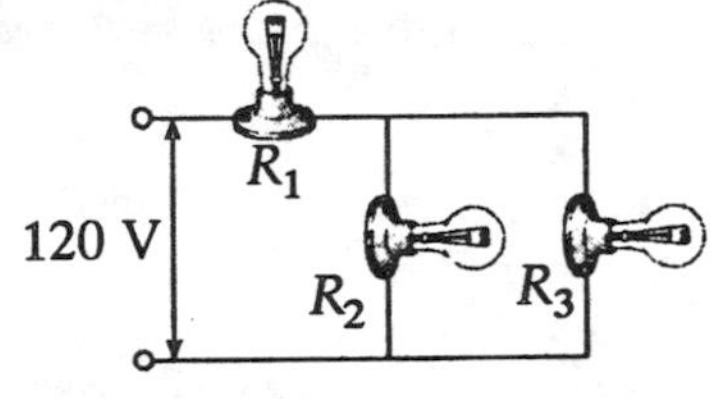

Figure P28.75

(a) The resistance of each bulb is

$$R = \frac{V}{I} = \frac{120\text{ V}}{0.500\text{ A}} = 240\ \Omega$$

As they are connected, R_2 and R_3 have equivalent resistance

$$\frac{1}{\dfrac{1}{240\ \Omega} + \dfrac{1}{240\ \Omega}} = 120\ \Omega$$

and the three together have equivalent resistance $240\ \Omega + 120\ \Omega = 360\ \Omega$.

The total current can be calculated as $I = \dfrac{V}{R} = \dfrac{120\text{ V}}{360\ \Omega} = 0.333\text{ A}$.

Thus the power dissipated is $P = VI = (120\text{ V})(0.333\text{ A}) = 40\text{ W}$ ◊

(b) For bulb R_1 $V = IR = (0.333\text{ A})(240\ \Omega) = 80.0\text{ V}$ ◊

The potential difference across both R_2 and R_3 is $120\text{ V} - 80.0\text{ V} = 40.0\text{ V}$ ◊

---

77. The value of a resistor R is to be determined using the ammeter-voltmeter setup shown in Figure P28.77. The ammeter has a resistance of 0.50 Ω, and the voltmeter has a resistance of 20 000.0 Ω. Within what range of actual values of R will the measured values be correct to within 5% if the measurement is made using the circuit shown in (a) Figure P28.77a and (b) Figure P28.77b?

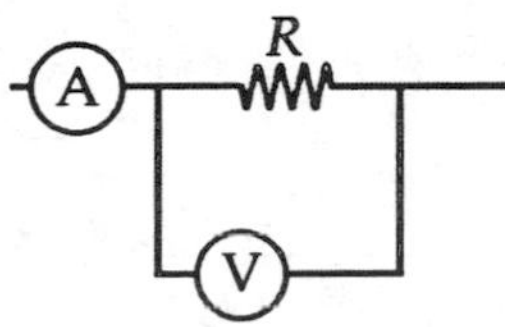

Figure P28.77 (a)

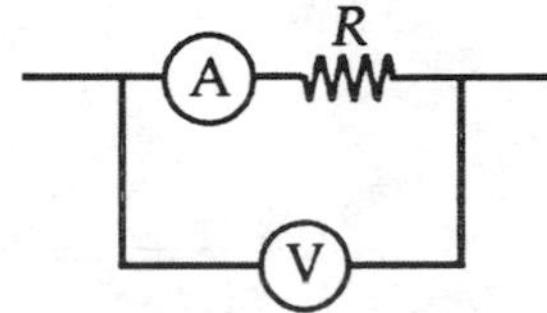

Figure P28.77 (b)

Solution

(a) In Figure P28.77 (a), at least a little current goes through the voltmeter, so the resistor carries less current than the ammeter says and the resistance computed by dividing the voltage by the inflated ammeter reading will be too small.

We require $\frac{V}{I} \geq 0.95R$ where I is the current through the ammeter.

Call I_R the current through the resistor. Then $I - I_R$ is the current in the voltmeter and the voltage rule says $V = I_R R = (I - I_R)\ 20\ 000\ \Omega$.

Thus $$I = \frac{I_R(R + 20000\ \Omega)}{20000\ \Omega}$$

and our requirement is $$\frac{I_R R}{\left(\frac{I_R(R + 20000\ \Omega)}{20000\ \Omega}\right)} \geq 0.95R$$

Solving, $$20000\ \Omega \geq 0.95(R + 20000\ \Omega) = 0.95R + 19000\ \Omega$$

and $R \leq \frac{1000\ \Omega}{0.95}$ or $R \leq 1.05\ \text{k}\Omega$ ◊

(b) In Figure 28.77b, the voltmeter reading is $I(0.5\ \Omega) + IR$, at least a little larger than the voltage across the resistor. So the resistance computed by dividing the inflated voltmeter reading by the ammeter reading will be too large. We require

$$\frac{V}{I} \leq 1.05R$$

$$\frac{I\,(0.5\ \Omega) + IR}{I} \leq 1.05R$$

$$0.5\ \Omega \leq 0.05R$$

$$R \geq 10.0\ \Omega$$ ◊

79. A battery has an emf $\mathcal{E}$ and internal resistance r. A variable resistor R is connected across the terminals of the battery. Find the value of R such that (a) the potential difference across the terminals is a maximum, (b) the current in the circuit is a maximum, (c) the power delivered to the resistor is a maximum.

Solution

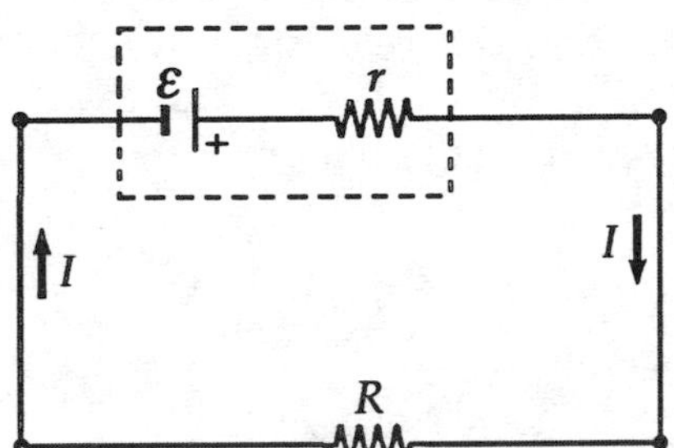

(a) $V_{\text{terminal}} = \mathcal{E} - Ir = \dfrac{\mathcal{E}R}{R+r}$

Therefore, $V_{\text{terminal}} \to \mathcal{E}$ as $R \to \infty$ ◊

(b) $I = \dfrac{\mathcal{E}}{R+r}$

Therefore, $I \to \dfrac{\mathcal{E}}{r}$ as $R \to 0$ ◊

(c) We note that the power is $P = I^2R = \dfrac{\mathcal{E}^2 R}{(R+r)^2}$

To maximize the power P as a function of R, we differentiate,

$$\frac{dP}{dR} = \mathcal{E}^2 R(-2)(R+r)^{-3} + \mathcal{E}^2(R+r)^{-2}$$

and require that $\dfrac{-2\mathcal{E}^2 R}{(R+r)^3} + \dfrac{\mathcal{E}^2}{(R+r)^2} = 0$

Then $2R = R + r$ and $R = r$ ◊

Making load resistance equal to source resistance to maximize power transfer is called impedance matching.

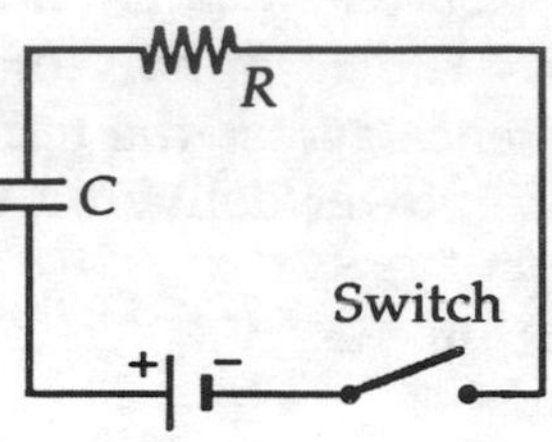

81. The values of the components in a simple series RC circuit containing a switch (See Figure 28.15) are $C = 1.00\ \mu\text{F}$, $R = 2.00 \times 10^6\ \Omega$, and $\mathcal{E} = 10.0$ V. At the instant 10.0 s after the switch is closed, calculate (a) the charge on the capacitor, (b) the current in the resistor, (c) the rate at which energy is being stored in the capacitor, and (d) the rate at which energy is being delivered by the battery.

Solution

(a) $q = CV\left(1 - e^{-t/RC}\right)$

$$q = (1.00 \times 10^{-6}\ \text{F})(10.0\ \text{V})\left[1 - e^{-10.0\ \text{s}/\left((2.00\times10^{6}\ \Omega)(1.00\times10^{-6}\ \text{F})\right)}\right]$$

$$q = 9.93 \times 10^{-6}\ \text{C} = 9.93\ \mu\text{C} \quad \Diamond$$

(b) $I = \dfrac{dq}{dt} = \dfrac{d}{dt}\left[CV\left(1 - e^{-t/RC}\right)\right]$

$$I = \left(\frac{V}{R}\right)e^{-t/RC} = \left(\frac{10.0\ \text{V}}{2.00 \times 10^{-6}\ \Omega}\right)e^{-(10.0/2.00)} = 3.37 \times 10^{-8}\ \text{A} \quad \Diamond$$

(c) Since the energy stored in the capacitor, $U = \dfrac{q^2}{2C}$,

the rate of storing energy $= \dfrac{dU}{dt} = \dfrac{q}{C}\dfrac{dq}{dt} = \left(\dfrac{q}{C}\right)I$

$$\text{Rate} = \left(\frac{9.90 \times 10^{-6}\ \text{C}}{1.00 \times 10^{-6}\ \text{F}}\right)(3.37 \times 10^{-8}\ \text{A}) = 3.34 \times 10^{-7}\ \text{W} \quad \Diamond$$

(d) $P_{\text{batt}} = I\mathcal{E} = (3.37 \times 10^{-8}\ \text{A})(10.0\ \text{V}) = 3.37 \times 10^{-7}\ \text{W} \quad \Diamond$

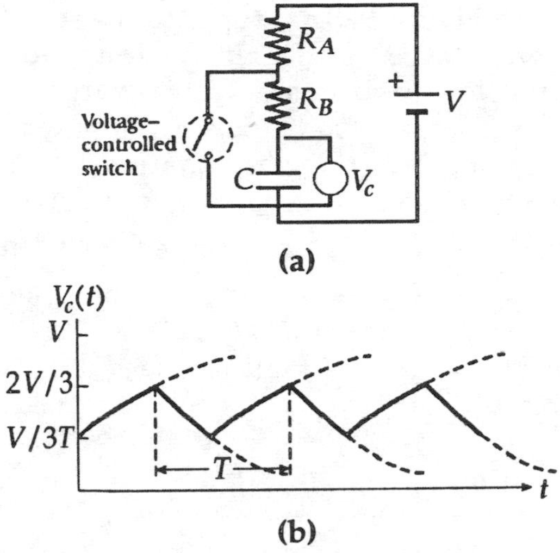

Figure P28.83

83. The switch in Figure P28.83 (a) closes when $V_c \geq 2V/3$ and opens when $V_c \leq V/3$. The voltmeter reads a voltage as plotted in Figure P28.83 (b). What is the period, T, of the waveform in terms of R_A, R_B, and C?

Solution The period comprises two parts: a fall time which we call t_1 and a rise time, t_2. To find the interval over which the V_c decreases, start at the point when the voltage has just reached 2/3 V and the switch has just closed. The voltage decays from 2/3 V toward zero with time constant R_BC, since R_B is the only resistance in the discharge circuit:

$$V(t) = \tfrac{2}{3}Ve^{-t/R_BC}$$

We want to know t_1 when $V(t)$ reaches $\frac{1}{3}V$. In four steps,

$$\tfrac{1}{3}V = \tfrac{2}{3}Ve^{-t_1/R_BC} \qquad e^{-t_1/R_BC} = \tfrac{1}{2} \qquad e^{t_1/R_BC} = 2 \qquad t_1 = R_BC\ln 2$$

Now we find the rise time. If V_c first starts from 0, it charges up to $V/3$ in a time t_0:

$$\tfrac{1}{3}V = V\left[1-e^{-t_0/(R_A+R_B)C}\right] \qquad e^{-t_0/(R_A+R_B)C} = \tfrac{2}{3} \qquad \text{and} \qquad t_0 = (R_A+R_B)C\ln\left(\tfrac{3}{2}\right)$$

If V_c first started from 0, it would charge up to $2V/3$ in a time (t_0 + t_2) according to

$$\tfrac{2}{3}V = V\left[1-e^{-(t_0+t_2)/(R_A+R_B)C}\right] \qquad \text{or} \qquad e^{-(t_0+t_2)/(R_A+R_B)C} = \tfrac{1}{3}.$$

Thus $\quad t_0 + t_2 = C(R_A + R_B)\ln 3$

So the rise time is

$$t_2 = t_0 + t_2 - t_0 = C(R_A+R_B)\left(\ln 3 - \ln\left(\tfrac{3}{2}\right)\right) = C(R_A+R_B)\ln\left(\frac{3}{(3/2)}\right) = C(R_A+R_B)\ln 2$$

The whole period is

$$T = t_1 + t_2 = R_BC\ln 2 + C(R_A+R_B)\ln 2 = C(R_A+2R_B)\ln 2 \quad \Diamond$$

87. In Figure P28.87, suppose the switch has been closed sufficiently long for the capacitor to become fully charged. Find (a) the steady-state current through each resistor and (b) the charge Q on the capacitor. (c) The switch is now opened at $t = 0$. Write an equation for the current i_{R_2} through R_2 as a function of time and (d) find the time that it takes for the charge on the capacitor to fall to one-fifth its initial value.

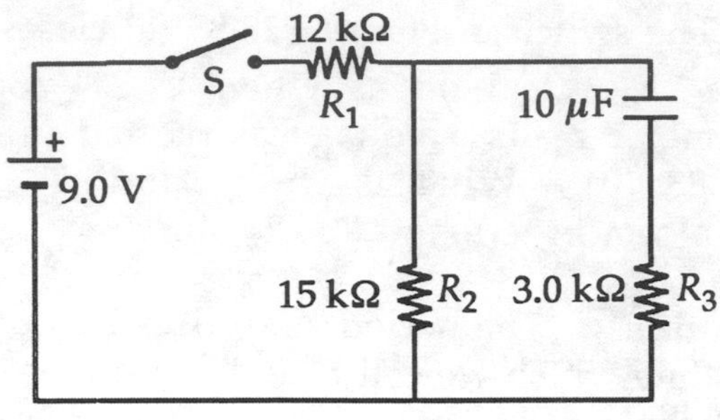

Figure P28.87

Solution

(a) When the capacitor is fully charged, no current flows in it or in the 3-kΩ resistor:

$$I_3 = 0 \quad \Diamond$$

So the same current flows in resistors 1 and 2: $I_1 = I_2$, as given by the voltage rule, $+9V - (12\text{ k}\Omega)I_1 - (15\text{ k}\Omega)I_1 = 0$.

$$I_1 = \frac{9\text{ V}}{27\text{ k}\Omega} = 0.333\text{ mA} = I_1 = I_2 \quad \Diamond$$

(b) For the left-hand mesh, the voltage rule gives for the capacitor voltage:

$$(+15\text{ k}\Omega)(0.333\text{ mA}) - V_c - (3.0\text{ k}\Omega)(0) = 0$$

$$V_c = 5.00\text{ V}$$

So its charge is $\quad Q = CV = (10\ \mu\text{F})\ 5\text{ V} = 50.0\ \mu\text{C} \quad \Diamond$

(c) At $t = 0$, the current in R_1 drops to zero. The capacitor, charged to 5 V with top plate positive, will drive the same current $I_2 = I_3$ counterclockwise around the left-hand mesh. At $t = 0$, its value is given by $+5\text{ V} - I_2(15\text{ k}\Omega) - I_2(3\text{ k}\Omega) = 0$.

$$I_2 = \frac{5\text{ V}}{18\text{ k}\Omega} = 0.278\text{ mA}$$

Thereafter, it decays as it drains the capacitor's charge, with time constant

$$R_{eq}C = (18\ \text{k}\Omega)\ 10\ \mu\text{F} = 180\ \text{ms}$$

So its equation is $I_2 = (0.278\ \text{mA})e^{-t/(180\ \text{ms})}$ ◊

(d) The charge decays according to the equation $q = Q_0 e^{-t/RC}$

Substituting known values into this yields $\frac{1}{5}(50\ \mu\text{C}) = (50\ \mu\text{C})\ e^{-t/(180\ \text{ms})}$

Solving for t, $0.200 = e^{-t/(180\ \text{ms})}$

$$\frac{-t}{180\ \text{ms}} = -1.61$$

and $t = (1.61)(180\ \text{ms}) = 290\ \text{ms}$ ◊

Chapter 29

Magnetic Fields

Chapter 29

MAGNETIC FIELDS

INTRODUCTION

Many historians of science believe that the compass, which uses a magnetic needle, was used in China as early as the 13th century B.C., its invention being of Arab or Indian origin. The early Greeks knew about magnetism as early as 800 B.C. They discovered that certain stones, now called magnetite (Fe_3O_4), attract pieces of iron. In 1269 Pierre de Maricourt mapped out the directions taken by a needle when it was placed at various points on the surface of a spherical natural magnet. He found that the directions formed lines that encircled the sphere and passed through two points diametrically opposite to each other, which he called the *poles* of the magnet. Subsequent experiments showed that every magnet, regardless of its shape, has two poles, called *north* and *south poles,* which exhibit forces on each other in a manner analogous to electric charges. That is, like poles repel each other and unlike poles attract each other.

NOTES FROM SELECTED CHAPTER SECTIONS

29.1 The Magnetic Field

Particles with charge q, moving with speed v in a magnetic field **B**, experience a magnetic force **F**:

- The magnetic force is proportional to the charge q and speed v of the particle.
- The magnitude and direction of the magnetic force depend on the velocity of the particle and on the magnitude and direction of the magnetic field.
- When a charged particle moves in a direction *parallel* to the magnetic field vector, the magnetic force **F** on the charge is *zero.*
- The magnetic force acts in a direction perpendicular to both **v** and **B**; that is, **F** is perpendicular to the plane formed by **v** and **B**.
- The magnetic force on a positive charge is in the direction opposite to the force on a negative charge moving in the same direction.
- If the velocity vector makes an angle θ with the magnetic field, the magnitude of the magnetic force is proportional to $\sin\theta$.

There are several important differences between electric and magnetic forces:

- The electric force is always in the direction of the electric field, whereas the magnetic force is perpendicular to the magnetic field.

- The electric force acts on a charged particle independent of the particle's velocity, whereas the magnetic force acts on a charged particle only when the particle is in motion.
- The electric force does work in displacing a charged particle, whereas the magnetic force associated with a steady magnetic field does *no* work when a particle is displaced.

29.4 Motion of a Charged Particle in a Magnetic Field

When a charged particle moves in an external magnetic field, the *work done* by the magnetic force on the particle is *zero*. The magnetic force changes the direction of the velocity vector but does not change its magnitude.

When a charged particle enters an external magnetic field along a direction perpendicular to the field, the particle will move in a circular path in a plane perpendicular to the magnetic field.

29.5 Applications of the Motion of Charged Particles in a Magnetic Field

The time required for one revolution of a charged particle in a *cyclotron* is independent of the speed (or radius) of the particle.

EQUATIONS AND CONCEPTS

The magnetic field (or magnetic induction) at some point in space is defined in terms of the *magnetic force* exerted on a moving positive electric charge at that point. The SI unit of the magnetic field is the tesla (T) or weber per square meter (Wb/m^2).

$$\mathbf{F} = q\mathbf{v} \times \mathbf{B} \tag{29.1}$$

The magnetic force will be of maximum magnitude when the charge moves along a direction perpendicular to the direction of the magnetic field. In general, the velocity vector may be directed along some direction other than 90.0° relative to the magnetic field. In this case, the magnetic force on the moving charge is less than its maximum value.

$$F_{max} = qvB\sin\theta$$

The SI unit of magnetic field intensity is the tesla (T).

$$1\,\mathrm{T} = 1\,\frac{\mathrm{N\cdot s}}{\mathrm{C\cdot m}}$$

Equation 29.2 can be written in a form which serves to define the magnitude of the magnetic field.

$$B \equiv \frac{F}{qv\sin\theta}$$

The cgs unit of magnetic field is the gauss (G).

$$1\ \text{T} = 10^4\ \text{G}$$

In order to determine the direction of the magnetic force, apply the right-hand rule:

> Hold your right hand with your fingers curling, first in the direction of **v**, and then in the direction of **B**, as shown. The magnetic force on a positive charge points in the direction of your thumb. If the charge is negative, then the direction of the force is reversed.

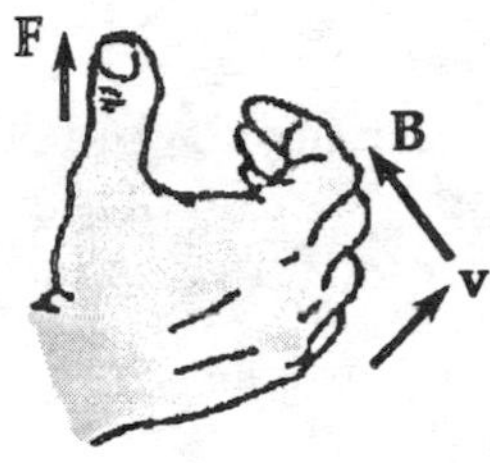

If a *straight* wire carrying a current is placed in an external magnetic field, a *magnetic force* will be exerted on the wire. The magnetic force on a wire of arbitrary shape is found by integrating over the length of the wire. In these equations the direction of **L** and $d\mathbf{s}$ is that of the current.

$$\mathbf{F} = I\,\mathbf{L}\times\mathbf{B} \qquad (29.3)$$

$$\mathbf{F} = I\int_a^b d\mathbf{s}\times\mathbf{B} \qquad (29.5)$$

The magnitude of the magnetic force on the conductor depends on the angle between the direction of the conductor and the direction of the field.

$$F = BIL\sin\theta$$

The magnetic force will be maximum when the conductor is directed perpendicular to the magnetic field.

$$F_{\text{max}} = BIL$$

When a closed conducting loop carrying a current is placed in an external magnetic field, there is a *net torque* exerted on the loop. In Equation 29.9, the area vector **A** is directed perpendicular to the area of the loop with a sense given by the right-hand rule. The magnitude of **A** is numerically equal to the area of the loop.

$$\tau = I\mathbf{A}\times\mathbf{B} \qquad (29.9)$$

The magnitude of the torque will depend on the angle between the direction of the magnetic field and the direction of the normal (or perpendicular) to the plane of the loop.

$$\tau = IAB\sin\theta$$

The magnitude of the torque will be maximum when the magnetic field is parallel to the plane of the loop.

$$\tau_{max} = IAB \quad (29.8)$$

The direction of rotation of the loop is such that the normal to the plane of the loop turns into a direction parallel to the magnetic field.

The torque on a current loop can also be expressed in terms of the magnetic moment of the loop.

$$\boldsymbol{\tau} = \boldsymbol{\mu} \times \mathbf{B} \quad (29.11)$$

where $\boldsymbol{\mu} = I\mathbf{A}$ (29.10)

When a charged particle enters the region of a uniform magnetic field with its velocity vector initially perpendicular to the direction of the field, the particle will move in a circular path in a plane perpendicular to the direction of the field. The *radius* of the circular path will be proportional to the linear momentum of the charged particle.

$$r = \frac{mv}{qB} \quad (29.12)$$

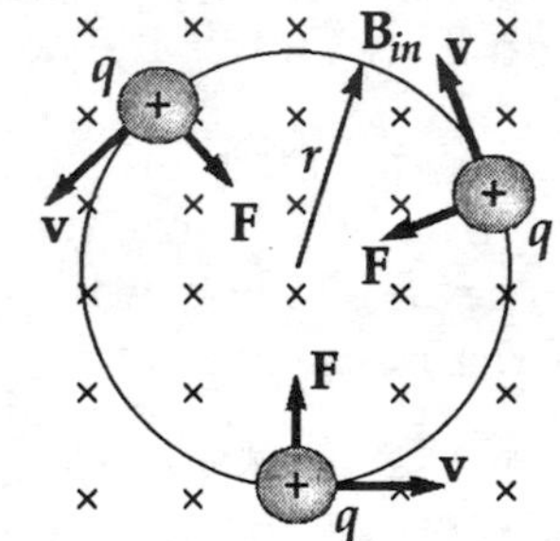

The *angular frequency* (or cyclotron frequency) of the particle will be proportional to the ratio of charge to mass. Note that the frequency, and hence the period of rotation do *not* depend on the radius of the path.

$$\omega = \frac{qB}{m} \quad (29.13)$$

$$T = \frac{2\pi m}{qB} \quad (29.14)$$

There are several important applications of the motion of charged particles in a magnetic field:

Velocity Selector—When a beam of charged particles is directed into a region where uniform electric and magnetic fields are perpendicular to each other and to the initial direction of the particle beam, those particles emerging along the initial beam direction will have a common velocity. Only those particles with a velocity $v = E/B$ will get through.

$$v = \frac{E}{B} \quad (29.16)$$

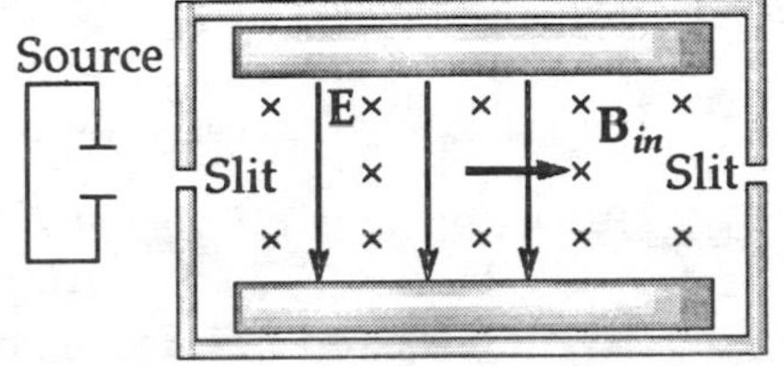

Mass Spectrometer—If an ion beam, after passing through a velocity selector, is directed perpendicularly into a second uniform magnetic field, the *ratio of charge to mass* for the isotopic species can be determined by measuring the radius of curvature of the beam in the field.

$$\frac{m}{q} = \frac{rB_0B}{E} \quad (29.17)$$

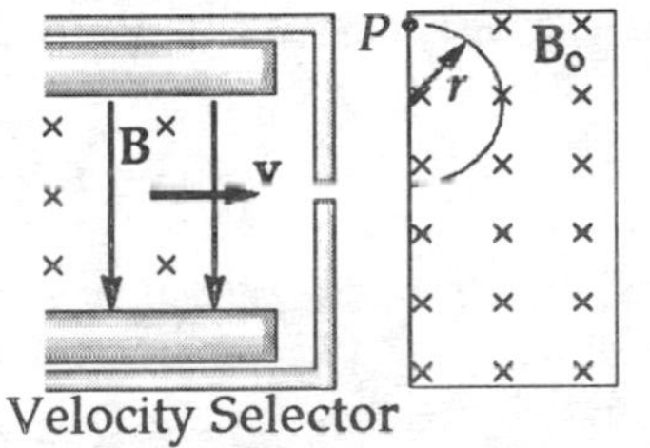

Velocity Selector

Cyclotron—The *maximum kinetic energy* acquired by an ion in a cyclotron depends on the radius of the dees and the intensity of the magnetic field. This relationship holds until the ion reaches relativistic energies (≅ 20 MeV). This is true for ions of proton mass or greater, but is not true for electrons.

$$K = \frac{q^2B^2R^2}{2m} \quad (29.18)$$

When the *Hall Coefficient* $(1/nq)$ is known for a calibrated sample, magnetic field strengths can be determined by accurate measurements of the *Hall voltage*, V_H. By measuring the voltage as + or –, one can find the sign of the charge carriers as well.

$$V_H = \frac{IBd}{nqA} \quad (29.21)$$

SUGGESTIONS, SKILLS, AND STRATEGIES

To remember the symbols for vectors that point away from you and towards you, think of a three-dimensional archery arrow as it turns to point first away from you (shown on the left), and then towards you, (shown on the right). The point on the arrow is represented by a dot; the feathers are represented by an 'x'.

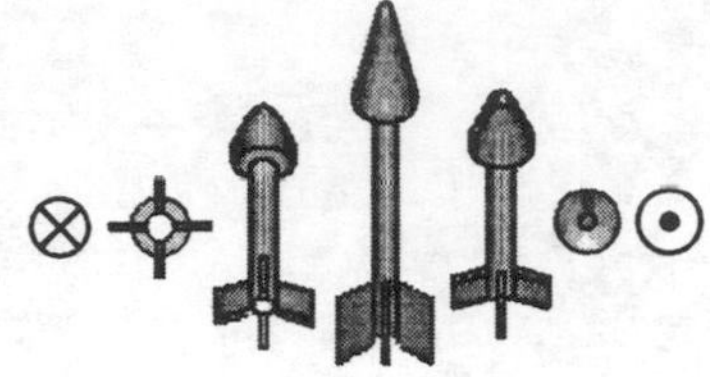

Equation 29.1, $\mathbf{F} = q\mathbf{v} \times \mathbf{B}$, serves as the definition of the magnetic field vector **B**. The direction of the magnetic force **F** is determined by the right-hand rule for the cross product which you have used before. This assumes that the charge q is a positive charge. If the vectors **v** and **B** are given in unit vector notation then $\mathbf{v} \times \mathbf{B}$ can be written as

$$\mathbf{v}\times\mathbf{B} = \mathbf{i}\left(v_y B_z - v_z B_y\right) + \mathbf{j}\left(v_z B_x - v_x B_z\right) + \mathbf{k}\left(v_x B_y - v_y B_x\right)$$

This means that the components of the magnetic force are:

$$F_x = q\left(v_y B_z - v_z B_y\right)$$

$$F_y = q\left(v_z B_x - v_x B_z\right)$$

$$F_z = q\left(v_x B_y - v_y B_x\right)$$

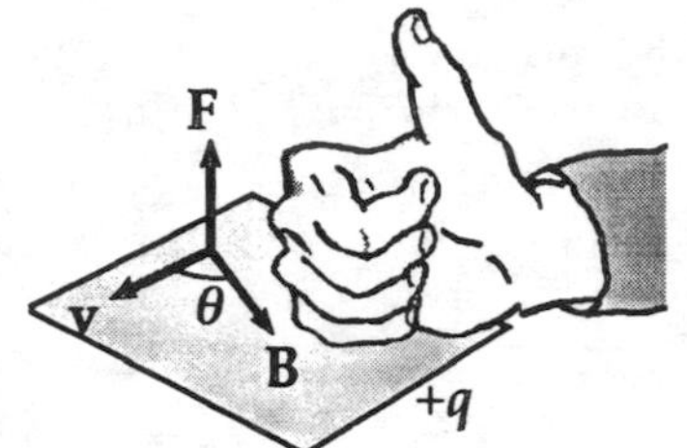

The right-hand rule and the vector cross product are also used to determine the direction of the torque and the direction of the resulting rotation for a closed current loop in a magnetic field. When the four fingers on your right hand circle the current loop in the direction of the current, the thumb will point in the direction of the area vector **A**. Applying this rule to the situation shown in the figure to the right, the vector **A** is directed out of the plane of the rectangular loop facing you as shown.

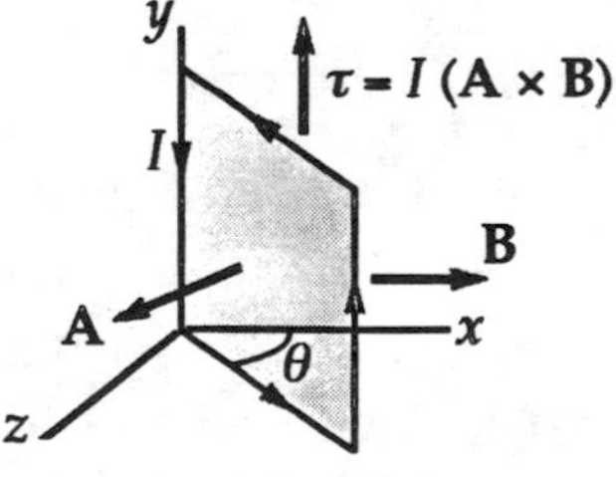

The resulting torque is parallel to the direction of $\mathbf{A} \times \mathbf{B}$ and is along the positive y axis as shown in the figure. This is consistent with the general rule for determining the direction of the cross product of two vectors. If the loop is considered to be hinged along the edge joining the y axis, then the rotation will be such that the angle θ decreases and the plane of the rectangular loop becomes parallel to the x-y plane. This is shown in the figure as a counterclockwise rotation about the y axis as seen from above.

REVIEW CHECKLIST

▷ Use the defining equation for a magnetic field **B** to determine the magnitude and direction of the magnetic force exerted on an electric charge moving in a region where there is a magnetic field. You should understand clearly the important differences between the forces exerted on electric charges by electric fields and those forces exerted on moving electric charges by magnetic fields.

▷ Calculate the magnitude and direction of the magnetic force on a current-carrying conductor when placed in an external magnetic field. You should be able to perform such calculations for either a straight conductor or one of arbitrary shape.

▷ Determine the magnitude and direction of the torque exerted on a closed current loop in an external magnetic field. You should understand how to correctly designate the direction of the area vector corresponding to a given current loop, and to incorporate the magnetic moment of the loop into the calculation of the torque on the loop.

▷ Calculate the period and radius of the circular orbit of a charged particle moving in a uniform magnetic field. Understand the essential features of the velocity selector and mass spectrometer and make appropriate quantitative calculations regarding the operation of these instruments.

SOLUTIONS TO SELECTED END-OF-CHAPTER PROBLEMS

1. Consider an electron near the equator. In which direction does it tend to deflect if its velocity is directed (a) downward, (b) northward, (c) westward, or (d) southeastward?

Solution The magnetic field near the equator is horizontally north.

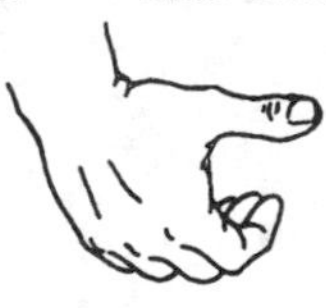

Down × North
is East

West × North
is Down

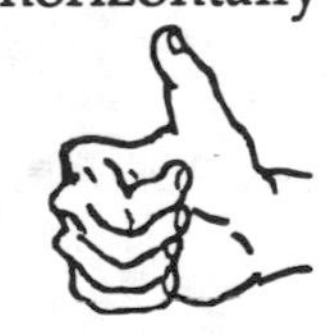

Southeast × North
is Up

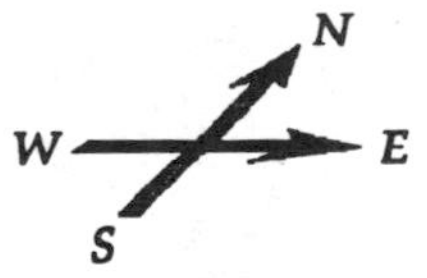

(a) If the velocity is down, $\mathbf{v} \times \mathbf{B}$ is east and $q\mathbf{v} \times \mathbf{B}$ for the negative electron is *west*. ◊

(b) If **v** is north, $\mathbf{v} \times \mathbf{B}$ is zero and the electron is not deflected. ◊

(c) If **v** is west, $\mathbf{v} \times \mathbf{B}$ = west × north = down, and the negative electron is deflected *up* . ◊

(d) If **v** is southeast, $\mathbf{v} \times \mathbf{B}$ = southeast × north = up, and the electron is deflected *down*. ◊

5. A proton moving at 4.0×10^6 m/s through a magnetic field of 1.7 T experiences a magnetic force of magnitude 8.2×10^{-13} N. What is the angle between the proton's velocity and the field?

Solution Since the magnitude of the force on a moving charge in a magnetic field is $F = qvB \sin \theta$,

$$\theta = \sin^{-1}\left[\frac{F}{qvB}\right]$$

$$\theta = \sin^{-1}\left[\frac{8.2\times10^{-13}\ \text{N}}{(1.60\times10^{-19}\ \text{C})(4.0\times10^{6}\ \text{m/s})(1.7\ \text{T})}\right] = 49° \quad \text{or} \quad 131° \quad \Diamond$$

---

9. A proton moves with a velocity of $\mathbf{v} = (2\mathbf{i} - 4\mathbf{j} + \mathbf{k})$ m/s in a region in which the magnetic field is $\mathbf{B} = (\mathbf{i} + 2\mathbf{j} - 3\mathbf{k})$ T. What is the magnitude of the magnetic force this charge experiences?

Solution

$$\mathbf{F} = q\mathbf{v}\times\mathbf{B}$$

$$= \left(1.60\times10^{-19}\ \text{C}\right)\left((2\mathbf{i}-4\mathbf{j}+\mathbf{k})\ \text{m/s}\right)\times\left((1\mathbf{i}+2\mathbf{j}-3\mathbf{k})\ \text{T}\right)$$

Since $1\ \text{C}\cdot\text{m}\cdot\text{T/s} = 1\ \text{N}$, we can write this in determinant form as:

$$\mathbf{F} = \left(1.60\times10^{-19}\ \text{N}\right)\begin{vmatrix} \mathbf{i} & \mathbf{j} & \mathbf{k} \\ 2 & -4 & 1 \\ 1 & 2 & -3 \end{vmatrix}$$

Expanding as described in Equation 11.14, we have

$$F = \left(1.60\times10^{-19}\ \text{N}\right)(10\mathbf{i}+7\mathbf{j}+8\mathbf{k}) = (16.0\mathbf{i}+11.2\mathbf{j}+12.8\mathbf{k})10^{-19}\ \text{N}$$

$$|\mathbf{F}| = \sqrt{16^2+11.2^2+12.8^2}\ 10^{-19}\text{N} = 23.4\times10^{-19}\ \text{N} \quad \Diamond$$

---

10. A proton moves perpendicular to a uniform magnetic field **B** at 1.00×10^7 m/s and experiences an acceleration of 2.00×10^{13} m/s^2 in the $+x$ direction when its velocity is in the +z direction. Determine the magnitude and direction of the field.

Solution

$$F = ma = (1.67 \times 10^{-27}\ \text{kg})(2.00 \times 10^{13}\ \text{m/s}^2)$$

$$F = 3.34 \times 10^{-14}\ \text{N} = qvB \sin 90°$$

$$B = \frac{F}{qv} = \frac{3.34 \times 10^{-14}\ \text{N}}{(1.60 \times 10^{-19}\ \text{C})(1.00 \times 10^7\ \text{m/s})}$$

$$B = 2.09 \times 10^{-2}\ \text{T} \quad \lozenge$$

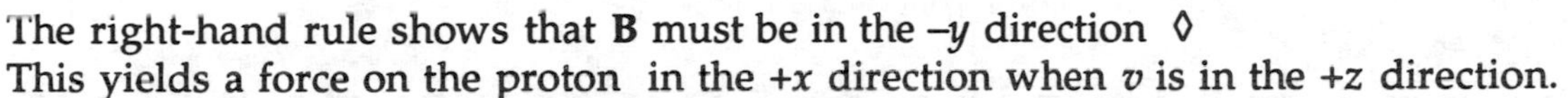

The right-hand rule shows that **B** must be in the $-y$ direction ◊
This yields a force on the proton in the $+x$ direction when v is in the +z direction.

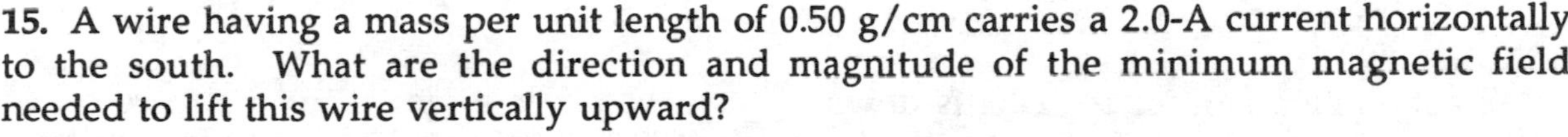

15. A wire having a mass per unit length of 0.50 g/cm carries a 2.0-A current horizontally to the south. What are the direction and magnitude of the minimum magnetic field needed to lift this wire vertically upward?

Solution

$$\frac{m}{\ell} = \left(0.50\ \frac{\text{g}}{\text{cm}}\right)\left(\frac{1\ \text{kg}}{1000\ \text{g}}\right)\left(\frac{100\ \text{cm}}{\text{m}}\right) = 5.0 \times 10^{-2}\ \text{kg/m}$$

I = 2 A south. To make **F** upward, **B** must be in a direction given by right-hand rule: eastward ◊

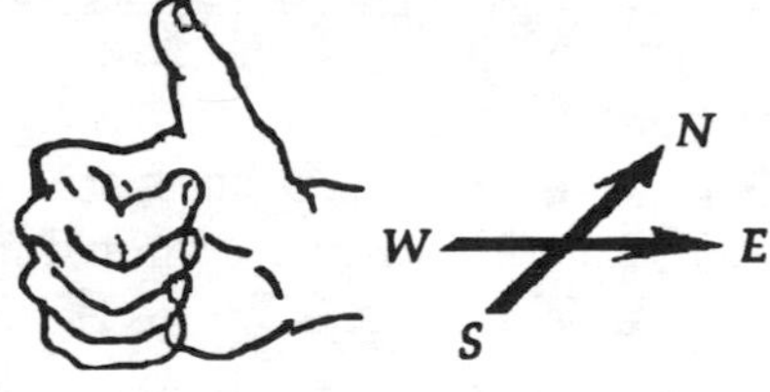

$F = I\ell B \sin\theta$: must counter balance $w = mg$

$$mg = I\ell B \sin\theta \qquad \text{or} \qquad \frac{m}{\ell} g = IB \sin\theta$$

$$(5.0 \times 10^{-2}\ \text{kg/m})(9.8\ \text{m/s}^2) = (2.0\ \text{A}) B \sin 90°$$

$$B = 0.245\ \text{T} \quad \lozenge$$

---

21. A strong magnet is placed under a horizontal conducting ring of radius r that carries a current I as in Figure P29.21. If the magnetic field makes an angle θ with the vertical at the ring's location, what are the magnitude and direction of the resultant force on the ring?

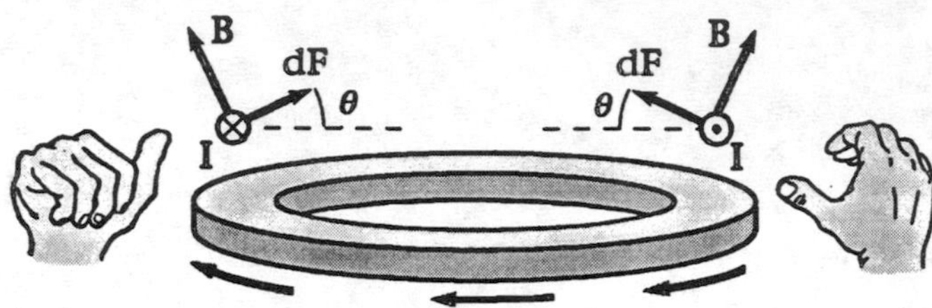

Figure P29.21

Solution The magnetic force on each bit of ring is radial inward and upward, at an angle θ above the radial line, according to:

$$|d\mathbf{F}| = I|d\mathbf{s} \times \mathbf{B}| = I\, ds\, B$$

The radially inward components tend to squeeze the ring, but cancel out as forces. The upward components $I\, ds\, B \sin\theta$ all add to

$$\mathbf{F} = I(2\pi r)B \sin\theta \text{ up } \Diamond.$$

The magnetic moment of the ring is down. This problem is a model for the force on a dipole in a nonuniform magnetic field, or for the force that one magnet exerts on another magnet.

---

23. A rectangular loop consists of $N = 100$ closely wrapped turns and has dimensions $a = 0.40$ m and $b = 0.30$ m. The loop is hinged along the y axis, and its plane makes an angle $\theta = 30°$ with the x axis (Fig. P29.23a). What is the magnitude of the torque exerted on the loop by a uniform magnetic field $B = 0.80$ T directed along the x axis when the current is $I = 1.2$ A in the direction shown? What is the expected direction of rotation of the loop?

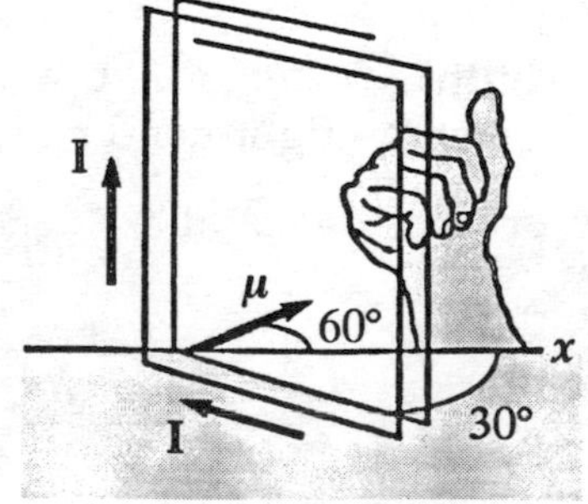

Figure P29.23

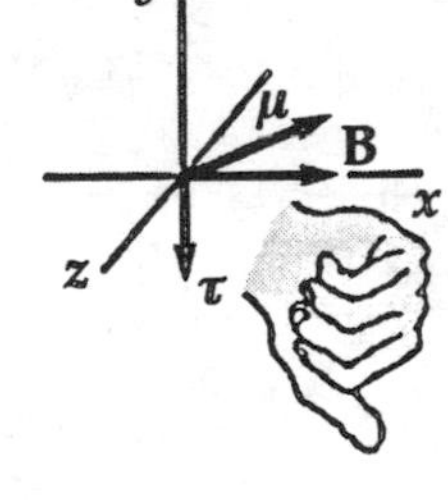

Solution The magnetic moment of the coil is $\mu = NIA$ perpendicular to its plane and making a 60° angle with the x axis as shown (Fig. P29.23b). The torque on the dipole is then $\tau = \mu \times \mathbf{B} = NAIB \sin\theta$ down, having a magnitude of

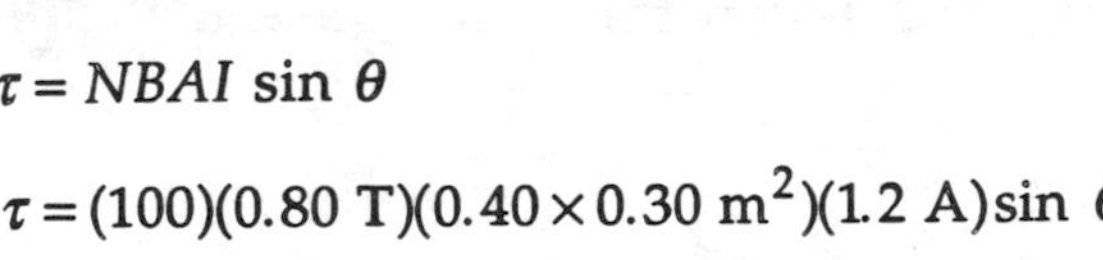

$$\tau = NBAI \sin \theta$$

$$\tau = (100)(0.80\text{ T})(0.40 \times 0.30\text{ m}^2)(1.2\text{ A})\sin 60°$$

$$\tau = 10\text{ N·m} \quad \lozenge$$

Note θ is the angle between the magnetic moment and the **B** field. Loop will rotate such to align the magnetic moment with the **B** field. Looking down along the y axis, loop will rotate in the clockwise direction.

---

33. A proton (charge $+e$, mass m_p), a deuteron (charge $+e$, mass $2m_p$), and an alpha particle, (charge $+2e$, mass $4m_p$) are accelerated through a common potential difference, V. The particles enter a uniform magnetic field, **B**, in a direction perpendicular to **B**. The proton moves in a circular path of radius r_p. Determine the values of the radii of the circular orbits for the deuteron, r_d, and the alpha particle, r_α, in terms of r_p.

Solution An electric field changes the speed of each particle according to

$$(K + U)_i = (K + U)_f. \quad qV = 1/2\, mv^2.$$

(We have assumed the particles start from rest.) The magnetic field changes their direction as described by $\Sigma\mathbf{F} = m\mathbf{a}$.

$$qvB \sin 90° = \frac{mv^2}{r}; \qquad \text{thus} \qquad r = \frac{mv}{qB} = \frac{m}{qB}\sqrt{\frac{2qV}{m}} = \frac{1}{B}\sqrt{\frac{2mV}{q}}$$

For the protons, $$r_p = \frac{1}{B}\sqrt{\frac{2m_pV}{e}}$$

For the deuterons, $$r_d = \frac{1}{B}\sqrt{\frac{2(2m_p)V}{e}} = \sqrt{2}\,r_p \quad \lozenge$$

For the alpha particles, $$r_\alpha = \frac{1}{B}\sqrt{\frac{2(4m_p)V}{2e}} = \sqrt{2}\,r_p \quad \lozenge$$

---

35. A cosmic-ray proton in interstellar space has an energy of 10 MeV and executes a circular orbit having a radius equal to that of Mercury's orbit around the Sun (5.8×10^{10} m). What is the magnetic field in that region of space?

Solution Think of the proton as having accelerated through a potential difference $V = 10^7$ V. Then its energy is

$$E = \tfrac{1}{2}mv^2 = eV$$

so its speed is $$v = \sqrt{\frac{2eV}{m}}$$

Now $\Sigma \mathbf{F} = ma$ becomes $$\frac{mv^2}{R} = evB \sin 90°$$

So $$B = \frac{mv}{eR} = \frac{m}{eR}\sqrt{\frac{2eV}{m}} = \frac{1}{R}\sqrt{\frac{2mV}{e}}$$

and $$B = \frac{1}{5.8 \times 10^{10}\ \text{m}}\sqrt{\frac{2(1.6727 \times 10^{-27}\ \text{kg})(10^7\ \text{V})}{1.60 \times 10^{-19}\ \text{C}}} = 7.88 \times 10^{-12}\ \text{T} \quad \Diamond$$

---

41. At the equator, near the surface of the Earth, the magnetic field is approximately 50 μT northward and the electric field is about 100 N/C downward (in other words, toward the ground). Find the gravitational, electric, and magnetic forces on a 100-eV electron moving eastward in a straight line in this environment.

Solution The gravitational force on the electron is

$$F_g = mg = 9.11 \times 10^{-31}\ \text{kg}\ (9.80\ \text{m/s}^2)\ \text{down} = 8.93 \times 10^{-30}\ \text{N down} \quad \Diamond$$

Likewise, the electric force can be easily calculated from

$$F_e = q\mathbf{E} = \left(-1.60 \times 10^{-19}\ \text{C}\right)(100\ \text{N/C})\ \text{down} = 1.60 \times 10^{-17}\ \text{N up} \quad \Diamond$$

The speed of the electron follows from

$$\tfrac{1}{2}mv^2 = 100\ \text{eV}\left(1.60 \times 10^{-19}\ \text{C/e}^-\right)\left(\frac{1\ \text{J}}{\text{V}\cdot\text{C}}\right)$$

Thus,

$$v = \sqrt{2\left(\frac{1.60\times10^{-17}\ \text{J}}{9.11\times10^{-31}\ \text{kg}}\right)\left(\frac{1\ \text{kg}\cdot\text{m}^2}{\text{J}\cdot\text{s}^2}\right)} = 5.93\times10^{6}\ \text{m/s}$$

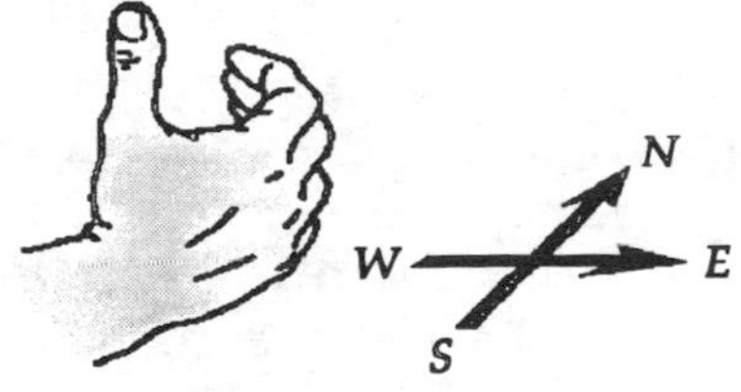

East × North = Up

From the velocity and magnetic field, we can calculate the force on the electron:

$$F_m = qv\times B = \left(-1.60\times10^{-19}\ \text{C}\right)\left(5.93\times10^{6}\ \text{m/s east}\right)\times\left(50\times10^{-6}\ \text{T north}\right)$$

Thus, the magnetic field is $= -4.74\times10^{-17}\ \text{N up} = 4.7\times10^{-17}\ \text{N down}$ ◊

---

47. A cyclotron designed to accelerate protons has a magnetic field of magnitude 0.45 T over a region of radius 1.2 m. What are (a) the cyclotron frequency and (b) the maximum speed acquired by the protons?

Solution

(a) The cyclotron frequency, $\omega = \frac{qB}{m}$

For protons, $\omega = \frac{\left(1.60\times10^{-19}\ \text{C}\right)(0.45\ \text{T})}{1.67\times10^{-27}\ \text{kg}} = 4.3\times10^{7}\ \text{rad/s}$ ◊

(b) $R = \frac{mv}{Bq}$ and $v = \frac{BqR}{m}$

$$v = \frac{(0.45\ \text{T})(1.60\times10^{-19}\ \text{C})(1.2\ \text{m})}{1.67\times10^{-27}\ \text{kg}} = 5.2\times10^{7}\ \text{m/s}\ ◊$$

---

49. The picture tube in a television uses magnetic deflection coils rather than electric deflection plates. Suppose an electron beam is accelerated through a 50.0-kV potential difference and then passes a distance of 1.00 cm through a uniform magnetic field produced by these coils for 1 cm. The screen is located 10.0 cm from the center of the coils and is 50.0 cm wide. When the field is turned off, the electron beam hits the center of the screen. What field strength is necessary to deflect the beam to the side of the screen?

Solution The beam is deflected by the angle

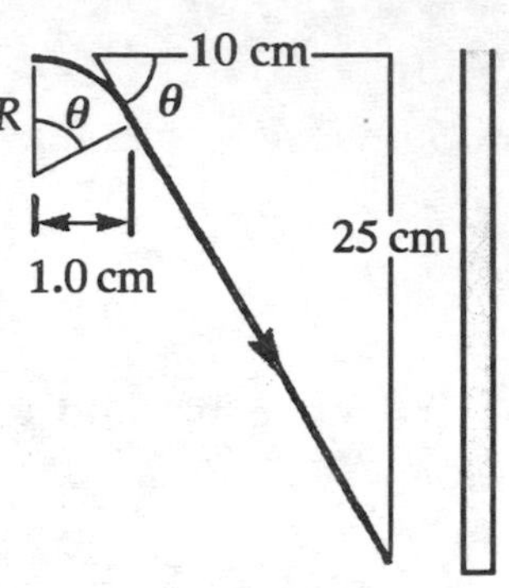

$$\theta = \tan^{-1}\frac{25}{10} = 68.2°$$

The two angles θ shown are equal because their sides are perpendicular, right side to right side and left side to left side. The radius of curvature of the electrons in the field is

$$R = \frac{1.0\text{ cm}}{\sin 68.2°} = 1.077\text{ cm}$$

Now, $\frac{1}{2}mv^2 = qV$ gives for the speed $v = \sqrt{\frac{2qV}{m}} = 1.33 \times 10^8\text{ m/s}$

where we choose to ignore the relativistic correction.

At last, $\Sigma\mathbf{F} = m\mathbf{a}$ becomes

$$\frac{mv^2}{R} = |q|vB\sin 90°$$

$$B = \frac{mv}{|q|R} = \frac{(9.11\times10^{-31}\text{ kg})(1.33\times10^{8}\text{ m/s})}{(1.60\times10^{-19}\text{ C})(1.077\times10^{-2}\text{ m})}$$

Solving, $B = 70.1\text{ mT}$ ◊

53. In an experiment designed to measure the Earth's magnetic field using the Hall effect, a copper bar 0.50 cm thick is positioned along an east-west direction. If a current of 8.0 A in the conductor results in a Hall voltage of 5.1×10^{-12} V, what is the magnitude of the Earth's magnetic field? (Assume that $n = 8.48 \times 10^{28}$ electrons/m^3 and that the plane of the bar is rotated to be perpendicular to the direction of **B**.)

Solution The Hall voltage is given by $V_H = \dfrac{IB}{nqt}$

Thus, $$B = \frac{nqtV_H}{I}$$

$$B = \frac{(8.48 \times 10^{28}\ e^- / m^3)(1.60 \times 10^{-19}\ C/e^-)(0.0050\ m)(5.10 \times 10^{-12}\ V)}{8\ A}$$

and $B = 4.32 \times 10^{-5}$ T or $B = 43.2\ \mu T$ ◊

55. The Hall effect can be used to measure n, the number of conduction electrons per unit volume for an unknown sample. The sample is 15 mm thick and when placed in a 1.8-T magnetic field produces a Hall voltage of 0.122 μV while carrying a 12-A current. What is the value of n?

Solution We apply Equation 29.22, where q stands for the absolute value of the charge of one corner.

$$V_H = \frac{IB}{nqt} \quad \text{or} \quad n = \frac{IB}{V_H q t}$$

$$n = \frac{(12\ C/s)(1.8\ N \cdot s / C \cdot m)}{(0.122 \times 10^{-6}\ N \cdot m / C)(1.60 \times 10^{-19}\ C)(15 \times 10^{-3}\ m)}$$

$$n = 7.38 \times 10^{28}\ \text{electrons}/m^3$$ ◊

57. A wire having a linear mass density of 1.0 g/cm is placed on a horizontal surface that has a coefficient of friction of 0.20. The wire carries a current of 1.5 A toward the east and moves horizontally to the north. What are the magnitude and direction of the smallest magnetic field that enables the wire to move in this fashion?

Solution The smallest field will exert the smallest force $\mathbf{F}_m$ required to slide the wire, which we can exert most effectively by making it act at angle θ above the horizontal to reduce friction by reducing the normal force. The wire is in equilibrium:

$$\Sigma \mathbf{F}_x = 0 \qquad F_m \cos\theta - \mu n = 0$$

$$\Sigma \mathbf{F}_y = 0 \qquad +n + F_m \sin\theta - mg = 0$$

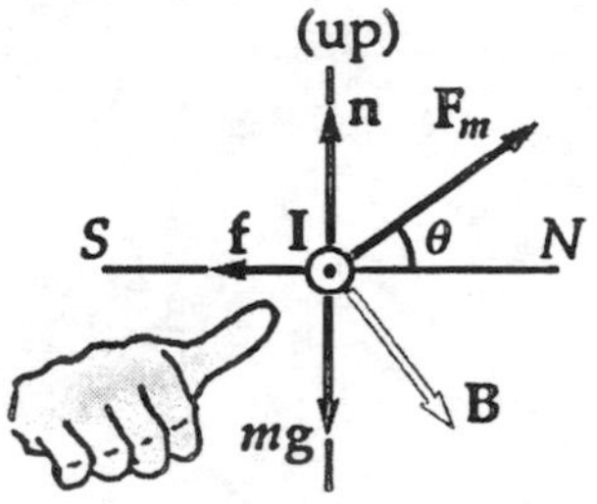

(up) n $\mathbf{F}_m$ S f I θ N mg B

Then $$F_m \cos\theta - \mu mg + \mu F_m \sin\theta = 0$$

$$F_m = \mu mg(\mu \sin\theta + \cos\theta)^{-1}$$

To minimize F_m we require $\dfrac{dF_m}{d\theta} = 0$

$$\frac{dF_m}{d\theta} = \frac{\mu mg(-1)(\mu\cos\theta - \sin\theta)}{(\mu \sin\theta + \cos\theta)^2} = 0$$

$$\mu = \frac{\sin\theta}{\cos\theta} = \tan\theta = 0.2 \qquad \theta = 11.3°$$

$$\frac{F_m}{\ell} = \frac{(m/\ell)\mu g}{(\mu\sin\theta + \cos\theta)} = \frac{(0.10\text{ kg/m})(0.20)(9.8\text{ m/s}^2)}{(0.20\sin 11.3° + \cos 11.3°)}$$

$$= 0.192\text{ N/m} = \frac{I\ell B \sin 90°}{\ell}$$

$$B = \frac{0.192\text{ N/m}}{1.5\text{ A}} = 0.13\text{ T} \quad \Diamond$$

The direction should be northward at (90 – 11.3) = 79° below the horizontal. ◊

58. Indicate the initial direction of the deflection of the charged particles as they enter the magnetic fields shown in Figure P29.58.

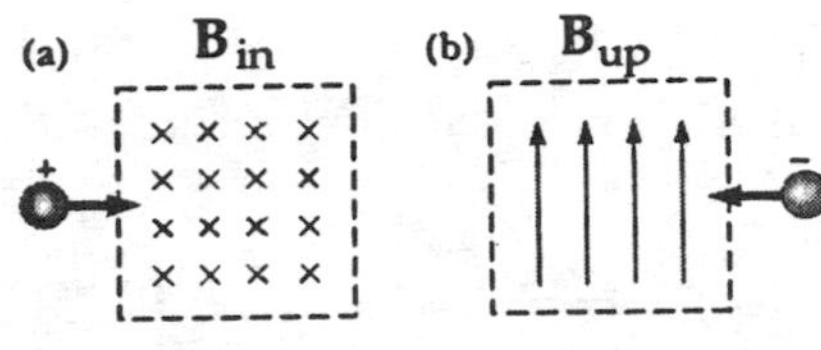

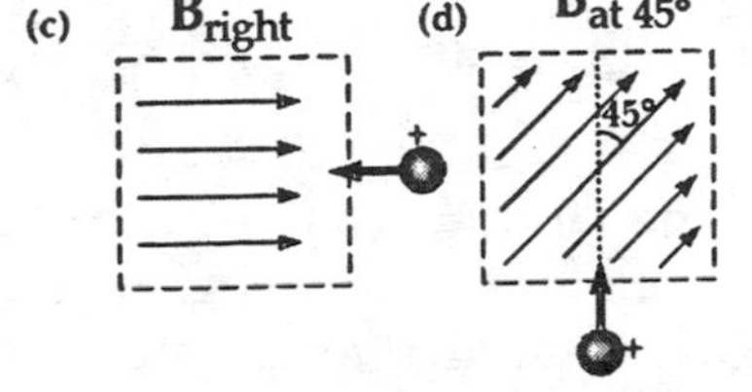

Figure P29.58

Solution

(a) By solution figure (a), $\mathbf{v} \times \mathbf{B}$ is (right)×(away) = up ◊

(b) By solution figure (b), $\mathbf{v} \times \mathbf{B}$ is (left)×(up) = away. Since the charge is negative, $q\mathbf{v} \times \mathbf{B}$ is toward you ◊

(c) $\mathbf{v} \times \mathbf{B}$ is zero since the angle between **v** and **B** is 180° and sin 180° = 0. There is no deflection. ◊

(d) $\mathbf{v} \times \mathbf{B}$ is (up) × (up and right), or away from you ◊

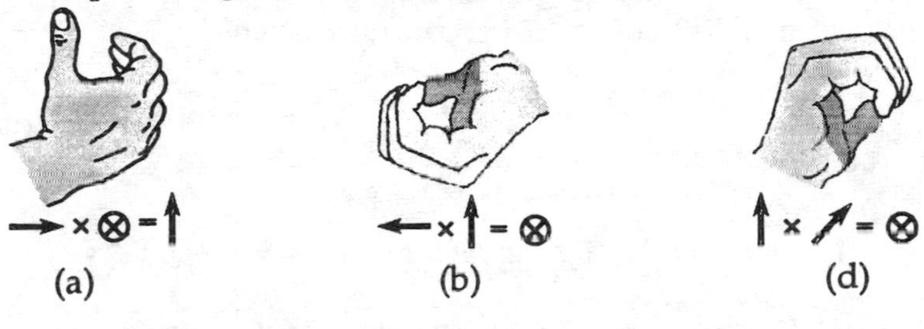

---

59. A positive charge $q = 3.2 \times 10^{-19}$ C moves with a velocity $\mathbf{v} = (2\mathbf{i} + 3\mathbf{j} - \mathbf{k})$ m/s through a region where both a uniform magnetic field and a uniform electric field exist. (a) Calculate the total force on the moving charge (in unit-vector notation) if $\mathbf{B} = (2\mathbf{i} + 4\mathbf{j} + \mathbf{k})$ T and $\mathbf{E} = (4\mathbf{i} - \mathbf{j} - 2\mathbf{k})$ V/m. (b) What angle does the force vector make with the positive x axis?

Solution The total force is the Lorentz force,

(a) $$\mathbf{F} = q\mathbf{E} + q(\mathbf{v} \times \mathbf{B}) = q(\mathbf{E} + \mathbf{v} \times \mathbf{B})$$

$$\mathbf{F} = q[(4\mathbf{i} - \mathbf{j} - 2\mathbf{k})\ \text{V/m} + (2\mathbf{i} + 3\mathbf{j} - \mathbf{k})\ \text{m/s} \times (2\mathbf{i} + 4\mathbf{j} + \mathbf{k})\ \text{T}]$$

$$\mathbf{F} = q\left[(4\mathbf{i} - \mathbf{j} - 2\mathbf{k})\ \text{V/m} + (7\mathbf{i} - 4\mathbf{j} + 2\mathbf{k})\ \frac{\text{m} \cdot \text{T}}{\text{s}}\right] = q\left([11\mathbf{i} - 5\mathbf{j}]\ \frac{\text{V}}{\text{m}}\right) = q\left([11\mathbf{i} - 5\mathbf{j}]\ \frac{\text{N}}{\text{C}}\right)$$

$$\mathbf{F} = (3.2 \times 10^{-19}\ \text{C})\left((11\mathbf{i} - 5\mathbf{j})\ \frac{\text{N}}{\text{C}}\right) = (3.5\mathbf{i} - 1.6\mathbf{j}) \times 10^{-18}\ \text{N} \quad ◊$$

(b) $\mathbf{F}\cdot\mathbf{i} = F\cos\theta = F_x$

$$\theta = \cos^{-1}\left(\frac{F_x}{F}\right) = \cos^{-1}\left(\frac{3.52}{3.87}\right) = 24° \quad \Diamond$$

---

61. The circuit in Figure P29.61 consists of wires at the top and bottom and identical metal springs at the left and right sides. The wire at the bottom has a mass of 10 g and is 5.0 cm long. The springs stretch 0.50 cm under the weight of the wire and the circuit has a total resistance of 12 Ω. When a magnetic field is turned on, directed out of the page, the springs stretch an additional 0.30 cm. What is the strength of the magnetic field? (The upper portion of the circuit is fixed.)

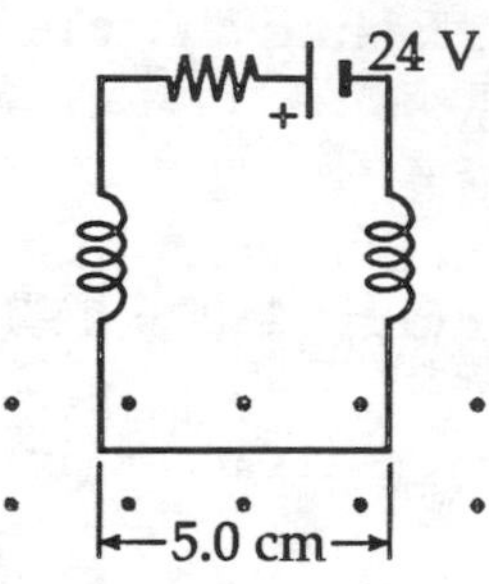

Figure P29.61

Solution The weight of the wire is

$$w = mg = \left(1.0\times10^{-2}\text{ kg}\right)\left(9.8\text{ m/s}^2\right) = 9.8\times10^{-2}\text{ N}$$

We find the spring constant of the pair of springs together, from

$$F_s = kx \qquad k = \frac{F_s}{x} = \frac{9.8\times10^{-2}\text{ N}}{0.50\times10^{-2}\text{ m}} = 19.6\text{ N/m}$$

The magnetic force is then $\quad (19.6\text{ N/m})(0.30\times10^{-2}\text{ m}) = 5.88\times10^{-2}\text{ N}$

The current is $\quad I = \frac{V}{R} = \frac{24\text{ V}}{12\ \Omega} = 2.0\text{ A}$

Then $\mathbf{F}_m = I\,\mathbf{L}\times\mathbf{B}$

$$\left(-5.88\times10^{-2}\mathbf{j}\text{ N}\right) = (2.0\text{ A})\left(5.0\times10^{-2}\text{ m}\right)(\mathbf{i}\times B\,\mathbf{k}) = \left(-1.0\times10^{-1}\,\mathbf{j}\,B\right)\frac{\text{C}\cdot\text{m}}{\text{s}}$$

$$B = 0.59\text{ T} \quad \Diamond$$

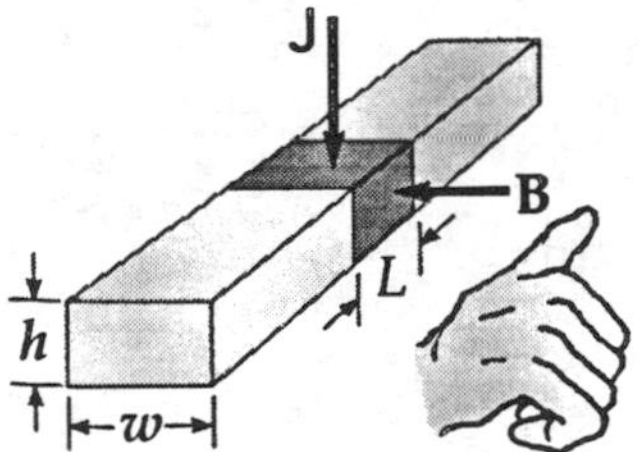

Figure 29.63

63. Sodium melts at 99°C. Liquid sodium, an excellent thermal conductor, is used in some nuclear reactors to remove thermal energy from the reactor core. The liquid sodium is moved through pipes by pumps that exploit the force on a moving charge in a magnetic field. The principle is as follows: Imagine the liquid metal to be in a pipe having a rectangular cross section of width w and height h. A uniform magnetic field perpendicular to the pipe affects a section of length L (Fig. P29.63). An electric current directed perpendicular to the pipe and to the magnetic field produces a current density J. (a) Explain why this arrangement produces on the liquid a force that is directed along the length of the pipe. (b) Show that the section of liquid in the magnetic field experiences a pressure increase JLB.

Solution

(a) By the right-hand rule, the electric current carried by the material experiences a force $I\,\mathbf{h} \times \mathbf{B}$ in the direction of $\mathbf{L}$. ◊

(b) The sodium, consisting of ions and electrons, flows along the pipe transporting no net charge. But inside the section of length L, electrons drift upward to constitute downward electric current $J\cdot$(area) $= JLw$. The current feels magnetic force $I\,\mathbf{h} \times \mathbf{B} = JLwhB \sin 90°$. This force along the pipe axis will make the fluid move, exerting pressure

$$\frac{F}{\text{area}} = \frac{JLwhB}{hw} = JLB \quad ◊$$

In the picture, the fluid moves away from you into the page.

69. Consider an electron orbiting a proton and maintained in a fixed circular path of radius $R = 5.29 \times 10^{-11}$ m by the Coulomb force. Treating the orbiting charge as a current loop, calculate the resulting torque when the system is in a magnetic field of 0.400 T directed perpendicular to the magnetic moment of the electron.

Solution

$$|\tau| = IAB$$

Define v to be the speed in the electron's circular path, and R to be the radius of the path.

The period of the motion is $T = \frac{2\pi R}{v}$,

and the effective current due to the orbiting electron is $I = \frac{q}{T}$.

The Coulomb force on the electron must equal the central force:

$$\frac{k_e q^2}{R^2} = \frac{mv^2}{R} \qquad \text{or} \qquad v = q\sqrt{\frac{k_e}{mR}}$$

Therefore,

$$T = 2\pi\sqrt{\frac{mR^3}{q^2 k_e}}$$

$$= 2\pi\sqrt{\frac{(9.1\times10^{-31}\ \text{kg})(5.29\times10^{-11}\ \text{m})^3}{(1.6\times10^{-19}\ \text{C})^2\left(8.99\times10^9\ \frac{\text{N}\cdot\text{m}^2}{\text{C}^2}\right)}} = 1.52\times10^{-16}\ \text{s}$$

and the torque

$$|\tau| = \left(\frac{q}{T}\right)AB = \frac{1.60\times10^{19}\ \text{C}}{1.52\times10^{-16}\ \text{s}}(\pi)(5.29\times10^{-11}\ \text{m})^2(0.4\ \text{T}) = 3.70\times10^{-24}\ \text{N}\cdot\text{m} \quad \lozenge$$

71. Protons having a kinetic energy of 5.00 MeV are moving in the positive x direction and enter a magnetic field $\mathbf{B} = (0.0500\ \mathrm{T})\mathbf{k}$ directed out of the plane of the page and extending from $x = 0$ to $x = 1.00$ m as in Figure P29.71. (a) Calculate the y component of the protons' momentum as they leave the magnetic field. (b) Find the angle α between the initial velocity vector of the proton beam and the velocity vector after the beam emerges from the field. (*Hint:* Neglect relativistic effects and note that 1 eV = 1.60×10^{-19} J.)

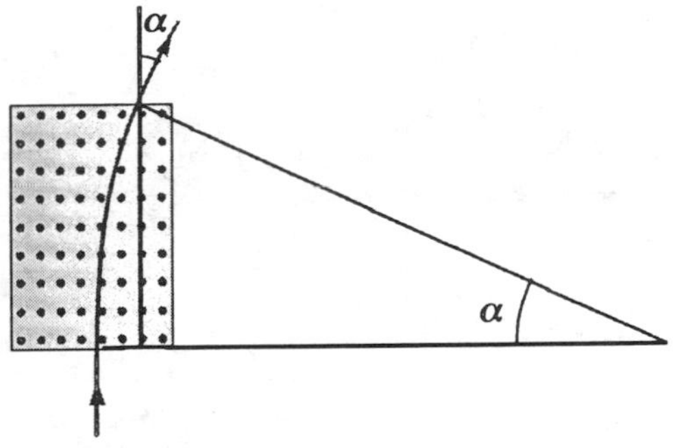

Figure P29.71 (modified)

Solution We first find the speed of each proton:

$$K = 5.00\ \mathrm{MeV} = \tfrac{1}{2}mv^2$$

$$v = \sqrt{\frac{2K}{m}} = \sqrt{\frac{2(5.00\times10^{6}\ \mathrm{eV})(1.60\times10^{-19}\ \mathrm{J/eV})}{1.67\times10^{-27}\ \mathrm{kg}}} = 3.10\times10^{7}\ \mathrm{m/s}$$

(This is only a little more than one-tenth the speed of light, so the correct relativistic answer would be nearly the same.) In the field, the trajectory of the protons is a section of a circle with radius given by

$$\Sigma\mathbf{F} = m\mathbf{a} \qquad qvB\sin 90° = \frac{mv^2}{r}$$

$$r = \frac{mv}{qB} = \frac{(1.67\times10^{-27}\ \mathrm{kg})(3.10\times10^{7}\ \mathrm{m/s})}{(1.60\times10^{-19}\ \mathrm{C})(0.0500\ \mathrm{N\cdot s/C\cdot m})} = 6.46\ \mathrm{m}$$

(b) It is easiest to do part (b) before (a):

$$\sin\alpha = \frac{1\ \mathrm{m}}{6.46\ \mathrm{m}} \qquad \alpha = 8.90° \quad \lozenge$$

(a) Each proton moves with constant speed, so

$$p_y = -mv\sin\alpha = (-1.67\times10^{-27}\ \mathrm{kg})(3.10\times10^{7}\ \mathrm{m/s})\left(\frac{1\ \mathrm{m}}{6.46\ \mathrm{m}}\right) = -8.00\times10^{-21}\ \mathrm{kg\cdot m/s} \quad \lozenge$$

Chapter 30

Sources of the Magnetic Field

Chapter 30

SOURCES OF THE MAGNETIC FIELD

INTRODUCTION

The preceding chapter treated a class of problems involving the magnetic force on a charged particle moving in a magnetic field. To complete the description of the magnetic interaction, this chapter deals with the origin of the magnetic field, namely, moving charges or electric currents. We begin by showing how to use the law of Biot and Savart to calculate the magnetic field produced at a point by a current element. Using this formalism and the superposition principle, we then calculate the total magnetic field due to a distribution of currents for several geometries. Next, we show how to determine the force between two current-carrying conductors, a calculation that leads to the definition of the ampere. We also introduce Ampère's law, which is very useful for calculating the magnetic field of highly symmetric configurations carrying steady currents. We apply Ampère's law to determine the magnetic field for several current configurations.

NOTES FROM SELECTED CHAPTER SECTIONS

30.1 The Biot-Savart Law

The *Biot-Savart law* says that if a wire carries a steady current I, the magnetic field $d\mathbf{B}$ at a point P associated with an element $d\mathbf{s}$ has the following properties:

- The vector $d\mathbf{B}$ is perpendicular both to $d\mathbf{s}$ (which is in the direction of the current) and to the unit vector $\hat{\mathbf{r}}$ directed from the element to the point P.
- The magnitude of $d\mathbf{B}$ is inversely proportional to r^2, where r is the distance from the element to the point P.
- The magnitude of $d\mathbf{B}$ is proportional to the current and to the length of the element, $d\mathbf{s}$.
- The magnitude of $d\mathbf{B}$ is proportional to $\sin\theta$, where θ is the angle between the vectors $d\mathbf{s}$ and $\hat{\mathbf{r}}$.

30.2 The Magnetic Force Between Two Parallel Conductors

Parallel conductors carrying currents in the *same direction attract* each other, whereas parallel conductors carrying currents in *opposite directions repel* each other.

The force between two parallel wires each carrying a current is used to define the ampere as follows:

If two long, parallel wires 1 m apart carry the same current and the force per unit length on each wire is 2×10^{-7} N/m, then the current is defined to be 1 A.

30.3 Ampère's Law

The direction of the magnetic field due to a current in a conductor is given by the right-hand rule:

If the wire is grasped in the right hand with the thumb in the direction of the current, the fingers will wrap (or curl) in the direction of **B**.

Ampère's law is valid only for *steady* currents and is useful only in those cases where the current configuration has a *high degree* of *symmetry*.

30.7 Gauss's Law in Magnetism

Gauss's law in magnetism states that the net magnetic flux through any closed surface is always zero.

30.8 Displacement Current and the Generalized Ampère's Law

Magnetic fields are produced both by conduction currents and by changing electric fields.

30.9 Magnetism in Matter

In order to describe the magnetic properties of materials, it is convenient to classify the material into three categories: paramagnetic, ferromagnetic, and diamagnetic. *Paramagnetic and ferromagnetic materials* are those that have atoms with permanent magnetic dipole moments. *Diamagnetic materials* are those whose atoms have no permanent magnetic dipole moments. For materials whose atoms have permanent magnetic moments, the diamagnetic contribution to the magnetism is usually overshadowed by a paramagnetic or ferromagnetic effect.

Paramagnetic substances have a positive but small susceptibility which is due to the presence of atoms (or ions) with *permanent* magnetic dipole moments. These dipoles interact only weakly with each other and are randomly oriented in the absence of an external magnetic field. Experimentally, one finds that the magnetization of a paramagnetic substance is proportional to the applied field and inversely proportional to the absolute temperature under a wide range of conditions.

A diamagnetic substance is one whose atoms have no permanent magnetic dipole moment. When an external magnetic field is applied to a diamagnetic substance, a weak magnetic dipole moment is *induced* in the direction opposite the applied field.

The effect called magnetic hysteresis shows that the magnetization of a ferromagnetic substance depends on the history of the substance as well as the strength of the applied field.

The magnetization curve is useful for another reason. *The area enclosed by the magnetization curve represents the work required to take the material through the hysteresis cycle.* Ferromagnetic substances contain atomic magnetic moments that tend to align parallel to each other even in a weak external magnetic field. Once the moments are aligned, the substance will remain magnetized after the external field is removed.

EQUATIONS AND CONCEPTS

The *Biot-Savart law* gives the magnetic field at a point in space due to an element of conductor $d\mathbf{s}$ which carries a current I and is at a distance r away from the point.

$$d\mathbf{B} = \frac{\mu_0}{4\pi}\frac{I d\mathbf{s} \times \hat{\mathbf{r}}}{r^2} \quad (30.4)$$

The permeability of free space is a constant.

$$\mu_0 = 4\pi \times 10^{-7}\ \mathrm{T{\cdot}m/A}$$

The *total magnetic field* is found by integrating the Biot-Savart law expression over the entire length of the conductor.

$$\mathbf{B} = \frac{\mu_0 I}{4\pi}\int \frac{d\mathbf{s} \times \hat{\mathbf{r}}}{r^2} \quad (30.5)$$

The magnetic field due to several important geometric arrangements of a current-carrying conductor can be calculated by use of the Biot-Savart law:

B at a distance a from a *long straight conductor,* carrying a current I.

$$B = \frac{\mu_0 I}{2\pi a} \qquad (30.7)$$

B at the center of an *arc of radius* R *which subtends an angle* θ *(in radians) at the center of the arc.*

$$B = \frac{\mu_0 I}{4\pi R}\theta \qquad (30.8)$$

B_x on the axis of a *circular loop* of radius R and at a *distance* x *from the plane* of the loop.

$$B_x = \frac{\mu_0 I R^2}{2\left(x^2 + R^2\right)^{3/2}} \qquad (30.9)$$

B at a distance a from a *straight wire* carrying a current I, where θ_1 and θ_2 are as shown in the figure below.

$$B = \frac{\mu_0 I}{4\pi a}\left(\cos\theta_1 - \cos\theta_2\right) \qquad (30.6)$$

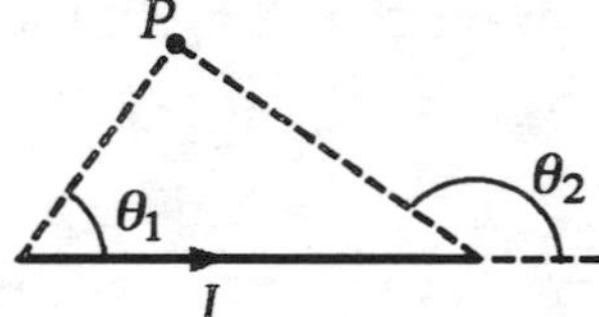

The magnitude of the *magnetic force per unit length* between very long parallel conductors depends on the distance a between the conductors and the magnitudes of the two currents.

$$\frac{F_1}{L} = \frac{\mu_0 I_1 I_2}{2\pi a} \qquad (30.13)$$

If the parallel currents I_1 and I_2 are in the same direction, the force between conductors will be one of attraction. Parallel conductors carrying currents in opposite directions will repel each other. In any case, the magnitude of the forces on the two conductors will be *equal.*

Ampère's law represents a relationship between the integral of the tangential component of the magnetic field around any closed path and the total current which threads the closed path.

$$\oint \mathbf{B}\cdot d\mathbf{s} = \mu_0 I \tag{30.15}$$

B inside a toroid having N turns and at a distance r from the center of the toroid.

$$B = \frac{\mu_0 N I}{2\pi r} \tag{30.18}$$

B near the center of a *solenoid* of n turns per unit length.

$$B = \mu_0 n I \tag{30.20}$$

The *magnetic flux* through a surface is the integral of the normal component of the field over the surface.

$$\Phi_B = \int \mathbf{B}\cdot d\mathbf{A} \tag{30.23}$$

The direction of the magnetic field due to current in a long wire is determined by using right-hand rule *B:*

> Hold the conductor in the right hand with the thumb pointing in the direction of the conventional current. The fingers will then wrap around the wire in the direction of the magnetic field lines. The magnetic field is tangent to the circular field lines at every point in the region around the conductor.

Comment on the direction of the magnetic field due to different current geometries.

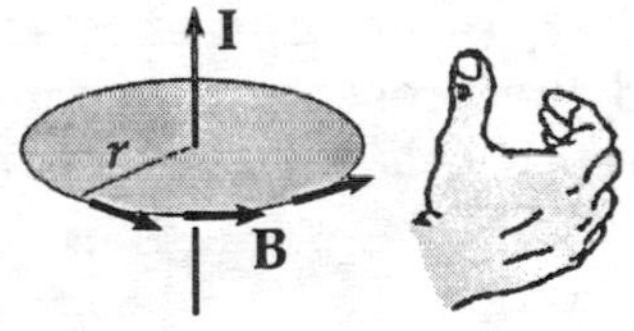

The direction of the magnetic field at the center of a current loop is perpendicular to the plane of the loop and directed in the sense given by the right-hand rule for B.

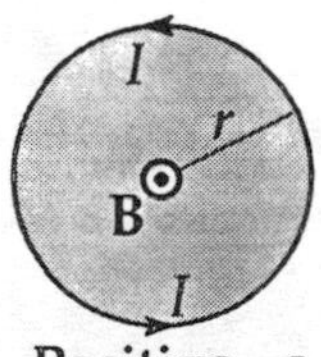

Positive q

Within a solenoid, the magnetic field is parallel to the axis of the solenoid and pointing in a sense determined by applying right-hand rule B to one of the coils.

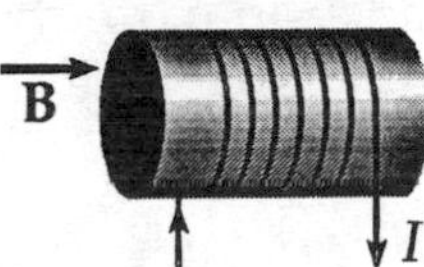

The magnetic moment μ of an orbiting electron is proportional to its orbital angular momentum L.	$\mu = \left(\frac{e}{2m}\right)L$ $L = 0,\ \hbar,\ 2\hbar,\ 3\hbar,\ \ldots$	(30.30)
The intrinsic magnetic moment μ_s associated with the spin of the electron is called the Bohr magneton.	$\mu_B = 9.27 \times 10^{-24}$ J/T	(30.32)
The magnetic state of a substance is described by a quantity called the magnetization vector, **M**. For paramagnetic and diamagnetic substances, the magnetization is proportional to the magnetic field strength.	$\mathbf{M} = \chi\mathbf{H}$	(30.36)
In a region where the total magnetic field is due to that of a current-carrying conductor and the presence of a magnetic substance, the *total field B* can be expressed in terms of **M** and **H**.	$\mathbf{B} = \mu_0\,(\mathbf{H} + \mathbf{M})$	(30.34)
The total field can also be expressed in terms of the permeability μ_m of the substance.	$\mathbf{B} = \mu_m\mathbf{H}$	(30.37)
The permeability of a magnetic substance is related to its magnetic susceptibility.	$\mu_m = \mu_0\,(1 + \chi)$	(30.38)
The magnetization of a paramagnetic substance is proportional to the applied field and inversely proportional to the absolute temperature. This is known as Curie's law.	$M = C\,\frac{B}{T}$	(30.39)

SUGGESTIONS, SKILLS, AND STRATEGIES

It is important to remember that the *Biot-Savart law*, given by Equation 30.4,

$$d\mathbf{B} = \frac{\mu_0 I}{4\pi}\frac{d\mathbf{s} \times \hat{\mathbf{r}}}{r^2}$$

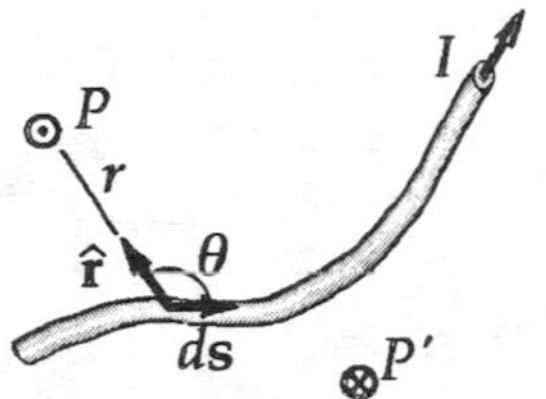

is a *vector* expression. The unit vector $\hat{\mathbf{r}}$ is directed from the element of conductor $d\mathbf{s}$ to the point P where the magnetic field is to be calculated, and r is the distance from $d\mathbf{s}$ to point P. For the arbitrary current element shown in the figure at right, the direction of $\mathbf{B}$ at point P, as determined by the right-hand rule for the cross product, is directed *out of the plane*; while the magnetic field at point P' due to the current in the element $d\mathbf{s}$ is directed *into the plane*. In order to find the *total* magnetic field at any point due to a conductor of finite sign, you must sum up the contributions from all current elements making up the conductor. This means that the total $\mathbf{B}$ field is expressed as an integral over the entire length of the conductor:

$$\mathbf{B} = \frac{\mu_0 I}{4\pi}\int \frac{d\mathbf{s} \times \hat{\mathbf{r}}}{r^2}$$

REVIEW CHECKLIST

- ▷ Use the Biot-Savart law to calculate the magnetic induction at a specified point in the vicinity of a current element, and by integration find the total magnetic field due to a number of important geometric arrangements. Your use of the Biot-Savart law must include a clear understanding of the *direction* of the magnetic field contribution relative to the direction of the current element which produces it and the direction of the vector which locates the point at which the field is to be calculated.

- ▷ Understand the basis for defining the ampere and the coulomb in terms of the magnetic force between parallel current-carrying conductors.

- ▷ Use Ampère's law to calculate the magnetic field due to steady current configurations which have a sufficiently high degree of symmetry such as a long straight conductor, a long solenoid, and a toroidal coil.

- ▷ Calculate the magnetic flux through a surface area placed in either a uniform or nonuniform magnetic field.

- ▷ Understand, via the generalized form of Ampère's law, that magnetic fields are produced both by *conduction currents* and by *changing electric fields*.

SOLUTIONS TO SELECTED END-OF-CHAPTER PROBLEMS

5. (a) A conductor in the shape of a square of edge length $\ell = 0.4$ m carries a current $I = 10$ A (Fig. P30.5). Calculate the magnitude and direction of the magnetic field at the center of the square. (b) If this conductor is formed into a single circular turn and carries the same current, what is the value of the magnetic field at the center?

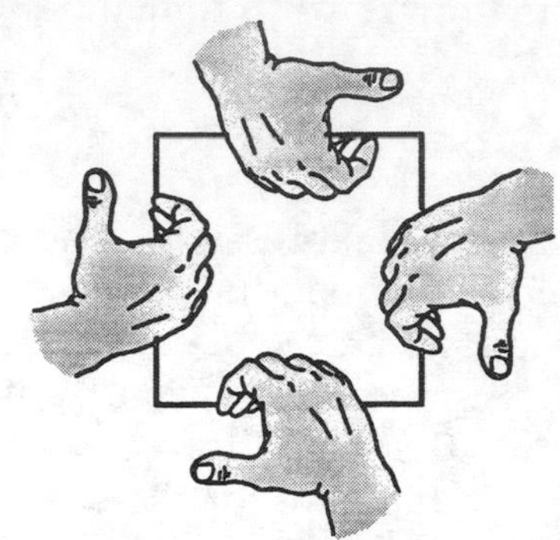

Figure P30.5

Solution

(a) We use Equation 30.6 for the field created by each side of the square. Each side creates field away from you, into the paper, in the same direction, so together they produce a field

$$B = \frac{4\mu_0 I}{4\pi a}\left(\cos\frac{\pi}{4} - \cos\frac{3\pi}{4}\right) = \frac{4\times10^{-6}\ \text{T}\cdot\text{m}}{0.20\ \text{m}}\left(\frac{\sqrt{2}}{2} + \frac{\sqrt{2}}{2}\right) = 2\sqrt{2}\times10^{-5}\ \text{T}$$

To three significant figures, **B** = 28.3 μT away from you into the paper ◊

(b) The new radius is found from $4\ell = 2\pi R$, so $R = 2\ell/\pi = 0.255$ m

$$B = \frac{\mu_0 I R^2}{2(x^2+R^2)^{3/2}} = \frac{\mu_0 I R^2}{2(0^2+R^2)^{3/2}} = \frac{\mu_0 I}{2R} = \frac{(4\pi\times10^{-7}\ \text{T}\cdot\text{m/A})(10\ \text{A})}{2(0.255\ \text{m})}$$

Caution! If you use your calculator, it may not understand the keystrokes:

[4] [×] [π] [EXP] [+/−] [7]

You may need to use

[4] [EXP] [+/−] [7] [×] [π]

Make sure you get $B = 2.47 \times 10^{-5}$ T = 24.7 μT away from you into the paper. ◊

A natural way to hold your right hand is with the fingers curving in the clockwise direction of the curving current. Then your extended thumb shows that the field is into the paper.

9. Determine the magnetic field at a point P located a distance x from the corner of an infinitely long wire bent at a right angle, as in Figure P30.9. The wire carries a steady current I.

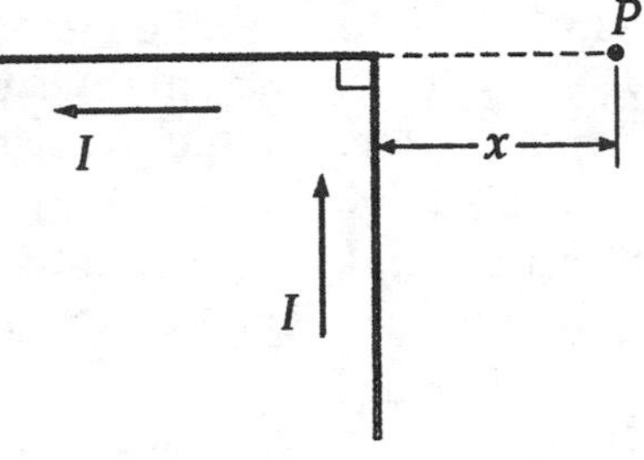

Figure P30.9

Solution The vertical section of wire constitutes one half of an infinitely long straight wire at distance x from P, so it creates field

$$B = \frac{1}{2}\left(\frac{\mu_0 I}{2\pi x}\right)$$

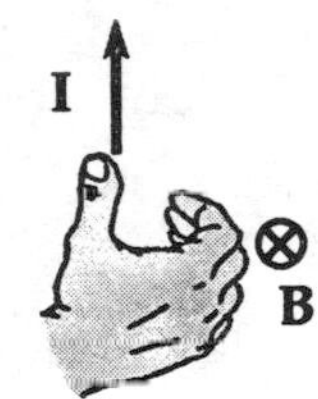

Hold your right hand with extended thumb in the direction of the current; the field is away from you, into the paper. For each bit of the horizontal section of wire $d\mathbf{s}$ is to the left and $\hat{\mathbf{r}}$ is to the right, so $d\mathbf{s} \times \hat{\mathbf{r}} = 0$. The horizontal current produces zero field at P. Thus,

$$\mathbf{B} = \frac{\mu_0 I}{4\pi x} \text{ into the paper} \quad \Diamond$$

---

15. Two long parallel conductors separated by 10.0 cm carry currents in the same direction. If $I_1 = 5.00$ A and $I_2 = 8.00$ A, what is the force per unit length exerted on each conductor by the other?

Solution Take the x direction to be the direction of the currents and the y axis to point from 8-A wire toward 5-A wire. At the location of 8-A current, the 5-A current creates a magnetic field

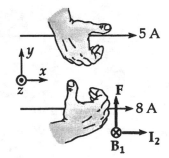

$$\mathbf{B}_1 = \frac{\mu_0 I_1}{2\pi a}(-\mathbf{k})$$

Then the force on I_2 is

$$\mathbf{F} = \ell \mathbf{I}_2 \times \mathbf{B} = \ell I_2 \frac{\mu_0 I_1}{2\pi a}(\mathbf{i} \times (-\mathbf{k})) = \frac{\mu_0 I_1 I_2 \ell}{2\pi a}\mathbf{j}$$

This is a force of attraction; substituting and solving, we find that the magnitude of the force per unit length on each wire is

$$\frac{F}{\ell} = \frac{\mu_0 I_1 I_2}{2\pi a} = 8.00 \times 10^{-5} \text{ N/m} \quad \Diamond$$

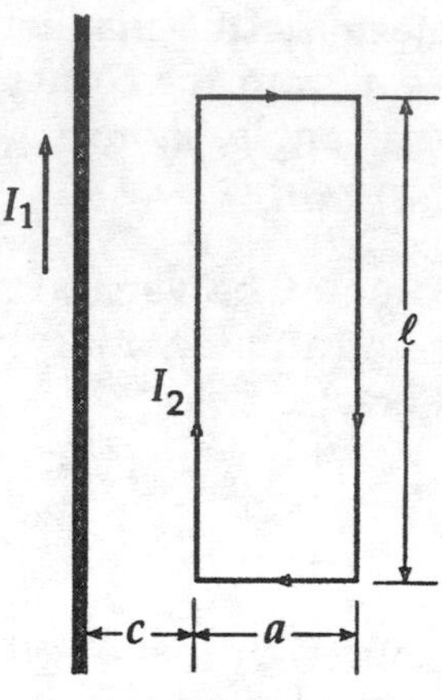

Figure P30.19

19. In Figure P30.19, the current in the long, straight wire is $I_1 = 5.00$ A and the wire lies in the plane of the rectangular loop, which carries 10.0 A. The dimensions are $c = 0.100$ m, $a = 0.150$ m, and $\ell = 0.450$ m. Find the magnitude and direction of the net force exerted on the loop by the magnetic field created by the wire.

Solution By symmetry the forces exerted on the segments of length a are equal and opposite and cancel. The magnetic field in the plane of I_2 to the right of I_1 is directed away from you into the plane. By the right-hand rule, $\mathbf{F} = I(\boldsymbol{\ell} \times \mathbf{B})$ is directed toward the *left* for the near side of the loop and directed toward the *right* for the side at $c + a$.

Thus, $$\mathbf{F} = \mathbf{F}_1 + \mathbf{F}_2 = \frac{\mu_0 I_1 I_2 \ell}{2\pi}\left(\frac{1}{c+a} - \frac{1}{c}\right)\mathbf{i}$$

$$\mathbf{F} = \frac{\mu_0 I_1 I_2 \ell}{2\pi}\left[\frac{-a}{c(c+a)}\right]\mathbf{i}$$

$$\mathbf{F} = \frac{(4\pi \times 10^{-7}\ \mathrm{N/A^2})(5.00\ \mathrm{A})(10.0\ \mathrm{A})(0.450\ \mathrm{m})}{2\pi}\left[\frac{-0.150\ \mathrm{m}}{(0.100\ \mathrm{m})(0.250\ \mathrm{m})}\right]\mathbf{i}$$

$$\mathbf{F} = (-2.70 \times 10^{-5}\mathbf{i})\ \mathrm{N} \quad \text{or} \quad F = 2.70 \times 10^{-5}\ \mathrm{N} \ \text{ toward the left} \quad \Diamond$$

---

24. What current is required in the windings of a long solenoid that has 1000 turns uniformly distributed over a length of 0.40 m in order to produce at the center of the solenoid a magnetic field of magnitude 1.0×10^{-4} T ?

Solution $$B = \mu_0 \frac{N}{\ell} I$$

$$I = \frac{B}{\mu_0 n} = \frac{(10^{-4}\ \mathrm{T \cdot A})(0.40\ \mathrm{m})}{(4\pi)(10^{-7}\ \mathrm{T \cdot m})(1000)} = 32\ \mathrm{mA} \quad \Diamond$$

---

27. The magnetic coils of a tokamak fusion reactor are in the shape of a toroid having an inner radius of 0.70 m and outer radius of 1.30 m. If the toroid has 900 turns of large-diameter wire, each of which carries a current of 14 kA, find the magnetic field strength along (a) the inner radius and (b) the outer radius.

Solution From Ampère's law, the magnetic field at a distance r from the center of the toroid is found to be

$$B = \frac{\mu_0 NI}{2\pi r}$$

(a) Along the inner radius,

$$B_i = \frac{(4\pi \times 10^{-7}\ \mathrm{T \cdot m / A})(900)(1.4 \times 10^{4}\ \mathrm{A})}{2\pi(0.70\ \mathrm{m})} = 3.60\ \mathrm{T} \quad \lozenge$$

(b) Along the outer radius, where $r = 1.3$ m, we find $B = 1.94\ \mathrm{T}$ ◊

---

31. A packed bundle of 100 long, straight, insulated wires forms a cylinder of radius $R = 0.50$ cm. (a) If each wire carries 2.0 A, what are the magnitude and direction of the magnetic force per unit length acting on a wire located 0.2 cm from the center of the bundle? (b) Would a wire on the outer edge of the bundle experience a force greater or smaller than the value calculated in part (a)?

Solution The force *on* one wire is exerted *by* the other ninety-nine, through the magnetic field they create.

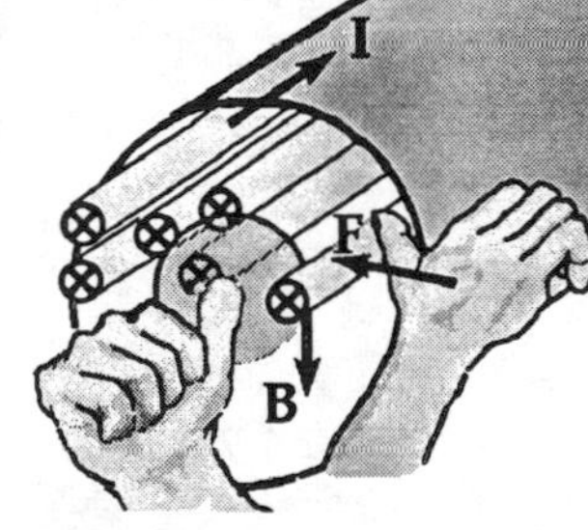

(a) According to Equation 30.17 in Example 30.4, the magnetic field at $r = 0.2$ cm from the center of the cable is:

$$B = \frac{\mu_0 I_0 r}{2\pi R^2} = \frac{(4\pi \times 10^{-7}\ \mathrm{T \cdot m / A})(99)(2.0\ \mathrm{A})(0.2 \times 10^{-2}\ \mathrm{m})}{2\pi\,(0.50 \times 10^{-2}\ \mathrm{m})^2}$$

$$B = 3.17 \times 10^{-3}\ \mathrm{T}$$

This field points tangent to a circle of radius 0.2 mm. It will exert a force $\mathbf{F} = I(\boldsymbol{\ell} \times \mathbf{B})$ toward the center of the bundle, on the hundredth wire:

$$\frac{F}{\ell} = IB\sin\theta = (2.0\ \mathrm{A})(3.17 \times 10^{-3}\ \mathrm{T})(\sin 90°) = 6.3\ \mathrm{mN / m}\ \text{toward the center}\ \lozenge$$

(b) As is shown in Figure 30.10 of the text, the field is strongest at the outer surface of the cable, so the force on one strand is greater here, by a factor of 5/2. ◊

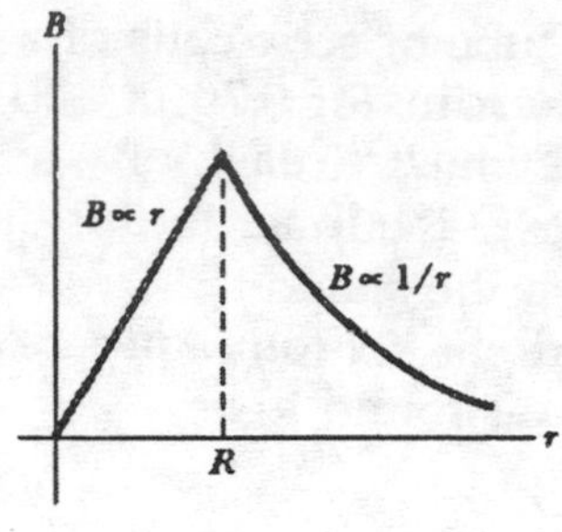

---

35. A solenoid has 500 turns, a length of 50.0 cm, and a radius of 5.0 cm. If it carries 4.0 A, calculate the magnetic field at an axial point located 15 cm from the center (that is, 10 cm from one end).

Solution We apply Equation 30.22,

$$B = \frac{\mu_0 NI}{2\ell}(\sin\phi_2 - \sin\phi_1)$$

where the angles are defined in the Fig. 30.17.

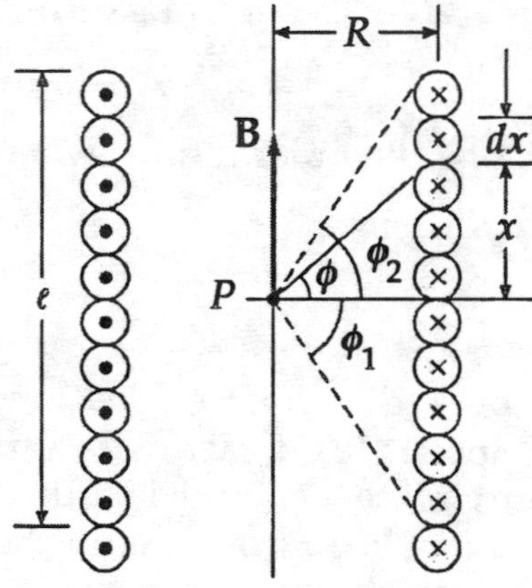

Figure 30.17

$$B = \frac{(4\pi\times10^{-7}\ \text{T}\cdot\text{m}/\text{A})(500)(4.0\ \text{A})}{2(0.50\ \text{m})}\left(\frac{10\ \text{cm}}{\sqrt{(10\ \text{cm})^2+(5\ \text{cm})^2}} - \frac{-40\ \text{cm}}{\sqrt{5^2+40^2}\ \text{cm}}\right)$$

$$= 2.51\times10^{-3}\ \text{T}\ (0.894 + 0.992)$$

$$= 4.7\ \text{mT along the solenoid axis}$$ ◊

The equation $B = \frac{\mu_0 NI}{\ell}$ for the field everywhere inside an infinitely long solenoid gives 5.03 mT, which is in error by only 6%.

---

37. A cube of edge length ℓ = 2.5 cm is positioned as shown in Figure P30.37. There is throughout the region a uniform magnetic field given by $\mathbf{B}$ = (5.0**i** + 4.0**j** + 3.0**k**) T. (a) Calculate the flux through the shaded face. (b) What is the total flux through the six faces?

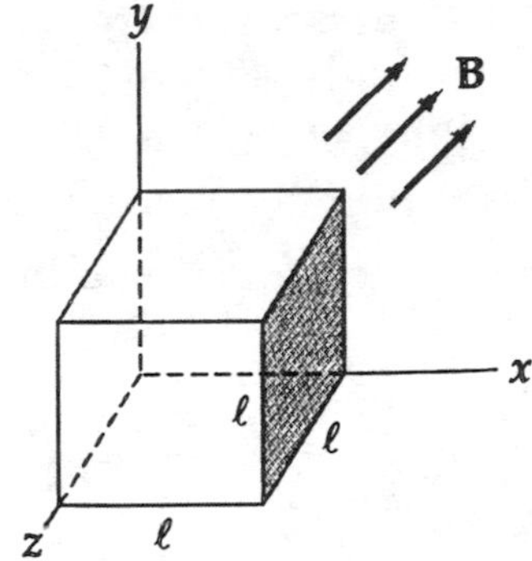

Figure P30.37

Solution

(a) $\Phi = \mathbf{B} \cdot \mathbf{A}$

$$\Phi = B_x A_x + B_y A_y + B_z A_z$$

$$\Phi = (5.0\text{ T})(0.025\text{ m})^2 = 3.13 \times 10^{-3}\text{ T}\cdot\text{m}^2 = 3.13\text{ mT}\cdot\text{m}^2 \quad \Diamond$$

(b) For a closed surface, $\oint \mathbf{B} \cdot d\,\mathbf{A} = 0$ so $\Phi = 0$ ◊

---

41. The applied voltage across the plates of a 4.00-μF capacitor varies in time according to the expression

$$V_{\text{app}} = (8.00\text{ V})(1 - e^{-t/4})$$

where t is in seconds. Calculate (a) the displacement current as a function of time and (b) the value of the current at t = 4.00 s.

Solution

(a) $I_d = C\dfrac{dV}{dt}$

$$I_d = C\frac{d}{dt}\left[8(1 - e^{-t/4})\right]$$

$$I_d = 8.00C\left(\tfrac{1}{4}e^{-t/4}\right) = 2.00(4 \times 10^{-6}\text{ F})e^{-t/4}\text{ V/s} = (8.00 \times 10^{-6}\text{ A})e^{-t/4} \quad \Diamond$$

(b) At $t = 4$ s,

$$I_d = (8.00 \times 10^{-6}\text{ A})e^{-1} = 2.94 \times 10^{-6}\text{ A} = 2.94\ \mu\text{A} \quad \Diamond$$

---

43. A 0.10 A current is charging a capacitor that has square plates, 5.0 cm on a side. If the plate separation is 4.0 mm, find (a) the time rate of change of electric flux between the plates and (b) the displacement current between the plates.

Solution The electric field in the space between the plates is $E = \dfrac{\sigma}{\epsilon_0} = \dfrac{Q}{\epsilon_0 A}$

The flux of this field is $\mathbf{E} \cdot \mathbf{A} = \left(\dfrac{Q}{\epsilon_0 A}\right) A \cos 0° = \dfrac{Q}{\epsilon_0}$

(a) The rate of change of flux is

$$\frac{d\Phi}{dt} = \frac{d}{dt}\frac{Q}{\epsilon_0} = \frac{1}{\epsilon_0}\frac{dQ}{dt} = \frac{I}{\epsilon_0} = \left(\frac{0.100 \text{ A}}{8.85 \times 10^{-12} \text{ C}^2/\text{N} \cdot \text{m}^2}\right)\left(\frac{1 \text{ C}}{\text{A} \cdot \text{s}}\right)$$

$$= 1.13 \times 10^{10} \text{ N} \cdot \text{m}^2/\text{C} \cdot \text{s} \quad \lozenge$$

(b) The displacement current is defined as

$$\epsilon_0 \frac{d\Phi}{dt} = \left(8.85 \times 10^{-12} \frac{\text{C}^2}{\text{N} \cdot \text{m}^2}\right)\left(1.13 \times 10^{10} \frac{\text{N} \cdot \text{m}^2}{\text{C} \cdot \text{s}}\right) = 0.100 \text{ A} \quad \lozenge$$

---

45. What is the relative permeability of a material that has a magnetic susceptibility of 10^{-4}?

Solution $\mu = \mu_0(1 + \chi)$

Relative permeability, $\dfrac{\mu}{\mu_0} = 1 + \chi = 1 + 10^{-4} = 1.0001 \quad \lozenge$

---

47. A toroid with a mean radius of 20 cm and 630 turns Fig. 30.29 is filled with powdered steel whose magnetic susceptibility χ is 100. If the current in the windings is 3.00 A, find B (assumed uniform) inside the toroid.

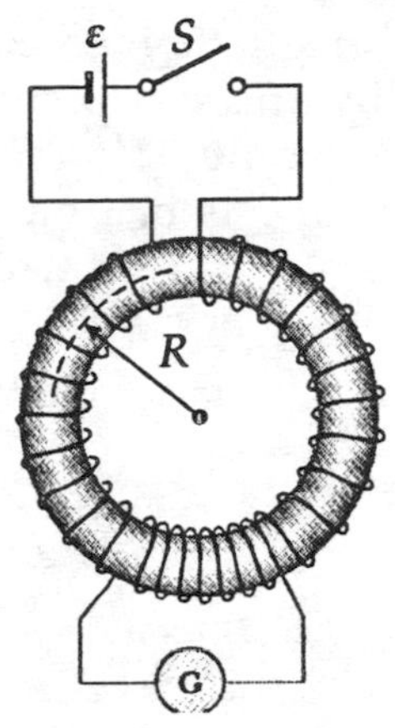

Figure 30.29

Solution

If the coil had a vacuum inside, the magnetic field would be given by Equation 30.18,

$$B = \frac{\mu_0 NI}{2\pi r}$$

That is, the magnetic field strength is $H = \frac{NI}{2\pi r}$

With the steel inside,

$$B = \mu_0(1+\chi)H = \mu_0(1+\chi)\frac{NI}{2\pi r} = \frac{(4\pi \times 10^{-7}\ \mathrm{T \cdot m/A})(101)(630)(3.00\ \mathrm{A})}{2\pi(0.20\ \mathrm{m})} = 0.191\ \mathrm{T} \quad \Diamond$$

---

51. A coil of 500 turns is wound on an iron ring ($\mu_m = 750\mu_0$) of 20-cm mean radius and 8.0-cm^2 cross-sectional area. Calculate the magnetic flux Φ in this Rowland ring when the current in the coil is 0.50 A.

Solution $\Phi = BA$

But $B = \mu_m nI$, so

$$\Phi = \mu_m nIA = (750)\left(4\pi \times 10^{-7}\ \frac{\mathrm{N}}{\mathrm{A}^2}\right)\left(\frac{500}{2\pi(0.20\ \mathrm{m})}\right)(0.50\ \mathrm{A})(8.0 \times 10^{-4}\ \mathrm{m}^2)$$

$$\Phi = 1.50 \times 10^{-4}\ \mathrm{Wb} = 150\ \mu\mathrm{Wb} \quad \Diamond$$

---

59. The magnetic moment of the Earth is approximately 8.7×10^{22} A·m^2. (a) If this were caused by the complete magnetization of a huge iron deposit, how many unpaired electrons would this correspond to? (b) At 2 unpaired electrons per iron atom, how many kilograms of iron would this correspond to? (The density of iron is 7900 kg/m^3, and there are approximately 8.5×10^{28} iron atoms/m^3.)

Solution The magnetic moment of each unpaired electron is the Bohr magneton,

$$9.27 \times 10^{-24}\ \mathrm{J/T} = \frac{9.27 \times 10^{-24}\ \mathrm{J}}{\mathrm{T}}\left(\frac{\mathrm{N \cdot m}}{\mathrm{J}}\right)\left(\frac{\mathrm{T \cdot C \cdot m}}{\mathrm{N \cdot s}}\right)\left(\frac{\mathrm{A \cdot s}}{\mathrm{C}}\right)$$

$$\mu_B = 9.27 \times 10^{-24}\ \mathrm{A \cdot m^2}$$

(a) The number of unpaired electrons is $N = \dfrac{8.7 \times 10^{22}\ \mathrm{A \cdot m^2}}{9.27 \times 10^{-24}\ \mathrm{A \cdot m^2}} = 9.4 \times 10^{45}$ ◊

Each iron atom has two unpaired electrons, so the number of iron atoms required is

$$\tfrac{1}{2}N = \tfrac{1}{2}(9.4 \times 10^{45}) = 4.7 \times 10^{45}.$$

(b) Mass $= \dfrac{4.7 \times 10^{45}\ (7900\ \mathrm{kg/m^3})}{8.5 \times 10^{28}\ \mathrm{atoms/m^3}} = 4.4 \times 10^{20}\ \mathrm{kg}$ ◊

---

61. Measurements of the magnetic field of a large tornado were made at the Geophysical Observatory in Tulsa, Oklahoma in 1962. If the tornado's field was $B = 1.5 \times 10^{-8}$ T pointing north when the tornado was 9.0 km east of the observatory, what current was carried up/down the funnel of the tornado?

Solution Consider the funnel as containing a long straight vertical current, to produce field $B = \mu_0 I / 2\pi r$, so

$$I = \frac{2\pi r B}{\mu_0} = \frac{2\pi(9000)(1.5 \times 10^{-8})}{4\pi \times 10^{-7}} = 680\ \mathrm{A}$$ ◊

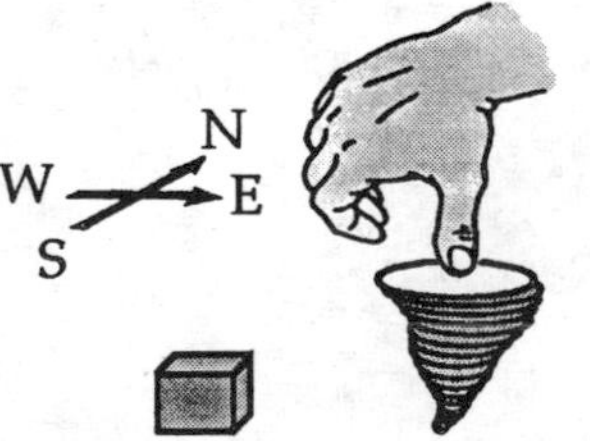

(Conventional) current is downward or negative charge flows upward.

---

65. A very long, thin strip of metal of width w carries a current I along its length as in Figure P30.65. Find the magnetic field in the plane of the strip (at an external point P) a distance b from one edge.

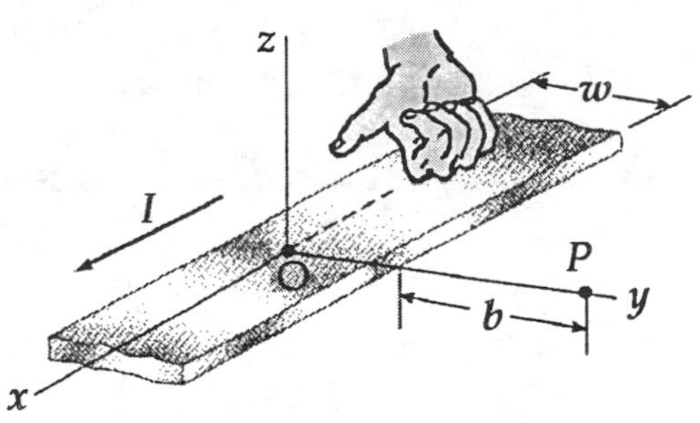

Figure P30.65

Solution Consider a long filament of the strip which has width dr and is a distance r from point P. The B field at a distance r from a long conductor is

$$B = \frac{\mu_0 I}{2\pi r}$$

Thus, the field due to the thin filament is

$$d\mathbf{B} = \frac{\mu_0\, dI}{2\pi r}\mathbf{k} \qquad \text{where} \qquad dI = I\left(\frac{dr}{w}\right) \qquad \text{so}$$

$$\mathbf{B} = \int_b^{b+w} \frac{\mu_0}{2\pi r}\left(I\frac{dr}{w}\right)\mathbf{k} = \frac{\mu_0 I}{2\pi w}\mathbf{k}\int_b^{b+w}\frac{dr}{r} = \frac{\mu_0 I}{2\pi w}\mathbf{k}\,\ln\left(1+\frac{w}{b}\right) \quad \Diamond$$

69A. A nonconducting ring of radius R is uniformly charged with a total positive charge q. The ring rotates at a constant angular speed ω about an axis through its center, perpendicular to the plane of the ring. What is the magnitude of the magnetic field on the axis of the ring a distance R from its center?

Solution

The time required for one revolution is given by $\omega = 2\pi/T$, or $T = 2\pi/\omega$. The spinning charged ring constitutes a loop carrying current $I = q/T = q\omega/2\pi$, so it creates magnetic field on its axis

$$B = \frac{\mu_0 I R^2}{2(x^2+R^2)^{3/2}} = \frac{\mu_0 q\omega R^2}{4\pi(R^2+R^2)^{3/2}} = \frac{\mu_0 q\omega}{8\sqrt{2}\pi R} \quad \Diamond$$

Related Calculation If $R = 0.10$ m, $q = 10\ \mu$C, and $\omega = 20$ rad/s, what is the magnitude of the magnetic field on the axis of the ring 0.050 m from its center?

Solution: In this case, the distance x is not equal to R, but to $R/2$. Thus,

$$B = \frac{\mu_0 I R^2}{2\left(\frac{R^2}{4} + R^2\right)^{3/2}} = \frac{\mu_0 q \omega}{5.59\pi R}$$

Substituting the given values, we find that $B = 1.43 \times 10^{-10}$ T ◊

---

73. A long cylindrical conductor of radius R carries a current I as in Figure P30.73. The current density J, however, is not uniform over the cross section of the conductor but is a function of the radius according to $J = br$, where b is a constant. Find an expression for the magnetic field B (a) at a distance $r_1 < R$ and (b) at a distance $r_2 > R$, measured from the axis.

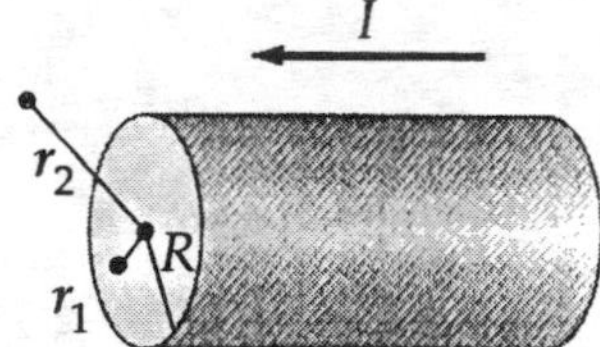

Figure P30.73

Solution Take a circle of radius r_1 or r_2 to apply $\oint \mathbf{B} \cdot d\mathbf{s} = \mu_0 I$ where for nonuniform current density $I = \int \mathbf{J} \cdot d\mathbf{A}$

In this case **B** is parallel to $d\mathbf{s}$ and **J** is parallel to $d\mathbf{A}$, so Ampère's law gives

$$\oint B\, ds = \mu_0 \int J\, dA$$

(a) When $r = r_1 < R$, $2\pi r_1 B = \mu_0 \int_0^{r_1} br(2\pi r dr)$ and $B = \frac{\mu_0 b r_1^2}{3}$ (inside) ◊

(b) When $r = r_2 > R$, $2\pi r_2 B = \mu_0 \int_0^{R} br(2\pi r dr)$ and $B = \frac{\mu_0 b R^3}{3 r_2}$ (outside) ◊

---

75. Two circular loops are parallel, coaxial, and almost in contact, 1.0 mm apart (Fig. P30.75). Each loop is 10 cm in radius. The top loop carries a clockwise current of 140 A. The bottom loop carries a counterclockwise current of 140 A. (a) Calculate the magnetic force that the bottom loop exerts on the top loop. (b) The upper loop has a mass of 0.021 kg. Calculate its acceleration, assuming that the only forces acting on it are the force in part (a) and its weight.

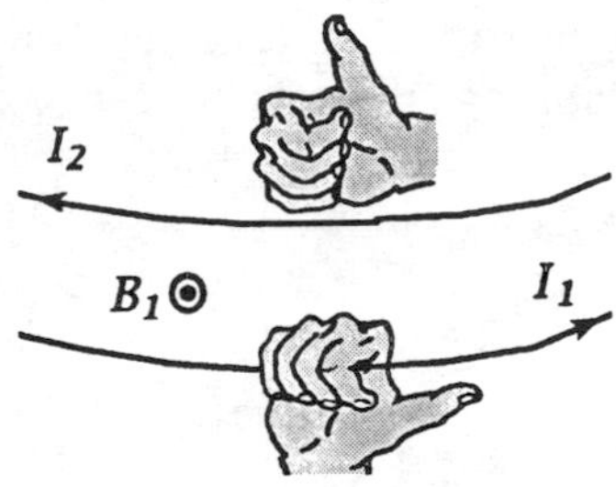

Figure P30.75

Solution

Note that 1 mm is small compared to 10 cm, so the lower wire creates a field $B_1 = \mu_0 I_1 / 2\pi a$, much as a long straight wire does. At the near point of the loop, the lower wire creates a magnetic field toward you in the space above it. The upper wire then experiences a force $I_2\mathbf{L} \times \mathbf{B}_1$ of repulsion, amounting to

(a) $F = \dfrac{\mu_0 I_1 I_2 L}{2\pi a}$ (Equation 30.13)

$$F = \frac{(4\pi \times 10^{-7})(140^2)(2\pi)(0.1)}{2\pi(1.0 \times 10^{-3})} = 2.5\text{ N up} \quad \Diamond$$

(b) $$a_{\text{loop}} = \frac{(2.46\text{ N}) - m_{\text{loop}}g}{m_{\text{loop}}} = 110\text{ m/s}^2 \text{ up} \quad \Diamond$$

---

81. A wire is formed into the shape of a square of edge length L (Fig. P.30.81). Show that when the current in the loop is I, the magnetic field at point P a distance x from the center of the square along its axis is

$$B = \frac{\mu_0 I L^2}{2\pi\left(x^2 + \dfrac{L^2}{4}\right)\sqrt{x^2 + \dfrac{L^2}{2}}}$$

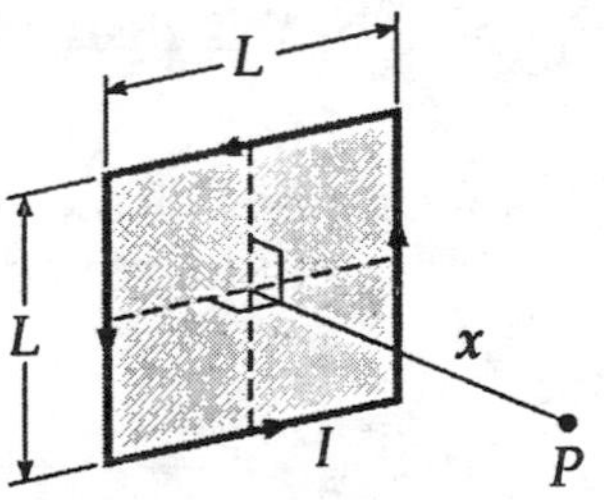

Figure P30.81

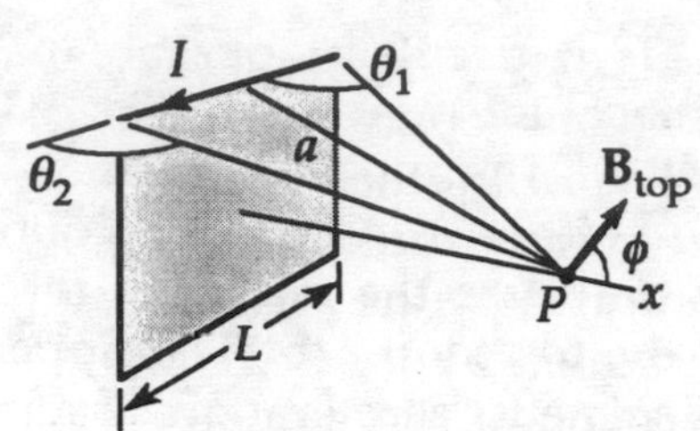

Solution Consider the top side of the square. The distance from its center to point P is

$$a = \sqrt{x^2 + \left(\frac{L}{2}\right)^2}$$

and Equation 30.6 describes the field it creates at point P.

The distance from one of the corners of the square to P is $\sqrt{x^2 + \left(\frac{L}{2}\right)^2 + \left(\frac{L}{2}\right)^2} = \sqrt{x^2 + \frac{L^2}{2}}$

Thus $\cos\theta_1 = \dfrac{L}{2\sqrt{x^2 + \frac{L^2}{2}}}$ and $\cos\theta_2 = -\cos\theta_1 = -\dfrac{L}{2\sqrt{x^2 + \frac{L^2}{2}}}$.

Then, $$B_{top} = \frac{\mu_0 I}{4\pi a}(\cos\theta_1 - \cos\theta_2),$$

$$B_{top} = \frac{\mu_0 I}{4\pi\sqrt{x^2 + \frac{L^2}{4}}}\left(\frac{L}{2\sqrt{x^2 + \frac{L^2}{2}}} + \frac{L}{2\sqrt{x^2 + \frac{L^2}{2}}}\right),$$

or $$B_{top} = \frac{\mu_0 I L}{4\pi\sqrt{x^2 + \frac{L^2}{4}}\sqrt{x^2 + \frac{L^2}{2}}}$$

The component of this field along the direction of x is $B_{top}\cos\phi = B_{top}\dfrac{L/2}{a}$.

Each of the four sides of the square produces this same size field along the direction of x, with the other components adding to zero. We assemble our results:

$$B = 4B_{top}\frac{L}{2a} = \frac{\mu_0 I L^2}{2\pi\left(x^2 + \frac{L^2}{4}\right)\sqrt{x^2 + \frac{L^2}{2}}}$$ away from the center of the square ◊

---

85. Four long, parallel conductors all carry 5.00 A. An end view of the conductors is shown in Figure P30.85. The current direction is out of the page at points *A* and *B* (indicated by the dots) and into the page at points *C* and *D* (indicated by the crosses). Calculate the magnitude and direction of the magnetic field at point *P*, located at the center of the square.

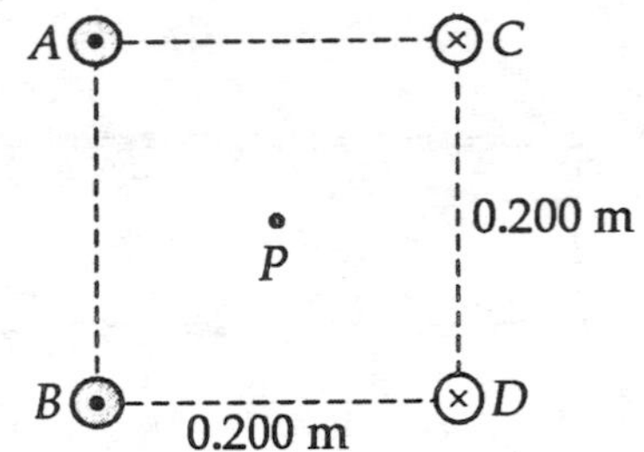

Figure P30.85

Solution

Each wire is distant from *P* by (0.200 m) cos 45° = 0.141 m.

Each wire produces a field at *P* of equal magnitude:

$$B_A = \frac{\mu_0 I}{2\pi a} = \frac{(2.00\times10^{-7}\ \text{T}\cdot\text{m/A})(5.00\ \text{A})}{(0.141\ \text{m})} = 7.07\ \mu\text{T}$$

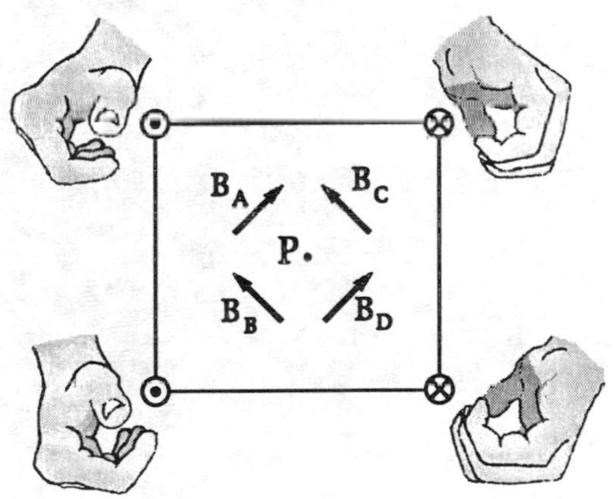

Carrying currents toward you, the left-hand wires produce fields at *P* of 7.07 μT, in the following directions:

A: to the top and right, at 45°

B: to the top and left, at 135°;

Carrying currents away from you, the wires to the right also produce fields at *P* of 7.07 *m*T, in the following directions:

C: upward and to the left, at 135°

D: upward and to the right, at 45°.

The total field is then

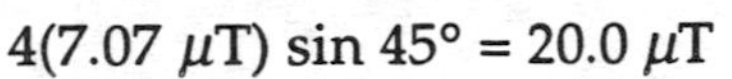

$4(7.07\ \mu\text{T})\sin 45° = 20.0\ \mu\text{T}$ toward the top of the page. ◊

Chapter 31

Faraday's Law

Chapter 31

FARADAY'S LAW

INTRODUCTION

Our studies so far have been concerned with electric fields due to stationary charges and magnetic fields produced by moving charges. This chapter deals with electric fields that originate from changing magnetic fields.

Experiments conducted by Michael Faraday in England in 1831 and independently by Joseph Henry in the United States that same year showed that an electric current could be induced in a circuit by a changing magnetic field. The results of these experiments led to a basic and important law of electromagnetism known as Faraday's law of induction. This law says that the magnitude of the emf induced in a circuit equals the time rate of change of the magnetic flux through the circuit.

As we shall see, an induced emf can be produced in several ways. For instance, an induced emf and an induced current can be produced in a closed loop of wire when the wire moves into a magnetic field. We shall describe such experiments along with a number of important applications that make use of the phenomenon of electromagnetic induction.

NOTES FROM SELECTED CHAPTER SECTIONS

31.1 Faraday's Law of Induction

The emf induced in a circuit is proportional to the time rate of change of magnetic flux through the circuit.

An emf can be induced in the circuit in several ways:

- The magnitude of the magnetic field can change as a function of time.
- The area of the circuit can change with time.
- The direction of the magnetic field relative to the circuit can change with time.
- Any combination of the above can change.

In particular, it is important to note that the magnitude of the induced emf depends on the *rate at which the magnetic field is changing.*

31.2 Motional emf

A potential difference will be maintained across a conductor moving in a magnetic field as long as the direction of motion through the field is not parallel to the field direction. If the motion is reversed, the polarity of the potential difference will also be reversed.

31.3 Lenz's Law

The polarity of the induced emf is such that it tends to produce a current that will create a magnetic flux to oppose the *change in flux* through the circuit.

31.7 Maxwell's Wonderful Equations

Maxwell's equations applied to free space are:

- Gauss's law (Eq. 31.12) which states that *the total electric flux through any closed surface equals the net charge inside that surface divided by* ϵ_0. This law describes how charge creates the electric field.
- Gauss's law for magnetism (Eq. 31.13) which states that *the net magnetic flux through a closed surface is zero.*
- Faraday's law of induction (Eq. 31.14) which states that *the line integral of the electric field around any closed path equals the rate of change of magnetic flux through any surface area bounded by the path.* This law describes how a changing magnetic field creates an electric field.
- The Ampère-Maxwell law (Eq. 31.15) which states that *the line integral of the magnetic field around any closed path is determined by the sum of the net conduction current through that path and the rate of change of electric flux through any surface bounded by that path.* This law describes how changing electric fields create a magnetic field.

These four equations, together with the Lorentz force law (Eq. 31.16), describe all electromagnetic phenomena.

EQUATIONS AND CONCEPTS

The total magnetic flux through a plane area, A, placed in a uniform magnetic field depends on the angle between the direction of the magnetic field and the direction perpendicular to the surface area.

$$\Phi \equiv B_{\perp}A = BA\cos\theta$$

$$\Phi_{max} = BA$$

The maximum flux through the area occurs when the magnetic field is perpendicular to the plane of the surface area. When the magnetic field is parallel to the plane of the surface area, the flux through the area is zero. The unit of magnetic flux is the weber, Wb.

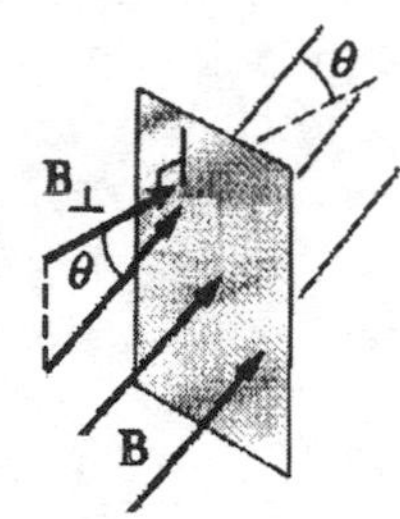

Faraday's law of induction states that the average emf induced in a circuit is proportional to the rate of change of magnetic flux through the circuit. The minus sign is included to indicate the polarity of the induced emf, which can be found by use of Lenz's law.

$$\mathcal{E} = -\frac{d\Phi_B}{dt} \tag{31.1}$$

Lenz's law states that the polarity of the induced emf (and the direction of the associated current in a closed circuit) produces a current whose magnetic field opposes the change in the flux through the loop. That is, the induced current tends to maintain the original flux through the circuit.

The magnetic flux threading a circuit is the integral of the normal component of the magnetic field over the area bounded by the circuit.

$$\Phi_B = \int \mathbf{B} \cdot d\mathbf{A} \tag{31.2}$$

A "motional" emf is induced in a conductor of length ℓ, moving with speed v, perpendicular to a magnetic field.

$$\mathcal{E} = -B\ell v \tag{31.5}$$

If the moving conductor is part of a complete circuit of resistance, R, a current will be induced in the circuit.

$$I = \frac{B\ell v}{R} \tag{31.6}$$

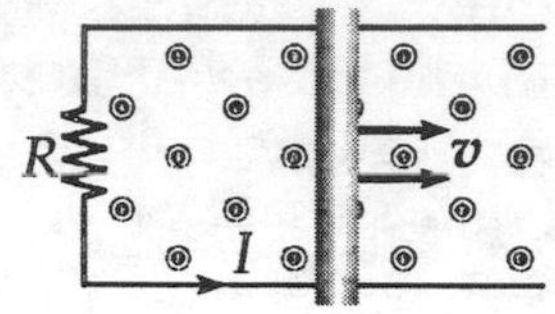

Faraday's law can be written in a more *general form* in terms of the integral of the electric field around a closed path. In this form the electric field is a *non-conservative, time-varying field.*

$$\oint \mathbf{E} \cdot d\mathbf{s} = -\frac{d\Phi_B}{dt} \tag{31.9}$$

or

$$\int \mathbf{E} \cdot d\mathbf{s} = -\frac{d}{dt} \int \mathbf{B} \cdot d\mathbf{A}$$

When a conducting loop of N turns and cross-sectional area, A, rotates with a constant angular velocity in a magnetic field, the emf induced in the loop will vary sinusoidally in time. For a given loop, the maximum value of the induced emf will be proportional to the angular velocity of the loop.

$$\mathcal{E} = NAB\omega \sin \omega t \tag{31.10}$$

$$\mathcal{E}_{max} = NAB\omega \tag{31.11}$$

Maxwell's equations as applied to free space (i.e. in the absence of any dielectric or magnetic material) are as follows (see Notes 31.7):

Gauss's law

$$\oint \mathbf{E} \cdot d\mathbf{A} = \frac{Q}{\epsilon_0} \tag{31.12}$$

Gauss's law for magnetism

$$\oint \mathbf{B} \cdot d\mathbf{A} = 0 \tag{31.13}$$

Faraday's law

$$\oint \mathbf{E} \cdot d\mathbf{s} = -\frac{d\Phi_B}{dt} \tag{31.14}$$

Ampère-Maxwell law

$$\oint \mathbf{B} \cdot d\mathbf{s} = \mu_0 I + \epsilon_0 \mu_0 \frac{d\Phi_E}{dt} \tag{31.15}$$

The force exerted on a particle of charge q due to the combined effect of the electric force and the magnetic field is the Lorentz force.

$$\mathbf{F} = q\mathbf{E} + q\mathbf{v} \times \mathbf{B} \qquad (31.16)$$

SUGGESTIONS, SKILLS, AND STRATEGIES

It is important to distinguish clearly between the *instantaneous value* of emf induced in a circuit and the *average value* of the emf induced in the circuit over a finite time interval.

To calculate the average induced emf, it is often useful to write Equation 31.3 as

$$\mathcal{E}_{\text{avg}} = -N\left(\frac{d\Phi_B}{dt}\right)_{\text{avg}} = -N\frac{\Delta\Phi_B}{\Delta t}$$

or

$$\mathcal{E}_{\text{avg}} = -N\left(\frac{\Phi_{m,f} - \Phi_{m,i}}{\Delta t}\right)$$

where the subscripts i and f refer to the magnetic flux through the circuit at the beginning and end of the time interval Δt. For a circuit (or loop) in a single plane, $\Phi_B = BA\cos\theta$, where θ is the angle between the direction of the normal to plane of the circuit (conducting loop) and the direction of the magnetic field.

Equation 31.4 can be used to calculate the *instantaneous value of an induced emf.* For a multi-turn loop, the induced emf is

$$\mathcal{E} = -N\frac{d}{dt}(BA\cos\theta)$$

where in a particular case B, A, θ, or any combination of those parameters can be time dependent while the others remain constant. The expression resulting from the differentiation is then evaluated using the values of B, A, and θ corresponding to the specified value.

REVIEW CHECKLIST

- ▷ Calculate the emf (or current) induced in a circuit when the magnetic flux through the circuit is changing in time. The variation in flux might be due to a change in (a) the area of the circuit, (b) the magnitude of the magnetic field, (c) the direction of the magnetic field, or (d) the orientation/location of the circuit in the magnetic field.
- ▷ Calculate the emf induced between the ends of a conducting bar as it moves through a region where there is a constant magnetic field (motional emf).
- ▷ Apply Lenz's law to determine the direction of an induced emf or current. You should also understand that Lenz's law is a consequence of the law of conservation of energy.
- ▷ Calculate the maximum and instantaneous values of the sinusoidal emf generated in a conducting loop rotating in a constant magnetic field.
- ▷ Calculate the electric field at various points in a charge-free region when the time variation of the magnetic field over the region is specified.

SOLUTIONS TO SELECTED END-OF-CHAPTER PROBLEMS

3. A powerful electromagnet has a field of 1.6 T and a cross-sectional area of 0.20 m^2. If we place a coil having 200 turns and a total resistance of 20 Ω around the electromagnet, and then turn off the power to the electromagnet in 20 ms, what is the current induced in the coil?

Solution The induced voltage is

$$\mathcal{E} = -N\frac{d(B\cdot A)}{dt} = -N\left(\frac{0 - B_i A\cos\theta}{t}\right)$$

$$= \frac{+200(1.6\text{ T})(0.20\text{ m}^2)(\cos 0°)}{20\times 10^{-3}\text{ s}}\left(\frac{1\text{ N}\cdot\text{s}}{\text{T}\cdot\text{C}\cdot\text{m}}\right)\left(\frac{1\text{ V}\cdot\text{C}}{\text{N}\cdot\text{m}}\right) = 3200\text{ V}$$

$$I = \frac{\mathcal{E}}{R} = \frac{3200\text{ V}}{20\ \Omega} = 160\text{ A} \quad \Diamond$$

The positive sign means that the current in the coil flows in the same direction as the current in the electromagnet.

11A. A long solenoid has n turns per meter and carries a current $I = I_0(1 - e^{-\alpha t})$. Inside the solenoid and coaxial with it is a loop that has a radius R and consists of a total of N turns of fine wire. What emf is induced in the loop by the changing current? (See Fig. P31.11)

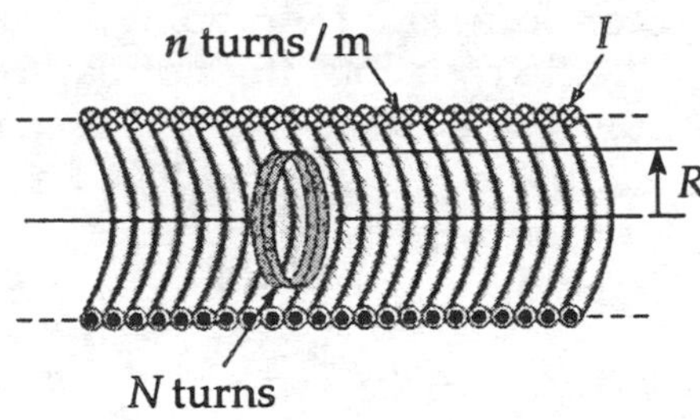

Figure P31.11

Solution The solenoid creates magnetic field $B = \mu_0 n I = \mu_0 n I_0(1 - e^{-\alpha t})$. The magnetic flux through one turn of the loop is

$$\Phi_m = \int B dA \cos\theta$$

$$\Phi_m = \mu_0 n I_0\left(1 - e^{-\alpha t}\right)\int dA$$

$$\Phi_m = \mu_0 n I_0\left(1 - e^{-\alpha t}\right)\pi R^2$$

The emf generated in the N-turn loop is

$$\mathcal{E} = -N\frac{d\Phi_m}{dt} = -N\mu_0 n I_0 \pi R^2 \alpha e^{-\alpha t} \quad \Diamond$$

The minus sign indicates that the emf will produce counterclockwise current in the smaller coil, opposite to the current in the solenoid.

Related Calculation: **11.** Find the induced emf if n = 400 turns per meter, I_0 = 30 A, α = 1.6 s^{-1}, R = 6.0 cm, and N = 250 turns.

Substituting these values into the answer for problem 11A,

$$\mathcal{E} = -(250)\left(4\pi \times 10^{-7}\ \text{N/A}^2\right)\left(400\ \text{m}^{-1}\right)(30\ \text{A})\pi(0.060\ \text{m})^2\left(1.6\ \text{s}^{-1}\right)e^{-1.6t}$$

$$\mathcal{E} = -68\, e^{-1.6t}\ \text{mV} \quad \Diamond$$

13. A coil formed by wrapping 50 turns of wire in the shape of a square is positioned in a magnetic field so that the normal to the plane of the coil makes an angle of 30° with the direction of the field. When the magnitude of the magnetic field is increased uniformly from 200 μT to 600 μT in 0.40 s, an emf of 80 mV is induced in the coil. What is the total length of the wire?

Solution $\mathcal{E} = -N\dfrac{d\Phi}{dt} = -N\dfrac{d}{dt}(BA\cos\theta) = -NA\cos\theta\dfrac{dB}{dt}$

$$|\mathcal{E}| = NA\cos\theta\left(\frac{\Delta B}{\Delta t}\right) \quad \text{and} \quad A = \frac{|\mathcal{E}|}{N\cos\theta\left(\dfrac{\Delta B}{\Delta t}\right)}$$

$$A = \frac{80\times10^{-3}\text{ V}}{(50)(\cos 30°)\left(\dfrac{600-200}{0.40}\right)\times10^{-6}\text{ T/s}} = 1.85\text{ m}^2$$

The edge length of the coil, $d = \sqrt{A}$, and the total length of the wire is $L = N(4d)$. Therefore,

$$L = (50)(4)\sqrt{1.85\text{ m}^2} = 270\text{ m} \quad \Diamond$$

---

14. A long, straight wire carries a current $I = I_0\sin(\omega t + \phi)$ and lies in the plane of a rectangular loop of N turns of wire, as shown in Figure P31.14. The quantities I_0, ω, and ϕ are all constants. Determine the emf induced in the loop by the magnetic field created by the current in the straight wire. Assume $I_0 = 50$ A, $\omega = 200\pi$ s^{-1}, $N = 100$, $a = b = 5.0$ cm, and $\ell = 20$ cm.

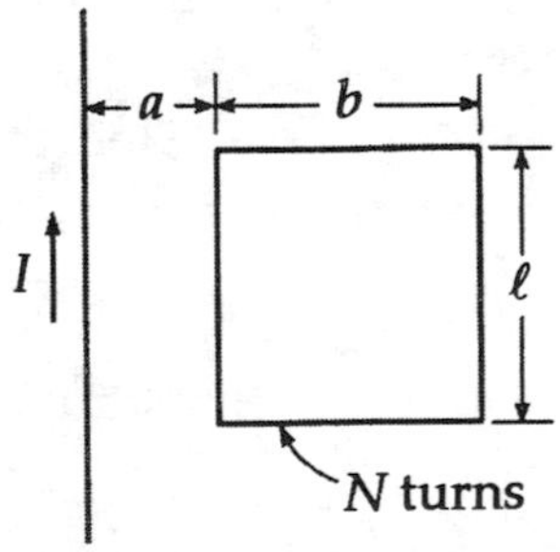

Figure P31.14

Solution The loop is the boundary of the rectangular loop. The magnetic field produced by the current in the straight wire is perpendicular to the plane of the loop at all points on its surface. The magnitude of the field is

$$B = \frac{\mu_0 I}{2\pi r}$$

Thus the flux linkage is

$$N\Phi_m = \frac{\mu_0 N I \ell}{2\pi}\int_a^{a+b}\frac{dr}{r} = \frac{\mu_0 N I_0 \ell}{2\pi}\ln\left(\frac{a+b}{a}\right)\sin(\omega t + \phi)$$

Finally, the induced emf is in absolute value

$$|\mathcal{E}| = N\frac{d\,\Phi_m}{dt} = \frac{\mu_0 N I_0 \ell\omega}{2\pi}\ln\left(\frac{a+b}{a}\right)\cos(\omega t + \phi)$$

$$= \frac{(4\pi\times 10^{-7}\ \mathrm{T\cdot m/A})(100)(50\ \mathrm{A})(0.20\ \mathrm{m})(200\pi\ \mathrm{s^{-1}})}{2\pi}\ln\left(\frac{5.0+5.0}{5.0}\right)\cos(\omega t + \phi)$$

$$|\mathcal{E}| = (87\ \mathrm{mV})\cos(200\pi t + \phi) \quad \Diamond$$

Related Commentary The term $\sin(\omega t + \phi)$ in the expression for the current in the straight wire does not change appreciably when ωt changes by 0.1 rad or less. Thus, the current does not change appreciably during a time interval

$$t < 0.1/(200\pi\ \mathrm{s^{-1}}) = 1.6 \times 10^{-4}\ \mathrm{s}.$$

We define a critical length,

$$ct = (3.0 \times 10^{8}\ \mathrm{m/s})(1.6 \times 10^{-4}\ \mathrm{s}) = 4.8 \times 10^{4}\ \mathrm{m},$$

equal to the distance to which field changes could be propagated during an interval of 1.6×10^{-4} s. This length is so much larger than any dimension of the loop or its distance from the wire that, although we consider the straight wire to be infinitely long, we can also safely ignore the field propagation effects in the vicinity of the loop. Moreover, the phase angle can be considered to be constant along the wire in the vicinity of the loop. If the frequency ω were much larger, say, $200\pi \times 10^{5}\ \mathrm{s^{-1}}$, the corresponding critical length would be only 48 cm. In this situation, propagation effects would be important and the above expression of $\mathcal{E}$ would require modification. As a "rule of thumb," we can consider field propagation effects for circuits of laboratory size to be negligible for frequencies, $f = \omega/2\pi$, that are less than about 10^{6} Hz.

19. In the arrangement shown in Figure P31.18, $R = 6.0\ \Omega$, and a 2.5-T magnetic field is directed into the paper. Let $\ell = 1.2$ m and neglect the mass of the bar. (a) Calculate the applied force required to move the bar to the right at a constant speed of 2.0 m/s. (b) At what rate is energy dissipated in the resistor?

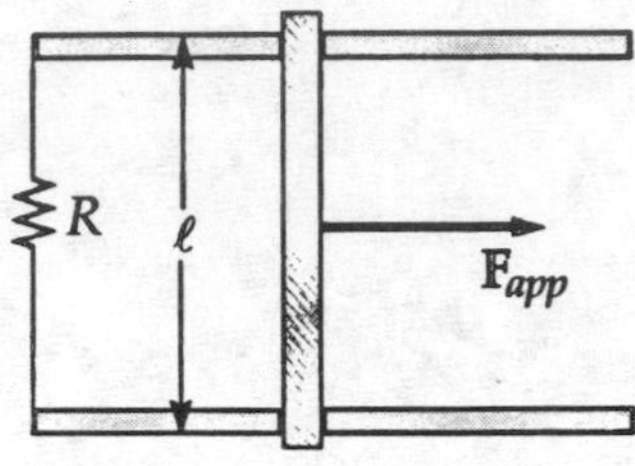

Figure P31.18

Solution

(a) At constant speed, the net force on the moving bar equals zero, or

$$|\mathbf{F}_{\text{app}}| = I|\boldsymbol{\ell} \times \mathbf{B}|$$

where the current in the bar $I = \dfrac{\mathcal{E}}{R}$ and $\mathcal{E} = BLv$. Therefore,

$$F_{\text{app}} = \left(\frac{B\ell v}{R}\right)LB = \frac{B^2\ell^2 v}{R} = \frac{(2.5\ \text{T})^2(1.2\ \text{m})^2(2.0\ \text{m/s})}{6.0\ \Omega} = 3.0\ \text{N} \quad \Diamond$$

(b) $P = F_{\text{app}}v = (3\ \text{N})(2.0\ \text{m/s}) = 6.0\ \text{W} \quad \Diamond$

---

23. A helicopter has blades of length 3.0 m, rotating at 2.0 rev/s around a central hub. If the vertical component of the Earth's magnetic field is 0.50×10^{-4} T, what is the emf induced between the blade tip and the center hub?

Solution

We suppose the length is measured out from the hub. Following Example 31.5,

$$\mathcal{E} = \tfrac{1}{2}B\omega\ell^2 = \tfrac{1}{2}(0.50 \times 10^{-4}\ \text{T})\left(\frac{2.0\ \text{rev}}{\text{s}}\right)\left(\frac{2\pi\ \text{rad}}{1.0\ \text{rev}}\right)(3.0\ \text{m})^2 = 2.8\ \text{mV} \quad \Diamond$$

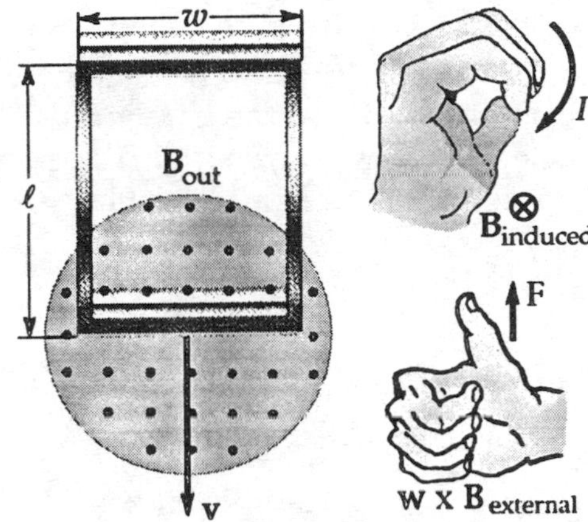

Figure P31.28

28. A conducting rectangular loop of mass M, resistance R, and dimensions w by ℓ falls from rest into a magnetic field **B** as in Figure P31.28. The loop accelerates until it reaches a terminal speed v_t. (a) Show that $v_t = \frac{MgR}{B^2w^2}$. (b) Why is v_t proportional to R? (c) Why is it inversely proportional to B^2?

Solution Let y represent the vertical dimension of the lower part of the loop where it is inside the strong magnetic field. As the loop falls, y increases; the loop encloses increasing flux toward you and has induced in it an emf to produce a current to make its own magnetic field away from you. This current is to the left in the bottom side of the loop, and feels an upward force in the external field.

(a) Symbolically, the flux is $\Phi = BA\cos\theta = Bwy\cos 0°$

The emf is $\mathcal{E} = -N\frac{d}{dt}(Bwy) = -Bw\frac{dy}{dt} = -Bwv$

The magnitude of the current is $I = \frac{|\mathcal{E}|}{R} = \frac{Bwv}{R}$

and the force is $\mathbf{F} = I\,\mathbf{w}\times\mathbf{B} = \left(\frac{Bwv}{R}\right)wB\,\sin 90° = \frac{B^2w^2v}{R}$ up

At terminal speed, the loop is in equilibrium:

$\Sigma F_y = 0$ becomes $\frac{+B^2w^2v_t}{R} - Mg = 0.$ Thus, $v_t = \frac{MgR}{B^2w^2}$ ◊

(b) The emf is directly proportional to v, but the current is inversely proportional to R. A large R means a small current at a given speed, so the loop must travel faster to get F_{mag} = weight.

(c) At a given speed, the current is directly proportional to the magnetic field. But the force is proportional to the product of the current and the field. For a small B, the speed must increase to compensate for both the small B and also the current, so $v \propto B^{-2}$.

29. A 0.15-kg wire in the shape of a closed rectangle 1.0 m wide and 1.5 m long has a total resistance of 0.75 Ω. The rectangle is allowed to fall through a magnetic field directed perpendicularly to the direction of motion of the rectangle (Fig. P31.28). The rectangle accelerates downward until it acquires a constant speed of 2.0 m/s with its top not yet in that region of the field. Calculate the magnitude of **B**.

Solution At constant speed, $F_g = F_B$ or $mg = ILB$

Therefore, $B = \dfrac{mg}{IL}$ but $I = \dfrac{\mathcal{E}}{R} = \dfrac{BLv}{R}$

so $$B = \sqrt{\frac{mgR}{vL^2}} = \sqrt{\frac{(0.15\text{ kg})(9.8\text{ m/s}^2)(0.75\ \Omega)}{(2.0\text{ m/s})(1.0\text{ m})^2}} = 0.74\text{ T} \quad \Diamond$$

---

33. A magnetic field directed into the page changes with time according to $B = (0.030t^2 + 1.4)$ T, where t is in seconds. The field has a circular cross section of radius $R = 2.5$ cm (Fig. P31.33). What are the magnitude and direction of the electric field at point P_1 when $t = 3.0$ s and $r_1 = 0.020$ m?

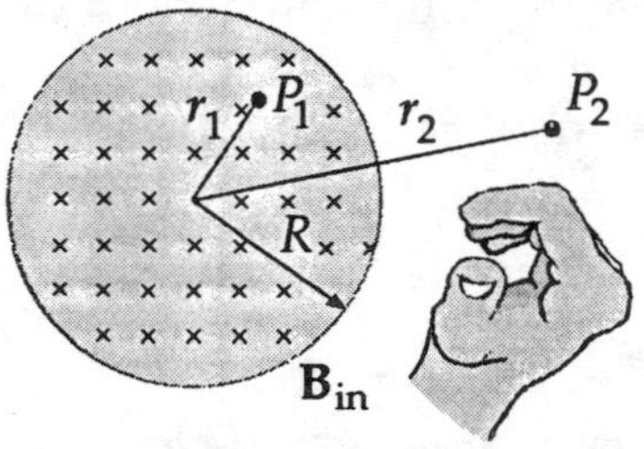

Figure P31.33

Solution $\oint \mathbf{E}\cdot d\mathbf{s} = -\dfrac{d\Phi}{dt}$

Consider a circular integration path of radius r_1:

$$E(2\pi r_1) = -\frac{d}{dt}(BA) = -A\left(\frac{dB}{dt}\right)$$

$$|E| = \frac{A}{2\pi r_1}\frac{d}{dt}(0.030t^2 + 1.4)\text{T} = \frac{\pi r_1^2}{2\pi r_1}(0.060t) = \frac{r_1}{2}(0.060t)$$

At $t = 3.0$ s, $$E = \left(\frac{0.020\text{ m}}{2}\right)(0.060\text{ T/sec})(3.0\text{ sec}) = 1.8\times10^{-3}\text{ N/C} \quad \Diamond$$

If there were a circle of wire of radius r_1, it would enclose increasing magnetic flux away from you. It would carry counterclockwise current to make its own magnetic field toward you, to oppose the change. Even without the wire and current, the counterclockwise electric field that would cause the current is lurking. At point P_1, it is upward and to the left, perpendicular to r_1. ◊

---

35. A long solenoid with 1000 turns/meter and radius 2.0 cm carries an oscillating current $I = (5.0\text{ A})\sin(100\pi t)$. What is the electric field induced at a radius $r = 1.0$ cm from the axis of the solenoid? What is the direction of this electric field when the current is increasing counterclockwise in the coil?

Solution In

$$\oint \mathbf{E}\cdot d\mathbf{s} = -\frac{d}{dt}\int \mathbf{B}\cdot d\mathbf{A}$$

consider a circle of radius 1.0 cm containing the point in question. The induced electric field is uniform in strength around this circle and everywhere tangent to it. The magnetic field is uniform within it:

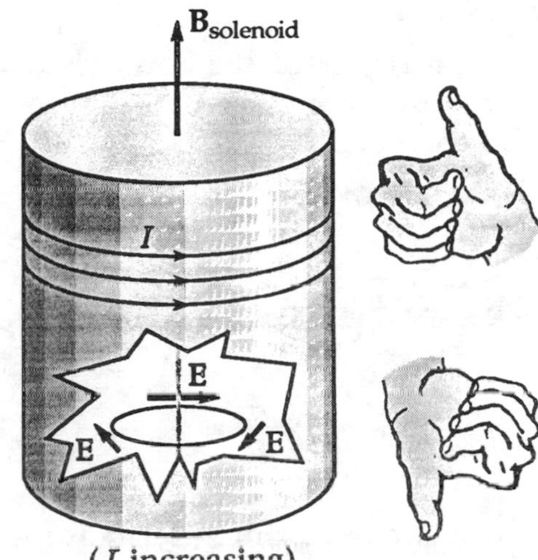

$$E\cos 0°\oint ds = -\frac{d}{dt}B\int dA$$

$$E2\pi r = -\frac{d}{dt}\left(\mu_0 n I \pi r^2\right)$$

$$E = -\left(\frac{\mu_0 n r}{2}\right)\frac{d}{dt}[(5.0\text{ A})\sin(100\pi t)]$$

$$E = -\left(\frac{\mu_0 n r}{2}\right)(5.0\text{ A})\left(100\pi\text{ s}^{-1}\right)\cos(100\pi t)$$

$$E = -\frac{1}{2}(4\pi\times 10^{-7}\text{ N/A}^2)(1.0\times 10^3\text{m}^{-1})(1.0\times 10^{-2}\text{ m})(5.0\text{ A})(100\pi\text{ s}^{-1})\cos(100\pi t)(1\text{ A}\cdot\text{s/C})$$

$$E = -(9.9\text{ mN/C})\cos(100\pi t) \quad \Diamond$$

Carrying counterclockwise current, the solenoid creates upward magnetic field in the picture. As the magnetic field increases, the electric field opposes the change by being clockwise ◊.

Note that if there *were* a wire of radius one centimeter, it would carry clockwise current to create a downward magnetic field of its own.

---

37. An aluminum ring of radius 5.0 cm and resistance $3.0 \times 10^{-4}\ \Omega$ is placed on top of a long air-core solenoid with 1000 turns per meter and radius 3.0 cm as in Figure P31.37. At the location of the ring, the magnetic field due to the current in the solenoid is one-half that at the center of the solenoid. If the current in the solenoid is increasing at a rate of 270 A/s, (a) what is the induced current in the ring? (b) At the center of the ring, what is the magnetic field produced by the induced current in the ring? (c) What is the direction of this field?

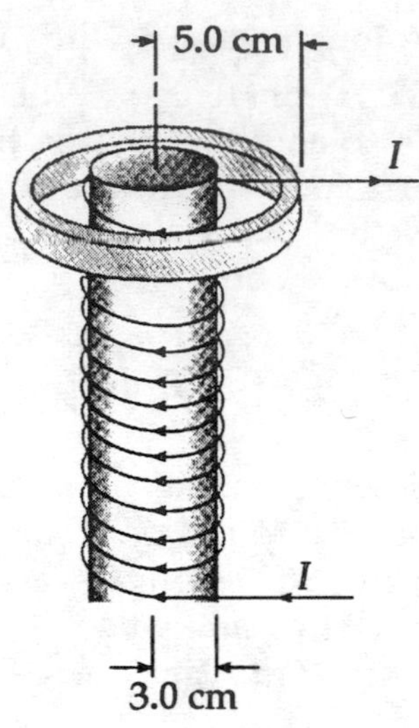

Figure P31.37

Solution

$$\mathcal{E} = -N\frac{d}{dt}(BA\cos\theta) = (-1)\frac{d}{dt}(0.50\ \mu_0 nIA\cos 0°)$$

$$\mathcal{E} = -0.50\ \mu_0 nA\frac{dI}{dt}$$

Note that A must be interpreted as the area of the solenoid, where the field is strong:

$$\mathcal{E} = -0.50(4\pi \times 10^{-7}\ \text{T}\cdot\text{m/A})(1000\ \text{turns/m})\left[\pi(0.030\ \text{m})^2\right](270\ \text{A/s})$$

Applying the conversions of $1\ \dfrac{\text{N}\cdot\text{s}}{\text{C}\cdot\text{m}\cdot\text{T}}$ and $1\ \dfrac{\text{V}\cdot\text{C}}{\text{N}\cdot\text{m}}$,

$$\mathcal{E} = -4.8 \times 10^{-4}\ \text{V}$$

(a) $I_{\text{ring}} = \dfrac{|\mathcal{E}|}{R} = \dfrac{0.00048}{0.00030} = 1.6\ \text{A}$ ◊

(b) $B_{\text{ring}} = \dfrac{\mu_0 I}{2R} = 2.0 \times 10^{-5}\ \text{T}$ ◊

(c) The coil's field points downward, and is increasing, so B_{ring} points upward. ◊

39. A square coil (20 cm × 20 cm) that consists of 100 turns of wire rotates about a vertical axis at 1500 rev/min, as indicated in Figure P31.39. The horizontal component of the Earth's magnetic field at the location of the coil is 2.0×10^{-5} T. Calculate the maximum emf induced in the coil by this field.

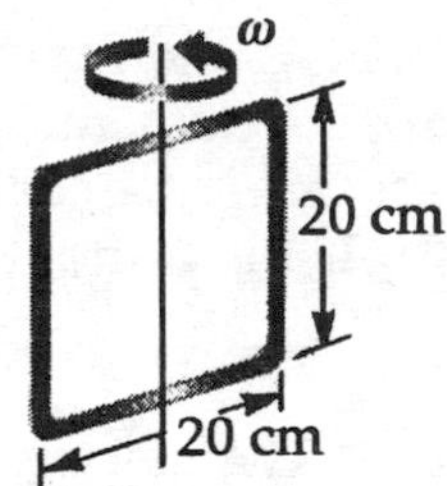

Figure P31.39

Solution Let θ represent the angle through which the coil turns, starting from $\theta = 0$ at an instant when the horizontal component of the Earth's field is perpendicular to the area.

Then
$$\mathcal{E} = -N\frac{d}{dt}(BA\cos\theta) = -NBA\frac{d}{dt}(\cos\omega t)$$

$$\mathcal{E} = +NBA\,\omega\sin\omega t$$

Here sin ωt oscillates between +1 and −1, so the spinning coil generates an alternating voltage having amplitude

$$\mathcal{E}_{max} = NBA\omega = NBA\,2\pi f$$

$$\mathcal{E}_{max} = 100(2.0\times10^{-5}\text{ T})(0.20\text{ m})^2(2\pi)(1500\text{ min}^{-1})\left(\frac{1\text{ min}}{60\text{ s}}\right) = 13\text{ mV}$$ ◊

---

41. A loop of area 0.10 m^2 is rotating at 60 rev/s with the axis of rotation perpendicular to a 0.20-T magnetic field. (a) If there are 1000 turns on the loop, what is the maximum voltage induced in it? (b) When the maximum induced voltage occurs, what is the orientation of the loop with respect to the magnetic field?

Solution For a loop rotating in a magnetic field,

(a) $\mathcal{E} = NBA\omega\sin\theta$ and $\mathcal{E}_{max} = NBA\omega$ so

$$\mathcal{E}_{max} = (1000)(0.20\text{ T})(0.10\text{ m}^2)(60\text{ rev/s})(2\pi\text{ rad}) = 7540\text{ V}$$ ◊

(b) $\mathcal{E} \to \mathcal{E}_{max}$ when $\sin\theta \to 1$ or $\theta = \pm\frac{\pi}{2}$

Therefore, at maximum emf, the plane of the loop is parallel to the field. ◊

---

45. (a) What is the maximum torque delivered by an electric motor if it has 80 turns of wire wrapped on a rectangular coil, of dimensions 2.5 cm by 4.0 cm? Assume that the motor uses 10 A of current and that a uniform 0.80-T magnetic field exists within the motor. (b) If the motor rotates at 3600 rev/min, what is the peak power produced by the motor?

Solution The electric motor is a turning coil in a magnetic field, just like a generator. Thus, it is appropriate to include this review problem from Chapter 29.

$$\boldsymbol{\tau} = \boldsymbol{\mu} \times \mathbf{B}$$

(a) $\tau_{max} = \mu B \sin 90° = NIAB$

Maximum $\tau_{max} = 80(10\text{ A})(2.5\times10^{-2}\text{ m})(4.0\times10^{-2}\text{ m})(0.80\text{ T}) = 0.640\text{ N}\cdot\text{m}$ ◊

(b) $P = \tau_{max}\omega = (0.640\text{ N}\cdot\text{m})\left(3600\ \dfrac{\text{rev}}{\text{min}}\right)\left(\dfrac{1\text{ min}}{60\text{ s}}\right)\left(\dfrac{2\pi\text{ rad}}{1\text{ rev}}\right) = 240\text{ W}$ ◊

---

49. A rectangular loop with resistance R has N turns, each of length ℓ and width w as shown in Figure P31.49. The loop moves into a uniform magnetic field **B** with velocity **v**. What are the magnitude and direction of the resultant force on the loop (a) as it enters the magnetic field, (b) as it moves within the field, (c) as it leaves the field?

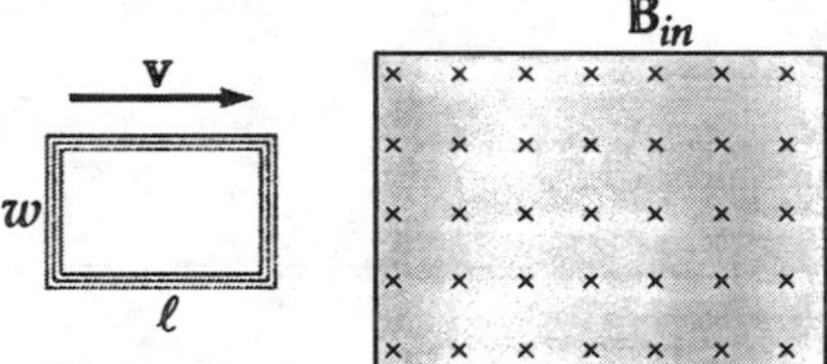

Figure P31.49

Solution

(a) Call x the distance that the leading edge has penetrated into the strong field region. The flux (Bwx away from you) through the coil increases in time so the voltage

$$\mathcal{E} = -N\frac{d}{dt}Bwx = -NBw\frac{dx}{dt} = -NBwv$$

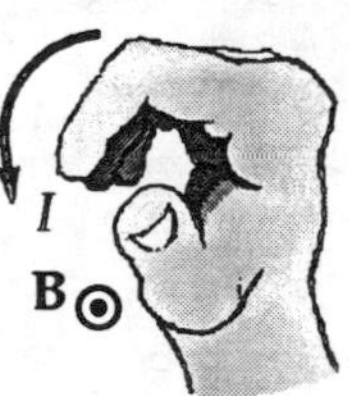

is induced in the coil, tending to produce counterclockwise current

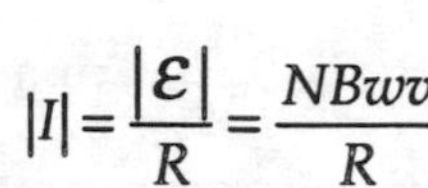

$$|I| = \frac{|\mathcal{E}|}{R} = \frac{NBwv}{R}$$

The current upward in the leading edge experiences a force

$$\mathbf{F} = NI\,\mathbf{L}\times\mathbf{B} = N\left(\frac{NBwv}{R}\right)w\,\mathbf{j}\times B(-\mathbf{k}) = \left(\frac{N^2B^2w^2v}{R}\right)(-\mathbf{i}) \quad \Diamond$$

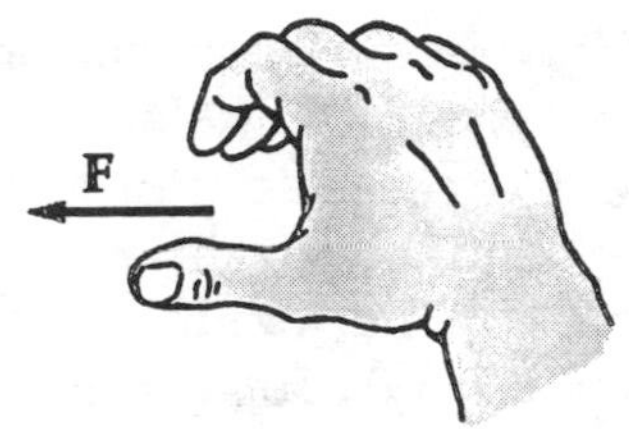

This retarding force models eddy-current damping.

(b) The flux through the coil is constant when it is wholly within the high-field region. The induced emf, induced current, and magnetic force are zero. ◊

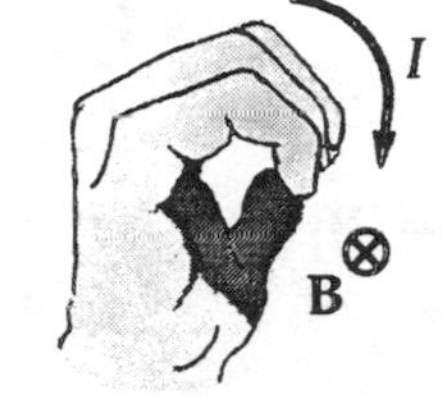

(c) As the coil leaves the field, the away-from-you flux it encloses decreases. The coil carries clockwise current to make some away-from-you field of its own. Again,

$$|I| = \frac{NBwv}{R}$$

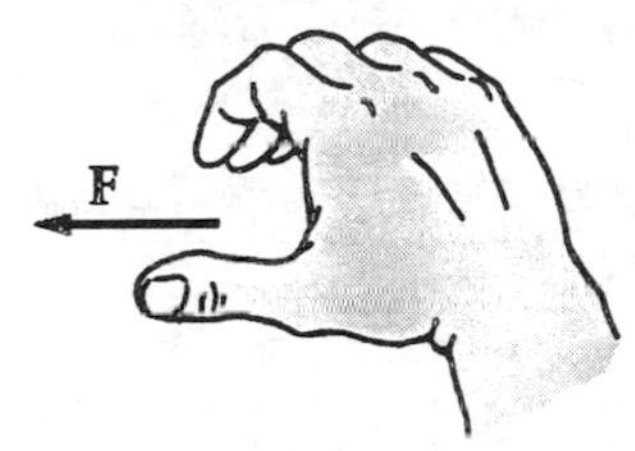

Now the trailing edge carries upward current to experience a force of

$$\mathbf{F} = \left(\frac{N^2B^2w^2v}{R}\right) \text{ to the left} \quad \Diamond$$

---

51. A proton moves through a uniform electric field $\mathbf{E} = 50\mathbf{j}$ V/m and a uniform magnetic field $\mathbf{B} = (0.20\mathbf{i} + 0.30\mathbf{j} + 0.40\mathbf{k})$ T. Determine the acceleration of the proton when it has a velocity $\mathbf{v} = 200\mathbf{i}$ m/s.

Solution $\mathbf{F} = m\mathbf{a} = q\mathbf{E} + q\mathbf{v}\times\mathbf{B}$

$$\mathbf{a} = \frac{e}{m}[\mathbf{E} + \mathbf{v}\times\mathbf{B}] \quad \text{where} \quad \mathbf{v}\times\mathbf{B} = \begin{vmatrix} \mathbf{i} & \mathbf{j} & \mathbf{k} \\ 200 & 0 & 0 \\ 0.20 & 0.30 & 0.40 \end{vmatrix} = -200(0.40)\mathbf{j} + 200(0.30)\mathbf{k}$$

$$\mathbf{a} = \frac{1.60\times10^{-19}}{1.67\times10^{-27}}[50\mathbf{j} - 80\mathbf{j} + 60\mathbf{k}] = 9.58\times10^{7}[-30\mathbf{j} + 60\mathbf{k}]$$

$$\mathbf{a} = \left(2.87\times10^{9}\right)(-\mathbf{j} + 2\mathbf{k})\ \text{m/s}^2 = (-2.9\times10^{9}\mathbf{j} + 5.8\times10^{9}\mathbf{k})\ \text{m/s}^2 \quad \Diamond$$

---

57. A rectangular coil of 60 turns, dimensions 0.10 m by 0.20 m and total resistance 10 Ω rotates with angular speed 30 rad/s about the y axis in a region where a 1.0-T magnetic field is directed along the x axis. The rotation is initiated so that the plane of the coil is perpendicular to the direction of **B** at $t = 0$. Calculate (a) the maximum induced emf in the coil, (b) the maximum rate of change of magnetic flux through the coil, (c) the induced emf at $t = 0.050$ s, and (d) the torque exerted on the loop by the magnetic field at the instant when the emf is a maximum.

Solution

Let θ represent the angle between the perpendicular to the coil and the magnetic field. Then $\theta = 0$ at $t = 0$ and $\theta = \omega t$ at all later times.

(a) $\mathcal{E} = -N\frac{d}{dt}(BA\cos\theta) = -NBA\frac{d}{dt}(\cos\omega t) = +NBA\omega\sin\omega t$

$\mathcal{E}_{max} = NBA\omega = 60(1.0\text{ T})(0.020\text{ m}^2)(30\text{ rad/s}) = 36.0\text{ V}$ ◊

(b) $\frac{d}{dt}(BA\cos\theta) = BA\omega\sin\omega t$

maximum $= BA\omega = 0.600\text{ T}\cdot\text{m}^2/\text{s}$ ◊

(c) $\mathcal{E} = NBA\omega\sin\omega t = (36\text{ V})\sin[(30\text{ rad/s})(0.050\text{ s})]$

$= (36\text{ V})\sin(1.5\text{ rad}) = (36\text{ V})\sin 85.9° = 36\text{ V}$ ◊

(d) $\boldsymbol{\tau} = \boldsymbol{\mu}\times\mathbf{B}$

The emf is maximal when $\theta = 90°$.

$$\tau_{max} = \mu B\sin 90° = NIAB = N\,\mathcal{E}_{max}\frac{AB}{R} = \frac{60(36\text{ V})(0.020\text{ m}^2)(1.0\text{ T})}{10\ \Omega} = 4.3\text{ N}\cdot\text{m}$$ ◊

65. A plane of a square loop of wire with edge length $a = 0.20$ m is perpendicular to the Earth's magnetic field at a point where $B = 15\ \mu$T, as in Figure P31.65. The total resistance of the loop and the wires connecting it to the galvanometer is 0.50 Ω. If the loop is suddenly collapsed by horizontal forces as shown, what total charge passes through the galvanometer?

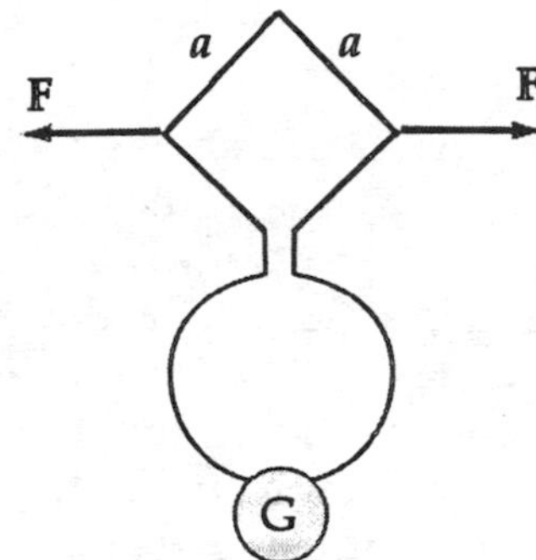

Figure P31.65

Solution

$$Q = \int I dt = \int \left(\frac{\mathcal{E}}{R}\right) dt$$

$$Q = \frac{1}{R}\int -\left(\frac{d\Phi}{dt}\right) dt = -\frac{1}{R}\int d\Phi$$

$$Q = -\frac{1}{R}\int d(BA) = -\frac{B}{R}\int_{A_1=a^2}^{A_2=0} dA = -\frac{B}{R}A\Big]_{A_1=a^2}^{A_2=0}$$

$$Q = \frac{Ba^2}{R} = \frac{(15\times10^{-6}\ \text{T})(0.20\ \text{m})^2}{0.50\ \Omega} = 1.20\times10^{-6}\ \text{C} = 1.20\ \mu\text{C} \quad \Diamond$$

---

67. The magnetic flux threading a metal ring varies with time t according to $\Phi_B = 3(at^3 - bt^2)$ T·m², with $a = 2.0\ \text{s}^{-3}$ and $b = 6.0\ \text{s}^{-2}$. The resistance of the ring is 3.0 Ω. Determine the maximum current induced in the ring during the interval from $t = 0$ to $t = 2.0$ s.

Solution $\Phi_m = (6t^3 - 18t^2)$ T·m²

$$\mathcal{E} = -\frac{d\Phi_m}{dt} = -18t^2 + 36t$$

Maximum $\mathcal{E}$ occurs when $\dfrac{d\mathcal{E}}{dt} = -36t + 36 = 0,$ which gives $t = 1.0$ s.

Thus, maximum current (at $t = 1.0$ s) is $I_{max} = \dfrac{\mathcal{E}}{R} = \dfrac{(-18+36)\text{V}}{3.0\ \Omega} = 6.0\ \text{A} \quad \Diamond$

---

71. Figure P31.71 shows a circular loop of radius r that has a resistance R spread uniformly throughout its length. The loop's plane is normal to the magnetic field **B** that decreases at a constant rate $dB/dt = -K$, where K is a positive constant. What are (a) the direction and (b) the value of the induced current? (c) Which point, a or b, is at the higher potential? Explain. (d) Discuss what force causes the current in the loop.

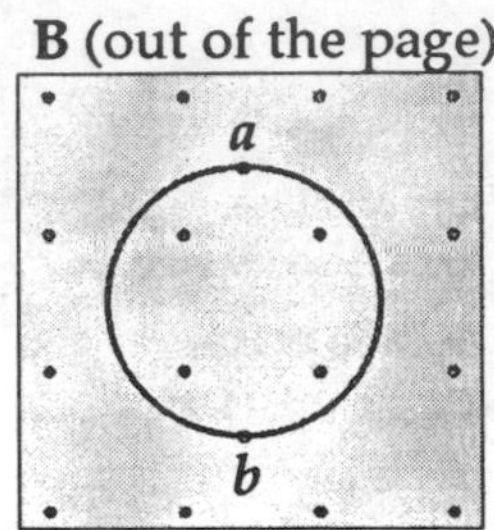

Figure P31.71

Solution

(a) The loop encloses decreasing flux toward you so it makes more flux toward you by carrying counterclockwise current. ◊

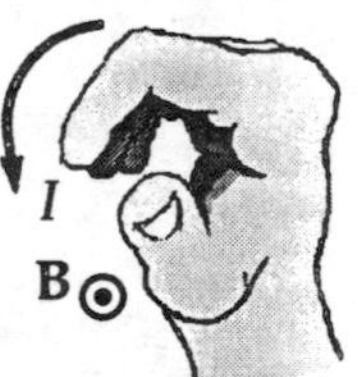

(b) $I = \frac{\mathcal{E}}{R} = -\frac{N}{R}\frac{d}{dt}(BA\cos\theta) = -\frac{A\cos 0°}{R}\frac{dB}{dt} = -\frac{\pi r^2}{R}(-K) = \pi r^2 \frac{K}{R}$ ◊

(c) A voltmeter connected to a and b reads zero. There is no potential function. None can be defined because the induced electric field is not a conservative field. The work it does when you carry a charge from a to b depends on your path. You can get arbitrarily large amounts of work out of the induced electric field by carrying a positive charge many times around the loop counterclockwise.

(d) The induced "electromotive force" is not a mechanical force. Charges in the loop are pushed on by the $q\mathbf{E}$ force of the induced electric field

$$E = \frac{\mathcal{E}}{2\pi r} = \frac{\pi r^2 K}{2\pi r} = \frac{rK}{2} \quad \text{counterclockwise} \quad ◊$$

Chapter 32

Inductance

Chapter 32

INDUCTANCE

INTRODUCTION

In the previous chapter, we saw that currents and emfs are induced in a circuit when the magnetic flux through the circuit changes with time. This electromagnetic induction has some practical consequences, which we describe in this chapter. First, we describe an effect known as *self-induction*, in which a time-varying current in a conductor induces in the conductor an emf that opposes the external emf that set up the current. Self-induction is the basis of the *inductor*, an electrical element that plays an important role in circuits that use time-varying currents. We discuss the energy stored in the magnetic field of an inductor and the energy density associated with a magnetic field.

Next, we study how an emf is induced in a circuit as a result of a changing magnetic flux produced by an external circuit, which is the basic principle of *mutual induction*. Finally, we examine the characteristics of circuits containing inductors, resistors, and capacitors in various combinations.

NOTES FROM SELECTED CHAPTER SECTIONS

32.1 Self-Inductance

The self-induced emf is always proportional to the *time rate of change* of current in the circuit.

The *inductance* of a device (an inductor) depends on its *geometry*.

32.2 *RL* Circuits

If a resistor and an inductor are connected in series to a battery, the current in the circuit will reach an *equilibrium* value ($\mathcal{E}/R$) after a time which is long compared to the *time constant* of the circuit, τ.

32.3 Energy in a Magnetic Field

In an *RL* circuit, the rate at which energy is supplied by the battery equals the sum of the rate at which heat is dissipated in the resistor and the rate at which energy is stored in the inductor. The *energy density* is proportional to the *square of the magnetic field*.

32.4 Mutual Inductance

If two coils are near each other, a time-varying current in one coil will give rise to an induced emf in the other. The *mutual inductance* depends on the geometry of the two circuits and their orientation with respect to each other. The emf induced by mutual induction in one coil is always proportional to the rate of change of current in the other.

32.5 Oscillations in an *LC* Circuit

The energy in an *LC* circuit continuously transfers between energy stored in the electric field of the capacitor and energy stored in the magnetic field of the inductor.

The angular frequency of the oscillations depends only on the values of inductance, L, and capacitance, C.

32.6 The *RLC* Circuit

The charge and current in an *RLC* circuit exhibit damped harmonic oscillations when the value of R in the circuit is small.

EQUATIONS AND CONCEPTS

When the current in a coil changes in time, a self-induced emf is present in the coil. The inductance, L, is a measure of the opposition of the coil to a change in the current.

$$\mathcal{E}_L = -L\frac{dI}{dt} \qquad (32.1)$$

The inductance of a given device, for example a coil, depends on its physical makeup—diameter, number of turns, type of material on which the wire is wound, and other geometric parameters. A circuit element which has a large inductance is called an inductor. The SI unit of inductance is the henry, H. A rate of change of current of 1 ampere per second in an inductor of 1 henry will produce a self-induced emf of 1 volt.

$$1\ \mathrm{H} = 1\frac{\mathrm{V\cdot s}}{\mathrm{A}} = 1\ \Omega\cdot\mathrm{s}$$

A coil, solenoid, toroid, coaxial cable, or other conducting device is characterized by a parameter called its *inductance*, *L*. The inductance can be calculated, knowing the current and magnetic flux.

$$L = \frac{N\Phi_B}{I} \quad (32.2)$$

The *inductance* of a particular circuit element can also be expressed as the ratio of the induced emf to the time rate of change of current in the circuit.

$$L = -\frac{\mathcal{E}_L}{dI/dt} \quad (32.3)$$

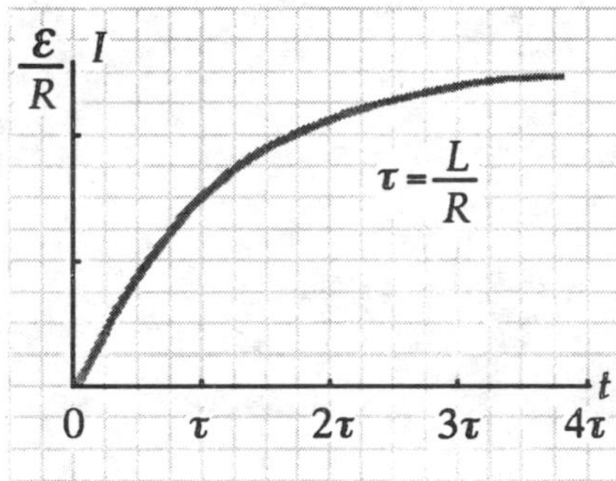

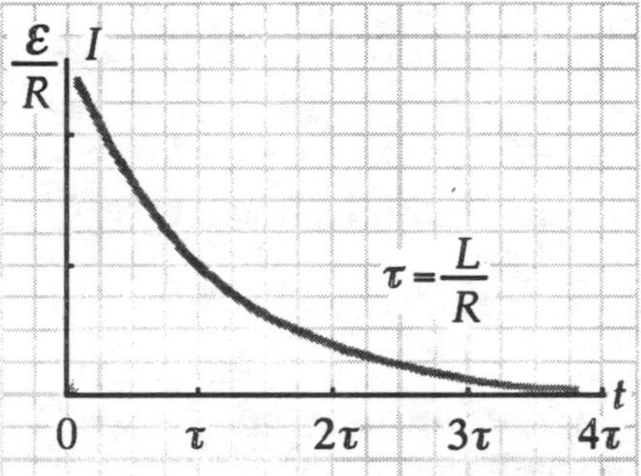

If the switch in the series circuit shown (which contains a battery, resistor, and inductor) is closed in position 1 at time $t = 0$, *current in the circuit* will increase in a characteristic fashion toward a maximum value of $\mathcal{E}/R$. This is shown in the first graph, above.

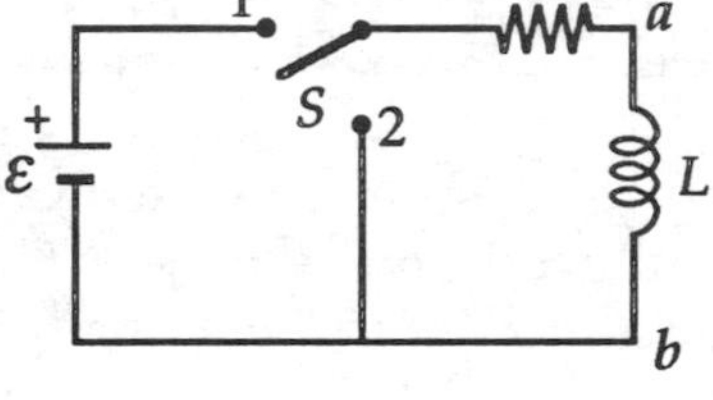

$$I(t) = \frac{\mathcal{E}}{R}(1 - e^{-t/\tau}) \quad (32.7)$$

Let the switch in the circuit shown in the figure be at position 1 with the current at its maximum value $I_0 = \mathcal{E}/R$. If the switch is thrown to position 2 at $t = 0$, the current *will decay* exponentially with time. The curve of the decay is shown in the second graph, above.

$$I(t) = \frac{\mathcal{E}}{R}e^{-t/\tau} \quad (32.10)$$

where $\frac{\mathcal{E}}{R} = I_0$

The *stored energy* U_B in the magnetic field of an inductor is proportional to the square of the current in the inductor.

$$U_B = \tfrac{1}{2}LI^2 \qquad (32.12)$$

It is often useful to express the energy in a magnetic field as *energy density* u_B; that is, energy per unit volume.

$$u_B = \frac{B^2}{2\mu_0} \qquad (32.14)$$

When two coils are nearby, an emf is produced by mutual induction in one coil which is proportional to the rate of change of current in the other. The *mutual inductance* M depends on the coil geometries and the relative positions of the coils.

$$\mathcal{E}_1 = -M\frac{dI_2}{dt}$$

$$\mathcal{E}_2 = -M\frac{dI_1}{dt}$$

The situation in which two closely wound coils are adjacent to each other can be used to define the *mutual inductance, M.* In this expression Φ_{21} is the magnetic flux through coil 2 due to the current in coil 1. The unit of mutual inductance is the henry, H. Note that M is a *shared property of the pair of coils.*

$$M_{21} \equiv \frac{N_2\Phi_{21}}{I_1} \qquad (32.15)$$

The figure shows a charged capacitor which can be connected by switch S to an inductor in a circuit with zero resistance. When the switch S is closed, the circuit exhibits a continual transfer of energy (back and forth) between the electric field of the capacitor and the magnetic field of the inductor. The process occurs at an angular frequency ω, called the *frequency of oscillation of the circuit.*

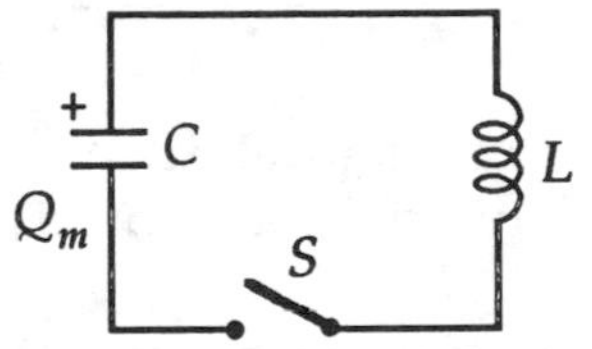

+
C
Q_m
S
L

$$\omega = \frac{1}{\sqrt{LC}} \qquad (32.22)$$

The *total energy* of this circuit can be expressed in terms of the charge on the capacitor and the current in the inductor at some arbitrary time (Equation 32.18). Alternatively, the total energy can be written in terms of the maximum charge on the capacitor and the maximum current in the inductor at a specified time t after the switch is closed (Eq. 32.26).

$$U = \frac{Q^2}{2C} + \tfrac{1}{2}LI^2 \qquad (32.18)$$

$$U = \frac{Q_{\text{max}}^2}{2C}\cos^2 \omega t + \frac{LI_{\text{max}}^2}{2}\sin^2 \omega t \qquad (32.26)$$

The *charge* on the capacitor and the *current* in the inductor vary sinusoidally in time and are 90° out of phase with each other. If the initial conditions are such that $I = 0$ and $Q = Q_{\text{max}}$ at $t = 0$, the charge and current vary with time according to Equations 32.24 and 32.25.

$$Q = Q_{\text{max}} \cos \omega t \qquad (32.24)$$

$$I = -I_{\text{max}} \sin \omega t \qquad (32.25)$$

SUGGESTIONS, SKILLS, AND STRATEGIES

Equation 32.2, $L = \frac{N\Phi_B}{I}$ and Equation 32.13, $U_B = \frac{1}{2}LI^2$, provide two different approaches for the calculation of the inductance L of a particular device.

In order to use Equation 32.2 to calculate L, take the following steps:

- Assume a current I to exist in the conductor for which you wish to calculate L (coil, solenoid, coaxial cable, or other device).
- Calculate the magnetic flux through the appropriate cross section using $\Phi_B = \int \mathbf{B} \cdot d\mathbf{A}$. Remember that in many cases, **B** will not be uniform over the area.
- Calculate L directly from the defining Equation 32.2.

In order to use Equation 32.13 to calculate L, take the following steps:

- Assume a current I in the conductor.
- Find an expression for **B** for the magnetic field produced by I.
- Use Equation 32.14, $u_B = \dfrac{B^2}{2\mu_0}$, and integrate this value of u_B over the appropriate volume to find the total energy stored in the magnetic field of the inductor $U_B = \int u\,dV$.
- Substitute this value of U_B into Equation 32.13 and solve for L.

REVIEW CHECKLIST

▷ Calculate the inductance of a device of suitable geometry.

▷ Calculate the magnitude and direction of the self-induced emf in a circuit containing one or more inductive elements when the current changes with time.

▷ Determine instantaneous values of the current in an LR circuit while the current is either increasing or decreasing with time.

▷ Calculate the total magnetic energy stored in a magnetic field. You should be able to perform this calculation if (1) you are given the values of the inductance of the device with which the field is associated and the current in the circuit, or (2) given the value of the magnetic field intensity throughout the region of space in which the magnetic field exists. In the latter case, you must integrate the expression for the energy density u_B over an appropriate volume.

▷ Calculate the emf induced by mutual inductance in one winding due to a time-varying current in a nearby inductor.

SOLUTIONS TO SELECTED END-OF-CHAPTER PROBLEMS

1. A 2.00-H inductor carries a steady current of 0.500 A. When the switch in the circuit is thrown open, the current disappears in 10.0 ms. What is the average induced emf in the inductor during this time?

Solution $\mathcal{E} = -L\dfrac{dI}{dt}$

$$\mathcal{E}_{ave} = -L\frac{I_f - I_0}{t} = (-2.00\text{ H})\left(\frac{0 - 0.500\text{ A}}{1.00\times 10^{-2}\text{ s}}\right)\left(1\ \frac{\text{V}\cdot\text{s}/\text{A}}{\text{H}}\right) = +100\text{ V} \quad \Diamond$$

---

3. A coiled telephone cord has 70 turns, a cross-sectional diameter of 1.3 cm, and an unstretched length of 60 cm. Determine an approximate value for the self-inductance of the unstretched cord.

Solution The spring of wire is a solenoid. We suppose that the plastic insulation has the same magnetic permeability as vacuum.

$$L = \frac{N^2\mu_0 A}{\ell} = \frac{(70)^2\left(4\pi\times 10^{-7}\ \dfrac{\text{T}\cdot\text{m}}{\text{A}}\right)\left(\dfrac{\pi(0.013\text{ m})^2}{4}\right)}{0.60\text{ m}} = 1.4\ \mu\text{H} \quad \Diamond$$

---

11. A 10.0-mH inductor carries a current $I = I_{max}\sin\omega t$, with $I_{max} = 5.00$ A and $\omega/2\pi = 60.0$ Hz. What is the back emf as a function of time?

Solution $\mathcal{E}_{back} = -\mathcal{E}_L = L\dfrac{dI}{dt} = L\dfrac{d}{dt}(I_0 \sin \omega t)$

$$\mathcal{E}_{back} = L\omega I_0 \cos \omega t = (0.0100\text{ H})(120\pi\text{ s}^{-1})(5.00\text{ A})\cos(120\pi t)$$

$$\mathcal{E}_{back} = (18.8\text{ V})\cos(377t) \quad \Diamond$$

16A. An inductor in the form of a solenoid contains N turns, has a length ℓ, and has cross-sectional area A. What uniform rate of decrease of current through the inductor induces an emf $\mathcal{E}$?

Solution $$L = \frac{\mu_0 N^2 A}{\ell} \quad \text{and} \quad \mathcal{E} = -L\frac{dI}{dt}$$

$$\frac{dI}{dt} = \frac{-\mathcal{E}}{L} = \frac{-\mathcal{E}\ell}{\mu_0 N^2 A} \quad \Diamond$$

Related Calculation Find the numerical rate of decrease in current that will induce an emf of 175 μV. Use a conductor that has 420 turns, a length of 16.0 cm, and a cross-sectional area of 3.00 cm^2.

Solution

Substituting the given values into the previous answer,

$$\frac{dI}{dt} = \frac{-\mathcal{E}\ell}{\mu_0 N^2 A} = \frac{(-175\times10^{-6}\ \text{V})(0.160\ \text{m})}{(4\pi\times10^{-7}\ \text{N}/\text{A}^2)(420)^2(3.00\times10^{-4}\ \text{m}^2)}$$

and $$\frac{dI}{dt} = -0.421\ \text{A}/\text{s} \quad \Diamond$$

---

21. A series RL circuit with L = 3.00 H and a series RC circuit with C = 3.00 μF have the same time constant. If the two circuits have the same resistance R, (a) what is the value of R and (b) what is the time constant?

Solution $$\frac{L}{R} = RC$$

(a) $$R = \sqrt{\frac{L}{C}} = \sqrt{\frac{3.00\ \Omega\cdot\text{s}}{3.00\times10^{-6}\ \text{s}/\Omega}} = 1.00\ \text{k}\Omega \quad \Diamond$$

(b) $$RC = (1000\ \Omega)(3.00\times10^{-6}\ \text{F}) = 3.00\ \text{ms} \quad \Diamond$$

---

23. A 12-V battery is about to be connected to a series circuit containing a 10-Ω resistor and a 2.0-H inductor. How long will it take the current to reach (a) 50% and (b) 90% of its final value?

Solution The time constant is $\tau = \frac{L}{R} = 0.2$ s:

(a) In $I = \frac{\mathcal{E}(1-e^{-t/\tau})}{R}$, the final value which the current approaches is

$$I = \frac{\mathcal{E}(1-e^{-\infty})}{R} = \frac{\mathcal{E}}{R}$$

We have at 50%, $(0.50)\frac{\mathcal{E}}{R} = \frac{\mathcal{E}(1-e^{t/0.2})}{R}$

Solving for t, $0.50 = 1 - e^{-t/0.20 \text{ s}}$

$$e^{-t/0.20} = 0.50$$

$$e^{t/0.20} = 2.0$$

$$\frac{t}{0.20} = \ln 2.0$$

Thus, $t = 0.14$ s ◊

(b) At 90%, $0.90 = 1 - e^{-t/\tau}$

and $t = \tau \ln\left[\frac{1}{1-0.90}\right]$

$$t = 0.20 \ln 10 = 0.46 \text{ sec}$$ ◊

25. A 140-mH inductor and a 4.9-Ω resistor are connected with a switch to a 6.0-V battery as shown in Figure P32.25. (a) If the switch is thrown to the left (connecting the battery), how much time elapses before the current reaches 220 mA? (b) What is the current in the inductor 10.0 s after the switch is closed? (c) Now the switch is quickly thrown from A to B. How much time elapses before the current falls to 160 mA?

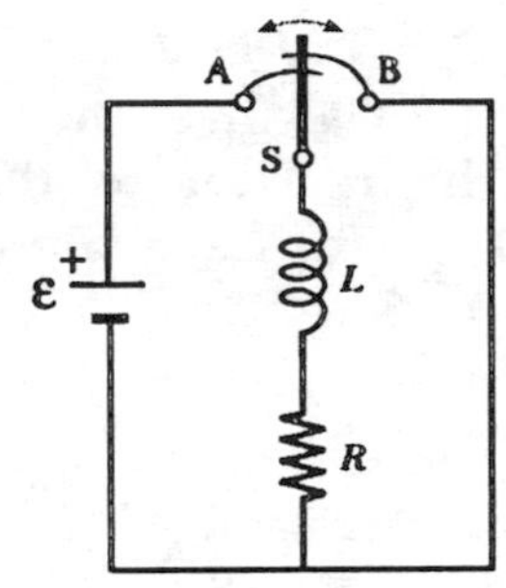

Figure P32.25

Solution The general *LR* equation is achieved by combining equations 32.7 and 32.8:

$$I = \frac{\mathcal{E}(1 - e^{-Rt/L})}{R}$$

(a) $$0.22\text{ A} = \frac{6.0\text{ V}}{4.9\ \Omega}(1 - e^{-4.9\ \Omega\, t/0.14\text{ H}})$$

$$0.180 = 1 - e^{-35t}$$

$$e^{35t} = 1.22$$

$$t = \left(\frac{\ln 1.22}{35}\right)\text{s} = 5.7\text{ ms} \quad \lozenge$$

Again referring to the general equation for answer (b) and (c),

(b) $$I = \frac{6.0\text{ V}}{4.9\ \Omega}\left(1 - e^{(-35)(10.0)}\right) = 1.2\text{ A} \quad \lozenge$$

(c) $$0.160\text{ A} = (1.22\text{ A})\, e^{(-4.9\ \Omega)t/0.14\text{ H}}$$

$$7.65 = e^{35t}$$

$$t = \left(\frac{\ln 7.65}{35}\right) = 58\text{ ms} \quad \lozenge$$

---

29. Let $L = 3.00$ H, $R = 8.00\ \Omega$, and $\mathcal{E} = 36.0$ V in Figure P32.26. (a) Calculate the ratio of the potential difference across the resistor to that across the inductor when $I = 2.00$ A. (b) Calculate the voltage across the inductor when $I = 4.50$ A.

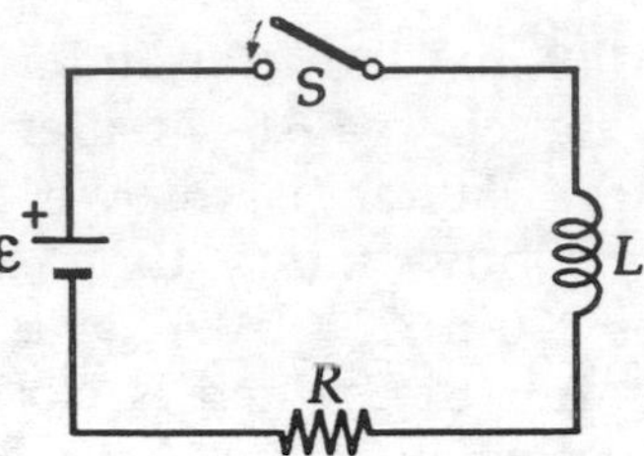

Figure P32.26

Solution

(a) When $I = 2.00$ A, the voltage across the resistor is

$$V_R = IR = (2.00\text{ A})(8.00\ \Omega) = 16.0\text{ V}$$

The voltage across the inductor is given by Kirchhoff's loop theorem:

$$36.0\text{ V} - 16.0\text{ V} - \mathcal{E}_L = 0 \qquad \text{and} \qquad \mathcal{E}_L = 20.0\text{ V}$$

So $$\frac{V_R}{\mathcal{E}_L} = \frac{16.0\text{ V}}{20.0\text{ V}} = 0.800 \quad \Diamond$$

(b) Similarly, $$+36.0\text{ V} - (4.50\text{ A})(8.00\ \Omega) - \mathcal{E}_L = 0 \quad \text{and} \quad \mathcal{E}_L = 0 \quad \Diamond$$

---

33. An air-core solenoid with 68 turns is 8.0 cm long and has a diameter of 1.2 cm. How much energy is stored in its magnetic field when it carries a current of 0.77 A?

Solution $$U_B = \tfrac{1}{2}LI^2$$

For a solenoid of length ℓ, $$L = \frac{\mu_0 N^2 A}{\ell}$$

Thus, $$U_B = \frac{\mu_0 N^2 A I^2}{2\ell} = \frac{\left(4\pi\times 10^{-7}\ \frac{\text{N}}{\text{A}^2}\right)(68)^2(\pi)(6.00\times 10^{-3}\text{ m})^2(0.77\text{ A})^2}{2(0.080\text{ m})} = 2.44\times 10^{-6}\text{ J} \quad \Diamond$$

---

35. A 10.0-V battery, a 5.00-Ω resistor, and a 10.0-H inductor are connected in series. After the current in the circuit has reached its maximum value, calculate (a) the power supplied by the battery, (b) the power dissipated in the resistor, (c) the power dissipated in the inductor, and (d) the energy stored in the magnetic field of the inductor.

Solution From Equation 32.7,

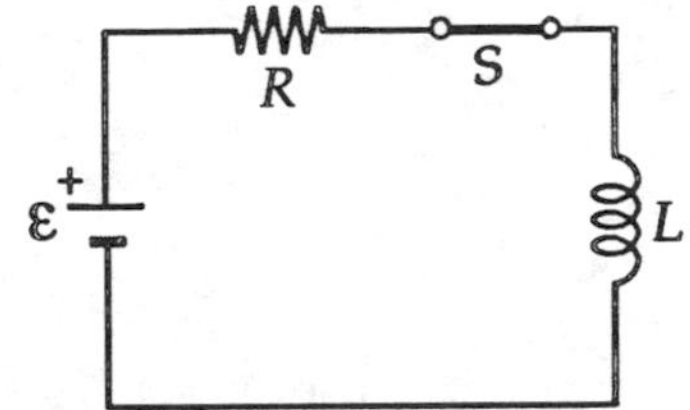

$$I = \frac{\mathcal{E}}{R}\left[1 - e^{-Rt/L}\right]$$

(a) The maximum current is reached after a long time t and is

$$I = \frac{\mathcal{E}}{R} = 2.00 \text{ A}$$

At that time, the inductor is fully energized and the battery output is

$$P = IV = (2.00 \text{ A})(10.0 \text{ V}) = 20.0 \text{ W} \quad \Diamond$$

(b) $P_{lost} = I^2R = (2.00 \text{ A})^2(5.00 \ \Omega) = 20 \text{ W}$ ◊

(c) $P_{inductor} = IV_{drop} = 0$ ◊

(d) $U_{stored} = \frac{I^2L}{2} = \frac{(10.0 \text{ H})(2.00 \text{ A})^2}{2} = 20.0 \text{ J}$ ◊

---

37. The magnetic field inside a superconducting solenoid is 4.5 T. The solenoid has an inner diameter of 6.2 cm and a length of 26 cm. Determine (a) the magnetic energy density in the field and (b) the energy stored in the magnetic field within the solenoid.

Solution

(a) The magnetic energy density is given by Equation 32.14,

$$u_B = \frac{B^2}{2\mu_0} = \frac{(4.5 \text{ T})^2}{2(1.26 \times 10^{-6} \text{ T} \cdot \text{m/A})}$$

$$u_B = 8.06 \times 10^6 \text{ J/m}^3 = 8.1 \text{ MJ/m}^3 \quad \Diamond$$

(b) The magnetic energy stored in the field equals u_B times the volume of the solenoid (the volume in which B is non-zero).

$$U_B = u_B V = u_B \pi r^2 h = (8.06 \times 10^6 \text{ J/m}^3)\pi(0.031 \text{ m})^2(0.26 \text{ m})$$

and $$U_B = 6.3 \text{ kJ} \quad \lozenge$$

---

39. On a clear day, there is a 100-V/m vertical electric field near the Earth's surface. At the same time, the Earth's magnetic field has a magnitude of 0.500×10^{-4} T. Compute the energy density of the two fields.

Solution $$u_E = \tfrac{1}{2}\epsilon_0 E^2 = \tfrac{1}{2}\left(8.85 \times 10^{-12} \frac{\text{C}^2}{\text{N} \cdot \text{m}^2}\right)\left(100 \frac{\text{N}}{\text{C}}\right)^2\left(\frac{\text{J}}{\text{N} \cdot \text{m}}\right) = 44.2 \text{ nJ/m}^3 \quad \lozenge$$

$$u_B = \frac{1}{2\mu_0}B^2 = \left(\frac{1}{2\left(4\pi \times 10^{-7} \text{ T} \cdot \text{m/A}\right)}\right)(5.00 \times 10^{-5} \text{ T})^2 = 995 \ \mu\text{T} \cdot \text{A/m}$$

Applying conversion factors of $1 \dfrac{\text{N} \cdot \text{s}}{\text{T} \cdot \text{C} \cdot \text{m}}$, $1 \dfrac{\text{C}}{\text{A} \cdot \text{s}}$, and $1 \dfrac{\text{J}}{\text{N} \cdot \text{m}}$,

$$u_B = 995 \ \mu\text{J/m}^3 \quad \lozenge$$

Energy is packed 22500 times more densely into the magnetic field.

---

47. An emf of 96.0 mV is induced in the windings of a coil when the current in a nearby coil is increasing at the rate of 1.20 A/s. What is the mutual inductance of the two coils?

Solution $$\mathcal{E}_2 = -M\left(\frac{dI_1}{dt}\right) \quad \text{and} \quad M = \frac{-\mathcal{E}_2}{dI_1/dt}$$

$$M = \frac{96.0 \times 10^{-3} \text{ V}}{1.20 \text{ A/s}} = 80.0 \times 10^{-3} \text{ H} = 80.0 \text{ mH} \quad \lozenge$$

---

51. Two solenoids A and B, spaced close to each other and sharing the same cylindrical axis, have 400 and 700 turns, respectively. A current of 3.5 A in coil A produces a flux of 300 μWb at the center of A and a flux of 90 μWb at the center of B. (a) Calculate the mutual inductance of the two solenoids. (b) What is the self-inductance of A? (c) What emf is induced in B when the current in A increases at the rate of 0.50 A/s?

Solution

(a) $M_{21} = \dfrac{N_2\Phi_{21}}{I_1} = \dfrac{700(90\times10^{-6}\text{ Wb})}{3.5\text{ A}} = 18\text{ mH}$ ◊

(b) $L = \dfrac{N\Phi}{I} = \dfrac{400(300\times10^{-6}\text{ Wb})}{3.5\text{ A}} = 34\text{ mH}$ ◊

(c) $\mathcal{E}_2 = -M_{21}\dfrac{dI_1}{dt} = -\left(18\times10^{-3}\text{ H}\right)(0.50\text{ A/s}) = -9.0\text{ mV}$ ◊

---

57. A fixed inductance L = 1.05 μH is used in series with a variable capacitor in the tuning section of a radio. What capacitance tunes the circuit into the signal from a station broadcasting at 6.30 MHz?

Solution

$$f_0 = \frac{1}{2\pi\sqrt{LC}}$$

Therefore

$$C = \frac{1}{(2\pi f_0)^2 L}$$

and

$$C = \frac{1}{\left[(2\pi)(6.30\times10^6\text{ Hz})\right]^2(1.05\times10^{-6}\text{ H})} = 608\text{ pF} \quad ◊$$

---

61. An LC circuit like that in Figure 32.11 consists of a 3.30-H inductor and an 840-pF capacitor, initially carrying a 105-μC charge. At t = 0 the switch is thrown closed. Compute the following quantities at t = 2.00 ms: (a) the energy stored in the capacitor, (b) the energy stored in the inductor, and (c) the total energy in the circuit.

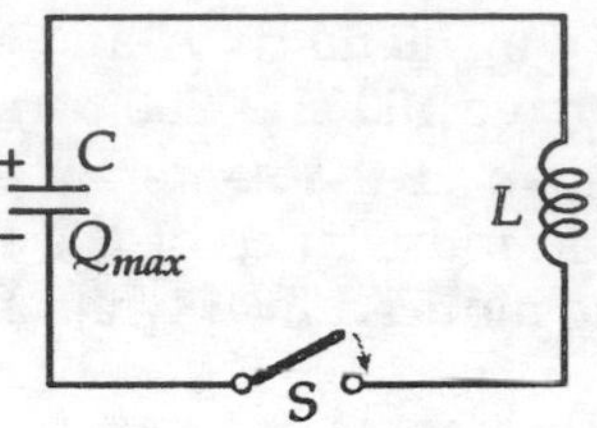

Figure P32.11

Solution At t =0 the capacitor charge is at its maximum value, so $\phi = 0$ in

$$Q = Q_{\max}\cos(\omega t + \phi) = Q_{\max}\cos\left(\frac{t}{\sqrt{LC}}\right)$$

$$= \left(105\times10^{-6}\ \text{C}\right)\cos\left(\frac{2.00\times10^{-3}\ \text{s}}{\sqrt{3.3\ \text{H}\ 840\ 10^{-12}\ \text{F}}}\right)$$

$$= \left(105\times10^{-6}\ \text{C}\right)(\cos 38.0\ \text{rad})$$

$$= 1.01\times10^{-4}\ \text{C}$$

(a) $U_C = \dfrac{Q^2}{2C} = \dfrac{(1.01\times10^{-4}\ \text{C})^2}{2(840\times10^{-12}\ \text{F})} = 6.03\ \text{J}$ ◊

(c) The constant total energy is that originally of the capacitor:

$$U = \frac{Q_{\max}^2}{2C} = \frac{(1.05\times10^{-4}\ \text{C})^2}{2(840\times10^{-12}\ \text{F})} = 6.56\ \text{J} \quad ◊$$

(b) $U_L = 6.56\ \text{J} - 6.03\ \text{J} = 0.529\ \text{J}$ ◊

We could also find this from

$$\tfrac{1}{2}LI^2 = \tfrac{1}{2}L\left(\frac{d}{dt}Q_{\max}\cos\omega t\right)^2 = \tfrac{1}{2}LQ_{\max}^2\omega^2\sin^2\omega t \quad ◊$$

63. In Figure 32.16, let $R = 7.60\ \Omega$, $L = 2.20$ mH, and $C = 1.80\ \mu$F. (a) Calculate the frequency of the damped oscillation of the circuit. (b) What is the critical resistance?

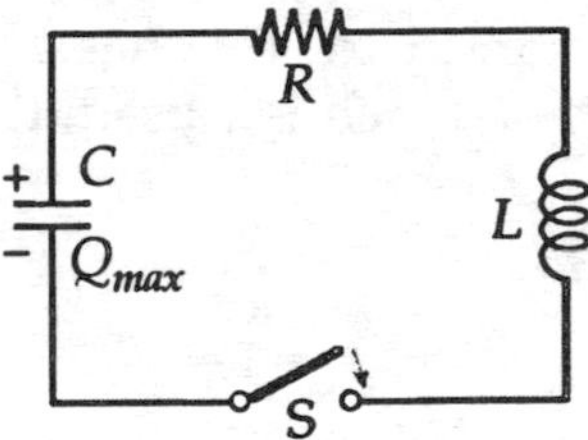

Figure P32.16

Solution

(a) $$\omega_d = \sqrt{\frac{1}{LC} - \left(\frac{R}{2L}\right)^2}$$

$$= \sqrt{\frac{1}{(2.20\times10^{-3})(1.80\times10^{-6})} - \left(\frac{7.6}{2(2.20\times10^{-3})}\right)^2}$$

$$\omega_d = 1.580\times10^4 \text{ rad/s} \qquad \text{and} \qquad f_d = \frac{\omega_d}{2\pi} = 2.51 \text{ kHz} \quad \lozenge$$

(b) $$R_C = \sqrt{\frac{4L}{C}} = \sqrt{\frac{4(2.20\times10^{-3}\text{ V}\cdot\text{s/A})}{(1.80\times10^{-6}\text{ A}\cdot\text{s/V})}} = 69.9 \text{ V/A} = 69.9\ \Omega \quad \lozenge$$

---

65. Consider an LC circuit in which $L = 500$ mH and $C = 0.100\ \mu$F. (a) What is the resonant frequency (ω_0)? (b) If a resistance of 1.00 kΩ is introduced into this circuit, what is the frequency of the (damped) oscillations? (c) What is the percent difference between the two frequencies?

Solution

(a) $$\omega_0 = \frac{1}{\sqrt{LC}} = \frac{1}{\sqrt{(0.50\text{ H})(1.00\times10^{-7}\text{ F})}} = 4.47\times10^3 \text{ rad/s} \quad \lozenge$$

(b) $$\omega_d = \sqrt{\frac{1}{LC} - \left(\frac{R}{2L}\right)^2} = \sqrt{\left(\frac{1}{(0.500\text{ H})(1.00\times10^{-7}\text{ F})}\right) - \left(\frac{1.00\times10^3\ \Omega}{(2)(0.500\text{ H})}\right)^2} = 4.36\times10^3 \text{ rad/s} \quad \lozenge$$

(c) $$\frac{\Delta\omega}{\omega_0} = \frac{4.47 - 4.36}{4.47} = 0.0253$$

Thus, the damped frequency is 2.53% lower than the undamped frequency. $\lozenge$

---

69. An inductor that has a resistance of 0.50 Ω is connected to a 5.0-V battery. One second after the switch is closed, the current in the circuit is 4.0 A. Calculate the inductance.

Solution $I = \frac{\mathcal{E}}{R}(1 - e^{-Rt/L})$

$$L = \frac{Rt}{\ln\left(\frac{\mathcal{E}}{\mathcal{E} - IR}\right)} = \frac{(0.50\ \Omega)(1.0\ \text{s})}{\ln\left(\frac{5.0}{5.0 - (4.0)(0.50)}\right)} = 0.98\ \text{H} \quad \Diamond$$

---

75. Assume that the switch in Figure P32.75 is initially in position 1. Show that if the switch is thrown from position 1 to position 2, all the energy stored in the magnetic field of the inductor is dissipated as thermal energy in the resistor.

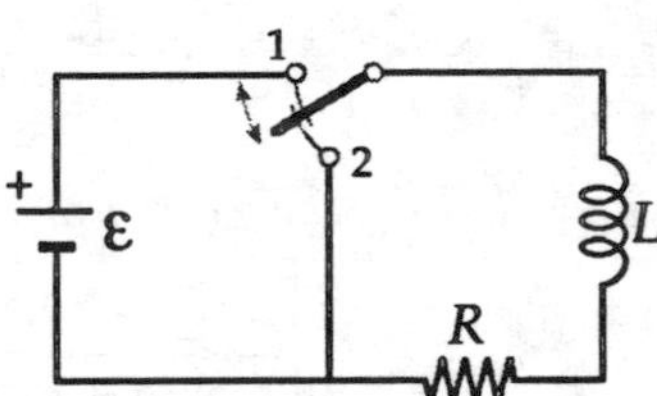

Figure P32.75

Solution If the inductor current has had a long time to grow, it will be nearly $\mathcal{E}/R$, and the stored energy will be

$$U_L = \tfrac{1}{2}LI^2 = \frac{L\mathcal{E}^2}{2R^2}$$

After the switch is thrown, $I = \frac{\mathcal{E}}{R}e^{-Rt/L}$

The heat power of the resistor is $P = \frac{dU}{dt} = I^2 R = \frac{\mathcal{E}^2}{R}e^{-2Rt/L}$

and the total heat generated is $U = \int_{\text{all } t} dU = \int_0^\infty P\,dt = \int_0^\infty \frac{\mathcal{E}^2}{R}e^{-2Rt/L}dt$

Substituting, $U = \frac{\mathcal{E}^2}{R}\left(-\frac{L}{2R}\right)\int_0^\infty e^{-2Rt/L}\left(-\frac{2R\,dt}{L}\right) = -\frac{L\mathcal{E}^2}{2R^2}e^{-2Rt/L}\Big]_0^\infty$

and $U = -\frac{L\mathcal{E}^2}{2R^2}(0 - 1) = +\frac{L\mathcal{E}^2}{2R^2} \quad \Diamond$

This is equal to the original energy of the inductor's magnetic field, as it must be according to energy conservation.

---

77. At $t = 0$, the switch in Figure P32.77 is thrown closed. By using Kirchhoff's rules for the instantaneous currents and voltages in this two-loop circuit, show that the current in the inductor is

$$I(t) = \frac{\mathcal{E}}{R_1}[1 - e^{-(R'/L)t}]$$

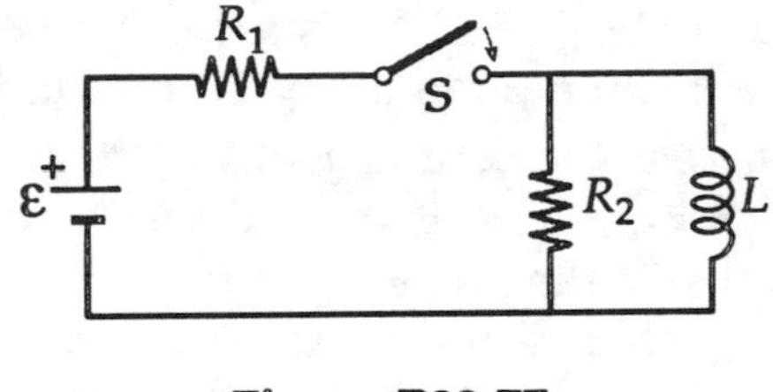

Figure P32.77

where $R' = R_1R_2/(R_1 + R_2)$.

Solution

Call I the downward current through the inductor and I_2 the downward current through R_2. Then $I + I_2$ is the current in R_1.

Left-hand loop: $$\mathcal{E} - (I + I_2)R_1 - I_2R_2 = 0$$

Outside loop: $$\mathcal{E} - (I + I_2)R_1 - L\frac{dI}{dt} = 0$$

Eliminate I_2, obtaining $$I\underbrace{\left(\frac{R_1R_2}{R_1 + R_2}\right)}_{R'} + L\frac{dI}{dt} = \underbrace{\left(\frac{R_2}{R_1 + R_2}\right)\mathcal{E}}_{\mathcal{E}'}$$

Thus, $$\mathcal{E}' - IR' - L\frac{dI}{dt} = 0$$

This is of the same form as Equation 32.6, so the reasoning on page 942 shows that the solution is the same form as Equation 32.7,

$$I = \frac{\mathcal{E}'}{R'}(1 - e^{-R't/L})$$

with $\frac{\mathcal{E}'}{R'} = \frac{\mathcal{E}R_2/(R_1 + R_2)}{R_1R_2/(R_1 + R_2)} = \frac{\mathcal{E}}{R_1}$: $$I(t) = \frac{\mathcal{E}}{R_1}[1 - e^{-(R't/L)}]$$

79. In Figure P32.79, the switch is closed at $t < 0$, and steady-state conditions are established. The switch is now thrown open at $t = 0$. (a) Find the initial voltage $\mathcal{E}_0$ across L just after $t = 0$. Which end of the coil is at the higher potential: a or b? (b) Make freehand graphs of the currents in R_1 and in R_2 as a function of time, treating the steady-state directions as positive. Show values before and after $t = 0$. (c) How long after $t = 0$ is the current in R_2 2.0 mA?

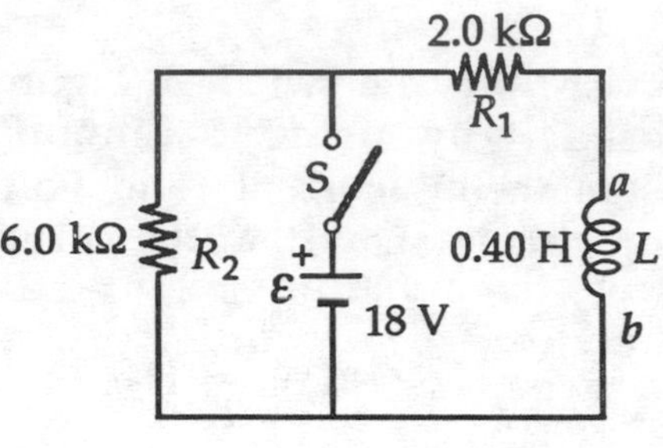

Figure P32.79

Solution Before $t = 0$, the current in R_2 is downward $\dfrac{18\text{ V}}{6\text{ k}\Omega} = 3.0\text{ mA}$

The current is clockwise in R_1 and the coil: $I = \dfrac{18\text{ V}}{2\text{ k}\Omega} = 9.0\text{ mA}$

(a) Just after $t = 0$, the current in the coil is 9.0 mA downward but decreasing. Noting that $\mathcal{E}_0 = V_b - V_a$, the outer loop gives (clockwise):

$$\mathcal{E}_0 - (9.0\text{ mA})(6\text{ k}\Omega) - (9.0\text{ mA})(2\text{ k}\Omega) = 0 \quad \text{and} \quad \mathcal{E}_0 = 72.0\text{ V} \quad \lozenge$$

Thus, point b is at the higher potential. ◊

(b) The currents in R_1 and R_2 are shown below. ◊

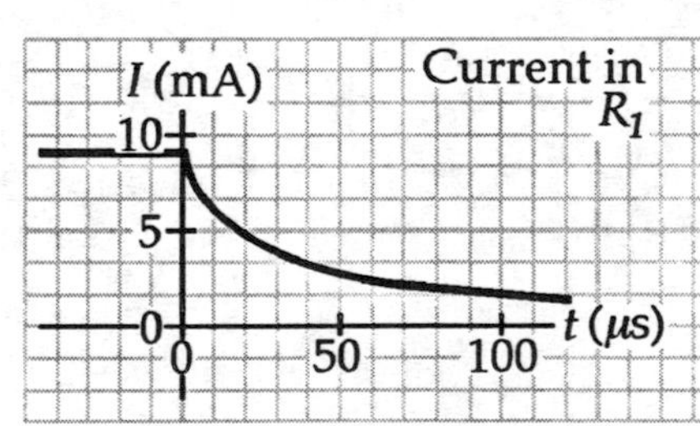

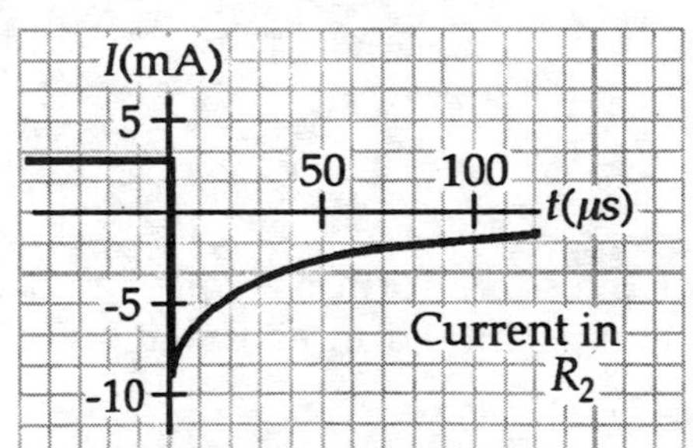

(c) $I = \dfrac{\mathcal{E}}{R}e^{-Rt/L}$ $\qquad 2.0 \times 10^{-3}\text{ A} = \dfrac{18\text{ V}}{2.0\text{ k}\Omega}e^{(-8.0\text{ k}\Omega)t/0.40\text{ H}}$

Resistance R_1 establishes the original value of the current in the outer loop, but the series combination of R_1 and R_2 establishes the decay constant.

$$t = \frac{0.40\text{ H}}{8.0\text{ k}\Omega}\ln 4.5 = 75\ \mu\text{s} \quad \lozenge$$

83. To prevent damage from arcing in an electric motor, a discharge resistor is sometimes placed in parallel with the armature. If the motor is suddenly unplugged while running, this resistor limits the voltage that appears across the armature coils. Consider a 12-V dc motor that has an armature that has a resistance of 7.5 Ω and an inductance of 450 mH. Assume the back emf in the armature coils is 10 V when the motor is running at normal speed. (The equivalent circuit for the armature is shown in Figure P32.83.) Calculate the maximum resistance R that limits the voltage across the armature to 80 V when the motor is unplugged.

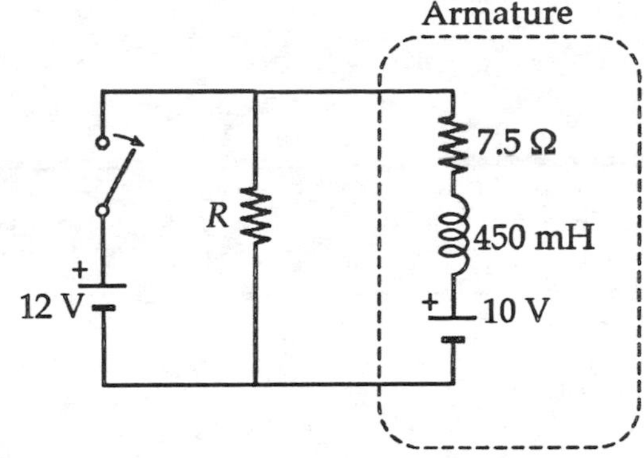

Figure P32.83

Solution The steady-state current when the switch is closed is found from Kirchhoff's loop rule:

$$+12\text{ V} - I(7.5\ \Omega) - 0 - 10\text{ V} = 0 \qquad I = \frac{2.0\text{ V}}{7.5\ \Omega} = 0.267\text{ A}$$

When the switch is opened, mechanical pigheadedness (inertia) keeps the armature spinning for a while, so the 10-V back emf is momentarily still there. Electrical pigheadedness (self-inductance) keeps the current at 0.267 A for an instant.

We require $80\text{ V} = R(0.267\text{ A})$ $\qquad R = 300\ \Omega$ ◊

Note that while the motor is running, this resistor turns

$$P = \frac{(12\text{ V})^2}{300\ \Omega} = 0.48\text{ W}$$

of power into heat (or wastes 0.48 W).

Chapter 33

Alternating Current Circuits

Chapter 33

ALTERNATING CURRENT CIRCUITS

INTRODUCTION

In this chapter, we describe alternating current (ac) currents. We investigate the characteristics of circuits containing familiar elements and driven by a sinusoidal voltage. Our discussion is limited to simple series circuits containing resistors, inductors, and capacitors, and we find that the ac current in each element is also sinusoidal but not necessarily in phase with the applied voltage. We conclude the chapter with two sections concerning the characteristics of *RC* filters, transformers, and power transmission.

NOTES FROM SELECTED CHAPTER SECTIONS

33.1 AC Sources and Phasors

Phasor diagrams are graphical constructions in which alternating quantities such as current and voltage are represented by rotating vectors called *phasors*. The maximum value of a time-varying quantity is represented by the *length* of the phasor; and the instantaneous value is the *projection* of the phasor onto the vertical axis. Phasors rotate counterclockwise with an angular velocity ω.

33.2 Resistors in an AC Circuit

If an ac circuit consists of a generator and a resistor, the current in the circuit is in phase with the voltage. That is, the current and voltage reach their maximum values at the same time. The *average* value of the current over *one complete cycle* is zero. The rms current refers to *root mean square,* which simply means the square root of the average value of the square of the current.

33.3 Inductors in an AC Circuit

If an ac circuit consists of a generator and an inductor, the current *lags behind* the voltage by 90°. That is, the voltage reaches its maximum value one-quarter of a period before the current reaches its peak value.

33.4 Capacitors in an AC Circuit

If an ac circuit consists of a generator and a capacitor, the current *leads* the voltage by 90°. That is, the current reaches its peak value one-quarter of a period before the voltage reaches its peak value.

33.5 The *RLC* Series Circuit

The ac current at all points in a series ac circuit has the same amplitude and phase. Therefore, the voltage across each element will have *different* amplitudes and phases: the voltage across the resistor is in phase with the current, the voltage across the inductor leads the current by 90°, and the voltage across the capacitor lags behind the current by 90°.

33.6 Power in an AC Circuit

The average power delivered by the generator is dissipated as heat in the resistor. There is no power loss in an ideal inductor or capacitor.

33.7 Resonance in a series *RLC* Circuit

The current in a series *RLC* circuit has its greatest value when the frequency of the generator equals ω_0; that is, when the "driving" frequency matches the resonance frequency.

33.9 The Transformer and Power Transmission

A transformer is a device designed to raise or lower an ac voltage and current without causing an appreciable change in the product IV. In its simplest form, it consists of a primary coil of N_1 turns and a secondary coil of N_2 turns, both wound on a common soft iron core. In an ideal transformer, the power delivered by the generator must equal the power dissipated in the load.

EQUATIONS AND CONCEPTS

A simple series alternating current circuit with a sinusoidal source of emf is shown in Figure 33.1. The rectangle ▭ used here represents the circuit element(s) which, in a particular case, may be

- a resistor R,
- a capacitor C,
- an inductor L, or
- some combination of two or more of the above components

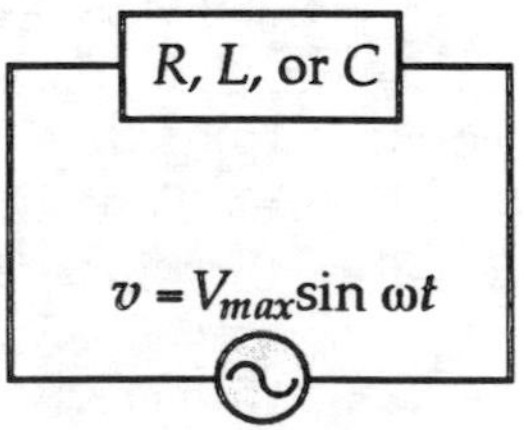

Figure 33.1

The *applied sinusoidal voltage* of frequency ω has a maximum value of V_{max}.

$$v = V_{max} \sin \omega t$$

When the circuit element is a *resistor* of value R, the current and voltage across the resistor are *in phase,* and I_{max} is the maximum current.

$$v_R = V_{max} \sin \omega t \qquad (33.2)$$

$$i_R = I_{max} \sin \omega t \qquad (33.3)$$

When the circuit element is an *inductor* of value L, the *current lags the voltage* across the inductor by 90°.

$$v_L = V_{max} \sin \omega t$$

$$i_L = \frac{V_{max}}{\omega L} \sin\left(\omega t - \frac{\pi}{2}\right) \qquad (33.8)$$

When the circuit element is a *capacitor* of value C, the *current leads the voltage* across the capacitor by 90°.

$$v_C = V_{max} \sin \omega t \qquad (33.12)$$

$$i_C = \omega C V_{max} \sin\left(\omega t + \frac{\pi}{2}\right) \qquad (33.15)$$

The maximum value of the current (or current amplitude) through each element is proportional to the amplitude of the ac voltage across the element. In the case of an inductor and a capacitor, the maximum value of the current depends also on the angular frequency of the source of emf.

$$I_{max} = \begin{cases} \dfrac{V_{max}}{R} & \text{Resistor} \\ \dfrac{V_{max}}{\omega L} = \dfrac{V_{max}}{X_L} & \text{Inductor} \quad (33.9) \\ \omega C V_{max} = \dfrac{V_{max}}{X_C} & \text{Capacitor} \quad (33.16) \end{cases}$$

In the general case, the ac circuit will contain a resistor, inductor, and capacitor in series with a sinusoidally varying voltage source as shown in Figure 33.2.

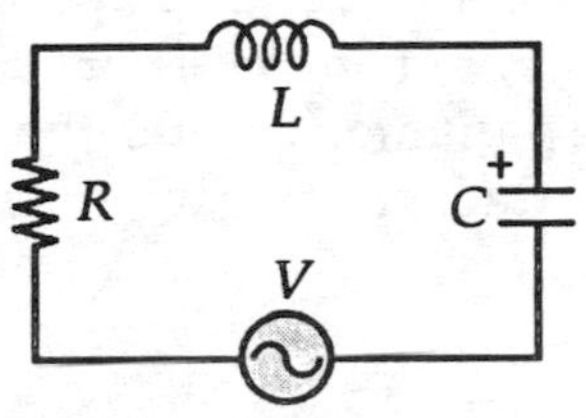

Figure 33.2

Since at any instant, the current has the same value (amplitude and phase) at every point in the circuit, it is convenient to express the phase relationship between the current and *instantaneous voltage* drops across R, L, and C relative to *the common current phase.*

$$v_R = I_{\text{max}} R \sin \omega t = V_R \sin \omega t \tag{33.19}$$

$$v_L = I_{\text{max}} X_L \sin\left(\omega t + \frac{\pi}{2}\right) \tag{33.20}$$

$$= V_L \cos \omega t$$

$$v_C = I_{\text{max}} X_C \sin\left(\omega t - \frac{\pi}{2}\right) \tag{33.21}$$

$$= -V_C \cos \omega t$$

Compare Equations 33.19, 33.20, and 33.21 to Equations 33.3, 33.8, and 33.15. You should convince yourself that *these two sets of equations express the same phase relationship between current and voltage.*

The *maximum voltage* across each circuit element can be written in the form of Ohm's law.

$$V_R = I_{\text{max}} R$$

$$V_L = I_{\text{max}} X_L$$

$$V_C = I_{\text{max}} X_C$$

In the previous set of equations, R is of course the resistance while X_L and X_C represent the *inductive reactance* and the *capacitive reactance* respectively. The reactances are frequency dependent. The inductive reactance increases with increasing frequency, while the capacitive reactance decreases with increasing frequency.

$$X_L = \omega L \tag{33.10}$$

$$X_C = \frac{1}{\omega C} \tag{33.17}$$

where $\omega = 2\pi f$

The *maximum current* in the circuit depends on the angular frequency ω of the source of emf, as well as the values of V_{max}, R, L, and C.

$$I_{max} = \frac{V_{max}}{\sqrt{R^2 + (X_L - X_C)^2}}$$

It is useful to define an operating parameter of the circuit called the *impedance*, Z, defined by Equation 33.23. V_{max} and I_{max} can be related in the form of Ohm's law, Equation 33.24. This requires that Z have the SI unit of ohm (Ω).

$$Z \equiv \sqrt{R^2 + (X_L - X_C)^2} \qquad (33.23)$$

$$V_{max} = I_{max} Z \qquad (33.24)$$

The applied voltage (across the source) and the current in the circuit will differ in phase by some angle ϕ, the *phase angle* of the circuit, given by Equation 33.25.

$$\tan \phi = \frac{X_L - X_C}{R} \qquad (33.25)$$

The *average power* delivered by a generator (source of emf) to an *RLC* series circuit is dissipated as heat in the resistor (there is zero power loss in ideal inductors and capacitors) and is directly proportional to cos ϕ, where ϕ is the phase angle. The quantity cos ϕ is called the *power factor* of the circuit. Since ϕ is frequency dependent (see Equation 33.25), the power factor also depends on frequency.

$$P_{av} = I_{rms} V_{rms} \cos \phi \qquad (33.29)$$

or

$$P_{av} = I_{rms}^2 R \qquad (33.30)$$

When measuring values of current and voltage in ac circuits, it is customary to use instruments which respond to *root-mean-square* (rms) values of these quantities rather than their maximum or instantaneous values.

$$V_{rms} = \frac{V_{max}}{\sqrt{2}}$$

$$I_{rms} = \frac{I_{max}}{\sqrt{2}}$$

You should notice that V_{rms} and I_{rms} are in the same ratio as V_{max} and I_{max}. Compare this equation to Equation 33.24.

$$V_{rms} = I_{rms}Z$$

From Equation 33.23 you can see that when the inductive reactance X_L equals the capacitive reactance X_C, the impedance of the circuit has its *minimum* value. Under these conditions, $Z = R$, and the current in the circuit will have its *maximum* value.

The condition that $Z = R$ occurs for a characteristic frequency of the circuit called the *resonance frequency*, ω_0, given by Equation 33.33. This is obtained from the condition $X_L = X_C$, where $X_L = \omega L$ and $X_C = \dfrac{1}{\omega C}$.

$$\omega_0 = \frac{1}{\sqrt{LC}} \qquad (33.33)$$

A simple transformer consists of a primary coil of N_1 turns and a secondary coil of N_2 turns wound on a common core. In Equation 33.41, V_1 represents the *voltage across the primary* (input voltage), while V_2 represents the *voltage across the secondary* (output voltage). Likewise, I_1 and I_2 represent the currents in the primary and secondary circuits. In the *ideal* transformer, the ratio of voltages is equal to the ratio of turns, and the ratio of currents is equal to the inverse of the ratio of turns.

$$V_2 = \frac{N_2}{N_1}V_1 \qquad (33.41)$$

$$I_1V_1 = I_2V_2 \qquad (33.42)$$

SUGGESTIONS, SKILLS, AND STRATEGIES

The *phasor diagram* is a very useful technique to use in the analysis of ac *RLC* circuits. In such a diagram, each of the rotating quantities V_R, V_L, V_C, and I_{max} is represented by a separate phasor (rotating vector). A 'phasor' diagram which describes the ac circuit of Figure 33.3(a) is shown in Figure 33.3(b). Each phasor has a length which is proportional to the magnitude of the voltage or current which it represents and rotates counterclockwise about the common origin with an angular frequency which equals the angular frequency of the alternating source, ω. The direction of the phasor which represents the current in the circuit is used as the *reference* direction to establish the correct phase differences among the phasors, which represent the voltage drops across the resistor, inductor, and capacitor. The *instantaneous values* v_R, v_L, v_C, and i are given by the *projection onto the vertical axis of the corresponding phasor.*

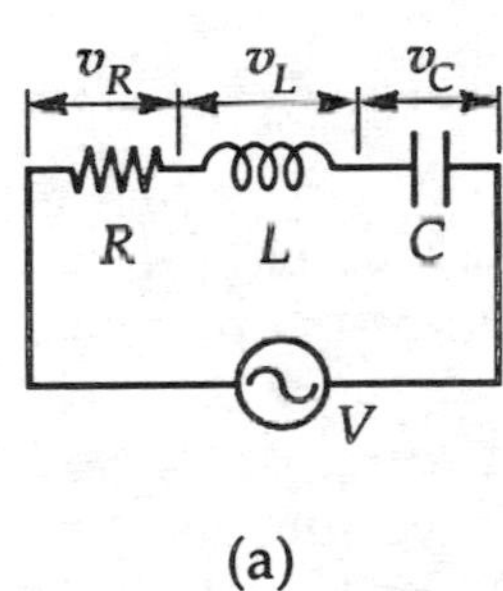

(a)

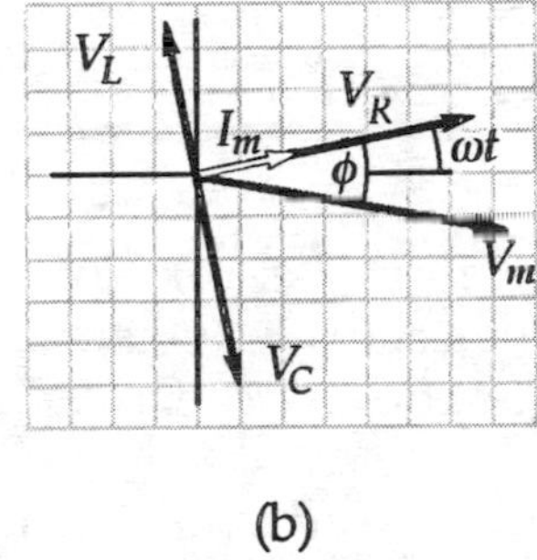

(b)

Figure 33.3

Consider the phasor diagram in Figure 33.3, where the maximum voltage across the resistor, V_R, is greater than the maximum voltage across the inductor, V_L. At the instant shown, the *instantaneous* value of the voltage across the inductor is greater than that across the resistor. Also notice that as time increases and the phasors rotate counterclockwise, maintaining their constant relative phase, V_R, V_L, and V_C (the voltage amplitudes) will remain constant in magnitude but the instantaneous values v_R, v_L, and v_C will vary sinusoidally with time. For the case shown in Figure 33.3(b), the phase angle ϕ is negative (this is because $X_C > X_L$ and therefore $V_C > V_L$); hence, the current in the circuit *leads* the applied voltage in phase.

The maximum voltage across each element in the circuit is the product of I_{max} and the resistance or reactance of that component. It is possible, therefore, to construct an *impedance triangle* for any series circuit as shown in Figure 33.4.

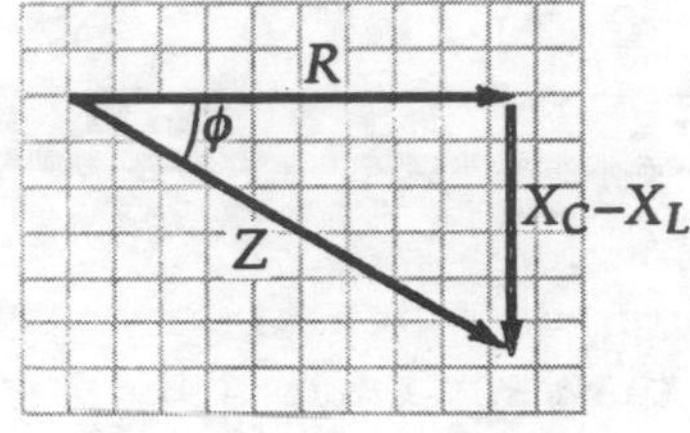

Figure 33.4

The following procedures are recommended when solving alternating current problems:

- The first step in analyzing alternating current circuits is to calculate as many of the unknown quantities such as X_L and X_C as possible. (Note that when calculating X_C, the capacitance should be expressed in farads, rather than, say, microfarads.)

- Apply the equation $V = IZ$ to that portion of the circuit of interest. That is, if you want to know the voltage drop across the combination of an inductor and a resistor, the equation reduces to $V = I\sqrt{R^2 + X_L^2}$.

REVIEW CHECKLIST

▷ Apply the formulas that give the reactance values in an ac circuit as a function of (i) capacitance, (ii) inductance, and (iii) frequency. Interpret the meaning of the terms *phase angle* and *power factor* in an ac circuit.

▷ Given an *RLC* series circuit in which values of resistance, inductance, capacitance, and the characteristics of the generator (source of emf) are known, calculate: (i) the instantaneous and rms voltage drop across each component, (ii) the instantaneous and rms current in the circuit, (iii) the phase angle by which the current leads or lags the voltage, (iv) the power expended in the circuit, and (v) the resonance frequency of the circuit.

▷ Understand the use of phasor diagrams for the description and analysis of ac circuits. Sketch circuit diagrams for high-pass and low-pass filter circuits and make calculations of the ratio of output to input voltage in each case.

▷ Understand the manner in which step-up and step-down transformers are used in the process of transmitting electrical power over large distances; and make calculations of primary to secondary voltage and current ratios for an ideal transformer.

SOLUTIONS TO SELECTED END-OF-CHAPTER PROBLEMS

1. Show that the rms value for the sawtooth voltage shown in Figure P33.1 is $V_{max} / \sqrt{3}$.

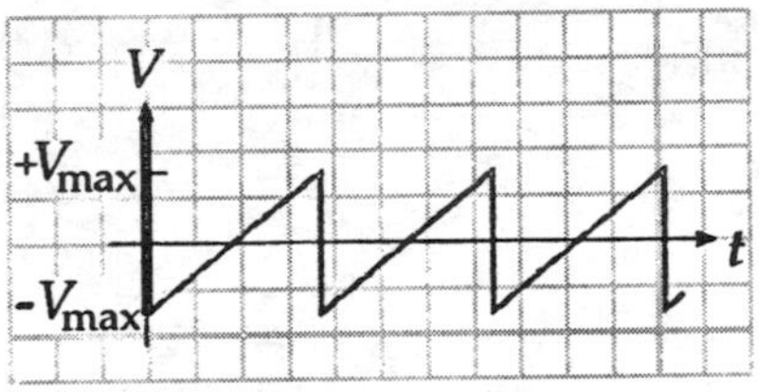

Figure P33.1

Solution Let T represent the period. During the first cycle, the equation of the graph is

$$v = -V_{max} + 2V_{max}t/T$$

The rms value is the effective heating value according to

$$V_{rms}^2 T = \int_0^T v^2\,dt$$

$$= \int_0^T \left(-V_{max} + \frac{2V_{max}t}{T}\right)^2 dt$$

$$= \int_0^T \left(V_{max}^2 - \frac{4V_{max}^2 t}{T} + \frac{4V_{max}^2 t^2}{T^2}\right) dt$$

$$= \left[V_{max}^2 t - \frac{4V_{max}^2 t^2}{2T} + \frac{4V_{max}^2 t^3}{3T^2}\right]_0^T$$

$$= V_{max}^2 T - 2V_{max}^2 T + \tfrac{4}{3}V_{max}^2 T$$

$$= \frac{V_{max}^2 T}{3}$$

Therefore, $$V_{rms} = \frac{V_{max}}{\sqrt{3}}$$ ◊

4A. Figure P33.4 shows three lamps connected to an ac (rms) voltage V. Lamps 1 and 2 have bulbs that each dissipate power P_1 and lamp 3 has a bulb that dissipates power P_2. Find the rms current and resistance of each bulb.

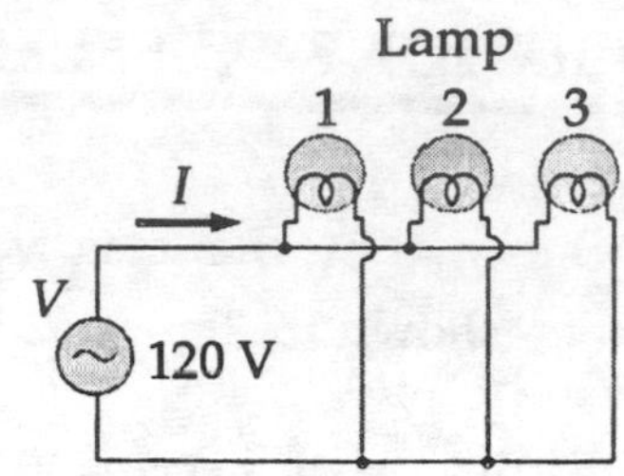

Figure P33.4

Solution $I_1 = I_2 = \frac{P_1}{V}$ ◊ $R_1 = R_2 = \frac{V}{I_1} = \frac{V^2}{P_1}$ ◊

$I_3 = \frac{P_2}{V}$ ◊ $R_3 = \frac{V^2}{P_2}$ ◊

Related Calculation Find the rms current and resistance of each bulb, if the power source is a 120-V ac (rms) household supply voltage, lamps 1 and 2 have 150-W bulbs, and lamp 3 has a 100-W bulb.

Solution Substituting the given values into our previous answers,

$$I_1 = I_2 = \frac{150 \text{ W}}{120 \text{ V}} = 1.25 \text{ A} \quad ◊ \qquad R_1 = R_2 = \frac{(120 \text{ V})^2}{150 \text{ W}} = 96.0\ \Omega \quad ◊$$

$$I_3 = \frac{100 \text{ W}}{120 \text{ V}} = 0.833 \text{ A} \quad ◊ \qquad R_3 = \frac{V^2}{P} = \frac{(120 \text{ V})^2}{100 \text{ W}} = 144\ \Omega \quad ◊$$

---

7. The current in the circuit shown in Figure 33.1 equals 60% of the maximum current at $t = 7.0$ ms, and the applied voltage is $v = V_{\text{max}} \sin \omega t$. What is the smallest frequency of the generator that gives this current?

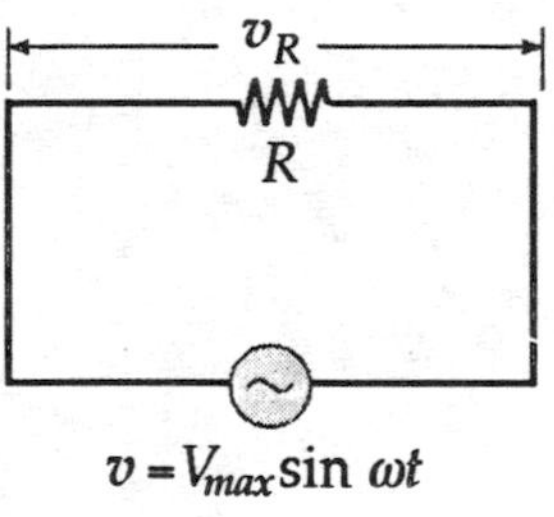

Figure 33.1

Solution $$i = \frac{v}{R} = \left(\frac{V_{\text{max}}}{R}\right) \sin \omega t$$

$$\frac{0.60 V_{\text{max}}}{R} = \left(\frac{V_{\text{max}}}{R}\right) \sin[\omega (7.0 \text{ ms})]$$

$$0.60 = \sin[\omega (7.0 \text{ ms})]$$

$$\omega (7.0 \text{ ms}) = 36.9° = 0.644 \text{ rad}$$

Therefore, $\omega = 91.9$ rad/s and $f = \frac{\omega}{2\pi} = 14.6$ cycle/s ◊

---

9. In a purely inductive ac circuit, as in Figure 33.4, V_{max} = 100 V. (a) If the maximum current is 7.5 A at 50 Hz, calculate the inductance L. (b) At what angular frequency ω is the maximum current 2.5 A?

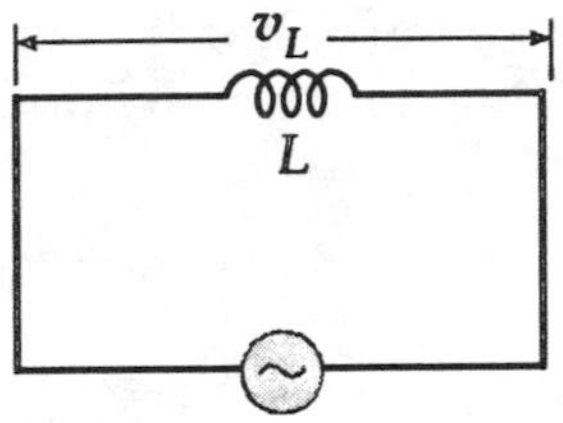

v_L

L

$v = V_{max} \sin \omega t$

Figure 33.4

Solution $\quad I_{max} = \dfrac{V_{max}}{X_L}$

(a) $$X_L = \frac{V_{max}}{I_{max}} = \frac{100\text{ V}}{7.5\text{ A}} = 13.3\ \Omega = \omega L$$

$$L = \frac{13.3\ \Omega}{\omega} = \frac{13.3\ \Omega}{2\pi\left(50\text{ s}^{-1}\right)} = 42.4\text{ mH} \quad \Diamond$$

(b) $$X_L = \frac{V_{max}}{I_{max}} = \frac{100\text{ V}}{2.5\text{ A}} = 40\ \Omega = \omega L$$

$$\omega = \frac{40\ \Omega}{42.4\text{ mH}} = 94\text{ rad/s} \quad \Diamond$$

A three-times-higher frequency makes the inductive reactance three times larger.

---

13. For the circuit shown in Figure 33.4 (above), V_{max} = 80.0 V, $\omega = 65\pi$ rad/s, and L = 70.0 mH. Calculate the current in the inductor at t = 15.5 ms.

Solution

$$X_L = \omega L = \left(65\pi\text{ s}^{-1}\right)\left(70.0\times10^{-3}\ \frac{\text{V}\cdot\text{s}}{\text{A}}\right) = 14.3\ \Omega$$

$$I_{max} = \frac{V_{max}}{X_L} = \frac{80.0\text{ V}}{14.3\ \Omega} = 5.60\text{ A}$$

$$I = -I_{max}\cos\omega t$$

$$= -(5.60\text{ A})\cos\left[\left(65\pi\text{ s}^{-1}\right)(0.0155\text{ s})\right]$$

$$= -(5.60\text{ A})\cos(3.17\text{ rad}) = +5.60\text{ A} \quad \Diamond$$

---

19. A 98.0-pF capacitor is connected to a 60.0-Hz power supply that produces a 20.0-V rms voltage. What is the maximum charge that appears on either of the capacitor plates?

Solution $$V_{max} = V_{rms}\sqrt{2} = (20.0 \text{ V})\sqrt{2} = 28.3 \text{ V}$$

$$Q_{max} = CV_{max} = \left(98.0 \times 10^{-12} \frac{\text{C}}{\text{V}}\right)(28.3 \text{ V}) = 2.77 \text{ nC} \quad \lozenge$$

---

21. What maximum current is delivered by an ac generator with V_{max} = 48 V and f = 90 Hz when connected across a 3.7-μF capacitor?

Solution $$I_{max} = \frac{V_{max}}{X_C} = V_{max}\omega C = V_{max}(2\pi f C)$$

$$I_{max} = (48 \text{ V})(2\pi)(90 \text{ Hz})(3.70 \times 10^{-6} \text{ F}) = 0.100 \text{ A} = 100 \text{ mA} \quad \lozenge$$

---

27. An *RLC* circuit consists of a 150-Ω resistor, a 21-μF capacitor, and a 460-mH inductor, connected in series with a 120-V, 60-Hz power supply. (a) What is the phase angle between the current and the applied voltage? (b) Which reaches its maximum earlier, the current or the voltage?

Solution The reactance of the inductor is $X_L = \omega L = 2\pi f L = 2\pi(60 \text{ s}^{-1})(0.460 \text{ H}) = 173 \ \Omega$

The reactance of the capacitor is $$X_C = \frac{1}{\omega C} = \frac{1}{2\pi f C} = \frac{1}{2\pi\left(60 \text{ s}^{-1}\right)\left(21 \times 10^{-6} \text{ F}\right)} = 126 \ \Omega$$

(a) $$\tan\phi = \frac{X_L - X_C}{R} = \frac{173 \ \Omega - 126 \ \Omega}{150 \ \Omega} = 0.314$$

$$\phi = 0.314 \text{ rad} = 18° \quad \lozenge$$

(b) Since $X_L > X_C$, ϕ is positive, so V_{app} leads the current. ◊

---

33. An inductor (L = 400 mH), a capacitor (C = 4.43 μF), and a resistor (R = 500 Ω) are connected in series. A 50.0-Hz ac generator produces a peak current of 250 mA in the circuit. (a) Calculate the required peak voltage V_{max}. (b) Determine the angle by which the current leads or lags the applied voltage.

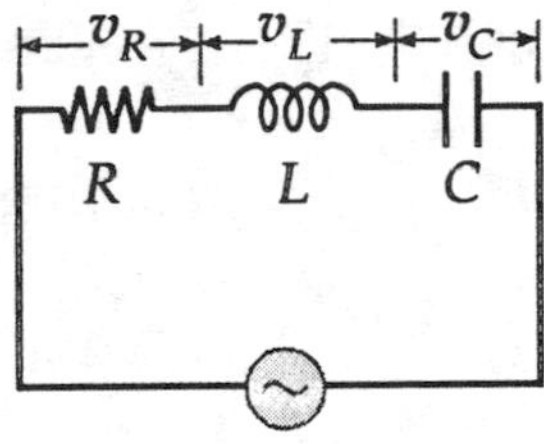

Solution

(a) We first find the impedence of the capacitor and the inductor:

$$X_L = \omega L = 2\pi(50.0\text{ Hz})(400 \times 10^{-3}\text{ H}) = 126\ \Omega$$

and $$X_C = \frac{1}{\omega C} = \frac{1}{(2\pi)(50.0\text{ Hz})(4.43 \times 10^{-6}\text{ F})} = 719\ \Omega$$

Then, we substitute these values into the equation for a series LRC circuit:

$$V_{max} = I_{max}Z - I_{max}\sqrt{R^2 + (X_L - X_C)^2}$$

Thus, $$Z = \sqrt{(500\ \Omega)^2 + (126\ \Omega - 719\ \Omega)^2} = 776\ \Omega$$

and $$V_{max} = I_{max}Z = (0.25\text{ A})(776\ \Omega) = 194\text{ V} \quad \Diamond$$

(b) $$\tan\phi = \frac{X_L - X_C}{R}$$

$$\phi = \tan^{-1}\left(\frac{126 - 719}{500}\right) = -49.9° \quad \Diamond$$

The current *leads* the voltage by 49.9°.

37. An ac voltage of the form $v = (100\text{ V})\sin(1000t)$ is applied to a series *RLC* circuit. If $R = 400\ \Omega$, $C = 5.0\ \mu\text{F}$, and $L = 0.50$ H, find the average power dissipated in the circuit.

Solution Comparing $v = (100\text{ V})\sin(1000t)$ with $v = V_{max}\sin\omega t$,

$$V_{max} = 100\text{ V} \quad \text{and} \quad \omega = 1000\text{ s}^{-1}$$

$$V_{rms} = \frac{100\text{ V}}{\sqrt{2}} = 70.7\text{ V}$$

$$X_C = \frac{1}{\omega C} = \frac{1}{(1000\text{ s}^{-1})(5.00\times10^{-6}\text{ F})} = 200\ \Omega$$

$$X_L = \omega L = \left(1000\text{ s}^{-1}\right)(0.50\text{ H}) = 500\ \Omega$$

$$Z = \sqrt{R^2 + (X_L - X_C)^2} = \sqrt{(400\ \Omega)^2 + (500\ \Omega - 200\ \Omega)^2} = 500\ \Omega$$

$$I_{rms} = \frac{V_{rms}}{Z} = \frac{70.7\text{ V}}{500\ \Omega} = 0.141\text{ A}$$

Only the resistor takes electric energy out of the circuit:

$$P = I_{rms}^2 R = (0.141\text{ A})(400\ \Omega) = 8.0\text{ W} \quad \Diamond$$

---

41. In a certain series *RLC* circuit, $I_{rms} = 9.0$ A, $V_{rms} = 180$ V, and the current leads the voltage by 37°. (a) What is the total resistance of the circuit? (b) Calculate the magnitude of the reactance of the circuit $(X_L - X_C)$.

Solution $P_{av} = I_{rms}V_{rms}\cos\phi = I_{rms}^2 R$

(a) $$R = \frac{V_{rms}\cos\phi}{I_{rms}} = \frac{(180\text{ V})\cos(-37°)}{9.0\text{ A}} = 16.0\ \Omega \quad \Diamond$$

(b) $$\tan\phi = \frac{X_L - X_C}{R}$$

$$X_L - X_C = R\tan\phi = (16.0\ \Omega)\tan(-37°) = -12.0\ \Omega \quad \text{or} \quad |X_L - X_C| = 12\ \Omega \quad \Diamond$$

---

45. An *RLC* circuit is used in a radio to tune into an FM station broadcasting at 99.7 MHz. The resistance in the circuit is 12.0 Ω, and the inductance is 1.40 μH. What capacitance should be used?

Solution The circuit is to be in resonance: $\omega L = \frac{1}{\omega C}$

$$C = \frac{1}{\omega^2 L} = \frac{1}{4\pi^2 f^2 L} = \frac{1}{4\pi^2 (99.7 \text{ MHz})^2 (1.40\ \mu\text{V}\cdot\text{s/A})} = 1.82 \text{ pF} \quad \Diamond$$

---

47. A coil of resistance 35.0 Ω and inductance 20.5 H is in series with a capacitor and a 200-V (rms), 100-Hz source. The rms current in the circuit is 4.00 A. (a) Calculate the capacitance in the circuit. (b) What is V_{rms} across the coil?

Solution We first calculate the impedance:

$$Z = \frac{V_{\text{rms}}}{I_{\text{rms}}} = \frac{200 \text{ V}}{4.00 \text{ A}} = 50.0\ \Omega = \sqrt{R^2 + (X_L - X_C)^2}$$

$$(50.0\ \Omega)^2 = (35.0\ \Omega)^2 + \left(2\pi\,(100 \text{ Hz})(20.5 \text{ H}) - X_C\right)^2$$

$$\pm 35.7\ \Omega = 12880\ \Omega - X_C$$

(a) Either $X_C = 12920\ \Omega = \frac{1}{2\pi(100 \text{ Hz})C}$ and $C = 123 \text{ nF}$ ◊

or $X_C = 12840\ \Omega = \frac{1}{2\pi(100 \text{ Hz})C}$ and $C = 124 \text{ nF}$ ◊

(b) $V_L = I_{\text{rms}} X_L = I_{\text{rms}} 2\pi f L = (4.00 \text{ A})(2\pi)(100 \text{ Hz})(20.5 \text{ V}\cdot\text{s/A}) = 51.5 \text{ kV}$ ◊

---

51. The *RC* high-pass filter shown in Figure 33.16 has a resistance $R = 0.50\ \Omega$. (a) What capacitance gives an output signal that has one-half the amplitude of a 300-Hz input signal? (b) What is the gain (V_{out}/V_{in}) for a 600-Hz signal?

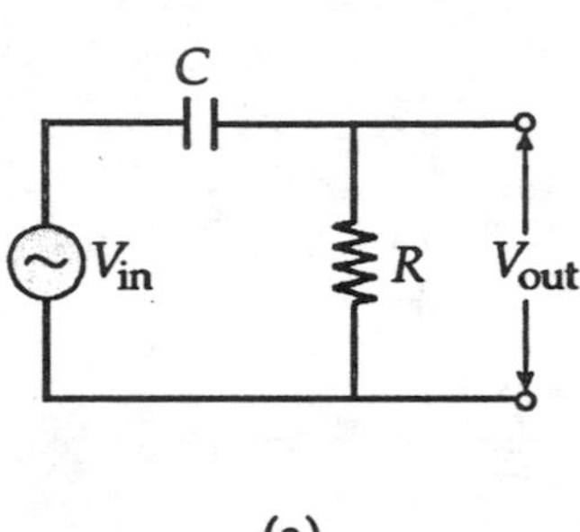

(a)

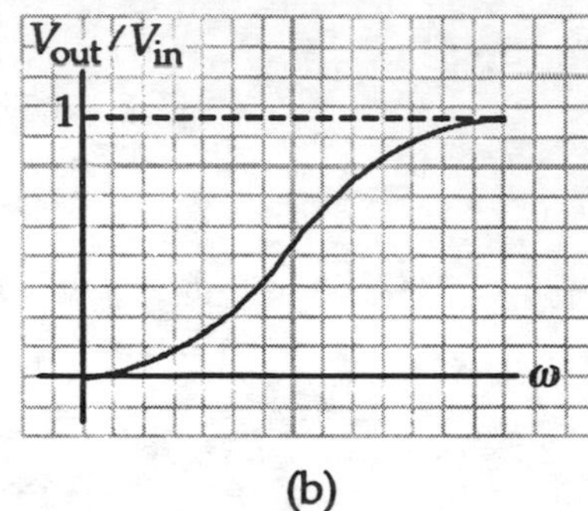

(b)

Figure 33.16

Solution

$$\frac{V_{out}}{V_{in}} = \frac{R}{\sqrt{R^2 + \left(\frac{1}{\omega C}\right)^2}}$$

(a) Solving for C gives

$$C = \frac{1}{\omega R\sqrt{\left(\frac{V_{in}}{V_{out}}\right)^2 - 1}}$$

$$= \frac{1}{(2\pi)(300\ \text{Hz})(0.50\ \Omega)\sqrt{(2)^2 - 1}} = 610\ \mu\text{F} \quad \Diamond$$

(b) Taking $\omega = 2\pi(600)$ rad/s, $R = 0.50\ \Omega$, and $C = 613\ \mu\text{F}$, we find

$$\frac{V_{out}}{V_{in}} = \frac{0.50\ \Omega}{\sqrt{(0.50\ \Omega)^2 + \left(\frac{1}{(1200\pi\ \text{rad/s})(613\ \mu\text{F})}\right)^2}} = 0.76 \quad \Diamond$$

57. Consider a low-pass filter followed by a high-pass filter, as shown in Figure P33.57. If $R = 1000\ \Omega$ and $C = 0.050\ \mu\text{F}$, determine $V_{\text{out}}/V_{\text{in}}$ for a 2.0-kHz input frequency.

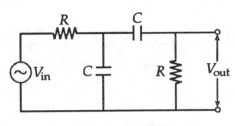

Figure P33.57

Solution Each capacitor has impedance

$$X_C = \frac{1}{\omega C} = \frac{1}{2\pi(2000/\text{s})(5.00\times10^{-8}\ \text{F})} = 1590\ \Omega$$
$\angle -90°$

The high-pass part of the circuit, shown in figures (a) and (b), has an impedance of

$$\sqrt{1590^2 + 1000^2}\angle\tan^{-1}\left(\frac{-1590}{1000}\right) \quad \text{or} \quad (1880\ \Omega)\angle -57.9°.$$

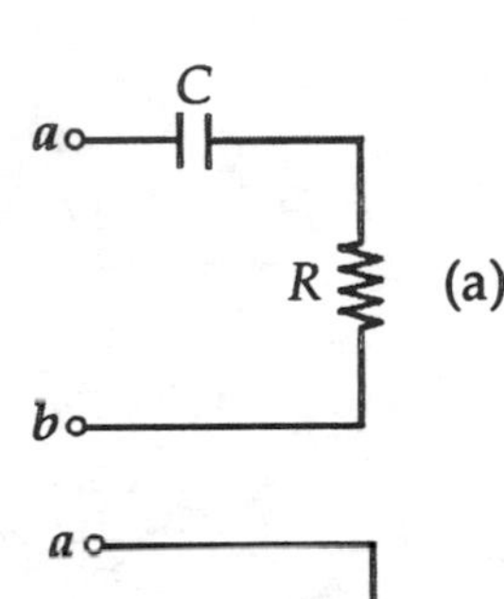

It alone has a voltage ratio of $\dfrac{V_{out}}{V_{ab}} = \dfrac{1000\ \Omega}{1880\ \Omega} = 0.532$

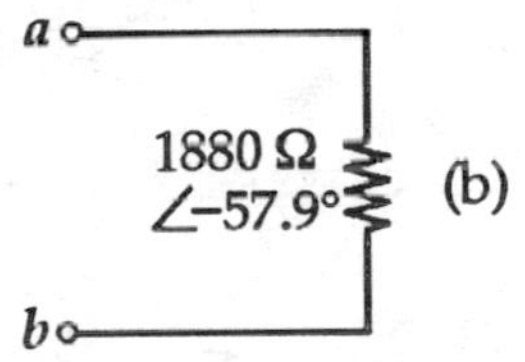

Now, to find the equivalent impedance across the segment in figure (c), we must add the reciprocal impedances

$$\left(6.28\times10^{-4}\ \frac{\text{A}}{\text{V}}\right)\angle 90° \quad + \quad \left(5.32\times10^{-4}\ \frac{\text{A}}{\text{V}}\right)\angle 57.9°$$

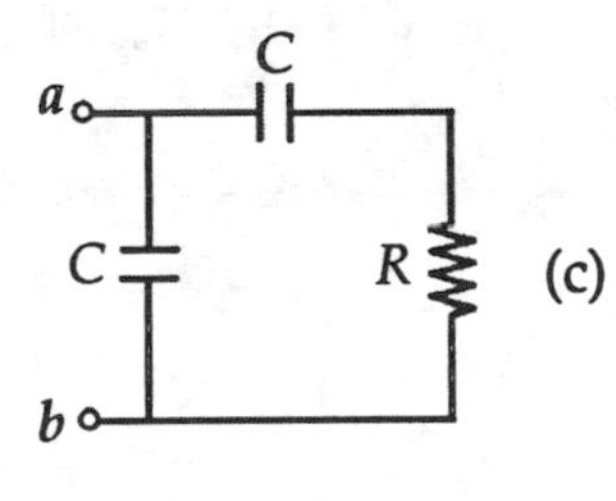

to obtain $$\left(1.115\times10^{-3}\ \frac{\text{A}}{\text{V}}\right)\angle 75.3°,$$

or an impedence of $(897\ \Omega)\angle -75.3°$, as in figure (d).

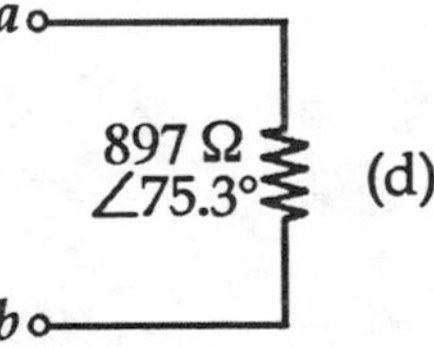

Last, the impedence seen by the input voltage is

$$(1000\ \Omega)\angle 0 + (897\ \Omega)\angle -75.3° = (1503\ \Omega)\angle -35.2°$$

The low-pass stage then has $\dfrac{V_{ab}}{V_{\text{in}}} = \dfrac{897\ \Omega}{1503\ \Omega} = 0.597$

and $$\frac{V_{\text{out}}}{V_{\text{in}}} = \left(\frac{V_{\text{out}}}{V_{ab}}\right)\left(\frac{V_{ab}}{V_{\text{in}}}\right) = (0.532)(0.597) = 0.317 \quad \Diamond$$

59. A transformer has $N_1 = 350$ turns and $N_2 = 2000$ turns. If the input voltage is $v(t) = (170\text{ V})\cos \omega t$, what rms voltage is developed across the secondary coil?

Solution

$$V_{1,\text{rms}} = \frac{170\text{ V}}{\sqrt{2}} = 120\text{ V}$$

$$V_2 = \frac{N_2}{N_1}V_1 = \frac{2000}{350}\,120\text{ V} = 687\text{ V} \quad \Diamond$$

65. A series *RLC* circuit consists of an 8.00-Ω resistor, a 5.00-μF capacitor, and a 50.0-mH inductor. A variable frequency source of amplitude 400 V (rms) is applied across the combination. Determine the power delivered to the circuit when the frequency is equal to one half the resonance frequency.

Solution The resonance frequency is

$$f = \frac{1}{2\pi\sqrt{LC}} = \frac{1}{2\pi\sqrt{(0.0500\text{ H})(5.00\times10^{-6}\text{ F})}} = 318\text{ Hz}$$

Operating at 159 Hz, we have

$$X_L = 2\pi f L = 2\pi(159\text{ Hz})(0.0500\text{ H}) = 50.0\ \Omega$$

$$X_C = \frac{1}{2\pi f C} = \frac{1}{2\pi(159\text{ Hz})(5.00\times10^{-6}\text{ F})} = 200\ \Omega$$

$$Z = \sqrt{R^2 + (X_L - X_C)^2} = \sqrt{8^2 + (50-200)^2}\ \Omega = 150\ \Omega$$

$$I = \frac{V}{Z} = \frac{400\text{ V}}{150\ \Omega} = 2.66\text{ A}$$

The power put in by the source is equal to the power taken out by the resistor:

$$I^2R = (2.66\text{ A})^2(8.00\ \Omega) = 56.7\text{ W} \quad \Diamond$$

69. As a way of determining the inductance of a coil used in a research project, a student first connects the coil to a 12-V battery and measures a current of 0.63 A. The student then connects the coil to a 24-V (rms), 60-Hz generator and measures an rms current of 0.57 A. What is the inductance?

Solution The battery sees only the resistance of the coil:

$$R = \frac{12\ \text{V}}{0.63\ \text{A}} = 19.0\ \Omega$$

With the alternating voltage,

$$Z = \frac{24\ \text{V}}{0.57\ \text{A}} = 42.1\ \Omega = \sqrt{R^2 + X_L{}^2}$$

$$X_L = \sqrt{42.1^2 - 19.0^2}\ \Omega = 37.6\ \Omega = 2\pi f L$$

$$L = \frac{37.6\ \Omega}{2\pi(60\ \text{Hz})} = 99.6\ \text{mH} \quad \Diamond$$

---

77. Consider a series *RLC* circuit having the following circuit parameters: $R = 200\ \Omega$, $L = 663$ mH, and $C = 26.5\ \mu$F. The applied voltage has an amplitude of 50.0 V and a frequency of 60.0 Hz. Find the following amplitudes: (a) The current i, including its phase constant ϕ relative to the applied voltage v; (b) the voltage V_R across the resistor and its phase relative to the current; (c) the voltage V_C across the capacitor and its phase relative to the current; and (d) the voltage V_L across the inductor and its phase relative to the current.

Solution We identify that

$$R = 200\ \Omega,\ L = 663\ \text{mH},\ C = 26.5\ \mu\text{F},\ \omega = 377\ \text{rad/s, and } V_{\text{max}} = 50\ \text{V}$$

So

$$\omega L = 250\ \Omega, \quad \text{and} \quad \left(\frac{1}{\omega C}\right) = 100\ \Omega$$

The impedance is

$$Z = \sqrt{R^2 + \left(\omega L - \frac{1}{\omega C}\right)^2} = \sqrt{(200\ \Omega)^2 + (250\ \Omega - 100\ \Omega)^2} = 250\ \Omega$$

(a) $I = \frac{V}{Z} = \frac{50\text{ V}}{250\ \Omega} = 0.200\text{ A}$ ◊

$\phi = \tan^{-1}\left(\frac{X_L - X_C}{R}\right) = 36.8°$ ◊ with V leading I

(b) $V_R = IR = 40.0\text{ V}$ at $\phi = 0°$ ◊

(c) $V_C = IX_C = (0.200\text{ A})(100\ \Omega) = 20.0\text{ V}$ at $\phi = -90.0°$ ◊

(d) $V_L = IX_L = (0.200\text{ A})(250\ \Omega) = 50.0\text{ V}$ at $\phi = 90.0°$ ◊

---

79. *Impedance matching:* A transformer may be used to provide maximum power transfer between two ac circuits that have different impedances. (a) Show that the ratio of turns N_1/N_2 needed to meet this condition is

$$\frac{N_1}{N_2} = \sqrt{\frac{Z_1}{Z_2}}$$

(b) Suppose you want to use a transformer as an impedance-matching device between an audio amplifier that has an output impedance of 8.00 kΩ and a speaker that has an input impedance of 8.00 Ω. What should your N_1/N_2 ratio be?

Solution $\frac{N_1}{N_2} = \frac{V_1}{V_2}$ $Z_1 = \frac{V_1}{I_1}$ $Z_2 = \frac{V_2}{I_2}$

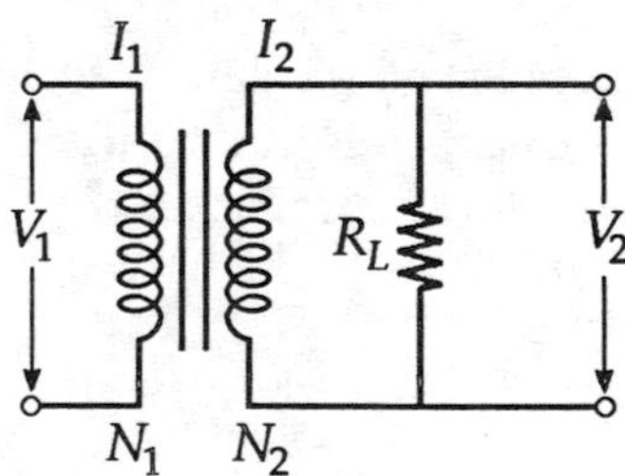

Thus, $\frac{N_1}{N_2} = \frac{Z_1 I_1}{Z_2 I_2}$

(a) Since $\frac{I_1}{I_2} = \frac{N_2}{N_1}$, we find $\frac{N_1}{N_2} = \sqrt{\frac{Z_1}{Z_2}}$ ◊

(b) $\frac{N_1}{N_2} = \sqrt{\frac{8000\ \Omega}{8.00\ \Omega}} = 31.6$ ◊

---

83. Consider the phase-shifter circuit shown in Figure P33.83. The input voltage is described by the expression v = (10 V) sin $200t$ (in SI units). Assuming that L = 500 mH, find (a) the value of R such that the output voltage v_{out} lags behind the input voltage by 30° and (b) the amplitude of the output voltage.

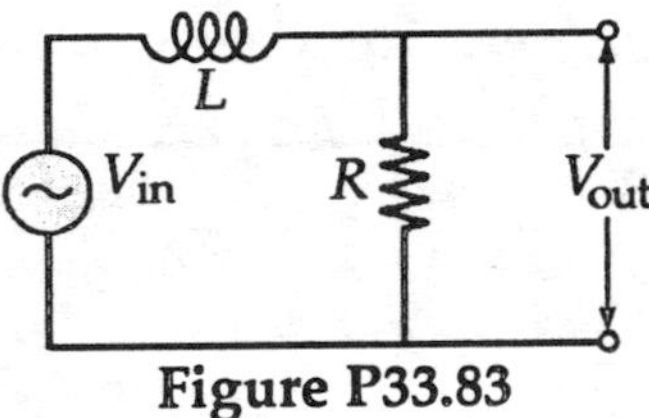

Figure P33.83

Solution

(a) We require $\tan 30° = \dfrac{X_L}{R}$

$$R = \frac{X_L}{\tan 30°} = \frac{\omega L}{\tan 30°} = \frac{(200 \text{ rad/s})(0.5 \text{ H})}{\tan 30°} = 173\ \Omega \quad \Diamond$$

(b) $$\frac{V_{\text{out}}}{V_{\text{in}}} = \frac{IR}{IZ} = \frac{R}{\sqrt{R^2 + X_L{}^2}}$$

$$V_{\text{out}} = 10\text{ V}\,\frac{173\ \Omega}{\sqrt{173^2 + 100^2}\ \Omega} = 8.66\text{ V} \quad \Diamond$$

---

87. A series RLC circuit in which R = 1.00 Ω, L = 1.00 mH, and C = 1.00 nF is connected to an ac generator delivering 1.00 V (rms). Make a careful plot of the power delivered to the circuit as a function of the frequency and verify that the half-width of the resonance peak is $R/2\pi L$.

Solution At resonance, $$\omega = \frac{1}{\sqrt{LC}} = \frac{1}{\sqrt{(1.00 \times 10^{-3}\text{ H})(1.00 \times 10^{-9}\text{ F})}} = 1.00 \times 10^{6}\text{ rad/s}$$

At that point $$Z = R = 1.00\ \Omega, \quad I = \frac{1.00\text{ V}}{1.00\ \Omega} = 1.00\text{ A},$$

and the power is $$I^2R = (1.0\text{ A})^2(1.00\ \Omega) = 1.00\text{ W}.$$

We compute the power at some other angular frequencies. Thus,

ω, 10^6 rad/s	ωL, Ω	$1/\omega C$, Ω	Z, Ω	I, Ω	$I^2R = P$, W
0.9990	999	1001	2.24	0.447	0.19984
0.9994	999.4	1000.6	1.56	0.640	0.40969
0.9995	999.5	1000.5	1.41	0.707	0.49988
0.9996	999.6	1000.4	1.28	0.781	0.60966
0.9998	999.8	1000.2	1.08	0.928	0.86205
1	1000	1000	1	1	1
1.0002	1000.2	999.8	1.08	0.928	0.86209
1.0004	1000.4	999.6	1.28	0.781	0.60985
1.0005	1000.5	999.5	1.41	0.707	0.50012
1.0006	1000.6	999.4	1.56	0.640	0.40998
1.001	1001	999	2.24	0.447	0.20016

The angular frequencies giving half the maximum power are

$$0.9995 \times 10^6 \text{ rad/s}$$

and 1.0005×10^6 rad/s,

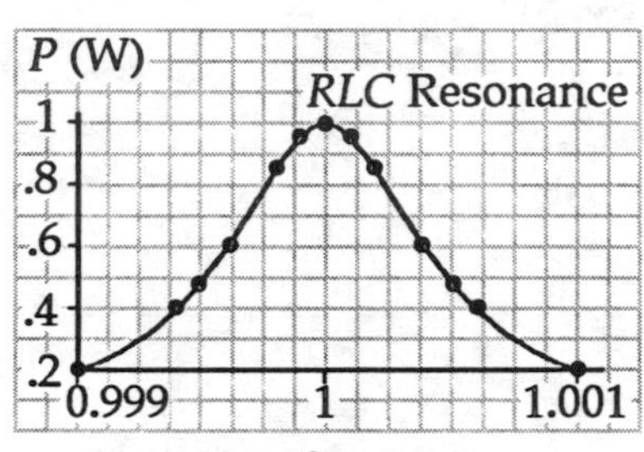

so the full width at half the maximum is

$$\Delta\omega = (1.0005 - 0.9995) \times 10^6 \text{ rad/s}$$

$$= 1.00 \times 10^3 \text{ rad/s}$$

Since $\Delta\omega = 2\pi\Delta f$, $\Delta f = 159$ Hz

and $$\frac{R}{2\pi L} = \frac{1.00\ \Omega}{2\pi\left(1.00 \times 10^{-3}\text{ H}\right)} = 159 \text{ Hz} \quad \Diamond$$

They agree.

Chapter 34

Electromagnetic Waves

ELECTROMAGNETIC WAVES

INTRODUCTION

The waves described in Chapters 16, 17, and 18 are mechanical waves. By definition, mechanical disturbances, such as sound waves, water waves, and waves on a string, require the presence of a medium. This chapter is concerned with the properties of electromagnetic waves that (unlike mechanical waves) can propagate through empty space.

In Section 31.7 we gave a brief description of Maxwell's equations, which form the theoretical basis of all electromagnetic phenomena. The consequences of Maxwell's equations are far-reaching and very dramatic for the history of physics. One of them, the Ampere-Maxwell law, predicts that a time-varying electric field produces a magnetic field just as a time-varying magnetic field produces an electric field (Faraday's law). From this generalization, Maxwell introduced the concept of displacement current, a new source of a magnetic field. Thus, Maxwell's theory provided the final important link between electric and magnetic fields.

NOTES FROM SELECTED CHAPTER SECTIONS

34.1 Maxwell's Equations and Hertz's Discoveries

Electromagnetic waves are generated by accelerating electric charges. The radiated waves consist of oscillating electric and magnetic fields, which are *at right angles to each other* and also *at right angles to the direction of wave propagation.*

The fundamental laws describing the behavior of electric and magnetic fields are Maxwell's equations. In this unified theory of electromagnetism, Maxwell showed that electromagnetic waves are a natural consequence of these fundamental laws.

The theory he developed is based upon the following four pieces of information:

- A charge creates an electric field. Electric fields originate on positive charges and terminate on negative charges. The electric field due to a point charge can be determined at a location by applying Coulomb's force law to a test charge placed at that location.

- Magnetic field lines always form closed loops; that is, they do not begin or end anywhere.

- A varying magnetic field induces an emf and hence an electric field. This is a statement of Faraday's law (Chapter 31).

- A moving charge (constituting a current) creates a magnetic field, as summarized in Ampère's law (Chapter 30).

- A varying electric field creates a magnetic field. This is Maxwell's addition to Ampère's law.

34.2 Plane Electromagnetic Waves

Following is a summary of the properties of electromagnetic waves:

- The solutions of Maxwell's third and fourth equations are wavelike, where both **E** and **B** satisfy the same wave equation.

- Electromagnetic waves travel through empty space with the speed of light,

$$c = \frac{1}{\sqrt{\mu_0 \epsilon_0}}$$

- The electric and magnetic field components of plane electromagnetic waves are perpendicular to each other and also perpendicular to the direction of wave propagation. The latter property can be summarized by saying that electromagnetic waves are transverse waves.

- The magnitudes of **E** and **B** in empty space are related by $E/B = c$.

- Electromagnetic waves obey the principle of superposition.

34.3 Energy Carried by Electromagnetic Waves

The magnitude of the Poynting vector represents the rate at which energy flows through a unit surface area perpendicular to the flow.

For an electromagnetic wave, the instantaneous energy density associated with the magnetic field equals the instantaneous energy density associated with the electric field. Hence, in a given volume, the energy is equally shared by the two fields.

34.4 Momentum and Radiation Pressure

Electromagnetic waves have momentum and exert pressure on surfaces on which they are incident. The pressure exerted by a normally incident wave on a *totally reflecting* surface is *double* that exerted on a surface which *completely absorbs* the incident wave.

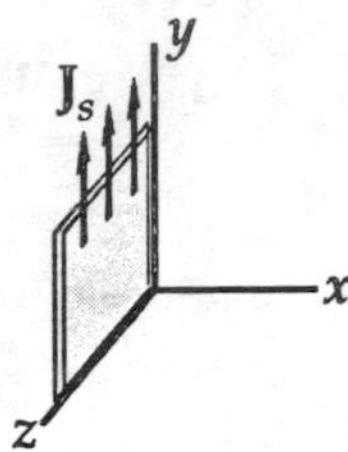

Figure 34.1

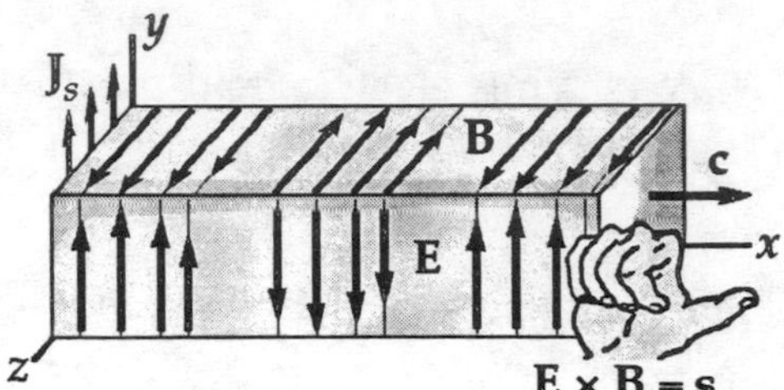

Figure 34.2 Linearly polarized electromagnetic wave due to an oscillating infinite sheet of current.

34.5 Radiation From an Infinite Current Sheet

An *infinite conducting sheet* provides the most convenient geometry for the description of the radiated electric and magnetic fields of a time-varying current. Consider an infinite sheet in the y-z plane (see Figure 34.1) carrying a sinusoidal *surface current per unit length* J_s of magnitude $J_s = J_0 \cos \omega t$.

The direction of **E**, **B**, and **c** are always orthogonal and are shown in Figure 34.2 relative to the direction of the surface current at a particular instant.

During the next half cycle, the directions of $\mathbf{J}_s$, **B**, and **E** will reverse. The direction of **c** will remain along the positive x axis. You should convince yourself that these directions are consistent with those required by the definition of the Poynting vector.

34.6 The Production of Electromagnetic Waves by an Antenna

The fundamental mechanism responsible for radiation by an antenna is the acceleration of a charged particle. Whenever a charged particle undergoes an acceleration, it must radiate energy. An alternating voltage applied to the wires of an antenna forces electric charges in the antenna to oscillate.

EQUATIONS AND CONCEPTS

Maxwell's equations are the fundamental laws governing the behavior of electric and magnetic fields. Electromagnetic waves are a natural consequence of these laws.

$$\oint \mathbf{E} \cdot d\mathbf{A} = \frac{Q}{\epsilon_0} \quad (34.1)$$

$$\oint \mathbf{B} \cdot d\mathbf{A} = 0 \quad (34.2)$$

You should notice that the integrals in Equations 34.1 and 34.2 are *surface integrals* in which the normal components of electric and magnetic fields are integrated over a *closed surface* while Equations 34.3 and 34.4 involve line integrals in which the tangential components of electric and magnetic fields are integrated around a *closed path.*

$$\oint \mathbf{E} \cdot d\mathbf{s} = -\frac{d\phi_B}{dt} \quad (34.3)$$

where $\phi_B = \mathbf{B} \cdot \mathbf{A}$

$$\oint \mathbf{B} \cdot d\mathbf{s} = \mu_0 I + \mu_0 \epsilon_0 \frac{d\phi_E}{dt} \quad (34.4)$$

where $\phi_E = \mathbf{E} \cdot \mathbf{A}$

Both **E** and **B** satisfy a differential equation which has the form of the general wave equation. These are the wave equations for electromagnetic waves in free space (where $Q = 0$ and $I = 0$). As stated here, they represent linearly polarized waves traveling with a speed c.

$$\frac{\partial^2 E}{\partial x^2} = \mu_0 \epsilon_0 \frac{\partial^2 E}{\partial t^2} \quad (34.8)$$

$$\frac{\partial^2 B}{\partial x^2} = \mu_0 \epsilon_0 \frac{\partial^2 B}{\partial t^2} \quad (34.9)$$

$$c = \frac{1}{\sqrt{\mu_0 \epsilon_0}} \quad (34.10)$$

The electric and magnetic fields vary in position and time as *sinusoidal transverse waves*. Their planes of vibration are perpendicular to each other and perpendicular to the direction of propagation.

$$E = E_{max} \cos(kx - \omega t) \quad (34.11)$$

$$B = B_{max} \cos(kx - \omega t) \quad (34.12)$$

The ratio of the magnitude of the electric field to the magnitude of the magnetic field is constant and equal to the speed of light c.

$$\frac{E_{max}}{B_{max}} = \frac{E}{B} = c \tag{34.13}$$

The *Poynting vector* **S** describes the energy flow associated with an electromagnetic wave. The direction of **S** is along the direction of propagation and the magnitude of **S** is the rate at which electromagnetic energy crosses a unit surface area perpendicular to the direction of **S**.

$$\mathbf{S} \equiv \frac{1}{\mu_0}\, \mathbf{E} \times \mathbf{B} \tag{34.18}$$

The *wave intensity* is the time average of the magnitude of the Poynting vector. E_{max} and B_{max} are the *maximum values* of the field magnitudes.

$$I = S_{av} = \frac{E_{max}^2}{2\mu_0 c} = \frac{c}{2\mu_0} B_{max}^2 \tag{34.20}$$

The electric and magnetic fields have *equal instantaneous energy densities.*

$$u_B = u_E = \tfrac{1}{2}\,\epsilon_0\, E^2 = \frac{B^2}{2\mu_0}$$

The total instantaneous energy density u is proportional to E^2 and B^2 while the *total average energy density* is proportional to E_{max}^2 and B_{max}^2. The average energy density is also proportional to the wave intensity.

$$u = \epsilon_0\, E^2 = \frac{B^2}{\mu_0}$$

$$u_{av} = \tfrac{1}{2}\,\epsilon_0\, E_{max}^2 = \frac{B_{max}^2}{2\mu_0} \tag{34.21}$$

$$I = S_{av} = c u_{av} \tag{34.22}$$

The *linear momentum p delivered to an absorbing surface* by an electromagnetic wave at normal incidence depends on the fraction of the total energy absorbed.

$$p = \frac{U}{c} \quad \text{(complete absorption)} \qquad (34.23)$$

$$p = \frac{2U}{c} \quad \text{(complete reflection)} \qquad (34.25)$$

An absorbing surface (at normal incidence) will experience a *radiation pressure* P which depends on the magnitude of the Poynting vector and the degree of absorption.

$$P = \frac{S}{c} \qquad (34.24)$$

$$P = \frac{2S}{c} \qquad (34.26)$$

The *magnetic field* due to an infinite current sheet in the y-z plane is in the x-z plane and varies in a sinusoidal fashion according to Equation 34.27. Note that the electromagnetic wave associated with this sheet of current propagates along the x axis as a linearly polarized wave. For small values of x, the magnetic field is independent of x.

$$B_z = -\frac{\mu_0 J_0}{2}\cos(kx - \omega t) \qquad (34.27)$$

$$B_z = -\frac{\mu_0}{2} J_0 \cos \omega t \quad (\text{for } x \cong 0)$$

A *radiated electric field* vibrates in the y-z plane according to Equation 34.28 and has the same space and time variations as the accompanying magnetic field.

$$E_y = -\frac{\mu_0 J_0 c}{2}\cos(kx - \omega t) \qquad (34.28)$$

There is an outgoing electromagnetic wave on each side of the infinite sheet. In each direction, the *rate of energy emission per unit area* (average intensity) is equal to the average of the Poynting vector.

$$S_{av} = \frac{\mu_0 J_0{}^2 c}{8} \qquad (34.30)$$

REVIEW CHECKLIST

▷ Describe the essential features of the apparatus and procedure used by Hertz in his experiments leading to the discovery and understanding of the source and nature of electromagnetic waves.

▷ For a properly described plane electromagnetic wave, calculate the values for the Poynting vector (magnitude), wave intensity, and instantaneous and average energy densities.

▷ Calculate the radiation pressure on a surface and the linear momentum delivered to a surface by an electromagnetic wave.

▷ Using the geometry of the infinite current sheet as an example, describe the relative directions and the space and time dependencies of the radiated electric and magnetic fields.

▷ Understand the production of electromagnetic waves and radiation of energy by an oscillating dipole. Use a diagram to show the relative directions for **E**, **B**, and **S** and account for the intensity of the radiated wave at points near the dipole and at distant points.

SOLUTIONS TO SELECTED END-OF-CHAPTER PROBLEMS

5. The magnetic field amplitude of an electromagnetic wave is 5.4×10^{-7} T. Calculate the electric field amplitude if the wave is traveling (a) in free space and (b) in a medium in which the speed of the wave is 0.8*c*.

Solution The fields in an electromagnetic wave are related by $\frac{E}{B} = c$.

(a) Substituting,

$$\frac{E}{5.4 \times 10^{-7}} = 3.0 \times 10^{8}$$

$$E = 160 \text{ V/m} \quad \Diamond$$

(b) If the speed of the wave is 0.8*c*,

$$\frac{E}{5.4 \times 10^{-7}} = 0.80\ (3.0 \times 10^{8})$$

and

$$E = 130 \text{ V/m} \quad \Diamond$$

---

9. Figure 34.3 shows a plane electromagnetic sinusoidal wave propagating in the x direction. The wavelength is 50.0 m, and the electric field vibrates in the xy plane with an amplitude of 22.0 V/m. Calculate (a) the sinusoidal frequency and (b) the magnitude and direction of **B** when the electric field has its maximum value in the negative y direction. (c) Write an expression for B in the form

$$B = B_{max}\cos(kx - \omega t)$$

with numerical values for B_{max}, k, and ω.

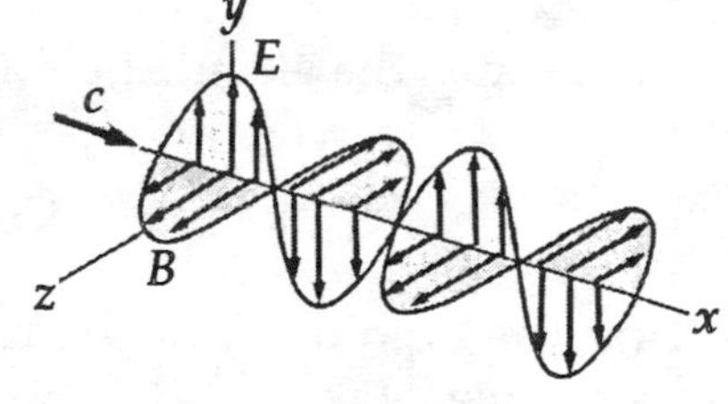

Figure P34.3

Solution

(a) $c = f\lambda$

$$f = \frac{c}{\lambda} = \frac{3.00\times10^{8}\text{ m/s}}{50.0\text{ m}} = 6.00\times10^{6}\text{ Hz} \quad \Diamond$$

(b) $c = \dfrac{E}{B}$

$$B = \frac{E}{c} = \frac{22.0\text{ V/m}}{3.00\times10^{8}\text{ m/s}} = 7.33\times10^{-8}\text{ T} = 73.3\text{ nT} \quad \Diamond$$

B is directed along *negative z direction* when **E** is in the negative y direction; therefore, $\mathbf{S} = \mathbf{E}\times\mathbf{B}/\mu_0$ will propagate in the direction of $(-\mathbf{j})\times(-\mathbf{k}) = +\mathbf{i}$.

(c) $B = B_{max}\cos(kx - \omega t)$

$$k = \frac{2\pi}{\lambda} \quad \text{and} \quad \omega = 2\pi f$$

Thus, $$B = B_{max}\cos\left[\left(\frac{2\pi}{\lambda}\right)x - (2\pi f)t\right]$$

$$B = (73.3\text{ nT})\cos\left[2\pi\left(\frac{x}{50.0} - 6.00\times10^{6}t\right)\right] \quad \Diamond$$

11. In SI units, the electric field in an electromagnetic wave is described by

$$E_y = 100 \sin(1.00 \times 10^7 x - \omega t)$$

Find (a) the amplitude of the corresponding magnetic wave, (b) the wavelength λ, and (c) the frequency f.

Solution

(a) $B = \dfrac{E}{c} = \dfrac{100 \text{ V/m}}{3.00 \times 10^8 \text{ m/s}} = 3.33 \times 10^{-7} \text{ T}$ ◊

(b) We compare the given wave function with $y = A\sin(kx - \omega t)$

to see that $k = 1.00 \times 10^7 \text{ m}^{-1}$.

Then, $\lambda = \dfrac{2\pi}{k} = \dfrac{2\pi}{1.00 \times 10^7 \text{ m}^{-1}} = 6.28 \times 10^{-7} \text{ m}$ ◊

(c) $f = \dfrac{c}{\lambda} = \dfrac{3.00 \times 10^8 \text{ m/s}}{6.28 \times 10^{-7} \text{ m}} = 4.78 \times 10^{14} \text{ Hz}$ ◊

---

15. What is the average magnitude of the Poynting vector 5.0 miles from a radio transmitter broadcasting isotropically with an average power of 250 kW?

Solution $S_{av} = \dfrac{P}{A} = \dfrac{P}{4\pi r^2}$

In meters, $r = (5.0 \text{ mi})(1609 \text{ m/mi}) = 8045 \text{ m}$

and the magnitude is $S = \dfrac{250 \times 10^3 \text{ W}}{(4\pi)(8.045 \times 10^3 \text{ m})^2} = 3.1 \times 10^{-4} \text{ W/m}^2$ ◊

---

19. The filament of an incandescent lamp has a 150-Ω resistance and carries a direct current of 1.0 A. The filament is 8.0 cm long and 0.90 mm in radius. (a) Calculate the Poynting vector at the surface of the filament. (b) Find the magnitude of the electric and magnetic fields at the surface of the filament.

Solution The rate at which the resistor converts electromagnetic energy into heat is

$$P = I^2R = 150 \text{ W},$$

and the surface area is $A = 2\pi rL = 2\pi(0.90 \times 10^{-3} \text{ m})(0.080 \text{ m}) = 4.52 \times 10^{-4} \text{ m}^2.$

(a) The Poynting vector is $S = \dfrac{P}{A} = 3.32 \times 10^5 \text{ W/m}^2$ ◊ (points radially inward)

(b) $B = \mu_0 \dfrac{I}{2\pi r} = \dfrac{\mu_0(1)}{2\pi(0.90 \times 10^{-3})} = 2.22 \times 10^{-4} \text{ T}$ ◊

$$E = \frac{\Delta V}{\Delta x} = \frac{IR}{L} = \frac{150 \text{ V}}{0.080 \text{ m}} = 1880 \text{ V/m} \quad ◊$$

Note: We could also calculate the Poynting vector from $S = \dfrac{EB}{\mu_0} = 3.32 \times 10^5 \text{ W/m}^2$ ◊

---

25. A radio wave transmits 25 W/m² of power per unit area. A flat surface of area A is perpendicular to the direction of propagation of the wave. Calculate the radiation pressure on it if the surface is a perfect absorber.

Solution For complete absorption, $P = \dfrac{S}{c}$

$$P = \frac{25 \text{ W/m}^2}{3 \times 10^8 \text{ m/s}} = 8.33 \times 10^{-8} \text{ N/m}^2 \quad ◊$$

---

27. A 15-mW helium-neon laser (λ = 632.8 nm) emits a beam of circular cross section whose diameter is 2.0 mm. (a) Find the maximum electric field in the beam. (b) What total energy is contained in a 1.0-m length of the beam? (c) Find the momentum carried by a 1.0-m length of the beam.

Solution The intensity of the light is the average magnitude of the Poynting vector:

$$I = \frac{P}{\pi r^2} = \frac{E_{max}^2}{2\mu_0 c}$$

(a) Therefore, the maximum electric field is

$$E_{max} = \sqrt{\frac{P(2\mu_0 c)}{\pi r^2}} = 1.90 \times 10^3 \text{ N/C} \quad \Diamond$$

(b) The power being 15 mW means that 15 mJ passes through a cross section of the beam in one second. This energy is uniformly spread through a beam length of 3×10^8 m, since that is how far the front end of the energy travels in one second. Thus, the energy in just a one-meter length is

$$\frac{15 \times 10^{-3} \text{ J/s}}{3.00 \times 10^8 \text{ m/s}}(1.0 \text{ m}) = 5.0 \times 10^{-11} \text{ J} \quad \Diamond$$

(c) $$p = \frac{U}{c} = \frac{5.0 \times 10^{-11} \text{ J}}{3.00 \times 10^8 \text{ m/s}} = 1.7 \times 10^{-19} \text{ kg} \cdot \text{m/s} \quad \Diamond$$

---

31. A rectangular surface of dimensions 120 cm × 40 cm is parallel to and 4.4 m from a much larger conducting sheet in which there is a sinusoidally varying surface current that has a maximum value of 10 A/m. (a) Calculate the average power incident on the smaller sheet. (b) What power per unit area is radiated by the larger sheet?

Solution Solving part (b) first, we have

(b) $$S_{av} = \frac{\mu_0 J_0{}^2 c}{8}$$

$$S_{av} = \frac{(4\pi \times 10^{-7} \text{ N/A}^2)(10 \text{ A/m})^2(3 \times 10^8 \text{ m/s})}{8} = 4.7 \text{ kW/m}^2 \quad \Diamond$$

(a) $$P = S_{av}A = \left(\frac{\mu_0 J_0{}^2 c}{8}\right)A$$

$$P = (4.71 \text{ kW/m}^2)(1.2 \text{ m})(0.40 \text{ m}) = 2.26 \times 10^3 \text{ W} = 2.26 \text{ kW} \quad \Diamond$$

---

35. A television set uses a dipole-receiving antenna for VHF channels and a loop antenna for UHF channels. The UHF antenna produces a voltage from the changing magnetic flux through the loop. (a) Using Faraday's law, derive an expression for the amplitude of the voltage that appears in a single-turn circular loop antenna with a radius r. The TV station broadcasts a signal with a frequency f, and the signal has an electric field amplitude E_{max} and magnetic field amplitude B_{max} at the receiving antenna's location. (b) If the electric field in the signal points vertically, what should be the orientation of the loop for best reception?

Solution We guess that the diameter of the loop is smaller than the wavelength, so we can approximate the magnetic field as uniform over the area of the loop while it oscillates in time as $B = B_{max} \cos \omega t$. The induced voltage is

$$\mathcal{E} = -\frac{d\Phi_B}{dt} = -\frac{d}{dt}(BA \cos \theta) = -A \frac{d}{dt}(B_{max} \cos \omega t \cos \theta)$$

$$\mathcal{E} = AB_{max}\omega(\sin \omega t \cos \theta)$$

$$\mathcal{E}(t) = 2\pi f B_{max} A \sin 2\pi f t \cos \theta = 2\pi^2 r^2 f B_{max} \cos \theta \sin 2\pi f t$$

(a) The amplitude of this emf is $\mathcal{E}_{max} = 2\pi^2 r^2 f B_{max} \cos \theta$,
where θ is the angle between the magnetic field and the normal to the loop. ◊

(b) If **E** is vertical, then **B** is horizontal, so the plane of the loop should be vertical, and the plane should point toward the transmitter. This will make $\theta = 0°$, so $\cos \theta$ takes on its maximum value.

---

38. Two radio-transmitting antennas are separated by half the broadcast wavelength and are driven in phase with each other. In which directions are (a) the strongest and (b) the weakest signals radiated?

Solution

(a) Along the perpendicular bisector of the line joining the antennas, you are equally distant from both. They oscillate in phase, so along this line you receive the two signals in phase. They interfere constructively to produce the strongest signal.

(b) Along the extended line joining the sources, the wave from the farther antenna must travel one-half wavelength farther to reach you, so you receive the waves 180° out of phase. They interfere destructively to produce the weakest signal.

---

39. What is the wavelength of an electromagnetic wave in free space that has a frequency of (a) 5.00×10^{19} Hz and (b) 4.00×10^{9} Hz?

Solution

(a) $$\lambda = \frac{c}{f} = \frac{3\times10^{8}\ \text{m/s}}{5\times10^{19}\ /\text{s}} = 6.00\ \text{pm} \quad \Diamond$$

This would be called an x-ray if it were emitted when an inner electron in an atom loses energy, or an electron in a vacuum tube. It would be called a gamma ray if it were radiated by an atomic nucleus.

(b) $$\lambda = \frac{c}{f} = \frac{3\times10^{8}\ \text{m/s}}{4\times10^{9}\ /\text{s}} = 7.50\ \text{cm} \quad \Diamond$$

According to Figure 34.17, this finger-length wave could be called a radio wave or a microwave.

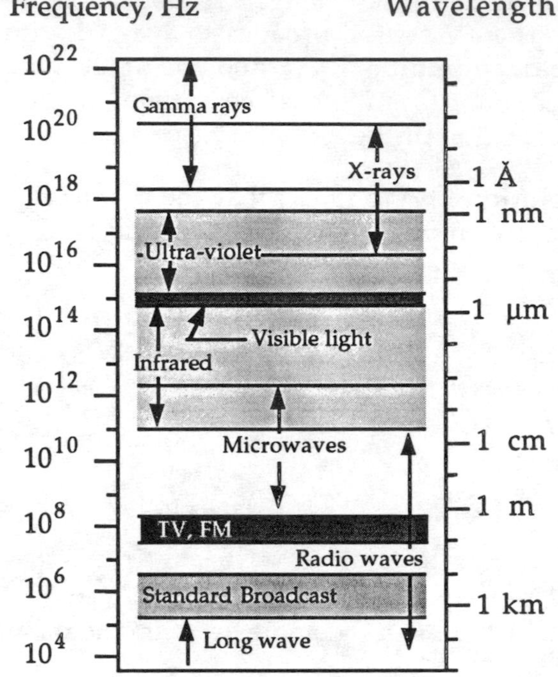

Figure 34.17

---

45. A community plans to build a facility to convert solar radiation to electrical power. They require 1.00 MW of power, and the system to be installed has an efficiency of 30% (that is, 30% of the solar energy incident on the surface is converted to electrical energy). What must be the effective area of a perfectly absorbing surface used in such an installation, assuming a constant energy flux of 1000 W/m²?

Solution At 30% efficiency, $P = 0.30SA$

$$A = \frac{P}{0.30S} = \frac{1\times10^{6}\ \text{W}}{(0.30)(1000\ \text{W}/\text{m}^2)} = 3330\ \text{m}^2 \quad \Diamond$$

This area is approximately 0.75 acres. ◊

---

47. A dish antenna having a diameter of 20 m receives (at normal incidence) a radio signal from a distant source, as shown in Figure P34.47. The radio signal is a continuous sinusoidal wave with amplitude $E_{max} = 0.20\ \mu\text{V/m}$. Assume the antenna absorbs all the radiation that falls on the dish. (a) What is the amplitude of the magnetic field in this wave? (b) What is the intensity of the radiation received by this antenna? (c) What is the power received by the antenna? (d) What force is exerted on the antenna by the radio waves?

Figure P34.47

Solution

(a) $B_{max} = \dfrac{E_{max}}{c} = 6.67 \times 10^{-16}\ \text{T}$ ◊

(b) $S_{av} = \dfrac{E_{max}^2}{2\mu_0 c} = 5.31 \times 10^{-17}\ \text{W/m}^2$ ◊

(c) $P_{av} = S_{av} A = 1.67 \times 10^{-14}\ \text{W}$ ◊

Do not mix up this power with the pressure $P = S/c$, which we use to find the force.

(d) $F = PA = \left(\dfrac{S_{av}}{c}\right) A = 5.56 \times 10^{-23}\ \text{N}$ ◊

(the weight of approximately 3000 Hydrogen atoms!)

---

51. In 1965, Penzias and Wilson discovered the cosmic microwave radiation left over from the Big Bang expansion of the Universe. The energy density of this radiation is $4.0 \times 10^{-14}\ \text{J/m}^3$. Determine the corresponding electric field amplitude.

Solution $u = \frac{1}{2} \epsilon_0 E_{max}^2$ (Eq. 34.21)

$$E_{max} = \sqrt{\frac{2u}{\epsilon_0}} = 95\ \text{mV/m} \quad ◊$$

---

57. A linearly polarized microwave of wavelength 1.50 cm is directed along the positive x axis. The electric field vector has a maximum value of 175 V/m and vibrates in the xy plane. (a) Assume that the magnetic field component of the wave can be written in the form $B = B_{max} \sin(kx - \omega t)$ and give values for B_{max}, k, and ω. Also, determine in which plane the magnetic field vector vibrates. (b) Calculate the magnitude of the Poynting vector for this wave. (c) What maximum radiation pressure would this wave exert if directed at normal incidence onto a perfectly reflecting sheet? (d) What maximum acceleration would be imparted to a 500-g sheet (perfectly reflecting and at normal incidence) of dimensions 1.0 m × 0.75 m?

Solution

(a) $B = B_{max} \sin(kx - \omega t)$

$$B_{max} = \frac{E_{max}}{c} = \frac{175 \text{ V/m}}{3.00 \times 10^{8} \text{ m/s}} = 5.83 \times 10^{-7} \text{ T} \quad \Diamond$$

$$k = \frac{2\pi}{\lambda} = \frac{2\pi}{0.015 \text{ m}} = 419 \text{ m}^{-1} \quad \Diamond$$

$$\omega = kc = (419 \text{ m}^{-1})(3.00 \times 10^{8} \text{ m/s}) = 1.26 \times 10^{11} \text{ rad/s} \quad \Diamond$$

(b) $$S_{av} = \frac{E_{max} B_{max}}{2\mu_0} = \frac{(175 \text{ V/m})(5.83 \times 10^{-7} \text{ T})}{2 \times 4\pi \times 10^{-7} \text{ N/A}^2} = 40.6 \text{ W/m}^2 \quad \Diamond$$

(c) For perfect reflection, $$P_r = \frac{2S}{c} = \frac{(2)(40.6 \text{ W/m}^2)}{3.00 \times 10^{8} \text{ m/s}} = 2.71 \times 10^{-7} \text{ N/m}^2 \quad \Diamond$$

(d) $$a = \frac{F}{m} = \frac{P_r A}{m} = \frac{(2.71 \times 10^{-7} \text{ N/m}^2)(0.75 \text{ m}^2)}{0.500 \text{ kg}} = 4.06 \times 10^{-7} \text{ m/s}^2 \quad \Diamond$$

59. An astronaut, stranded in space "at rest" 10 m from his spacecraft, has a mass (including equipment) of 110 kg. Having a 100-W light source that forms a directed beam, he decides to use the beam as a photon rocket to propel himself continuously toward the spacecraft. (a) Calculate how long it takes him to reach the spacecraft by this method. (b) Suppose, instead, he decides to throw the light source away in a direction opposite the spacecraft. If the mass of the light source is 3.0 kg and, after being thrown, moves at 12 m/s *relative to the recoiling astronaut*, how long does he take to reach the spacecraft?

Solution

(a) The light exerts back on him radiation pressure $P = \frac{F}{A} = \frac{S}{c}$

So, it exerts back on him a force of

$$F = \frac{SA}{c} = \frac{\text{Power}}{c} = \frac{100\text{ J/s}}{3.00 \times 10^8\text{ m/s}} = 3.33 \times 10^{-7}\text{ N} = (110\text{ kg})\,a$$

to give him an acceleration of $a = 3.03 \times 10^{-9}\text{ m/s}^2$

Then he moves according to $x = \frac{1}{2}at^2$

$$t = \sqrt{\frac{2x}{a}} = 8.12 \times 10^4\text{ s} = 22.6\text{ h} \quad \Diamond$$

(b) Momentum is conserved between an original picture of the astronaut at rest; and a final picture of a 107-kg astronaut moving at speed v and a 3-kg flashlight moving in the opposite direction at speed 12 m/s – v relative to the spacecraft. Thus,

$$0 = (107\text{ kg})v - (3.0\text{ kg})(12\text{ m/s} - v)$$

$$0 = (107\text{ kg})v - 36\text{ kg}\cdot\text{m/s} + (3.0\text{ kg})v$$

$$v = \frac{36}{110} = 0.327\text{ m/s}$$

$$t = \frac{10\text{ m}}{0.327\text{ m/s}} = 31\text{ s} \quad \Diamond$$

Chapter 35

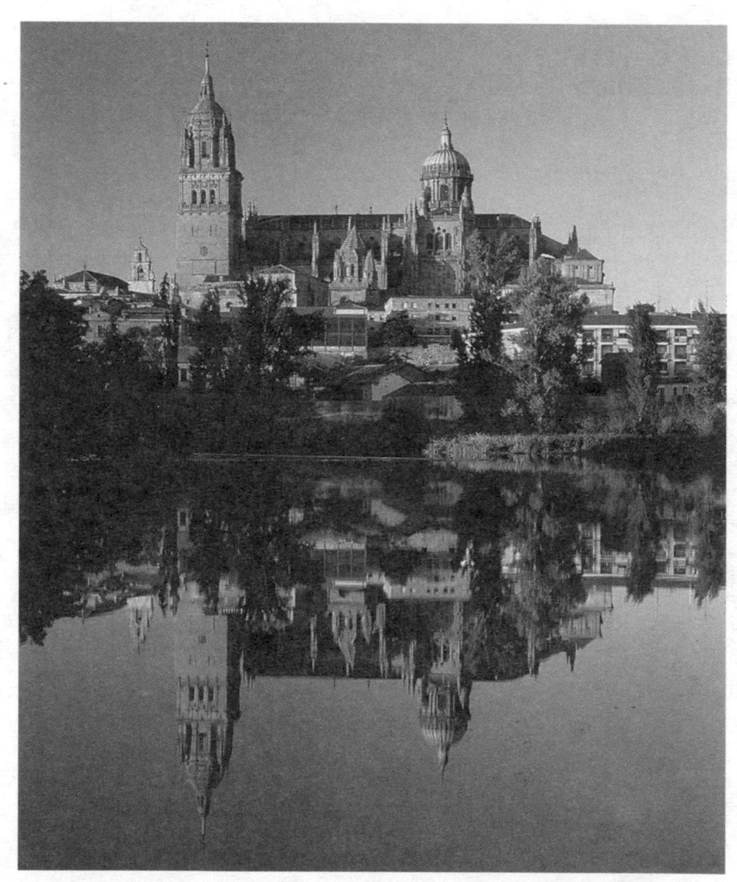

The Nature of Light and the Laws of Geometric Optics

Chapter 35

THE NATURE OF LIGHT AND THE LAWS OF GEOMETRIC OPTICS

INTRODUCTION

The chief architect of the particle theory of light was Newton. With this theory he provided simple explanations of some known experimental facts concerning the nature of light, namely the laws of reflection and refraction.

In 1678 a Dutch physicist and astronomer, Christian Huygens (1629 - 1695), showed that a wave theory of light could also explain the laws of reflection and refraction. The wave theory did not receive immediate acceptance for several reasons. All the waves known at the time (sound, water, and so on) traveled through some sort of medium, but light from the Sun could travel to Earth through empty space. Furthermore, it was argued that if light were some form of wave, it would bend around obstacles; hence, we should be able to see around corners. It is now known that light does indeed bend around the edges of objects. This phenomenon, known as *diffraction*, is not easy to observe because light waves have such short wavelengths. For more than a century most scientists rejected the wave theory and adhered to Newton's particle theory. This was, for the most part, due to Newton's great reputation as a scientist.

NOTES FROM SELECTED CHAPTER SECTIONS

35.3 The Ray Approximation in Geometric Optics

In the ray approximation, it is assumed that a wave travels through a medium along a straight line in the direction of its rays. A ray for a given wave is a straight line perpendicular to the wave front. This approximation also neglects diffraction effects.

35.4 Reflection and Refraction

A line drawn perpendicular to a surface at the point where an incident ray strikes the surface is called the *normal line*. Angles of reflection and refraction are measured relative to the normal.

When an incident ray undergoes partial reflection and partial refraction, the incident, reflected and refracted rays are all *in the same plane*.

The *path of a light ray through a refracting surface is reversible*.

As light travels from one medium into another, the *frequency does not change*.

35.5 Dispersion and Prisms

For a given material, the index of refraction is a function of the wavelength of light passing through the material. This effect is called dispersion. In particular, when light passes through a prism, a given ray is refracted at two surfaces and emerges bent away from its original direction by an angle of deviation, δ. Due to dispersion, δ is different for different wavelengths.

35.6 Huygens' Principle

Every point on a given wave front can be considered as a point source for a *secondary wavelet*. At some later time, the new position of the wave front is determined by the surface tangent to the set of secondary wavelets.

35.7 Total Internal Reflection

Total internal reflection, illustrated in Figure 35.1, is possible only when light rays traveling in one medium are incident on an interface bounding a second medium of *lesser* index of refraction than the first.

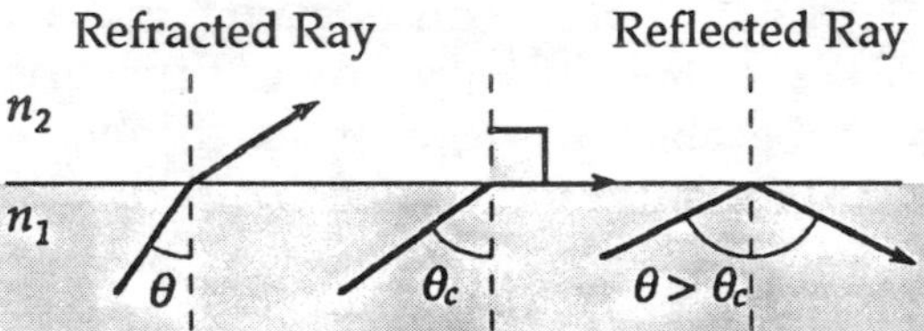

Figure 35.1 Total internal reflection of light occurs at angles of incidence $\theta_1 \geq \theta_c$ where $n_1 > n_2$

35.8 Fermat's Principle

A light ray traveling between any two points will follow a path which requires the *least time*.

EQUATIONS AND CONCEPTS

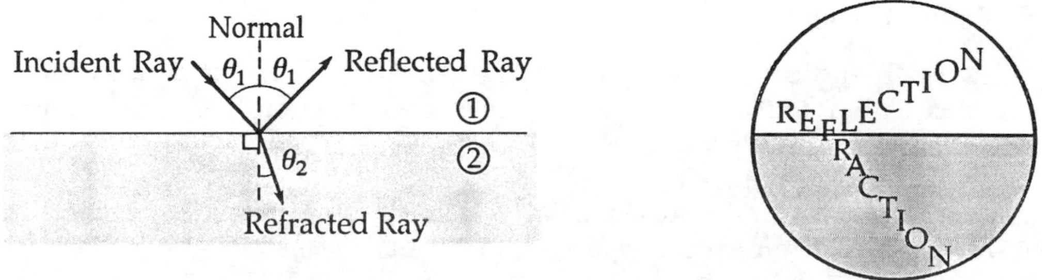

Figure 35.2 The reflection and refraction of light waves at an interface. The direction of the waves are perpendicular to the wave fronts.

Consider the situation in Figure 35.2, in which a light ray is incident obliquely on a smooth, planar surface which forms the boundary between two transparent media of different optical densities. A portion of the ray will be reflected back into the original medium, while the remaining fraction will be transmitted into the second medium.

The energy of a photon is proportional to the frequency of the associated electromagnetic wave.

$$E = hf \qquad (35.1)$$

The ratio between the energy and the frequency is known as Planck's constant.

$$h = 6.63 \times 10^{-34}\ \mathrm{J \cdot s}$$

The law of reflection states that the angle of incidence (the angle measured between the incident ray and the normal line) equals the angle of reflection (the angle measured between the reflected ray and the normal line).

$$\theta_1' = \theta_1 \qquad (35.2)$$

This is one form of the statement of Snell's law. The angle of refraction (measured relative to the normal line) depends on the angle of incidence, and also on the ratio of the speeds of light in the two media on either side of the refracting surface.

$$\frac{\sin\theta_2}{\sin\theta_1} = \frac{v_2}{v_1} = \text{constant} \qquad (35.3)$$

This is the most widely used and most practical form of Snell's law. This equation involves a parameter called the index of refraction, the value of which is characteristic of a particular medium. The index of refraction is defined in Equation 35.4 and Equation 35.7.

$$n_1 \sin\theta_1 = n_2 \sin\theta_2 \qquad (35.8)$$

Each transparent medium is characterized by a dimensionless number, the index of refraction which equals the ratio of the speed of light in the medium to the speed of light in vacuum.

$$n = \frac{c}{v} \tag{35.4}$$

The frequency of a wave is characteristic of the source. Therefore, as light travels from one medium into another of different index of refraction, the frequency remains constant but the wavelength changes. The index of refraction of a given medium can be expressed as the ratio of the wavelength of light in vacuum to the wavelength in that medium.

$$n = \frac{\lambda_0}{\lambda_n} \tag{35.7}$$

For angles of incidence equal to or greater than the critical angle, the incident ray will be totally internally reflected back into the first medium.

$$\sin\theta_c = \frac{n_2}{n_1} \tag{35.10}$$

$$n_1 > n_2$$

Total internal reflection is possible only when a light ray is directed from a medium of high index of refraction into a medium of lower index of refraction.

REVIEW CHECKLIST

- ▷ Understand Huygens' principle and the use of this technique to construct the subsequent position and shape of a given wave front.
- ▷ Describe the methods used by Roemer and Fizeau for the measurement of c and make calculations using sets of typical values for the quantities involved.
- ▷ Determine the directions of the reflected and refracted rays when a light ray is incident obliquely on the interface between two optical media.
- ▷ Understand the manner in which Fermat's principle of least time can be used as a basis of a derivation of the laws of reflection and refraction.
- ▷ Understand the conditions under which total internal reflection can occur in a medium and determine the critical angle for a given pair of adjacent media.

SOLUTIONS TO SELECTED END-OF-CHAPTER PROBLEMS

5. In an experiment to measure the speed of light using the apparatus of Fizeau (Fig. P35.5), the distance between light source and mirror was 11.45 km and the wheel had 720 notches. The experimentally determined value of c was 2.998×10^8 m/s. Calculate the minimum angular speed of the wheel for this experiment.

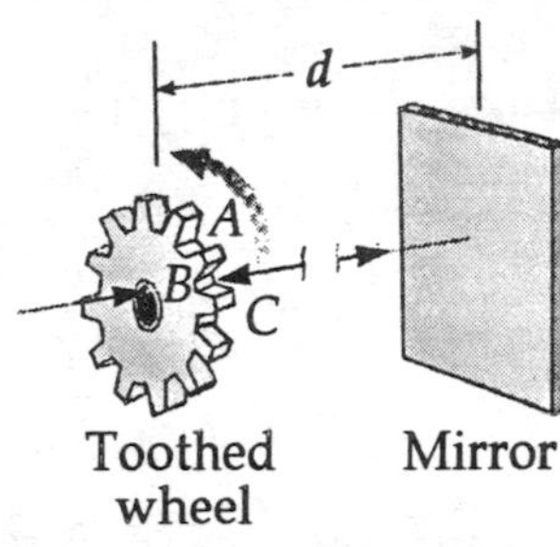

Figure P35.5

Solution The distance down to the mirror and back is $2 \times 11.45 \times 10^3 \text{ m} = 2.290 \times 10^4 \text{ m}$. The round-trip time is expected to be

$$t = \frac{x}{v} = \frac{2.290 \times 10^4 \text{ m}}{2.998 \times 10^8 \text{ m/s}} = 7.638 \times 10^{-5} \text{ s}$$

In this time the wheel should turn by $\frac{1}{2}\left(\frac{1}{720}\right)$ rev, to move a tooth into the place of a notch. Its angular speed should be

$$\omega = \frac{\theta}{t} = \frac{1}{2}\left(\frac{1 \text{ rev}}{720}\right)\left(\frac{1}{7.638 \times 10^{-5} \text{ s}}\right) = 9.091 \text{ rev/s} \quad \Diamond$$

Higher angular speeds must be available so that the experimenter can home in on the special dark setting from both sides. The demonstration is most convincing if the wheel turns at twice this angular speed, to replace a notch with another notch during the light's travel time, restoring the returning light to full brightness.

---

11. An underwater scuba diver sees the Sun at an apparent angle of 45° from the vertical. Where is the Sun?

Solution Refraction happens as sunlight in air crosses into water. The interface is horizontal, so the normal is vertical:

$n_1 \sin\theta_1 = n_2 \sin\theta_2$ gives $\sin\theta_1 = 1.33 \sin(45°)$

$$\sin\theta_1 = (1.33)(0.707) = 0.940$$

The sunlight is at $\theta_1 = 70.5°$ to the vertical, so the Sun is 19.5° above the horizon. $\Diamond$

---

15. A ray of light strikes a flat block of glass (n = 1.50) of thickness 2.00 cm at an angle of 30.0° with the normal. Trace the light beam through the glass, and find the angles of incidence and refraction at each surface.

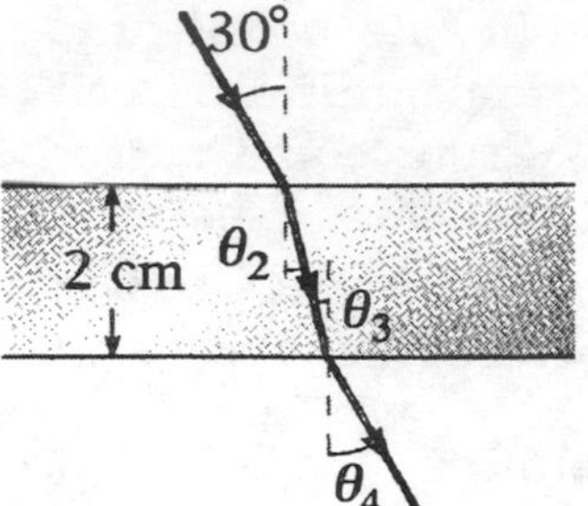

Solution At entry: $n_1 \sin\theta_1 = n_2 \sin\theta_2$

$$1.00 \sin 30° = 1.50 \sin\theta_2$$

and $$\theta_2 = \sin^{-1}\left(\frac{0.500}{1.50}\right) = 19.5° \quad \Diamond$$

To do geometrical optics, you must remember some geometry. The surfaces of entry and exit are parallel so their normals are parallel. Then angle θ_2 of refraction at entry and the angle θ_3 of incidence at exit are alternate interior angles formed by the ray as a transversal cutting parallel lines.

So $$\theta_3 = \theta_2 = 19.5° \quad \Diamond$$

At the exit, $$n_2 \sin\theta_3 = n_1 \sin\theta_4$$

$$1.5 \sin 19.5° = 1 \sin\theta_4 \quad \text{and} \quad \theta_4 = 30.0° \quad \Diamond$$

The exiting ray in air is parallel to the original ray in air. Thus, a car windshield of uniform thickness will not distort, but shows the driver the actual direction to every object outside.

---

29. A prism that has an apex angle of 50.0° is made of cubic zirconia, with n = 2.20. What is its angle of minimum deviation?

Solution From Equation 35.9,

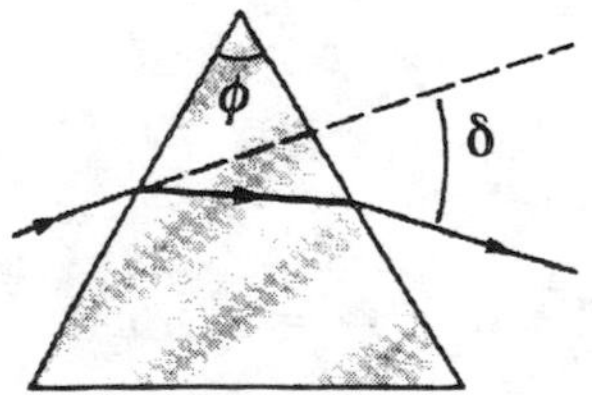

$$n = \frac{\sin\left(\frac{\phi + \delta_{\min}}{2}\right)}{\sin\left(\frac{\phi}{2}\right)}$$

Solving for $\delta_{\min}$, $$\delta_{\min} = 2 \sin^{-1}\left(n \sin\frac{\phi}{2}\right) - \phi$$

$$\delta_{\min} = 2 \sin^{-1}(2.20 \sin 25.0°) - 50.0° = 86.8° \quad \Diamond$$

---

30. A triangular glass prism with apex angle 60.0° has an index of refraction $n = 1.50$. (a) What is the smallest angle of incidence θ_1 for which a light ray can emerge from the other side? (See Figure P35.30.) (b) For what angle of incidence θ_1 does the light ray leave at the same angle θ_1?

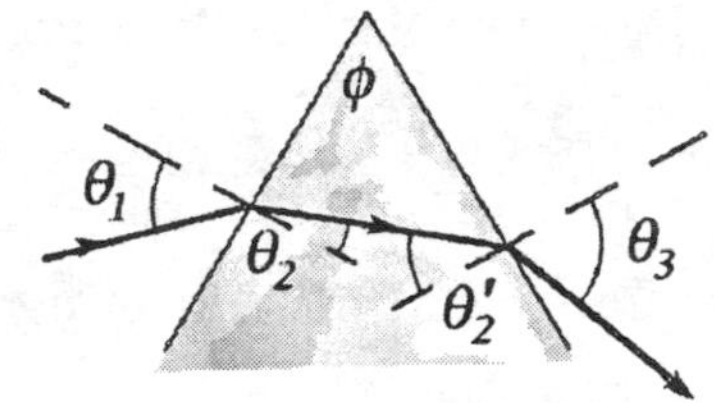

Figure 35.19

Solution

Call the angles of incidence and refraction, at the surfaces of entry and exit, θ_1, θ_2, θ_2', and θ_3, in order as shown. The apex angle (ϕ = 60°) is the angle between the surfaces of entry and exit. The ray in the glass forms a triangle with these surfaces, in which the interior angles must add to 180°.

Thus, $$(90-\theta_2)+\phi+(90-\theta_2')=180$$

Then, $$\theta_2+\theta_2'=\phi$$ (a general rule for light going through prisms)

(a) The photograph on page 1040 of text shows the effect nicely. At the first refraction, $\sin\theta_1 = n\sin\theta_2$. Total internal reflection occurs for

$$1.50\ \sin\theta_2' = 1.00\ \sin\ 90°$$

$$\theta_2' = \sin^{-1}\left(\frac{1}{1.5}\right) = \sin^{-1}(0.667) = 41.8°$$

$$\theta_2 = 60° - \theta_2' = 18.26°$$

Since $\sin\theta_1 = n\sin\theta_2$, $\theta_1 = \sin^{-1}(n\sin\theta_2) = 27.9°$ ◊

(b) When light passes symmetrically through the prism, $\theta_1 = \theta_3$ requires $\theta_2 = \theta_2'$.

With $\theta_2 + \theta_2' = 60°$, $\theta_2 = 30°$.

And since $1.00\sin\theta_1 = 1.50\ \sin\ 30°$, $\theta_1 = 48.6°$. ◊

This symmetric passage is easy to identify experimentally, since it happens to result in minimum deviation of the light.

31. The index of refraction for violet light in silica flint glass is 1.66, and that for red light is 1.62. What is the angular dispersion of visible light passing through a prism of apex angle 60.0° if the angle of incidence is 50.0° (Fig. P35.31a)?

Solution For the incoming ray, $\sin\theta_2 = \dfrac{\sin\theta_1}{n}$

Using the figure to the right,

$$(\theta_2)_{\text{violet}} = \sin^{-1}\left(\frac{\sin 50°}{1.66}\right) = 27.48°$$

$$(\theta_2)_{\text{red}} = \sin^{-1}\left(\frac{\sin 50°}{1.62}\right) = 28.22°$$

θ_1 = incident angle = 50°
θ_2 = angle of refraction
θ_2' = 2nd angle of refraction
θ_3 = angle of transmission

For the outgoing ray,

$$\theta_2' = 60° - \theta_2 \quad \text{and} \quad \sin\theta_3 = n\sin\theta_2'$$

$$(\theta_3)_{\text{violet}} = \sin^{-1}[1.66 \sin 32.52°] = 63.17°$$

$$(\theta_3)_{\text{red}} = \sin^{-1}[1.62 \sin 31.78°] = 58.56°$$

The dispersion is $\Delta\theta_3 = 63.17° - 58.56° = 4.61°$ ◊

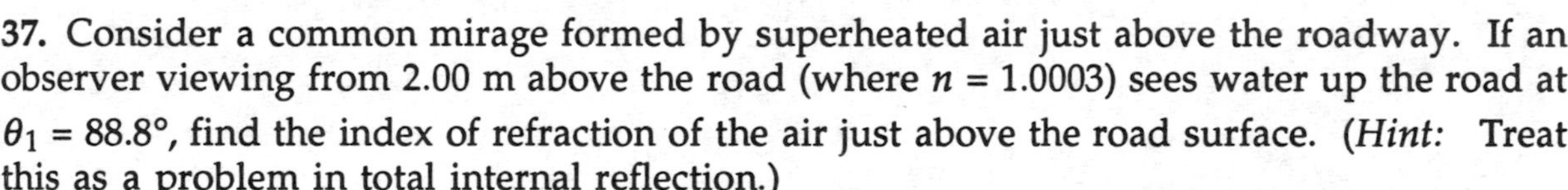

37. Consider a common mirage formed by superheated air just above the roadway. If an observer viewing from 2.00 m above the road (where $n = 1.0003$) sees water up the road at $\theta_1 = 88.8°$, find the index of refraction of the air just above the road surface. (*Hint:* Treat this as a problem in total internal reflection.)

Solution Think of the air as in two discrete layers, the first medium being cooler air with $n_1 = 1.0003$ and the second medium being hot air with a lower index, which reflects light from the sky because refraction is impossible according to $n_1 \sin\theta_1 \geq n_2 \sin 90°$.

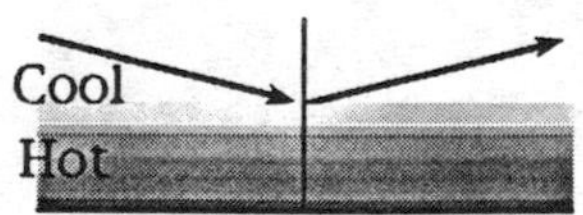

$$1.0003 \sin 88.8° \geq n_2$$

$$n_2 \leq 1.00008 \quad ◊$$

41. A specimen of glass has an index of refraction of 1.61 for the wavelength corresponding to the prominent bright line in the sodium spectrum. If an equiangular prism is made from this glass, what angle of incidence results in minimum deviation of the sodium line?

Solution Equation 35.9 gives the index of refraction in terms of the angle of minimum deviation and the apex angle of the prism:

$$n = \frac{\sin\left(\dfrac{\phi + \delta_{\min}}{2}\right)}{\sin\left(\dfrac{\phi}{2}\right)}$$

This gives $$\delta_{\min} = 2\left[\sin^{-1}\left(n\sin\frac{\phi}{2}\right) - \frac{\phi}{2}\right] = 2\left[\sin^{-1}(1.61\ \sin\pi/6) - \pi/6\right] = 0.824\ \text{rad}$$

It can be shown from the geometry of the arrangement (see Example 35.7 of the text) that

$$\theta = \tfrac{1}{2}(\phi + \delta_{\min}) \quad \text{or} \quad \theta = \tfrac{1}{2}\left(\frac{\pi}{3} + 0.824\right)\text{rad} = 0.674\ \text{rad} = 53.6° \quad \Diamond$$

---

43. A small underwater pool light is 1.00 m below the surface. The light emerging from the water forms a circle on the water surface. What is the diameter of this circle?

Solution At the edge of the circle, the light is totally internally reflected:

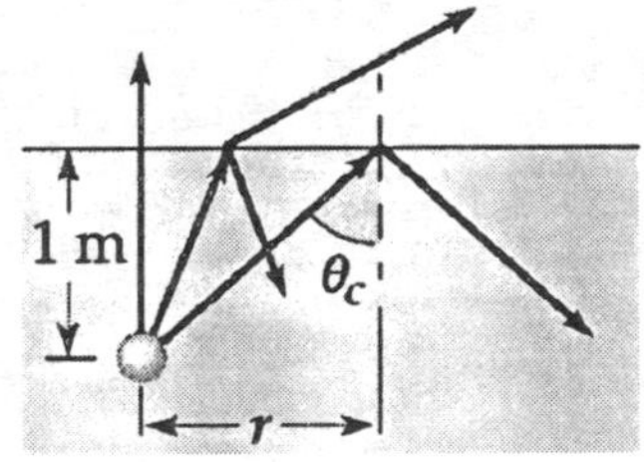

$$n_1 \sin\theta_c = n_2 \sin 90°$$

$$\left(\frac{4}{3}\right)\sin\theta_c = 1$$

$$\theta_c = \sin^{-1}(0.750) = 48.6°$$

The radius then satisfies $$\tan\theta_c = \frac{r}{(1.00\ \text{m})}$$

So the diameter is $$d = 2\ \tan\theta_c = 2\ \tan\ 48.6° = 2.27\ \text{m} \quad \Diamond$$

---

45. A drinking glass is 4.00 cm wide at the bottom, as shown in Figure P35.45. When an observer's eye is placed as shown, the observer sees the edge of the bottom of the glass. When this glass is filled with water, the observer sees the center of the bottom of the glass. Find the height of the glass.

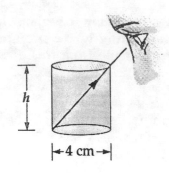

Figure P35.45

Solution $\tan\theta_1 = \dfrac{4}{h}$ and $\tan\theta_2 = \dfrac{2}{h}$

$$\tan^2\theta_1 = 4\ \tan^2\theta_2$$

$$\frac{\sin^2\theta_1}{(1-\sin^2\theta_1)} = \frac{4\ \sin^2\theta_2}{(1-\sin^2\theta_2)} \qquad (1)$$

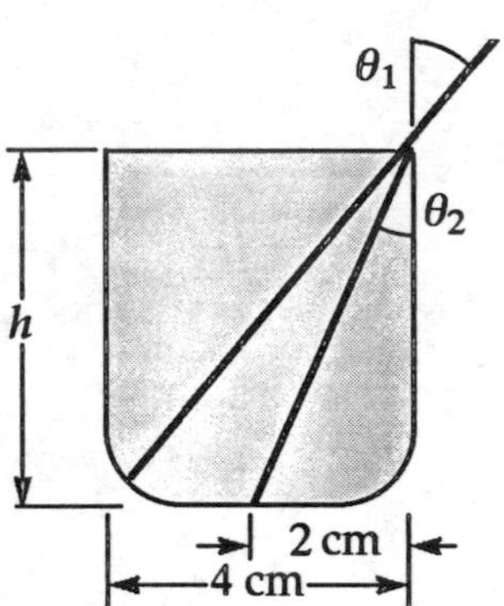

$$n_1\sin\theta_1 = n_2\sin\theta_2 \quad \text{or} \quad \sin\theta_1 = 1.333\ \sin\theta_2$$

Squaring both sides,

$$\sin^2\theta_1 = 1.778\ \sin^2\theta_2$$

Substitute this value into (1) which yields $\dfrac{1.778\ \sin^2\theta_2}{(1-1.778\ \sin^2\theta_2)} = \dfrac{4\ \sin^2\theta_2}{(1-\sin^2\theta_2)}$

Defining $x = \sin^2\theta_2$ and solving for x, $\dfrac{0.444}{(1-1.778x)} = \dfrac{1}{(1-x)}$

$$0.444 - 0.444x = 1 - 1.778x$$

and $x = 0.417 = \sin^2\theta_2$

Thus $\theta_2 = \sin^{-1}\sqrt{0.417} = 40.2°$

and $h = \dfrac{(2.00\text{ cm})}{\tan\theta_2} = \dfrac{(2.00\text{ cm})}{\tan(40.2°)} = 2.37\text{ cm}$ ◊

47. Derive the law of reflection (Eq. 35.2) from Fermat's principle of least time. (See the procedure outlined in Section 35.8 for the derivation of the law of refraction from Fermat's principle.)

Solution To derive the law of *reflection*, locate point O so that the time of travel from point A to point B will be minimum. Remembering that $\sec\theta = 1/\cos\theta$,

The *total* light path is $\quad L = a\sec\theta_1 + b\sec\theta_2$

and the time of travel is $\quad t = \left(\dfrac{1}{v}\right)(a\sec\theta_1 + b\sec\theta_2)$

For the minimum time, $dt = 0$. Therefore, if point O is displaced by dx, then

$$dt = \left(\frac{1}{v}\right)(a\sec\theta_1\tan\theta_1\,d\theta_1 + b\sec\theta_2\tan\theta_2\,d\theta_2) = 0 \qquad (1)$$

Also, $$c + d = a\tan\theta_1 + b\tan\theta_2 = \text{constant}$$

so $$a\sec^2\theta_1\,d\theta_1 + b\sec^2\theta_2\,d\theta_2 = 0 \qquad (2)$$

Combining (1) and (2), we find $\quad \theta_1 = \theta_2 \quad \Diamond$

---

49. A light ray of wavelength 589 nm is incident at an angle θ on the top surface of a block of polystyrene, as shown in Figure P35.49. (a) Find the maximum value of θ for which the refracted ray undergoes total internal reflection at the left vertical face of the block. Repeat the calculation for the case in which the polystyrene block is immersed in (b) water and (c) carbon disulfide.

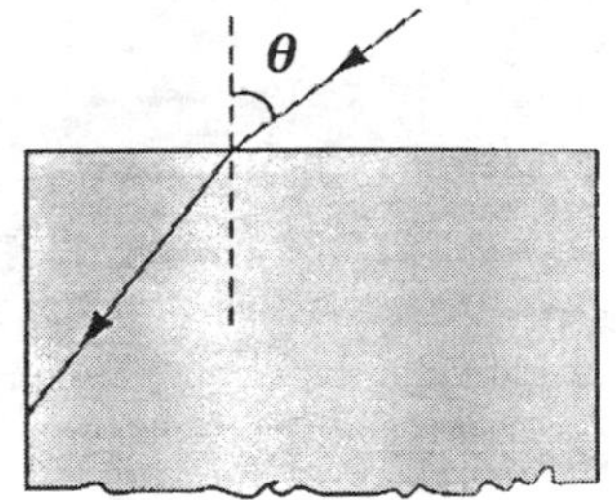

Figure P35.49

Solution For polystyrene *surrounded by air*, internal reflection requires

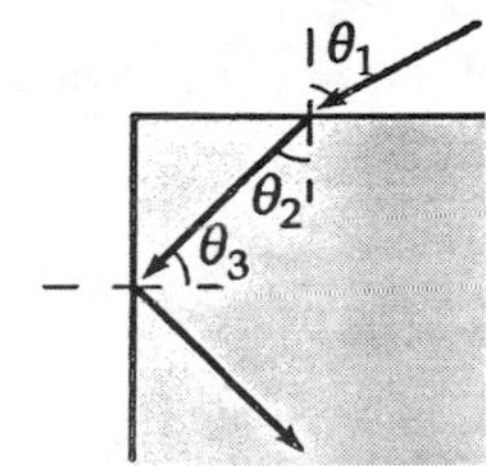

(a) $$\theta_3 = \sin^{-1}\left(\frac{1}{1.49}\right) = 42.2°$$

and then from the geometry, $\quad \theta_2 = 90° - \theta_3 = 47.8°$

From Snell's law, this would require that $\theta_1 > 90°$. Therefore, the real maximum value for θ_1 is 90°: total internal reflection always happens.

(b) For polystyrene *surrounded by water,* we have

$$\theta_3 = \sin^{-1}\left(\frac{1.33}{1.49}\right) = 63.2° \quad \text{and} \quad \theta_2 = 26.8°$$

and from Snell's law $\theta_1 = 30.3°$ ◊

(c) *This is not possible* since the beam is initially traveling in a medium of lower index of refraction.

---

51. A hiker stands on a mountain peak near sunset and observes a rainbow caused by water droplets in the air about 8 km away. The valley is 2 km below the mountain peak and entirely flat. What fraction of the complete circular arc of the rainbow is visible to the hiker?

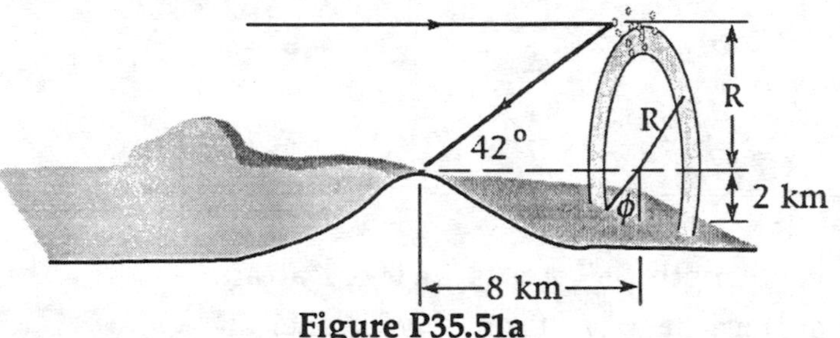

Figure P35.51a

Solution Horizontal light rays from the setting Sun pass above the hiker. The light rays are twice refracted and once reflected, as in Figure 35.30, by just the certain special raindrops at 40° to 42° from the hiker's shadow, and reach the hiker as the rainbow. The hiker sees more of the violet inner edge, so we consider the red outer edge. The radius R of the circle of droplets is

$$R = (8 \text{ km})(\sin 42°) = 5.35 \text{ km}$$

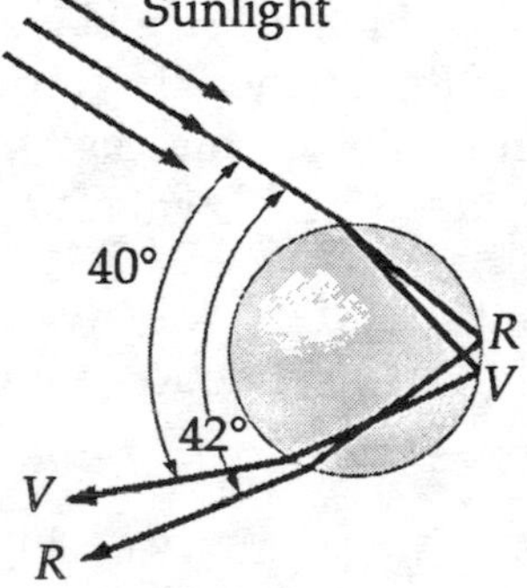

Figure 35.30

Then the angle ϕ, between the vertical and the radius where the bow touches the ground, is given by

$$\cos\phi = \frac{2\text{ km}}{R} = \frac{2\text{ km}}{5.35\text{ km}} = 0.374 \quad \text{or} \quad \phi = 68.1°$$

The angle filled by the visible bow is $360° - (2 \times 68.1°) = 223.8°$, so the visible bow is

$$\frac{223.8°}{360°} = 62.2\% \text{ of a circle} \quad \Diamond$$

This striking view motivated Charles Wilson's 1906 invention of the cloud chamber, a standard tool of nuclear physics. Look for a full-circle rainbow around your shadow when you fly in an airplane.

---

53. A light ray is incident on a prism and refracted at the first surface as shown in Figure P35.53. Let Φ represent the apex angle of the prism and n its index of refraction. Find in terms of n and Φ the smallest allowed value of the angle of incidence at the first surface for which the refracted ray does not undergo internal reflection at the second surface.

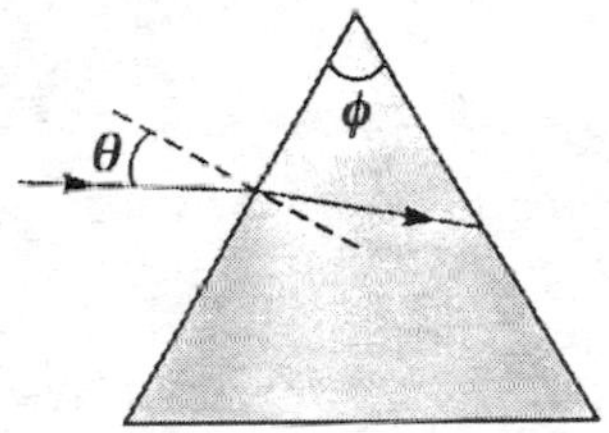

Figure P35.53

Solution Refer to Figure P35.53b.

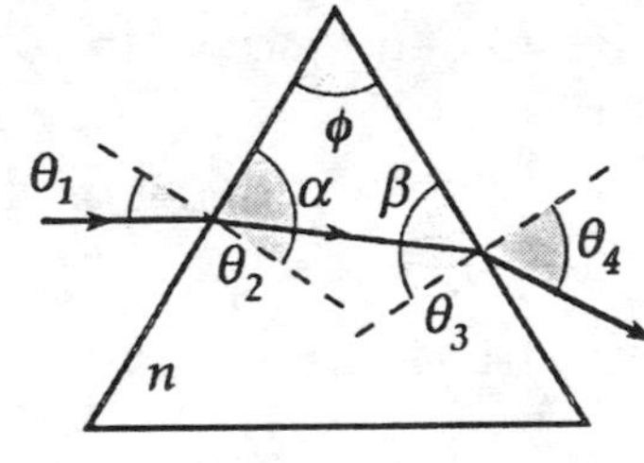

We see that $\theta_2 + \alpha = 90°$ and $\theta_3 + \beta = 90°$

so $\theta_2 + \theta_3 + \alpha + \beta = 180°$

Also, from the figure we see $\alpha + \beta + \phi = 180°$

Therefore, $\phi = \theta_2 + \theta_3$

By applying Snell's law at the first and second refracting surfaces, we find

$$\theta_2 = \sin^{-1}\left(\frac{\sin\theta_1}{n}\right) \quad \text{and} \quad \theta_3 = \sin^{-1}\left(\frac{\sin\theta_4}{n}\right)$$

Substituting these values into the expression for ϕ,

$$\phi = \sin^{-1}\left(\frac{\sin\theta_1}{n}\right) + \sin^{-1}\left(\frac{\sin\theta_4}{n}\right)$$

The limiting condition for internal reflection at the second surface is $\theta_4 \to 90°$. Under these conditions, we have

$$\sin\theta_1 = n\sin\left[\phi - \sin^{-1}\left(\frac{1}{n}\right)\right]$$

or $$\theta_1 = \sin^{-1}\left[\sqrt{n^2 - 1}\sin\phi - \cos\phi\right] \quad \lozenge$$

---

55. A laser beam strikes one end of a slab of material, as shown in Figure P35.55. The index of refraction of the slab is 1.48. Determine the number of internal reflections of the beam before it emerges from the opposite end of the slab.

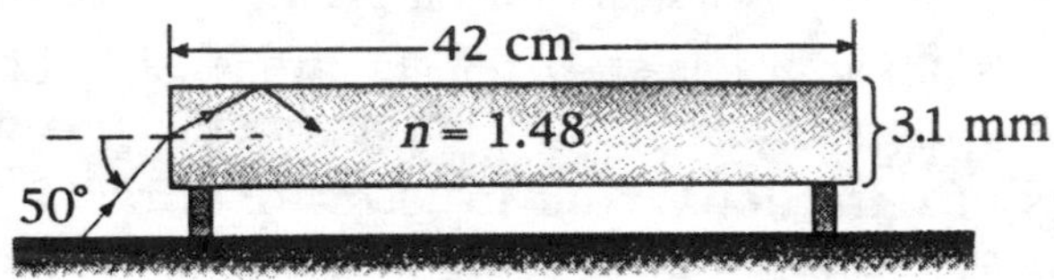

Figure P35.55

Solution On entrance, $\sin 50° = 1.48 \sin \theta_2$

$$\theta_2 = 31.17°$$

The beam strikes the top face at horizontal coordinate $\quad x_1 = \dfrac{1.55 \text{ mm}}{\tan 31.17°} = 2.562 \text{ mm}$

Thereafter, the beam strikes a face every $\quad 2x_1 = 5.124$ mm.

Since the slab is 420 mm long, the beam makes $\dfrac{420 - 2.562}{5.124} = 81$ more reflections;

Therefore, the total is $\quad$ 82 reflections. $\lozenge$

---

57. The light beam in Figure P35.57 strikes surface 2 at the critical angle. Determine the angle of incidence θ_1.

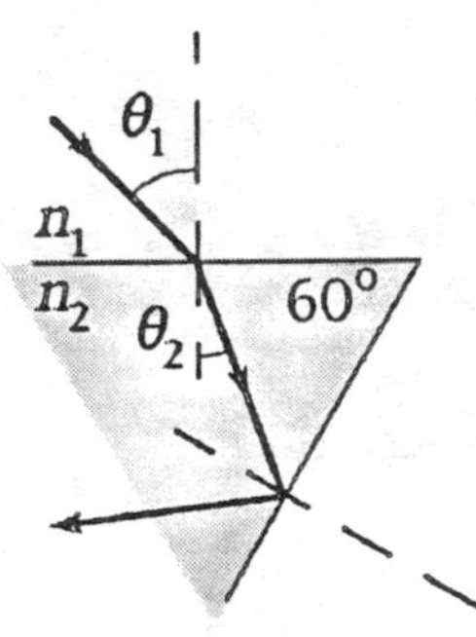

Figure P35.57

Solution Call the index of refraction of the prism material n_2. Use n_1 to represent the surrounding stuff. At surface 2,

$$n_2 \sin 42° = n_1 \sin 90°.$$

So $$\frac{n_2}{n_1} = \frac{1}{\sin 42°} = 1.49$$

Call the angle of refraction θ_2 at the surface 1. The ray inside the prism forms with surfaces 1 and 2 a triangle whose interior angles add to 180°. Thus,

$$(90° - \theta_2) + 60° + (90° - 42°) = 180°$$

$$60° = \theta_2 + 42°$$

$$\theta_2 = 18°$$

At surface 1, $$n_1 \sin\theta_1 = n_2 \sin 18°$$

$$\sin\theta_1 = \left(\frac{n_2}{n_1}\right) \sin 18° \qquad \sin\theta_1 = \frac{\sin 18°}{\sin 42°}$$

$$\theta_1 = 27.5° \quad \Diamond$$

Chapter 36

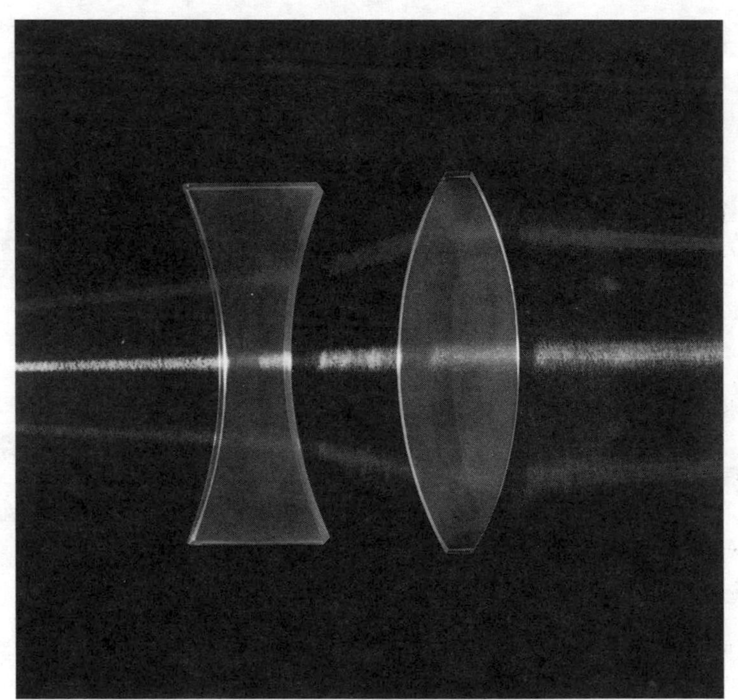

Geometric Optics

Chapter 36

GEOMETRIC OPTICS

INTRODUCTION

This chapter is concerned with the images formed when spherical waves fall on flat and spherical surfaces. We find that images can be formed by reflection or by refraction and that mirrors and lenses work because of this reflection and refraction. Such devices, commonly used in optical instruments and systems, are described in detail. In this chapter, we continue to use the ray approximation and to assume that light travels in straight lines, both steps valid because here we are studying the field called *geometric optics*. In subsequent chapters, we shall concern ourselves with interference and diffraction effects, or the field of *wave optics*.

NOTES FROM SELECTED CHAPTER SECTIONS

36.1 Images Formed by Flat Mirrors

The image formed by a plane mirror has the following properties:

- The image is as far behind the mirror as the object is in front.
- The image is unmagnified, virtual, and upright. (By upright, we mean that, if the object arrow points upward as in Figure 36.2, so does the image arrow.)
- The image has right-left reversal.

36.2 Images Formed by Spherical Mirrors

A spherical mirror is a reflecting surface which has the shape of a segment of a sphere.

Real images are formed at a point when *reflected light actually passes through the point*.

Virtual images are formed at a point when light rays *appear to diverge from the point*.

The point of intersection of any two of the following rays in a ray diagram for mirrors locates the image:

- The first ray is drawn from the top of the object parallel to the optical axis and is reflected back through the focal point, F.
- The second ray is drawn from the top of the object to the vertex of the mirror and is reflected with the angle of incidence equal to the angle of reflection.
- The third ray is drawn from the top of the object through the center of curvature, C, which is reflected back on itself.

36.4 Thin Lenses

The following three rays form the ray diagram for a thin lens:

- The first ray is drawn parallel to the optic axis. After being refracted by the lens, this ray passes through (or appears to come from) one of the focal points.
- The second ray is drawn through the center of the lens. This ray continues in a straight line.
- The third ray is drawn through the focal point F, and emerges from the lens parallel to the optic axis.

36.5 Lens Aberrations

Aberrations are responsible for the formation of imperfect images by lenses and mirrors. Spherical aberration is due to the variation in focal points for parallel incident rays that strike the lens at various distances from the optical axis. Chromatic aberration arises from the fact that light of different wavelengths focuses at different points when refracted by a lens.

36.6 The Camera

The f-number of a lens is the ratio of the focal length to the diameter. The smaller the f-number, the faster the lens.

36.7 The Eye

The near point represents the closest distance for which the lens will produce a sharp image on the retina. This distance usually increases with age and has an average value of around 25 cm.

The *power* of a lens in *diopters* equals the inverse of the focal length in meters.

EQUATIONS AND CONCEPTS

The *mirror equation* is used to locate the position of an image formed by reflection of paraxial rays. The focal point of a spherical mirror is located midway between the center of curvature and the vertex of the mirror.

$$\frac{1}{p}+\frac{1}{q}=\frac{1}{f} \qquad (36.6)$$

$$f=\frac{R}{2} \qquad (36.5)$$

In using the equations related to the image-forming properties of spherical mirrors, spherical refracting surfaces, and thin lenses, you must be very careful to use the correct algebraic sign for each physical quantity. The sign conventions appropriate for the form of the equations stated here are summarized in the SKILLS section.

The *lateral magnification* of a spherical mirror can be stated either as a ratio of image size to object size or in terms of the ratio of image distance to object distance.

$$M=\frac{h'}{h}=-\frac{q}{p} \qquad (36.2)$$

A magnified image of an object can be formed by a single spherical refracting surface of radius R which separates two media whose indices of refraction are n_1 and n_2.

$$\frac{n_1}{p}+\frac{n_2}{q}=\frac{n_2-n_1}{R} \qquad (36.8)$$

$$M=\frac{h'}{h}=-\frac{n_1 q}{n_2 p}$$

A special case is that of the virtual image formed by a *planar refracting surface* $(R=\infty)$.

$$\frac{n_1}{p}=-\frac{n_2}{q}$$

$$q=-\frac{n_2}{n_1}p \qquad (36.9)$$

The *thin lens* is an important component in many optical instruments. The location of the image formed by a given object is determined by the characteristic properties of the lens (index of refraction n and radii of curvature R). If the lens is surrounded by a medium other than air, the index of refraction given in Equation 36.10 must be the *index of the lens relative to the surrounding medium.* This means, for example, that a hollow biconvex lens ("air lens"), if immersed in water, would have a *negative* focal length. The lateral magnification of a thin lens has the same form as that of a spherical mirror (Eq. 36.2).

$$\frac{1}{p}+\frac{1}{q}=(n-1)\left(\frac{1}{R_1}-\frac{1}{R_2}\right) \quad (36.10)$$

$$\frac{1}{p}+\frac{1}{q}=\frac{1}{f} \quad (36.12)$$

Thin lenses are often used in combination. A special case occurs when two thin lenses are in contact. The *focal length of the combination* given by Equation 36.13 will be *less* than that of either lens individually.

$$\frac{1}{f}=\frac{1}{f_1}+\frac{1}{f_2} \quad (36.13)$$

The use of the lens and mirrors in the design and operation of optical instruments can be illustrated by several relatively simple examples:

Camera: The *light intensity* I incident on the film per unit area is inversely proportional to the square of a ratio of the diameter of the lens to its focal length. The *f-number* equals the ratio of the focal length to the lens diameter.

$$I \propto \frac{1}{(f/D)^2} \quad (36.15)$$

$$f\text{-number} = \frac{f}{D} \quad (36.14)$$

Eye: The power of a lens in diopters is the reciprocal of the focal length measured in meters (including the correct algebraic sign).

$$P = \frac{1}{f}$$

Simple magnifier: When an object is at the near point (25 cm), the angle subtended by the object at the eye is θ. When a converging lens of focal length f is placed between the eye and the object, an image which subtends an angle θ_0 can be formed at the near point. The *angular magnification* or magnifying power of the lens can be expressed in alternate forms.

$$m \equiv \frac{\theta}{\theta_0} \qquad (36.16)$$

$$m = 1 + \frac{25 \text{ cm}}{f} \qquad (36.18)$$

Compound microscope: This instrument contains an objective lens of short focal length f_0 and an eye piece of focal length f_e. The two lenses are separated by a distance L. When an object is located just beyond the focal point of the objective, the two lenses in combination form an enlarged, virtual and inverted image of lateral magnification M.

$$M = -\frac{L}{f_0}\left(\frac{25 \text{ cm}}{f_e}\right) \qquad (36.20)$$

Astronomical telescope: Two converging lenses are separated by a distance equal to the sum of their focal lengths. The angular magnification is equal to the ratio of the two focal lengths.

$$m = -\frac{f_0}{f_e} \qquad (36.21)$$

SUGGESTIONS, SKILLS, AND STRATEGIES

A major portion of this chapter is devoted to the development and presentation of equations which can be used to determine the location and nature of images formed by various optical components acting either singly or in combination. It is essential that these equations be used with the correct algebraic sign associated with each quantity involved. You must understand clearly the sign conventions for mirrors, refracting surfaces, and lenses. The following discussion represents a review of these sign conventions.

SIGN CONVENTIONS FOR MIRRORS

Equations: $\frac{1}{p}+\frac{1}{q}=\frac{1}{f}=\frac{2}{R}$ $\qquad$ $M=\frac{h'}{h}=-\frac{q}{p}$

The front side of the mirror is the region on which light rays are incident and reflected.

p is + if the object is in front of the mirror (real object).
p is – if the object is in back of the mirror (virtual object).

q is + if the image is in front of the mirror (real image).
q is – if the image is in back of the mirror (virtual image).

Both f and R are + if the center of curvature is in front of the mirror (concave mirror).

Both f and R are – if the center of curvature is in back of the mirror (convex mirror).

If M is positive, the image is erect.
If M is negative, the image is inverted.

You should check the sign conventions as stated against the situations described in Figure 36.1.

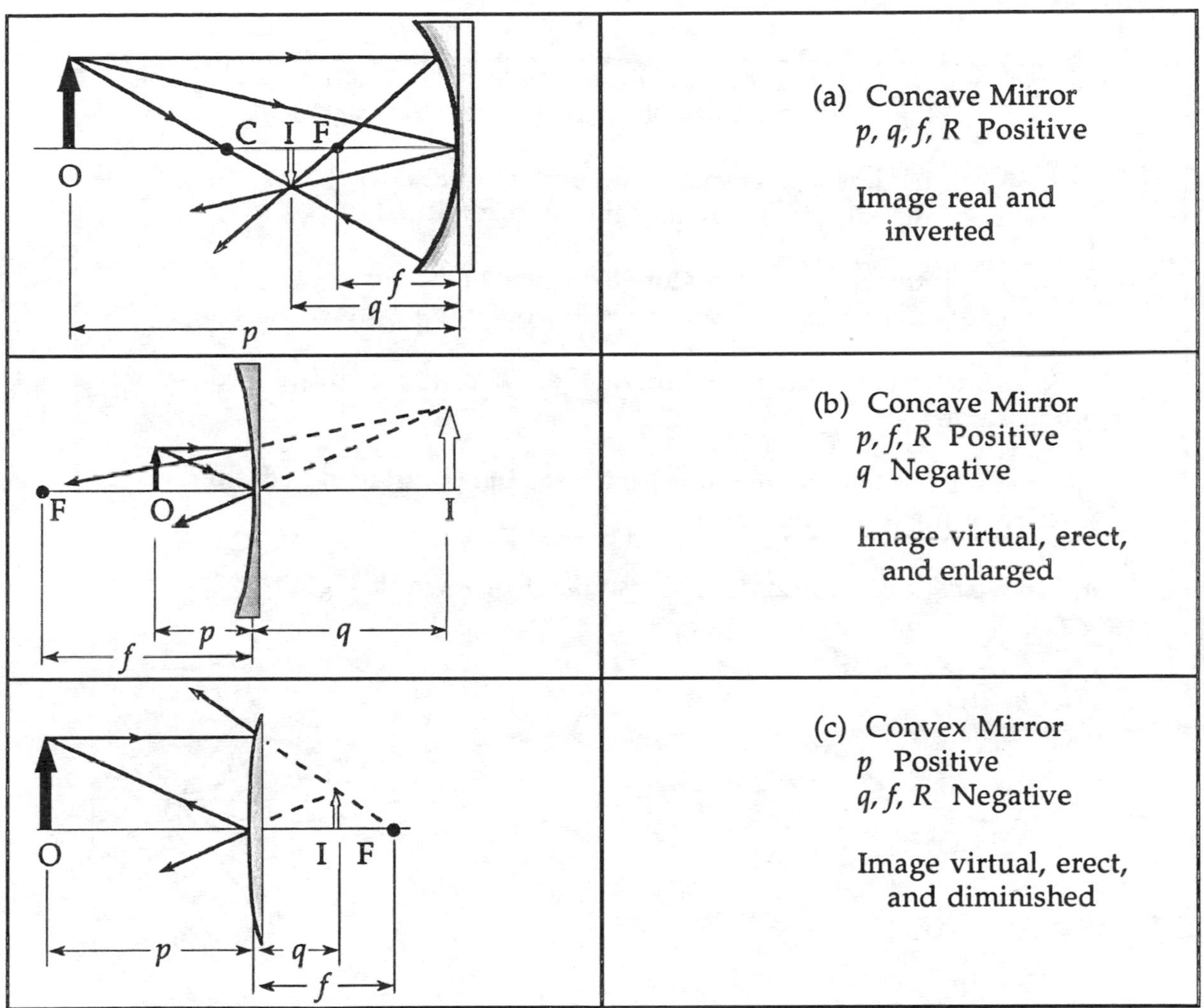

Figure 36.1 Figures describing sign conventions for mirrors.

SIGN CONVENTIONS FOR REFRACTING SURFACES

Equations: $\frac{n_1}{p}+\frac{n_2}{q}=\frac{n_2-n_1}{R}$ $\qquad M=\frac{h'}{h}=-\frac{q}{p}$

In the following table, the *front* side of the surface is the side *from which the light is incident*.

p is + if the object is in front of the surface (real object).
p is – if the object is in back of the surface (virtual object).

q is + if the image is in back of the surface (real image).
q is – if the image is in front of the surface (virtual image).

R is + if the center of curvature is in back of the surface.
R is – if the center of curvature is in front of the surface.

n_1 refers to the index of the medium on the side of the interface from which the light comes.

n_2 is the index of the medium into which the light is transmitted after refraction at the interface.

Review the above sign conventions for the situations shown in Figure 36.2.

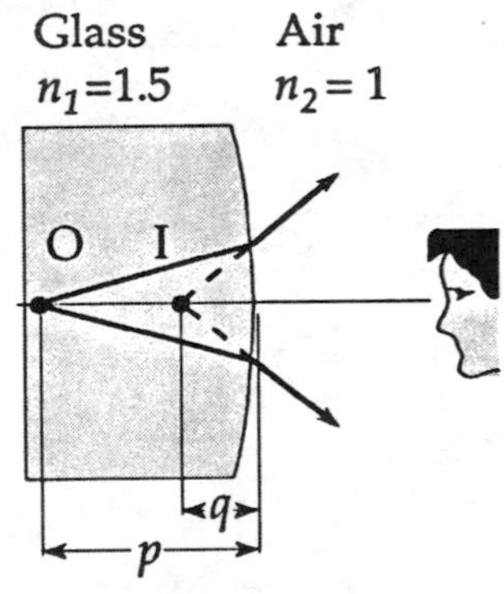

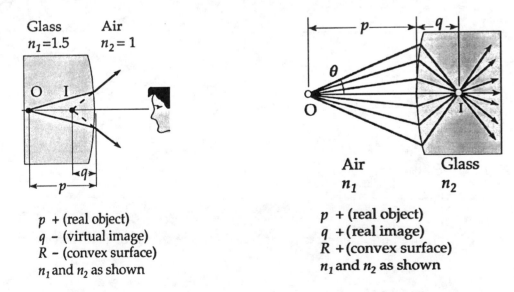

p + (real object)
q - (virtual image)
R - (convex surface)
n_1 and n_2 as shown

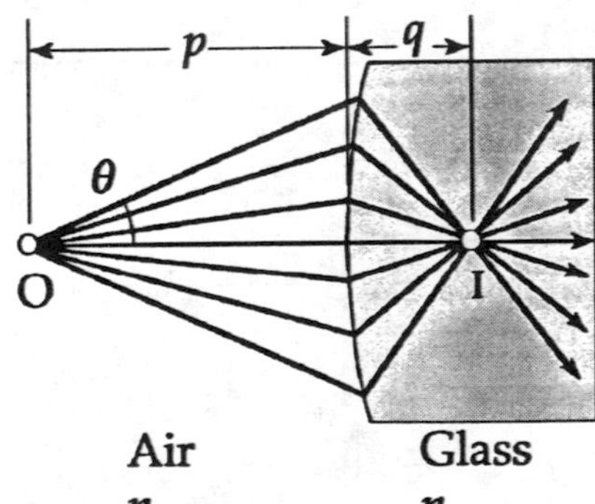

p + (real object)
q + (real image)
R + (convex surface)
n_1 and n_2 as shown

Figure 36.2 Figures describing sign conventions for refracting surfaces.

SIGN CONVENTIONS FOR THIN LENSES

Equations: $\frac{1}{p}+\frac{1}{q}=\frac{1}{f}=(n-1)\left(\frac{1}{R_1}-\frac{1}{R_2}\right)$ $\qquad M=\frac{h'}{h}=-\frac{q}{p}$

In the following table, the *front* of the lens is the *side from which the light is incident.*

p is + if the object is in front of the lens. p is – if the object is in back of the lens.
q is + if the image is in back of the lens. q is – if the image is in front of the lens.
f is + if the lens is convex. f is – if the lens is concave.
R_1 and R_2 are + if the center of curvature is in back of the lens.
R_1 and R_2 are – if the center of curvature is in front of the lens.

The sign conventions for thin lenses are illustrated by the examples shown in Figure 36.3.

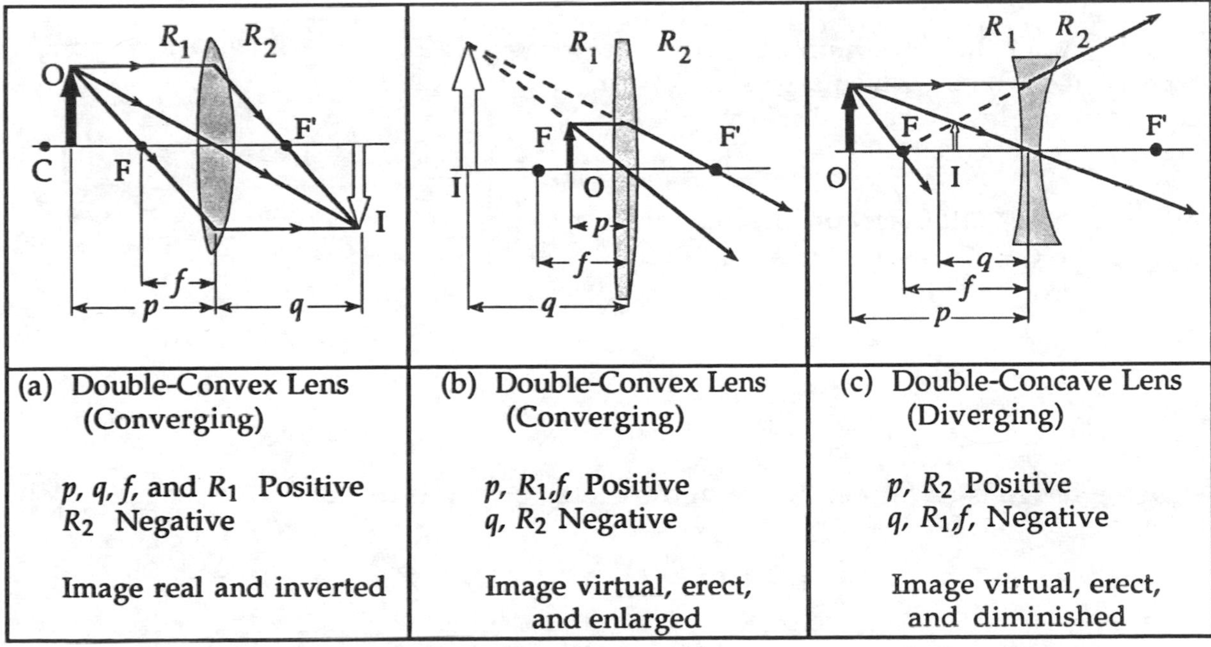

R1
R2
O
C
F
F'
I
f
p
q
(a) Double-Convex Lens (Converging)
p, q, f, and R1 Positive
R2 Negative
Image real and inverted
(b) Double-Convex Lens (Converging)
p, R1,f, Positive
q, R2 Negative
Image virtual, erect, and enlarged
(c) Double-Concave Lens (Diverging)
p, R2 Positive
q, R1,f, Negative
Image virtual, erect, and diminished

Figure 36.3 Figures describing the sign conventions for various thin lenses.

REVIEW CHECKLIST

▷ Identify the following properties which characterize an image formed by a lens or mirror system with respect to an object: position, magnification, orientation (i.e. inverted, erect or right-left reversal) and whether real or virtual.

▷ Understand the relationship of the algebraic signs associated with calculated quantities to the nature of the image and object: real or virtual, erect or inverted.

▷ Calculate the location of the image of a specified object as formed by a plane mirror, spherical mirror, plane refracting surface, spherical refracting surface, thin lens, or a combination of two or more of these devices. Determine the magnification and character of the image in each case.

▷ Construct ray diagrams to determine the location and nature of the image of a given object when the geometrical characteristics of the optical device (lens or mirror) are known.

SOLUTIONS TO SELECTED END-OF-CHAPTER PROBLEMS

3. Determine the minimum height of a vertical flat mirror in which a person 5'10" in height can see his or her full image. (A ray diagram would be helpful.)

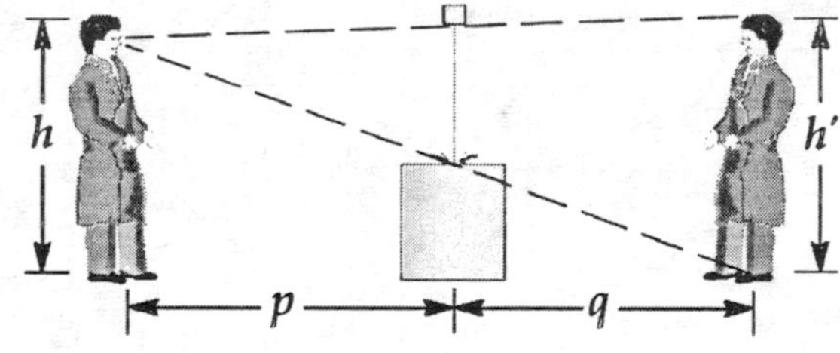

Solution The flatness of the mirror is described $R=\infty$, $f=\infty$, and $1/f=0$. By our general mirror equation,

$$\frac{1}{p}+\frac{1}{q}=\frac{1}{f}, \quad \text{or} \quad q=-p$$

Thus, the image is as far behind the mirror as the person is in front. The magnification is then

$$M=\frac{-q}{p}=1=\frac{h'}{h}, \quad \text{so} \quad h'=h=70"$$

The height of the mirror is defined by the triangle from the person's eyes to the top and bottom of his image, as shown. From the geometry of the triangle, we see that the mirror must be:

$$h'\left(\frac{p}{p-q}\right)=h'\left(\frac{p}{2p}\right)=\frac{h'}{2}$$

Thus, the mirror must be at least 35" high. ◊

---

7. A concave mirror has a radius of curvature of 60 cm. Calculate the image position and magnification of an object placed in front of the mirror at distances of (a) 90 cm and (b) 20 cm. (c) Draw ray diagrams to obtain the image in each case.

Solution

(a) $\frac{1}{p}+\frac{1}{q}=\frac{2}{R}$ or $\frac{1}{90\text{ cm}}+\frac{1}{q}=\frac{2}{60\text{ cm}}$

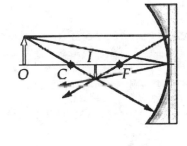

$$\frac{1}{q}=\frac{2}{60\text{ cm}}-\frac{1}{90\text{ cm}}=0.022\text{ cm}^{-1}$$

$$q=45\text{ cm}$$

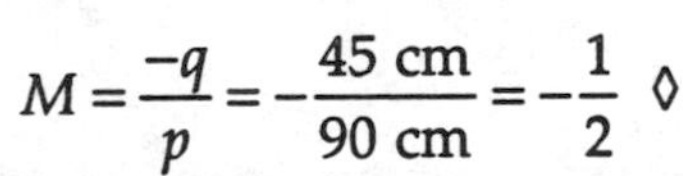

$$M=\frac{-q}{p}=-\frac{45\text{ cm}}{90\text{ cm}}=-\frac{1}{2}\quad ◊$$

(b) $\frac{1}{p}+\frac{1}{q}=\frac{2}{R}$ or $\frac{1}{20\text{ cm}}+\frac{1}{q}=\frac{2}{60\text{ cm}}$

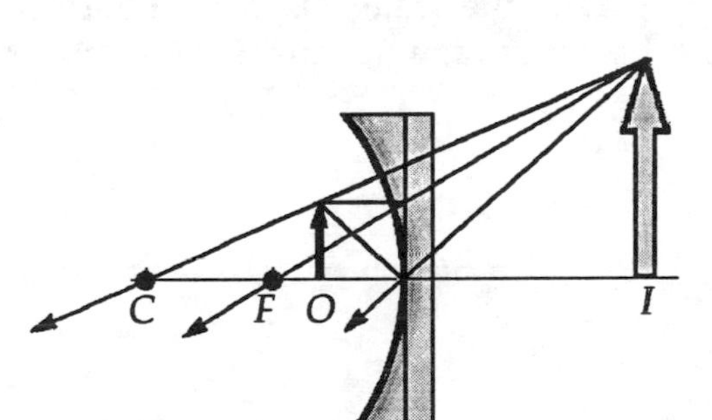

◊

$$\frac{1}{q}=\frac{2}{60\text{ cm}}-\frac{1}{20\text{ cm}}=-0.0167\text{ cm}^{-1}$$

$$q=-60\text{ cm}$$

$$M=-\frac{q}{p}=-\frac{-60\text{ cm}}{20\text{ cm}}=3\quad ◊$$

---

13. A spherical mirror is to be used to form, on a screen located 5.0 m from the object, an image five times the size of the object. (a) Describe the type of mirror required. (b) Where should the mirror be positioned relative to the object?

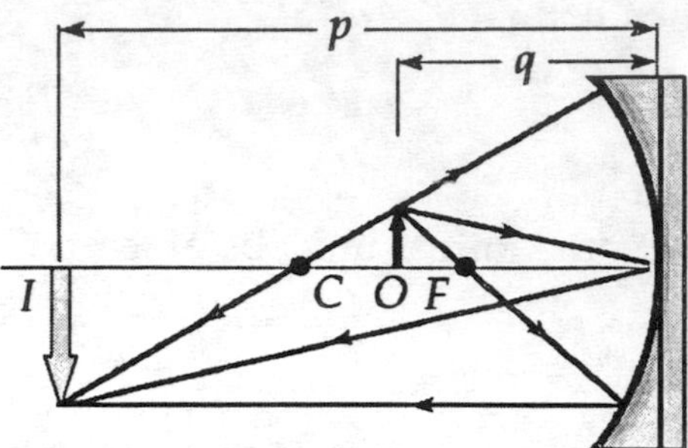

Solution

We are given that $q = (p + 5.0 \text{ m})$.

Since the image must be real,

$$M = -5.0 = -\frac{q}{p}, \text{ or } q = 5.0p$$

(b) With these two equations, the distance of the mirror from the object is:

$$p + 5.0 = 5.0p \quad \text{and} \quad p = 1.25 \text{ m} \quad \Diamond$$

(a) We can solve for the focal length of the mirror from

$$\frac{1}{f} = \frac{1}{p} + \frac{1}{q} = \frac{1}{1.25 \text{ m}} + \frac{1}{6.25 \text{ m}} \qquad \text{so that} \qquad f = 1.04 \text{ m}$$

Noting that the image is real, inverted and enlarged, we can say that the mirror must be concave, and must have a radius of

$$R = 2f = 2.08 \text{ m} \quad \Diamond$$

15. A spherical convex mirror has a radius of 40.0 cm. Determine the position of the virtual image and magnification for object distances of (a) 30.0 cm and (b) 60.0 cm. (c) Are the images upright or inverted?

Solution The convex mirror is described by $f = R/2 = (-40.0\text{ cm})/2 = -20.0\text{ cm}$.

(a) $\frac{1}{p}+\frac{1}{q}=\frac{1}{f}$, so $\frac{1}{30.0\text{ cm}}+\frac{1}{q}=\frac{1}{-20.0\text{ cm}}$ and $q=-12.0\text{ cm}$ ◊

$$M=\frac{-q}{p}=-\frac{-12.0\text{ cm}}{30.0\text{ cm}}=+0.400$$ ◊

(c) The image is behind the mirror, upright, virtual, and diminished. ◊

(b) $\frac{1}{60.0\text{ cm}}+\frac{1}{q}=-\frac{1}{20.0\text{ cm}}$

$q=-15.0\text{ cm}$ ◊

$$M=\frac{-q}{p}=-\frac{-15.0\text{ cm}}{60.0\text{ cm}}=+0.250$$ ◊

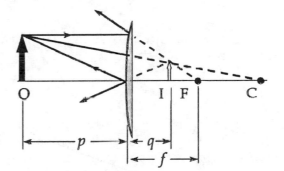

(c) The image is behind the mirror, upright, virtual, and diminished. ◊

The principle ray diagram is an essential complement to the numerical description of the image. Draw the rays into this diagram:

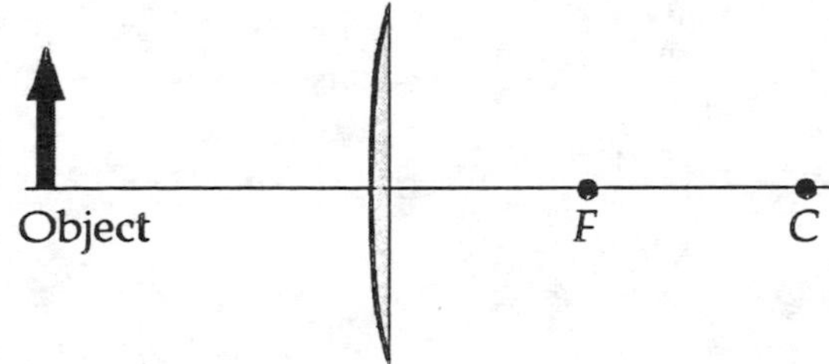

Use your diagram to figure out that the image is behind the mirror, upright, virtual and diminished. ◊

17. A smooth block of ice ($n = 1.309$) rests on the floor with one face parallel to the floor. The block has a vertical thickness of 50.0 cm. Find the location of the image of a pattern in the floor covering as formed by rays that are nearly perpendicular to the block.

Solution The upper surface of the block is a single refracting surface with zero curvature, and with infinite radius of curvature.

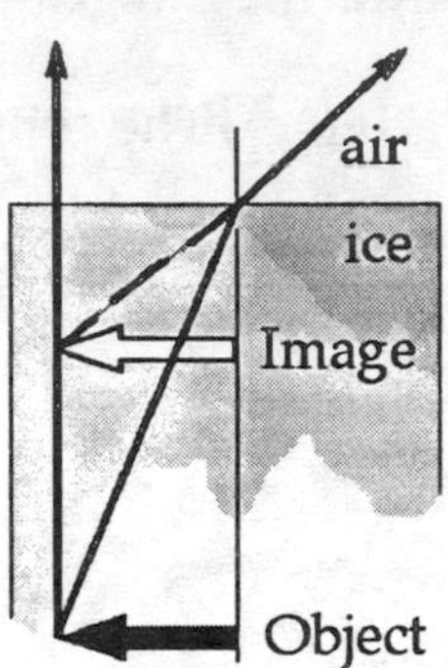

$$\frac{n_1}{p} + \frac{n_2}{q} = \frac{n_2 - n_1}{R}$$

$$\frac{1.309}{50.0 \text{ cm}} + \frac{1}{q} = \frac{1.00 - 1.309}{\infty} = 0$$

$$q = \frac{-50 \text{ cm}}{1.309} = -38.2 \text{ cm}$$

The floor appears to be 38.2 cm below the upper surface of the ice. The image is virtual, upright, and actual size. ◊

---

21. A glass sphere ($n = 1.50$) of radius 15 cm has a tiny air bubble located 5.0 cm from the center. The sphere is viewed along a direction parallel to the radius containing the bubble. What is the apparent depth of the bubble below the surface of the sphere?

Solution

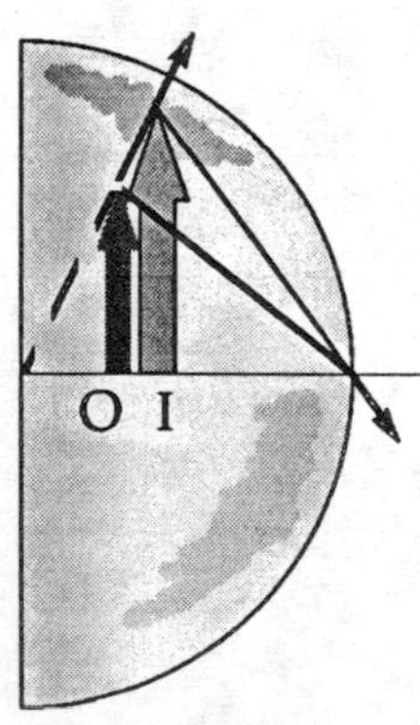

From Equation 36.8, $$\frac{n_1}{p} + \frac{n_2}{q} = \frac{n_2 - n_1}{R}$$

We solve for q to find $$q = \frac{n_2 R p}{p(n_2 - n_1) - n_1 R}$$

In this case,

$$n_1 = 1.5 \qquad n_2 = 1.0 \qquad p = 10 \text{ cm}$$

Thus the apparent depth is

$$q = \frac{(1.0)(-15 \text{ cm})(10 \text{ cm})}{(10 \text{ cm})(1.0 - 1.5) - (1.5)(-15 \text{ cm})} = 8.6 \text{ cm} \quad ◊$$

The image is virtual, upright, and enlarged.

---

23. A glass hemisphere is used as a paperweight with its flat face resting on a stack of papers. The radius of the circular cross section is 4.0 cm, and the index of refraction of the glass is 1.55. The center of the hemisphere is directly over a letter "O" that is 2.5 mm in diameter. What is the diameter of the image of the letter as seen looking along a vertical radius?

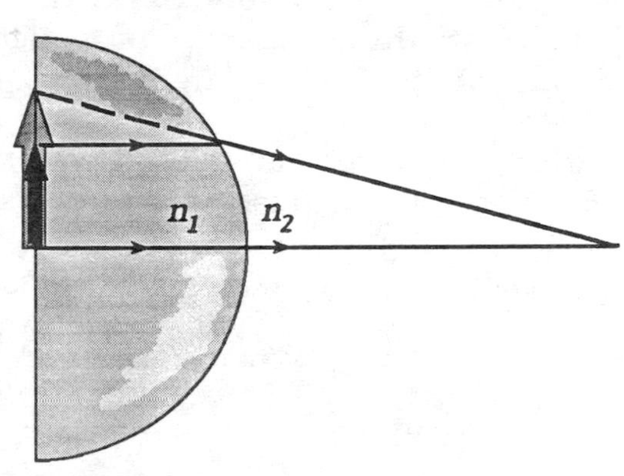

Solution The image forms by refraction at a single surface:

$$\frac{n_1}{p}+\frac{n_2}{q}=\frac{n_2-n_1}{R} \qquad \text{or} \qquad \frac{1.55}{4.0\text{ cm}}+\frac{1.00}{q}=\frac{1.00-1.55}{-4.0\text{ cm}}$$

Thus,
$$q=\frac{1}{[0.55/4.0\text{ cm}-1.55/4.0\text{ cm}]}=-4.0\text{ cm}$$

The virtual image is at the same location as the object.

To find the image diameter h', consider the refraction of the ray from the top end of O through V. We have

$$n_1\sin\theta_1=n_2\sin\theta_2 \qquad \text{where} \qquad \sin\theta_1\cong\tan\theta_1=h/p$$

$$\text{and} \qquad \sin\theta_2\cong\tan\theta_2=\frac{h'}{|q|}=-\frac{h'}{q}$$

Then,
$$n_1\frac{h}{p}=-n_2\frac{h'}{q}$$

$$h'=-\frac{n_1qh}{n_2p}=\frac{-1.55(-4.0\text{ cm})(2.5\text{ mm})}{1(4.0\text{ cm})}=3.9\text{ mm} \quad \lozenge$$

The image is virtual, upright, and enlarged.

---

27. The left face of a biconvex lens has a radius of curvature of 12 cm, and the right face has a radius of curvature of 18 cm. The index of refraction of the glass is 1.44. (a) Calculate the focal length of the lens. (b) Calculate the focal length if the radii of curvature of the two faces are interchanged.

Solution Convex outward on both sides, the lens has the centers of curvature of its surfaces on opposite sides. The second surface has negative radius:

(a) $$\frac{1}{f}=(n-1)\left[\frac{1}{R_1}-\frac{1}{R_2}\right]=(1.44-1.00)\left[\frac{1}{12\text{ cm}}-\frac{1}{-18\text{ cm}}\right]$$

$f = 16$ cm ◊

(b) $$\frac{1}{f}=(0.44)\left[\frac{1}{18\text{ cm}}-\frac{1}{-12\text{ cm}}\right]$$

$f = 16$ cm ◊

Reversing the lens does not change what it does to the light.

---

30. A person looks at a gem with a jeweler's microscope—a converging lens that has a focal length of 12.5 cm. The microscope forms a virtual image 30.0 cm from the lens. (a) Determine the magnification. Is the image upright or inverted? (b) Construct a ray diagram for this arrangement.

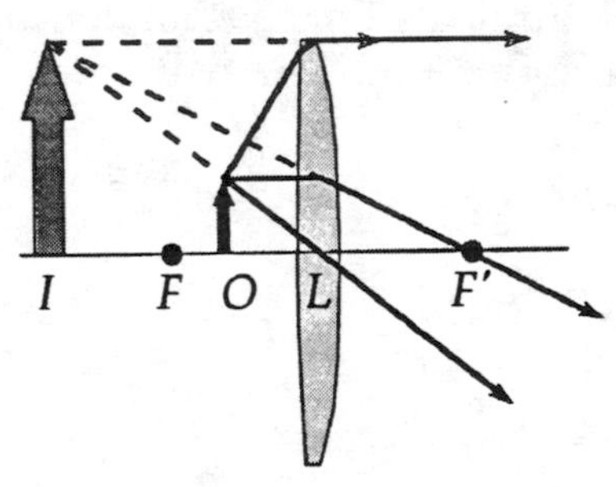

Solution $$\frac{1}{p}+\frac{1}{q}=\frac{1}{f}$$

Thus, $$\frac{1}{p}+\frac{1}{-30\text{ cm}}=\frac{1}{12.5\text{ cm}}$$

$p = 8.82$ cm

$$M=-\frac{q}{p}=-\frac{(-30)}{8.82}=3.40,\quad \text{upright}$$ ◊

---

33. An object is located 20 cm to the left of a diverging lens having a focal length f = –32 cm. Determine (a) the location and (b) the magnification of the image. (c) Construct a ray diagram for this arrangement.

Solution $\quad \frac{1}{p}+\frac{1}{q}=\frac{1}{f} \quad$ or $\quad \frac{1}{20\text{ cm}}+\frac{1}{q}=\frac{1}{-32\text{ cm}}$

(a) So $\quad q=-\left(\frac{1}{20\text{ cm}}+\frac{1}{32\text{ cm}}\right)^{-1}=-12.3\text{ cm}$ ◊

(b) $\quad M=-\frac{q}{p}=-\frac{-12.3\text{ cm}}{20\text{ cm}}=0.62$ ◊

(c) The image is virtual, upright, and diminished. ◊

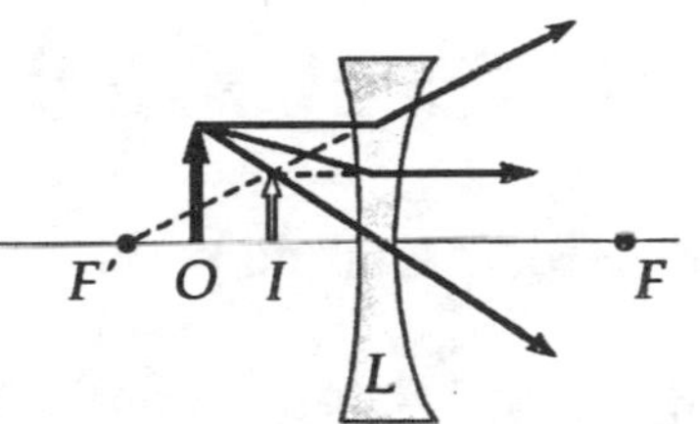

---

37. The nickel's image in Figure P36.37 has twice the diameter of the nickel and is 2.84 cm from the lens. Determine the focal length of the lens.

Figure P36.37

Solution

Looking through the lens, you see the image beyond the lens. Therefore, the image is virtual, with q = –2.84 cm.

Now, $\quad M=\frac{h'}{h}=2=-\frac{q}{p},$

so $\quad p=-\frac{q}{2}=1.42\text{ cm}$

Thus, $\quad f=\left[\frac{1}{p}+\frac{1}{q}\right]^{-1}=\left[\frac{1}{1.42\text{ cm}}+\frac{1}{-2.84\text{ cm}}\right]^{-1}$

and $\quad f=2.84\text{ cm}$ ◊

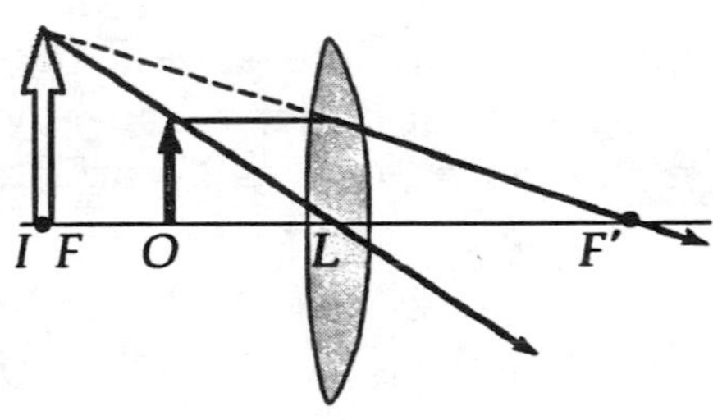

---

39. A camera is being used with correct exposure at $f/4$ and a shutter speed of (1/16) s. In order to photograph a fast-moving subject, the shutter speed is changed to (1/28) s. Find the new f-number setting needed to maintain satisfactory exposure.

Solution Call the f-number N_f. In available-light photography, the incoming light intensity I is set by the subject. The photographer must get enough light energy ($E = IAt$) onto the film for a reasonable exposure. Here A is the aperture area, and

$$A = \pi \frac{D^2}{4}$$

The exposure in each case must be the same, so $IA_1t_1 = IA_2t_2$

Substituting for A and eliminating I, this becomes $D_1^{\,2}t_1 = D_2^{\,2}t_2$

The fraction, focal length divided by fnumber, is the diameter D of the lens aperture. Therefore, we may write $D = f/N_f$, and substitute.

$$(f/4)^2(1/16 \text{ s}) = (D)_2^{\,2}(1/128 \text{ s})$$

$$D_2 = (f/4)\sqrt{128/16} = f/1.41$$

Thus the new image will be taken with a setting of $f/1.41$, and an f-number of 1.41 ◊

---

41. A nearsighted woman cannot see objects clearly beyond 25 cm (the far point.) If she has no astigmatism and contact lenses are prescribed, what are the power and type of lens required to correct her vision?

Solution The lens should take parallel light rays from a very distant object ($p = \infty$) and make them diverge from a virtual image at the woman's far point, which is 25 cm beyond the lens, at $q = -25$ cm. Thus,

$$\frac{1}{f} = \frac{1}{p} + \frac{1}{q} = \frac{1}{\infty} + \frac{1}{-25 \text{ cm}}$$

Hence, the power of the lens is $P = \frac{1}{f} = -\frac{1}{0.25 \text{ m}} = -4.0$ diopter; this is a diverging lens. ◊

---

47. A philatelist examines the printing detail on a stamp using a convex lens of focal length 10.0 cm as a simple magnifier. The lens is held close to the eye and the lens-to-object distance is adjusted so that the virtual image is formed at the normal near point (25 cm). Calculate the magnification.

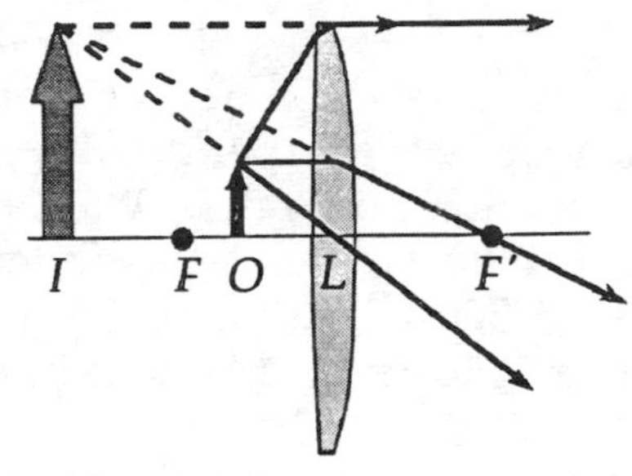

Solution The image is virtual: $q = -25$ cm. The distances from stamp to magnifying glass is p in

$$\frac{1}{p} + \frac{1}{q} = \frac{1}{f}$$

Therefore, $$\frac{1}{p} + \frac{1}{-25 \text{ cm}} = \frac{1}{10 \text{ cm}}$$

and $$p = 7.14 \text{ cm}$$

Then, $$M = -q/p = -\frac{-25 \text{ cm}}{7.14 \text{ cm}} = 3.50 \ \lozenge$$

---

49. The Yerkes refracting telescope has a 1.0-m diameter objective lens of focal length 20 m and an eyepiece of focal length 2.5 cm. (a) Determine the magnification of the planet Mars as seen through this telescope. (b) Are the martian polar caps right side up or upside down?

Solution

(a) The angular magnification is

$$m = -\frac{f_0}{f_e} = \frac{-2.0 \text{ m}}{0.025 \text{ m}} = -800 \quad \lozenge$$

(b) The minus sign means that the image is upside down. ◊

---

59. A parallel beam of light enters a glass hemisphere perpendicular to the flat face, as shown in Figure P36.59. The radius is $R = 6.00$ cm, and the index of refraction is $n = 1.560$. Determine the point at which the beam is focused. (Assume paraxial rays.)

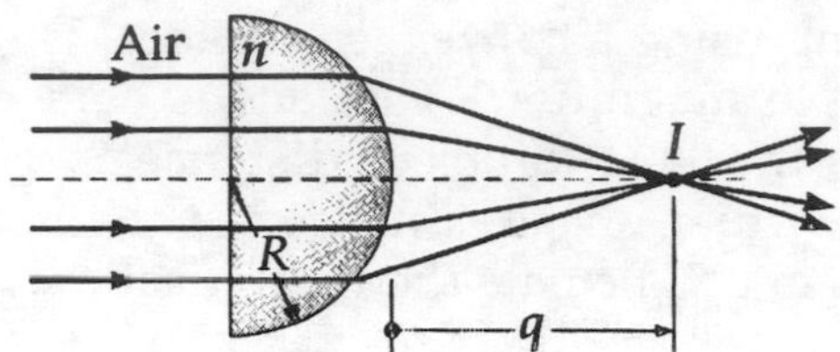

Figure P36.59

Solution A hemisphere is too thick to be described as a thin lens. The light is undeviated on entry into the flat face. Because of this, we instead consider the light's exit from the second surface, for which $R = -6.00$ cm. The incident rays are parallel, as described by $p = \infty$.

Then $$\frac{n_1}{p}+\frac{n_2}{q}=\frac{n_2-n_1}{R}$$ becomes $$0+\frac{1}{q}=\frac{(1-1.56)}{-6.00\text{ cm}}$$

and $$q = 10.7\text{ cm} \quad \Diamond$$

---

60. A thin lens having refractive index n is immersed in a liquid having index n'. Show that the focal length f of the lens is

$$\frac{1}{f}=\left(\frac{n}{n'}-1\right)\left(\frac{1}{R_1}-\frac{1}{R_2}\right)$$

Solution At the first surface, $$\frac{n'}{p_1}+\frac{n}{q_1}=\frac{n-n'}{R_1} \qquad (1)$$

At the second interface, $$\frac{n}{p_2}+\frac{n'}{q_2}=\frac{n'-n}{R_2} \qquad (2)$$

Since, for a thin lens, $p_2 = -q_1$, $$\frac{-n}{q_1}+\frac{n'}{q_2}=\frac{n'-n}{R_2} \qquad (3)$$

Adding (3) to (1) and dropping the subscripts on p and q,

$$\frac{n'}{p} + \frac{n'}{q} = (n - n')\left(\frac{1}{R_1} - \frac{1}{R_2}\right)$$

To find the focal length, let $p = \infty$.

$$0 + \frac{n'}{q} = (n - n')\left(\frac{1}{R_1} - \frac{1}{R_2}\right)$$

$$\frac{1}{q} = \frac{1}{f} = \left(\frac{n}{n'} - 1\right)\left(\frac{1}{R_1} - \frac{1}{R_2}\right) \quad \Diamond$$

---

65. An object is placed 12.0 cm to the left of a diverging lens of focal length –6.0 cm. A converging lens of focal length 12.0 cm is placed a distance d to the right of the diverging lens. Find the distance d so that the final image is at infinity. Draw a ray diagram for this case.

Solution From Equation 36.12, $q_1 = \dfrac{f_1 p_1}{p_1 - f_1} = \dfrac{(-6.0\text{ cm})(12.0\text{ cm})}{12.0\text{ cm} - (-6.0\text{ cm})} = -4.0\text{ cm}$

When we require that $q_2 \to \infty$, Equation 36.12 becomes $p_2 = f_2$; in this case,

$$p_2 = d - (-4.0\text{ cm})$$

Therefore, $d + 4.0\text{ cm} = f_2 = 12.0\text{ cm}$, and $d = 8.0\text{ cm}$ $\Diamond$

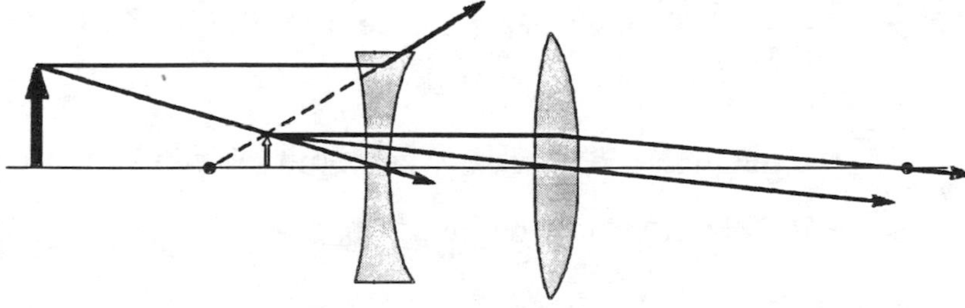

---

69. A colored marble is dropped into a large tank filled with benzene (n = 1.50). (a) What is the depth of the tank if the apparent depth of the marble when viewed from directly above the tank is 35 cm? (b) If the marble has a diameter of 1.5 cm, what is its apparent diameter when viewed from directly above the tank?

Solution $\frac{n_1}{p}+\frac{n_2}{q}=\frac{n_2-n_1}{R}$

(a) When $R = \infty$, $$p=-q\left(\frac{n_1}{n_2}\right)$$

Therefore, $$p=-(-35\text{ cm})\left(\frac{1.5}{1.0}\right)=52.5\text{ cm}$$

(b) $M=-\frac{n_1 q}{n_2 p}=\frac{h'}{h}$, or $h'=-h\left(\frac{n_1 q}{n_2 p}\right)$

Thus, $$h'=(-1.5\text{ cm})\frac{(1.5)(-35\text{ cm})}{(1)(52.5\text{ cm})}=1.5\text{ cm} \quad \Diamond$$

---

71. The disk of the Sun subtends an angle of 0.50° at the Earth. What are the position and diameter of the solar image formed by a concave spherical mirror of radius 3.0 m?

Solution For the mirror, $f = R/2$ = +1.5 m. In addition, because the distance to the Sun is so much larger than any other figures, we can take $p = \infty$.

The mirror equation, $\frac{1}{p}+\frac{1}{q}=\frac{1}{f}$, then gives a distance from the mirror of $q = f$ = 1.5 m. ◊

Now, in $M=-\frac{q}{p}=\frac{h'}{h}$, the magnification is nearly zero, but we can be more precise: h/p is the angular diameter of the object. Thus the image diameter is

$$h'=-\frac{hq}{p}=(-0.50°)\left(\frac{\pi}{180}\text{ rad / deg}\right)(1.50\text{ m})=-1.31\text{ cm} \quad \Diamond$$

---

75. In a darkened room, a burning candle is placed 1.5 m from a white wall. A lens is placed between candle and wall at a location that causes a larger, inverted image to form on the wall. When the lens is moved 90 cm toward the wall, another image of the candle is formed. Find (a) the two object distances that produce the images stated above and (b) the focal length of the lens. (c) Characterize the second image.

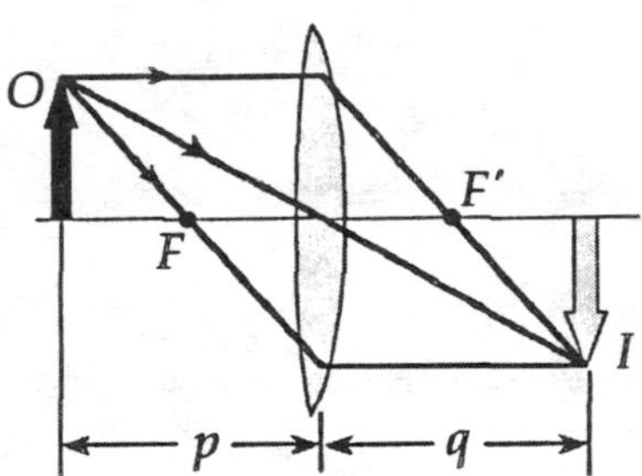

Solution In the original situation, $p_1 + q_1 = 1.5$ m.
In the final situation, $p_2 = p_1 + 0.90$ m, and $q_2 = q_1 - 0.90$ m.

Our lens equation is $$\frac{1}{p_1}+\frac{1}{q_1}=\frac{1}{f}=\frac{1}{p_2}+\frac{1}{q_2}$$

Substituting, we have $$\frac{1}{p_1}+\frac{1}{1.5\text{ m}-p_1}=\frac{1}{p_1+0.90}+\frac{1}{0.6-p_1}$$

Adding the fractions, $$\frac{1.5\text{ m}-p_1+p_1}{p_1(1.5\text{ m}-p_1)}=\frac{0.6-p_1+p_1+0.90}{(p_1+0.90)(0.6-p_1)}$$

Simplified, this becomes $$p_1(1.5\text{ m}-p_1)=(p_1+0.90)(0.6-p_1)$$

(a) Thus, $$p_1=\frac{0.54}{1.8}\text{ m }=0.30\text{ m} \quad \Diamond$$

$$p_2=p_1+0.90\ =1.2\text{ m} \quad \Diamond$$

(b) $\dfrac{1}{f}=\dfrac{1}{0.30\text{ m}}+\dfrac{1}{1.5\text{ m}-0.30\text{ m}}$ and $f=0.24$ m ◊

(c) The second image is real, inverted, and diminished, with $M=\dfrac{-q_2}{p_2}=-0.25$ ◊

Chapter 37

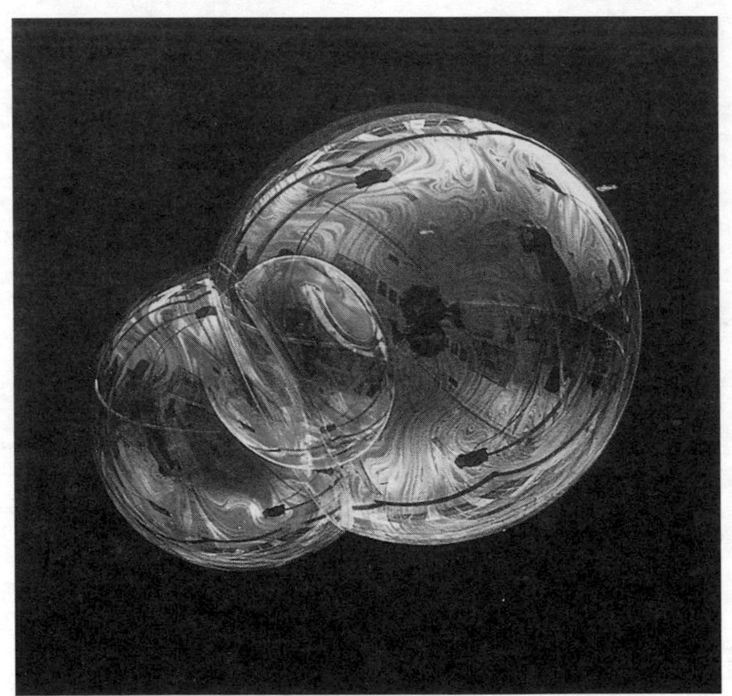

Interference of Light Waves

Chapter 37

INTERFERENCE OF LIGHT WAVES

INTRODUCTION

In the previous chapter on geometric optics, we used light rays to examine what happens when light passes through a lens or reflects from a mirror. The next two chapters are concerned with wave optics, which deals with the interference, diffraction, and polarization of light. These phenomena cannot be adequately explained with the ray optics of Chapter 36, but we describe how treating light as waves rather than as rays leads to a satisfying description of such phenomena.

NOTES FROM SELECTED CHAPTER SECTIONS

37.1 Conditions for Interference

In order to observe *sustained* interference in light waves, the following conditions must be met:

- The sources must be coherent; they must maintain a *constant phase* with respect to each other.
- The sources must be *monochromatic*— of a *single wavelength*.
- The *superposition principle* must apply.

37.2 Young's Double-Slit Experiment

A schematic diagram illustrating the geometry used in Young's double-slit experiment is shown in the figure below. The two slits S_1 and S_2 serve as coherent monochromatic sources. The *path difference* $\delta = r_2 - r_1 = d\sin\theta$.

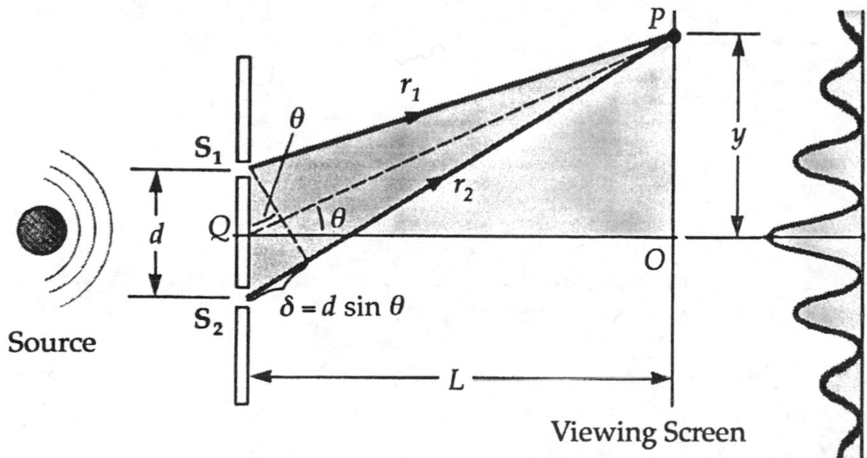

Figure 37.1

37.4 Phasor Addition of Waves

The technique of phasor addition of waves is described in the SKILLS section of this chapter.

37.5 Change of Phase due to Reflection

An electromagnetic wave undergoes a *phase change of 180°* upon reflection from a medium that is *optically more dense* than the one in which it was traveling. There is also a 180° phase change upon reflection from a *conducting surface.*

37.6 Interference in Thin Films

Interference effects in thin films depend on the difference in length of path traveled by the interfering waves as well as any phase changes which may occur due to reflection. It can be shown by the Lloyd's mirror experiment that when light is reflected from a surface of index of refraction greater than the index of the medium in which it is initially traveling, the reflected ray undergoes a 180° (or π radians) phase change. In analyzing interference effects, this can be considered equivalent to the gain or loss of a half wavelength in path difference. Therefore, there are two different cases to consider: (a) a film surrounded by a common medium and (b) a thin film located between two different media. These cases are illustrated in the following figures.

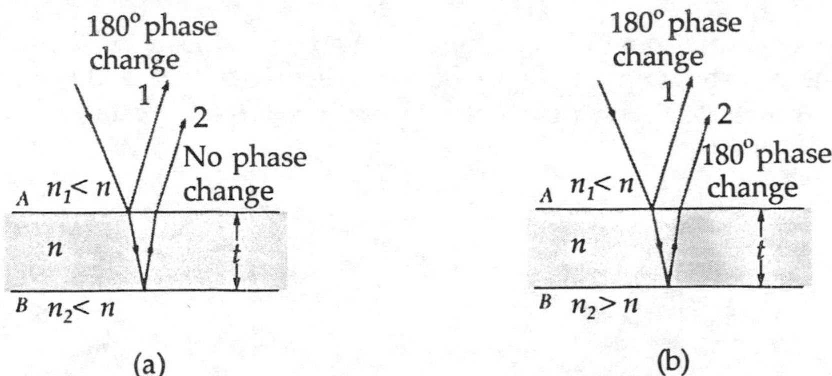

Figure 37.2 (a) Interference of light resulting from reflections at two surfaces of a thin film of thickness t, index of refraction n. (b) A thin film between two different media.

In case (a), where a phase change occurs at the top surface, the reflected rays will be in phase (constructive interference) if the thickness of the film is an odd number of quarter wavelengths—so that the path lengths will differ by an odd number of half wavelengths.

In case (b), the phase changes due to reflection at *both* the top and bottom surfaces are offsetting and, therefore, constructive interference for the reflected rays will occur when the film thickness is an integer number of half wavelengths—and, therefore, the path difference will be a whole number of wavelengths.

EQUATIONS AND CONCEPTS

In the arrangement used for the Young's double-slit experiment (Figure 37.1), two slits separated by a distance, d, serve as monochromatic coherent sources. The light intensity at any point on the screen is the resultant of light reaching the screen from both slits. Also, light from the two slits reaching any point on the screen (except the center) travel unequal path lengths. This difference in length of path is called the path difference.

$$\delta = r_2 - r_1 = d\sin\theta \qquad (37.1)$$

$$y = L\tan\theta$$

Bright fringes (constructive interference) will appear at points on the screen for which the path difference is equal to an integral multiple of the wavelength. The positions of bright fringes can also be located by calculating their vertical distance from the center of the screen (y). In each case, the number m is called the order number of the fringe. The central bright fringe ($\theta = 0$, $m = 0$) is called the zeroth-order maximum.

$$\delta = d\sin\theta = m\lambda \qquad (37.2)$$

for $m = 0, \pm 1, \pm 2, \ldots$

$$y_{\text{bright}} \cong \frac{\lambda L}{d} m \quad \text{(for small } \theta\text{)} \qquad (37.5)$$

Dark fringes (destructive interference) will appear at points on the screen which correspond to path differences of an odd multiple of half wavelengths. For these points of destructive interference, waves which leave the two slits in phase arrive at the screen 180° out of phase.

$$\delta = d\sin\theta = \left(m + \tfrac{1}{2}\right)\lambda \qquad (37.3)$$

for $m = 0, \pm 1, \pm 2, \ldots$

$$y_{\text{dark}} \cong \frac{\lambda L}{d}\left(m + \tfrac{1}{2}\right) \quad \text{(for small } \theta\text{)} \qquad (37.6)$$

The wavelength of light in a medium, λ_n, is less than the wavelength in free space, λ.

$$\lambda_n = \frac{\lambda}{n}$$

Two waves which leave the slits initially in phase arrive at the screen *out of phase* by an amount which depends on the path difference.

$$\phi = \frac{2\pi}{\lambda} d \sin\theta \qquad (37.8)$$

$$\phi = \frac{2\pi}{\lambda} \delta$$

Two waves initially in phase and of equal amplitude (E_0) will produce a resultant amplitude at some point on the screen which depends on the phase difference (and therefore on the path difference).

$$E_p = 2E_0 \cos\left(\frac{\phi}{2}\right) \sin\left(\omega t + \frac{\phi}{2}\right) \qquad (37.10)$$

The average light intensity (I_{av}) at any point P on the screen is proportional to the square of the amplitude of the resultant wave. *The average intensity can be written:*

Where $I_0 = 4E_0^2$,

- as a function of phase difference ϕ;

$$I_{av} = I_0 \cos^2\left(\frac{\phi}{2}\right) \qquad (37.11)$$

- as a function of the angle (θ) subtended by the screen point at the source midpoint; or

$$I_{av} = I_0 \cos^2\left(\frac{\pi d \sin\theta}{\lambda}\right) \qquad (37.12)$$

- as a function of the vertical distance (y) from the center of the screen.

$$I_{av} = I_0 \cos^2\left(\frac{\pi d y}{\lambda L}\right) \text{ (for small } \theta\text{)} \qquad (37.13)$$

The intensity pattern observed on the screen will vary as the number of equally spaced sources is increased; however, the *positions of the principle maxima remain the same.*

In thin-film interference (Figure 37.2), the wavelength of light in the film, λ_n, is not the same as the wavelength in the surrounding medium.

$$\lambda_n = \frac{\lambda}{n} \qquad (37.14)$$

The conditions for interference in thin films can be stated in terms of the thickness (t) and index of refraction of the film. The conditions expressed by Equations 37.16 and 37.17 are valid when the film is surrounded by a common medium.

$$2nt = \left(m + \tfrac{1}{2}\right)\lambda \quad (m = 0, 1, 2, \ldots) \qquad (37.16)$$

constructive interference

$$2nt = m\lambda \quad (m = 0, 1, 2, \ldots) \qquad (37.17)$$

destructive interference

SUGGESTIONS, SKILLS, AND STRATEGIES

The following features should be kept in mind while working thin-film interference problems:

- Identify the thin film from which interference effects are being observed.

- The type of interference that occurs in a specific problem is determined by the phase relationship between that portion of the wave reflected at the upper surface of the film and that portion reflected at the lower surface of the film.

- Phase differences between the two portions of the wave occur because of differences in the distances traveled by the two portions and by phase changes occurring upon reflection.

  The wave reflected from the lower surface of the film has to travel a distance equal to twice the thickness of the film before it returns to the upper surface of the film where it interferes with that portion of the wave reflected at the upper surface. When *distances alone are considered,* if this extra distance is equal to an integral multiple of λ the interference will be constructive; if the extra distance equals $\frac{1}{2}\lambda$, $\frac{3}{2}\lambda$, and so forth, destructive interference will occur.

  Reflections may change the results above which are based solely on the distance traveled. When a wave traveling in a particular medium reflects off a surface having a higher index of refraction than the one it is in, a 180° phase shift occurs. This has the same effect as if the wave lost $\frac{1}{2}\lambda$. These losses must be considered in addition to those losses that occur because of the extra distance one wave travels over another.

- When distance and phase changes upon reflection are both taken into account, the interference will be constructive if the waves are out of phase by an integral multiple of λ Destructive interference will occur when the phase difference is $\frac{1}{2}\lambda$, $\frac{3}{2}\lambda$, and so forth.

The technique of phasor addition offers a convenient alternative to the algebraic method for finding the resultant wave amplitude at some point on a screen. This is especially true when a large number of waves are to be combined. The method of phasor addition is outlined in the following steps and is illustrated in Figure 37.3 for the case of two equal amplitude waves differing in phase by an angle of ϕ.

- Draw the phasors representing the waves end to end. The angle between successive phasors is equal to the phase angle between the waves from successive source slits. The length of each phasor is proportional to the magnitude of the wave it represents.

- The resultant wave is the vector sum (vector from the tail of the first phasor to the head of the last one) of the individual phasors.

- The phase angle (α) of the resultant wave is the angle between the direction of the resultant phasor and the direction of the first phasor.

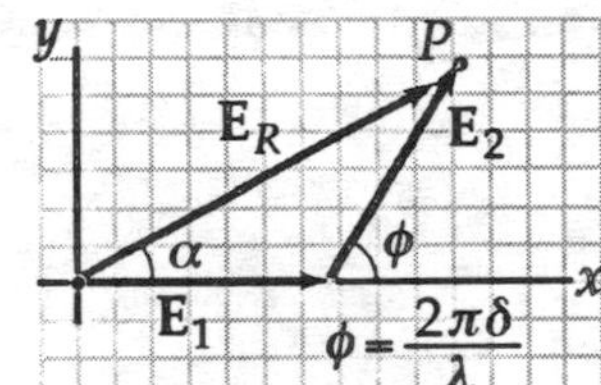

Figure 37.3

REVIEW CHECKLIST

▷ Describe Young's double-slit experiment to demonstrate the wave nature of light. Account for the phase difference between light waves from the two sources as they arrive at a given point on the screen. State the conditions for constructive and destructive interference in terms of each of the following: path difference, phase difference, distance from the center of the screen, and angle subtended by the observation point at the source mid-point.

▷ Outline the manner in which the superposition principle leads to the correct expression for the intensity distribution on a distant screen due to two coherent sources of equal intensity.

▷ Describe the use of the phasor diagram method to determine the amplitude and phase of the wave which is the resultant of two or three coherent sources.

▷ Account for the conditions of constructive and destructive interference in thin films considering both path difference and any expected phase changes due to reflection.

SOLUTIONS TO SELECTED END-OF-CHAPTER PROBLEMS

1. A pair of narrow, parallel slits separated by 0.25 mm is illuminated by green light (λ = 546.1 nm). The interference pattern is observed on a screen 1.2 m away from the plane of the slits. Calculate the distance (a) from the central maximum to the first bright region on either side of the central maximum and (b) between the first and second dark bands.

Solution When using Equations 37.2 and 37.4 as in this problem, you may have difficulty keeping track of four different symbols for distances. Three are shown in the drawing to the right. Note that:

y is the unknown distance from the bright fringe at the center of the interference pattern ($m = 0$), to the adjacent bright fringe ($m = 1$);

λ (not shown), is the wavelength of the light used.

Since for *very small* θ,

$$\sin\theta \cong \tan\theta \cong \theta \quad \text{and} \quad \tan\theta = y/L:$$

Equation 37.2, $\sin\theta = \dfrac{m\lambda}{d}$ becomes Equation 37.5, $y_{\text{bright}} \cong \dfrac{\lambda L}{d} m$

(a) $$y_{\text{bright}} = \frac{(546\times10^{-9}\ \text{m})(1.20\ \text{m})}{0.250\times10^{-3}\ \text{m}}(1) = 2.62\times10^{-3}\ \text{m} = 2.62\ \text{mm} \quad \Diamond$$

(b) If you have trouble remembering whether Equation 37.5 or Equation 37.6 applies to a given situation, you can instead remember that the first bright band is in the center, and dark bands are halfway between bright bands. Thus, Equation 37.5 applies with m = 0, 1, 2, . . . for bright bands, and with m = 0.5, 1.5, 2.5, . . . for dark bands. The dark band version of Equation 37.5 is simply Equation 37.6:

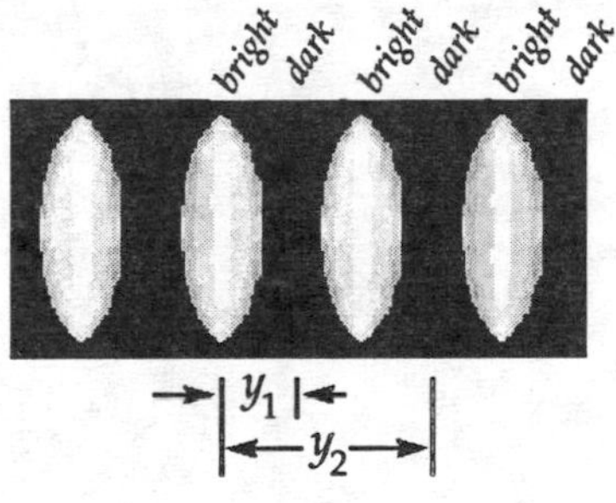

$$y_{\text{dark}} = \frac{\lambda L}{d}\left(m + \tfrac{1}{2}\right)$$

$$\Delta y_{\text{dark}} = \left(1 + \tfrac{1}{2}\right)\frac{\lambda L}{d} - \left(0 + \tfrac{1}{2}\right)\frac{\lambda L}{d} = \frac{\lambda L}{d} = 2.62\ \text{mm} \quad \Diamond$$

10. Two radio antennas separated by 300 m as in Figure P37.10 simultaneously transmit identical signals at the same wavelength. A radio in a car traveling due north receives the signals. (a) If the car is at the position of the second maximum, what is the wavelength of the signals? (b) How much farther must the car travel to encounter the next minimum in reception? (*Caution:* Do not use small-angle approximations in this problem.)

Solution Note, with the conditions given, the small angle approximation *does not work well.* That is, sin θ, tan θ, and θ are significantly different. The approach to be used is outlined below.

(a) At the $m = 2$ maximum,

$$\tan\theta = \frac{400 \text{ m}}{1000 \text{ m}} = 0.400$$

and $\theta = 21.8°$ so

$$\lambda = \frac{d\sin\theta}{m} = \frac{(300 \text{ m})\sin 21.8°}{2}$$

$\lambda = 55.7$ m ◊

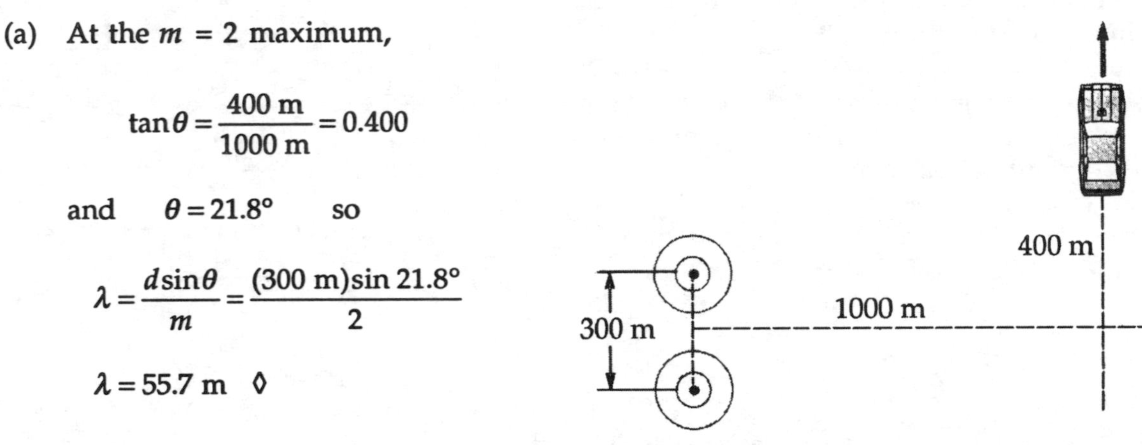

Figure P37.10

(b) The next minimum encountered is the $m = 2$ minimum, and at that point,

$$d\sin\theta = \left(m + \tfrac{1}{2}\right)\lambda \quad \text{which becomes} \quad d\sin\theta = \tfrac{5}{2}\lambda$$

or $$\sin\theta = \frac{5\lambda}{2d} = \frac{5(55.7 \text{ m})}{2(300 \text{ m})} = 0.464 \quad \text{and} \quad \theta = 27.7°$$

so $$y = (1000 \text{ m})\tan 27.7° = 524 \text{ m}$$

Therefore, the car must travel an additional 124 m. ◊

13. In Figure 37.1 of this solution manual, let L = 120 cm and d = 0.25 cm. The slits are illuminated with coherent 600-nm light. Calculate the distance y above the central maximum for which the average intensity on the screen is 75% of the maximum.

Solution For small θ, $$I_{avg} = I_0 \cos^2\left(\frac{\pi d \sin\theta}{\lambda}\right)$$

From the drawing, we see that $$\sin\theta \cong \frac{y}{L}$$

Substituting, we have $$y = \frac{\lambda L}{\pi d}\cos^{-1}\sqrt{\frac{I_{avg}}{I_0}}$$

In addition, we are given $I_{avg} = 0.75 I_0$

Thus, $$y = \frac{(6.0\times10^{-7}\text{ m})(1.20\text{ m})}{\pi(2.5\times10^{-3}\text{ m})}\cos^{-1}\sqrt{0.75}$$

$$y = 4.8\times10^{-5}\text{ m} = 0.048\text{ mm}$$ ◊

---

18. In Figure 37.1 of this solution manual, let L = 1.2 m and d = 0.12 mm, and assume that the slit system is illuminated with monochromatic 500-nm light. Calculate the phase difference between the two wave fronts arriving at P when (a) θ = 0.50° and (b) y = 5.0 mm. (c) What is the value of θ for which (1) the phase difference is 0.333 rad and (2) the path difference is $\lambda/4$?

Solution

(a) The path difference is $\delta = d\sin\theta = (0.12\times10^{-3}\text{ m})\sin 0.5° = 1.05\times10^{-6}\text{ m}$

The phase difference is $$\phi = \frac{2\pi\delta}{\lambda} = \frac{2\pi(1.05\times10^{-6}\text{ m})}{(500\times10^{-9}\text{ m})} = 13.2\text{ rad} = 0.59\text{ rad}$$ ◊

This is equivalent to 34° ◊

(b) $\tan\theta = \dfrac{y}{L} = \dfrac{5.0\times10^{-3}\text{ m}}{1.2\text{ m}} \cong \sin\theta$

$$\phi = \frac{2\pi d\sin\theta}{\lambda} = \frac{2\pi\left(0.12\times10^{-3}\text{ m}\right)\left(4.17\times10^{-3}\right)}{500\times10^{-9}\text{ m}} = 6.28\text{ rad},\quad \text{or (equivalently) } \phi = 0 \quad \Diamond$$

(c) $\phi = 0.333 = \dfrac{2\pi d\sin\theta}{\lambda}$;

$$\theta = \sin^{-1}\left(\frac{\lambda\phi}{2\pi d}\right) = \sin^{-1}\left[\frac{\left(500\times10^{-9}\text{ m}\right)(0.333)}{2\pi\left(1.2\times10^{-4}\text{ m}\right)}\right] = 1.3\times10^{-2}\text{ deg} \quad \Diamond$$

$\dfrac{\lambda}{4} = d\sin\theta$;

$$\theta = \sin^{-1}\left(\frac{\lambda}{4d}\right) = \sin^{-1}\left[\frac{\left(500\times10^{-9}\text{ m}\right)}{4\left(1.2\times10^{-4}\text{ m}\right)}\right] = 5.8\times10^{-2}\text{ deg} \quad \Diamond$$

---

22. Monochromatic light (λ = 632.8 nm) is incident on two parallel slits separated by 0.20 nm. What is the distance to the first maximum and its intensity (relative to the central maximum) on a screen 2.0 m beyond the slits?

Solution From Equation 37.2, $d\sin\theta = m\lambda$

Because $\sin\theta \cong \dfrac{y}{L}$, and $m = 1$, $\quad y_{\text{bright}} = \dfrac{\lambda L}{d} = \dfrac{\left(632.8\times10^{-9}\text{ m}\right)(2.0\text{ m})}{2.0\times10^{-4}\text{ m}}$

$$y_{\text{bright}} = 6.33\text{ mm} \quad \Diamond$$

From Equation 37.13, the relative intensity will be $\dfrac{I_1}{I_0} = \cos^2\left(\dfrac{\pi d}{\lambda L}y_{\text{bright}}\right)$

But since $y_{\text{bright}} = \dfrac{\lambda L}{d}$, $\quad \dfrac{I_1}{I_0} = \cos^2(\pi) = 1$

So the intensity will be approximately the same as at the central maximum. $\Diamond$

---

23. Two narrow parallel slits separated by 0.85 mm are illuminated by 600-nm light, and the viewing screen is 2.80 m away from the slits. (a) What is the phase difference between the two interfering waves on a screen at a point 2.50 mm from the central bright fringe? (b) What is the ratio of the intensity at this point to the intensity at the center of a bright fringe?

Solution From Equation 37.8,

(a) $$\phi = \frac{2\pi d}{\lambda}\sin\theta = \frac{2\pi d}{\lambda}\left(\frac{y}{\sqrt{y^2+D^2}}\right) \cong \frac{2\pi y d}{\lambda D}$$

$$\phi = \frac{2\pi\left(0.85\times10^{-3}\text{ m}\right)\left(2.50\times10^{-3}\text{ m}\right)}{\left(600\times10^{-9}\text{ m}\right)(2.8\text{ m})} = 7.95\text{ rad} = 96° \quad \Diamond$$

(b) $$\frac{I}{I_{\max}} = \frac{\cos^2\left(\frac{\pi d}{\lambda}\sin\theta\right)}{\cos^2\left(\frac{\pi d}{\lambda}\sin\theta_{\max}\right)} = \frac{\cos^2\left(\frac{\phi}{2}\right)}{\cos^2(m\pi)} = \cos^2\left(\frac{\phi}{2}\right) = 0.45 \quad \Diamond$$

---

27. Determine the resultant of the two waves $E_1 = 6.0 \sin(100\,\pi t)$ and $E_2 = 8.0 \sin(100\,\pi t + \pi/2)$.

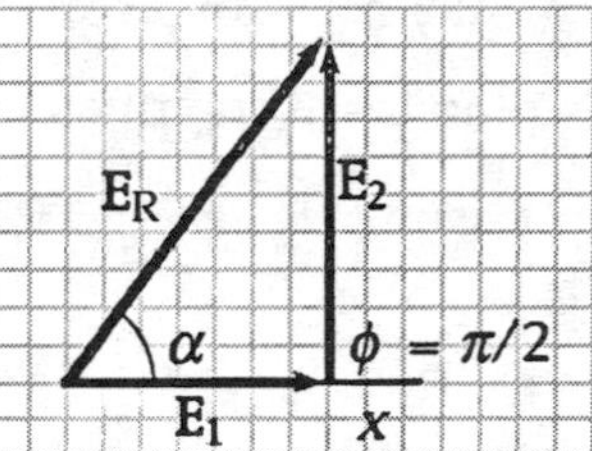

Solution Let the x axis lie along $\mathbf{E}_1$ in phase space. Its component form is then $\mathbf{E}_1 = 6.0\mathbf{i} + 0\mathbf{j}$. The components of $\mathbf{E}_2$ are

$$\mathbf{E}_2 = 8.0\cos\left(\frac{\pi}{2}\right)\mathbf{i} + 8.0\sin\left(\frac{\pi}{2}\right)\mathbf{j} = 8.0\mathbf{j}$$

The resultant is then $\mathbf{E}_R = 6.0\mathbf{i} + 8.0\mathbf{j}$, with an amplitude of $\sqrt{6.0^2+8.0^2} = 10$.

The phase, from $\tan\alpha = \dfrac{8.0}{6.0}$, is $\alpha = 0.927$ rad.

Thus, $$\mathbf{E}_R = 10\sin(100\pi t + 0.927) \quad \Diamond$$

---

31. Consider N coherent sources described by

$$E_1 = E_0 \sin(\omega t + \phi), \qquad E_2 = E_0 \sin(\omega t + 2\phi),$$

$$E_3 = E_0 \sin(\omega t + 3\phi), \ldots, \qquad E_N = E_0 \sin(\omega t + N\phi).$$

Find the minimum value of ϕ for which $E_R = E_1 + E_2 + E_3 + \ldots + E_N$ is zero.

Solution Take $\phi = \dfrac{360°}{N}$, and $E_R = \Sigma E_0 \sin(\omega t + N\phi) = 0$

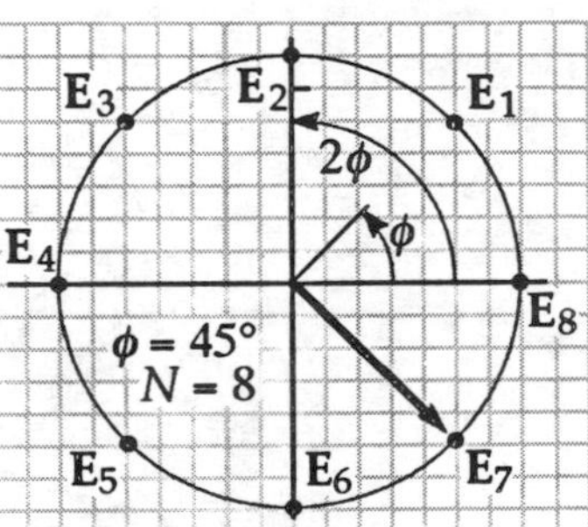

where N defines the number of coherent sources. In essence, the set of electric field components completes a full 360° circle, returning to zero.

In this situation, when the vectors are added, the sources are symmetric about $E_R = 0$, and thus sum to zero.

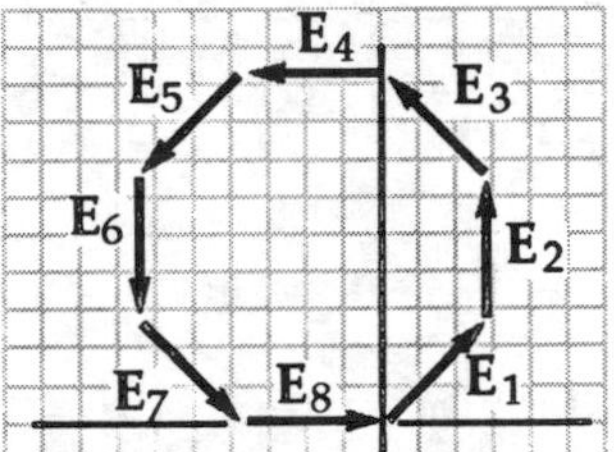

Alternatively, one can note that the vectors must create a regular polygon when summed, in order to return to the origin. (The figures to the right may help.)

If N has an integer divisor m, then the choice $\phi = \dfrac{360°}{m}$ will also make all the phasors add to zero. ◊

---

37. A thin layer of liquid methylene iodide (n = 1.756) is sandwiched between two flat parallel plates of glass. What must be the thickness of the liquid layer if normally incident 600-nm light is to be strongly reflected?

Solution A phase change of π occurs upon reflection at the first interface but not at the second. The total phase shift of the second reflected wave relative to the first is then:

$$\phi = 2\pi = \frac{2nt}{\lambda}(2\pi) + \pi$$

$$t = \frac{\lambda}{4n} = \frac{(600 \times 10^{-9}\ \text{m})}{4(1.756)} = 85.4 \times 10^{-9}\ \text{m} \qquad ◊$$

---

39. An oil film (n = 1.45) floating on water is illuminated by white light at normal incidence. The film is 280 nm thick. Find (a) the dominant observed color in the reflected light and (b) the dominant color in the transmitted light. Explain your reasoning.

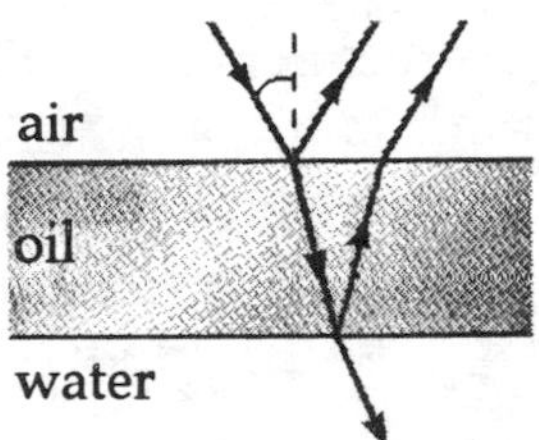

Solution The light reflected from the top of the oil film undergoes phase reversal. Since 1.45 > 1.33, the light reflected from the bottom undergoes no reversal. For constructive interference of reflected light, we then have

$$2nt = \left(m + \tfrac{1}{2}\right)\lambda \qquad \text{or} \qquad \lambda = \frac{2nt}{\left(m+\frac{1}{2}\right)} = \frac{2(1.45)(280\text{ nm})}{\left(m+\frac{1}{2}\right)}$$

(a) For $m = 0$, $\lambda = 1624$ nm (infrared)
$m = 1$, $\lambda = 541$ nm (green)
$m = 2$, $\lambda = 325$ nm (ultraviolet)

Infrared and ultraviolet are invisible to the human eye, so the dominant color is green. ◊

(b) Subtracting green from white gives red + violet = purple. ◊

We could also note that the reflected light contains little red and violet according to the condition for destructive interference $2nt = m\lambda$.

For $m = 1$, $\lambda = 812$ nm (near infrared)
$m = 2$, $\lambda = 406$ nm (violet)

So these colors make it through to be in the transmitted beam.

---

43. An air wedge is formed between two glass plates separated at one edge by a very fine wire as in Figure P37.43. When the wedge is illuminated from above by 600-nm light, 30 dark fringes are observed. Calculate the radius of the wire.

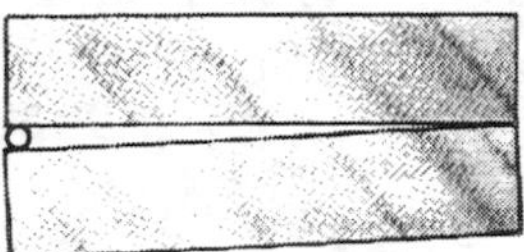

Figure P37.43

Solution Light reflecting from the bottom surface of the top plate undergoes no phase shift, while light reflecting from the top surface of the bottom plate is shifted by π, and also has to travel an extra distance $2t$, where t is the thickness of the air wedge.

For destructive interference, $2t = m\lambda \quad (m = 0, 1, 2, 3, \ldots)$

The first dark fringe appears where $m = 0$ at the line of contact between the plates. The thirtieth dark fringe gives for the diameter of the wire $2t = 29\lambda$, and $t = 14.5\lambda$.

$$r = t/2 = 7.25(600 \times 10^{-9}\text{ m}) = 4.35\ \mu\text{m} \quad \lozenge$$

---

45. Mirror M_1 in Figure P37.45 is displaced a distance ΔL. During this displacement, 250 fringes (formation of successive dark or bright bands) are counted. The light being used has a wavelength of 632.8 nm. Calculate the displacement ΔL.

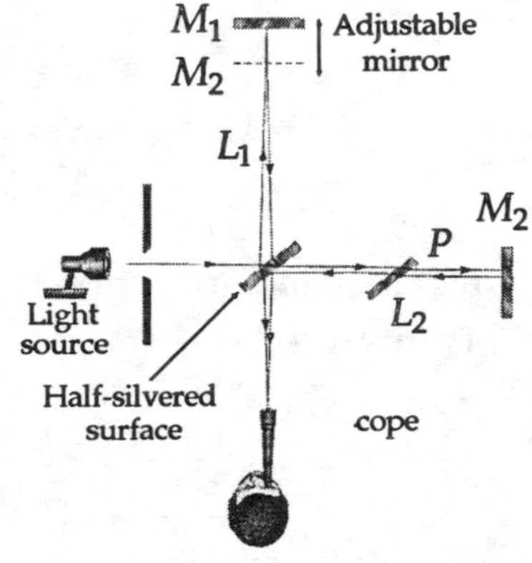

Figure P37.45

Solution When the mirror on one arm is displaced by ΔL, the path difference increases by $2\Delta L$. A shift resulting in the formation of successive dark (or bright) fringes requires a path length change of one-half wavelength.

Therefore, $2\Delta L = \dfrac{m\lambda}{2}$, where in this case $m = 250$.

$$\Delta L = \frac{m\lambda}{4} = \frac{(250)(6.328 \times 10^{-7}\text{ m})}{4} = 3.955 \times 10^{-5}\text{ m} \quad \lozenge$$

---

51. Astronomers observed a 60-MHz radio source both directly and by reflection from the sea. If the receiving dish is 20 m above sea level, what is the angle of the radio source above the horizon at first maximum?

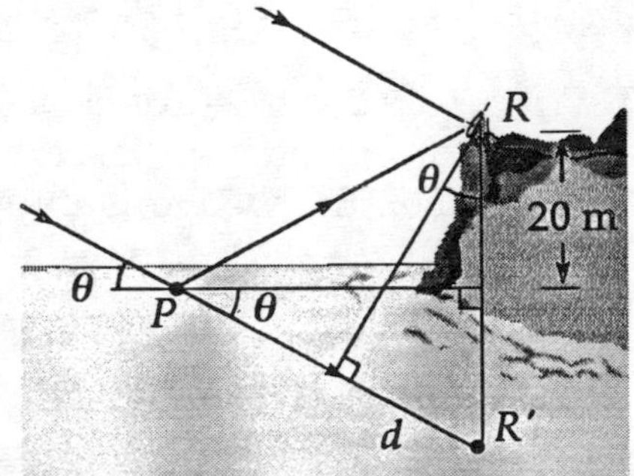

Solution One radio wave reaches the receiver R directly from the distant source at an angle θ above the horizontal. The other wave reflects from the water at P, there undergoing phase reversal.

Constructive interference first occurs for a path difference of $d = \lambda/2$. The angles θ in the figure are equal because they each form part of a right triangle with a shared angle at R'.

It is equally far from P to R as from P to R', the mirror image of the telescope.

So the path difference is $$d = 2(20\text{ m})\sin\theta = (40\text{ m})\sin\theta$$

The wavelength is $$\lambda = \frac{c}{f} = \frac{3.00\times 10^8\text{ m/s}}{60\times 10^6\text{ Hz}} = 5.0\text{ m}$$

Applying $(40\text{ m})\sin\theta = \lambda/2$, $$\sin\theta = \frac{5.0\text{ m}}{80\text{ m}} \quad\text{and}\quad \theta = 3.6° \quad ◊$$

---

53. Measurements are made of the intensity distribution in a Young's interference pattern (Fig. P37.53). At a particular value of y, it is found that $I/I_0 = 0.81$ when 600-nm light is used. What wavelength of light should be used to reduce the relative intensity at the same location to 64%?

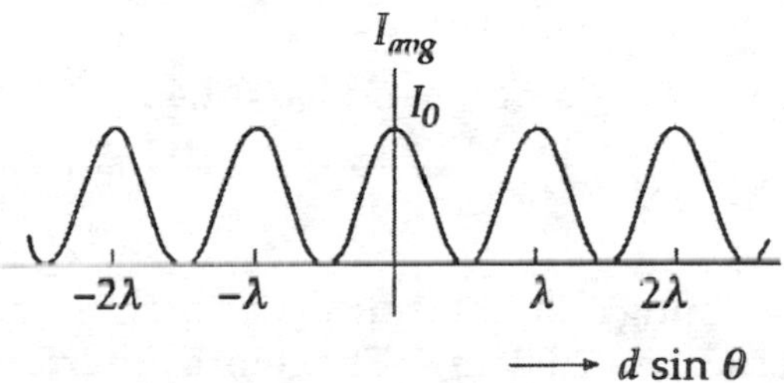

Figure P37.53

Solution

From Equation 37.13, $$\frac{I_1}{I_0} = \cos^2\left(\frac{\pi y d}{\lambda_1 L}\right) = 0.81$$

Therefore, $$\frac{\pi y d}{L} = \lambda_1 \cos^{-1}\sqrt{0.81} = (600\text{ nm})\cos^{-1}(0.90) = 270.6\text{ nm}$$

For the second wavelength, $$\frac{I_2}{I_0} = \cos^2\left(\frac{\pi y d}{\lambda_2 L}\right) = 0.64$$

and $$\lambda_2 = \frac{\pi y d / L}{\cos^{-1}\sqrt{0.64}}$$

But we know the numerator: $$\lambda_2 = \frac{270.6\text{ nm}}{\cos^{-1}\sqrt{0.64}} = 420\text{ nm} \quad ◊$$

---

55. In a Young's interference experiment, the two slits are separated by 0.15 mm, and the incident light includes light of wavelengths λ_1 = 540 nm and λ_2 = 450 nm. The overlapping interference patterns are formed on a screen 1.40 m from the slits. Calculate the minimum distance from the center of the screen to the point where a bright line of the λ_1 light coincides with a bright line of the λ_2 light.

Solution For Young's experiment, use Equation 37.2:

$$d\sin\theta = m_1\lambda_1 = m_2\lambda_2$$

$$\frac{\lambda_1}{\lambda_2} = \frac{540\text{ nm}}{450\text{ nm}} = \frac{m_2}{m_1} = \frac{6}{5}$$

Therefore,
$$\sin\theta = \frac{6\lambda_2}{d} = \frac{6(450\text{ nm})}{0.15\text{ mm}} = 0.018$$

Since $\sin\theta \cong \tan\theta$ and $L = 1.4$ m, $x = L\tan\theta = (0.018)(1.4\text{ m}) = 2.52\text{ cm}$ ◊

---

57. Young's double-slit experiment is performed with 589-nm light and a slits-to-screen distance of 2.00 m. The tenth interference minimum is observed 7.26 mm from the central maximum. Determine the spacing of the slits.

Solution In the equation, $d\sin\theta = \left(m + \frac{1}{2}\right)\lambda$, the first minimum is described by $m = 0$ and the tenth by $m = 9$.

So, $\sin\theta = \frac{\lambda}{d}\left(9 + \frac{1}{2}\right)$ Also, $\tan\theta = \frac{y}{L}$

For small θ, $\sin\theta \cong \tan\theta$. Thus,

$$d = \frac{9.5\lambda}{\sin\theta} = \frac{9.5\lambda L}{y} = \frac{9.5(5890\times 10^{-10}\text{ m})(2.0\text{ m})}{7.26\times 10^{-3}\text{ m}} = 1.54\times 10^{-3}\text{ m}$$

$d = 1.54$ mm ◊

---

59. A hair is placed at one edge between two flat glass plates 8.00 cm long. When this arrangement is illuminated with 600-nm light, 121 dark bands are counted, starting at the point of contact of the two plates. How thick is the hair?

Solution Light reflecting from the bottom surface of the top plate undergoes no phase shift, while light reflecting from the top surface of the bottom plate is phase shifted by π upon reflection and also has to travel extra distance $2t$, where t is the thickness of the air wedge. For destructive interference, $2t = m\lambda$ ($m = 0, 1, 2, 3, \ldots$). Therefore, for 121 dark bands, $2t = 120\lambda$.

$$t = 60\lambda = 3.60 \times 10^{-5}\text{ m} \quad \lozenge$$

---

63. The condition for constructive interference by reflection from a thin film in air as developed in Section 37.6 assumes nearly normal incidence. (a) Show that if the light is incident on the film at an angle $\theta_1 \gg 0$ (relative to the normal) then the condition for constructive interference is $2nt\cos\theta_2 = \left(m+\frac{1}{2}\right)\lambda$, where θ_2 is the angle of refraction. (b) Calculate the minimum thickness for constructive interference if 590-nm light is incident at an angle of 30° on a film that has an index of refraction of 1.38.

Solution

(a) In the figure, before they head off together in parallel paths, one beam travels distance y and suffers phase reversal on reflection; the other beam travels distance $2x$ inside the film, where its wavelength is λ/n. This makes its optical path length $2nx$; the condition for constructive interference is

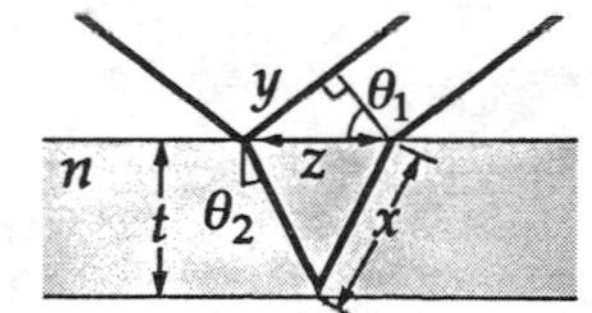

$$2nx - y = \left(m+\tfrac{1}{2}\right)\lambda$$

From the figure we have

$$\cos\theta_2 = \frac{t}{x}, \qquad \tan\theta_2 = \frac{z}{2t}, \qquad \text{and} \qquad \sin\theta_1 = \frac{y}{z}$$

Therefore, using Snell's law,

$$y = z\sin\theta_1 = (\sin\theta_1)2t\tan\theta_2 = \frac{(n\sin\theta_2)2t\sin\theta_2}{\cos\theta_2}$$

Then the condition for constructive interference becomes

$$\frac{2tn}{\cos\theta_2} - \frac{2nt\sin^2\theta_2}{\cos\theta_2} = \left(m+\tfrac{1}{2}\right)\lambda$$

Since $1-\sin^2\theta_2 = \cos^2\theta_2,$ this becomes $2nt\cos\theta_2 = \left(m+\tfrac{1}{2}\right)\lambda$ ◊

(b) $t = \left(\frac{1}{\cos\theta_2}\right)\left(\frac{\lambda}{2n}\right)\left(m+\tfrac{1}{2}\right)$ and for minimum thickness choose $m = 0$.

Also, from Snell's law, $\sin\theta_2 = \frac{\sin\theta_1}{n} = \frac{\sin 30°}{1.38} = 0.362$

$\theta_2 = 21.2°$ and $t = \frac{5.9\times10^{-7}\text{ m}}{(\cos 21.2°)4(1.38)} = 1.15\times10^{-7}\text{ m}$ ◊

---

64. Use phasor addition to find the resultant amplitude and phase constant when the following three harmonic functions are combined:

$$E_1 = \sin\left(\omega t + \frac{\pi}{6}\right) \qquad E_2 = 3.0\sin\left(\omega t + \frac{7\pi}{2}\right) \qquad E_3 = 6.0\sin\left(\omega t + \frac{4\pi}{3}\right)$$

Solution The phase diagram is shown at the right. From the diagram, and using vector addition, we find:

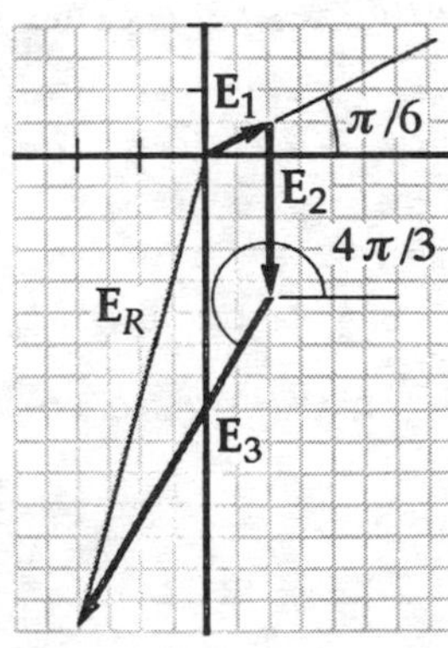

resultant amplitude = 7.99 ◊

and phase constant = 255° ◊

$$E_R = 7.99\sin(\omega t + 255°)$$

If you need to review vector addition, remember that you can convert vectors to and from their components using

$$R_x = R\cos\theta \qquad R_y = R\sin\theta \qquad \theta = \tan^{-1}\left(R_y/R_x\right) \qquad R = \sqrt{R_x^{\,2} + R_y^{\,2}}$$

and add the vector components to find the resultant vector.

70A. Consider the double-slit arrangement shown in Figure P37.70, where the slit separation is d and the slit-to-screen distance is L. A sheet of transparent plastic having an index of refraction n and thickness t is placed over the upper slit. As a result, the central maximum of the interference pattern moves upward a distance y'. Find y'.

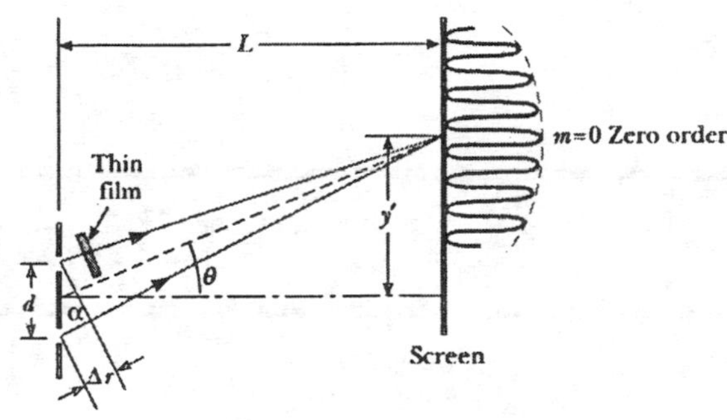

Figure P37.70

Solution Call t the thickness of the film. The central maximum corresponds to zero phase difference. Thus, the added distance Δr traveled by the light from the lower slit must introduce a phase difference equal to that introduced by the plastic film. The phase difference ϕ is

$$\phi = 2\pi\left(\frac{t}{\lambda_a}\right)(n-1)$$

The corresponding difference in *path length* Δr is

$$\Delta r = \phi\left(\frac{\lambda_a}{2\pi}\right) = 2\pi\left(\frac{t}{\lambda_a}\right)(n-1)\left(\frac{\lambda_a}{2\pi}\right) = t(n-1)$$

Note that the wavelength of the light does not appear in this equation. In Figure P37.70, the two rays from the slits are essentially parallel, so the angle θ may be expressed as

$$\tan\,\theta = \frac{\Delta r}{d} = \frac{y'}{L}$$

Eliminating Δr by substitution,

$$\frac{y'}{L} = \frac{t(n-1)}{d} \quad \text{gives} \quad y' = \frac{t(n-1)L}{d} \quad \Diamond$$

Chapter 38

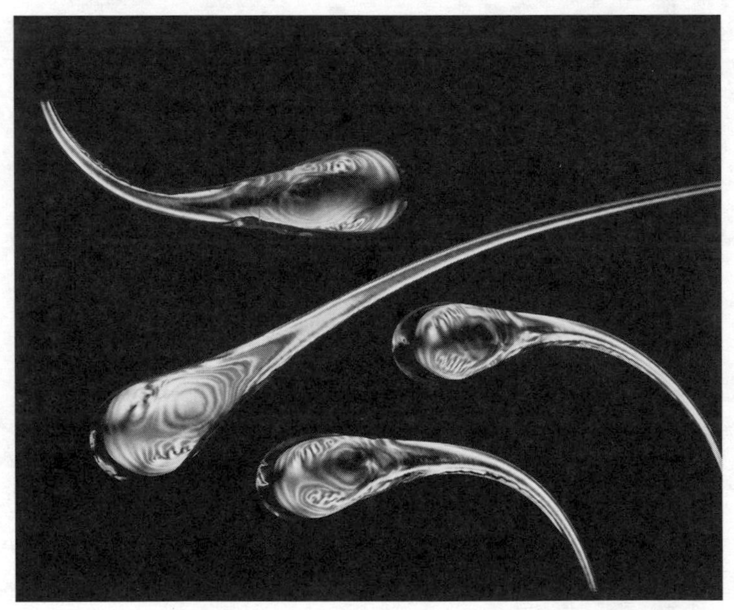

Diffraction and Polarization

DIFFRACTION AND POLARIZATION

INTRODUCTION

When light waves pass through a small aperture, a diffraction pattern is observed rather than a sharp spot of light, showing that light spreads beyond the aperture into regions where a shadow would be expected if light traveled in straight lines. Other waves, such as sound waves and water waves, also have this property of being able to bend around corners. This phenomenon, known as diffraction, can be regarded as interference from a great number of coherent wave sources. In other words, diffraction and interference are basically equivalent.

In Chapter 34, we learned that electromagnetic waves are transverse. That is, the electric and magnetic field vectors are perpendicular to the direction of propagation. In this chapter, we see that under certain conditions, light waves can be polarized in various ways, such as by passing light through polarizing sheets.

NOTES FROM SELECTED CHAPTER SECTIONS

38.1 Introduction to Diffraction

Fraunhofer diffraction occurs when light rays reaching a point (on a screen) from a diffracting source (edge, slit, etc.) are approximately parallel.

Fresnel diffraction occurs when the observing screen is a *finite* distance from the slit (or edge) and the light rays are not rendered parallel by a lens.

38.2 Single-Slit Diffraction

According to *Huygens' principle,* each portion of a slit acts as a source of waves; therefore, light from one portion of a slit can interfere with light from another portion.

In Figure 38.1 (a) on the following page, parallel light rays are shown incident on a slit of width a. In order to analyze the resulting Fraunhofer diffraction pattern, it is convenient to subdivide the width of the slit into many strips of equal width. Each substrip can be considered as a source represented by a phasor.

The positions of the minima in a single-slit diffraction pattern are determined by the conditions stated in Equation 38.1. The *secondary maxima* (of progressively diminished intensity) *lie approximately halfway between the minima.*

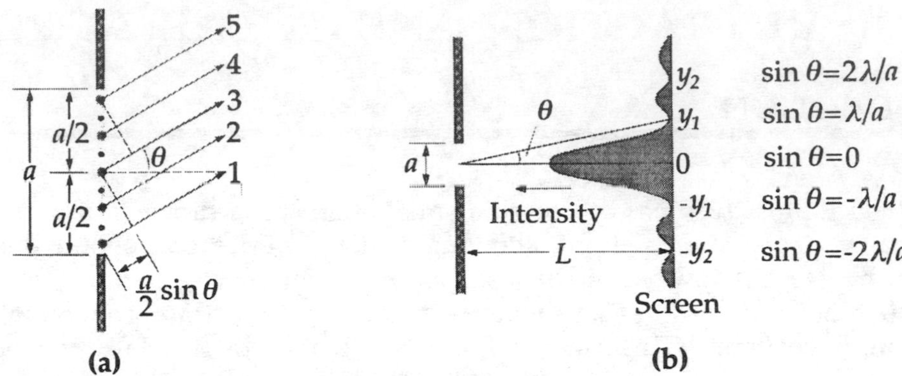

Figure 38.1 (a) Diffraction of light by a narrow slit of width *a*, and (b) the resulting intensity distribution.

38.3 Resolution of Single-Slit and Circular Apertures

When the central maximum of one image falls on the first minimum of another image, the images are said to be just resolved. This limiting condition of resolution is known as *Rayleigh's criterion.*

38.6 Polarization of Light Waves

If the electric field vector of the light wave emitted by the atoms or molecules of a light source vibrate in the *same direction at all times,* the wave is *plane polarized* or *linearly polarized.*

If the tip of the **E** vector rotates around a circle over time, the wave is *circularly polarized.*

If $\mathbf{E}_x$ and $\mathbf{E}_y$ are not equal in magnitude and differ in phase by 90°, the resulting wave is *elliptically polarized* —the tip of the **E** vector moves around in an ellipse.

Polarization by selective absorption is illustrated in Figure 38.2 on the following page. A beam of unpolarized light is incident on a sheet of polarizing material. The electric field vectors of the transmitted light will have nonzero components only along a direction parallel to the axis of polarization of the polarizer (indicated by straight lines drawn on the disk.) The light emerging from the polarizer is plane polarized and is allowed to fall on a second sheet of polarizing material (called the analyzer) such that there is an angle θ between the plane of polarization of the incident beam and the transmission axis of the analyzer.

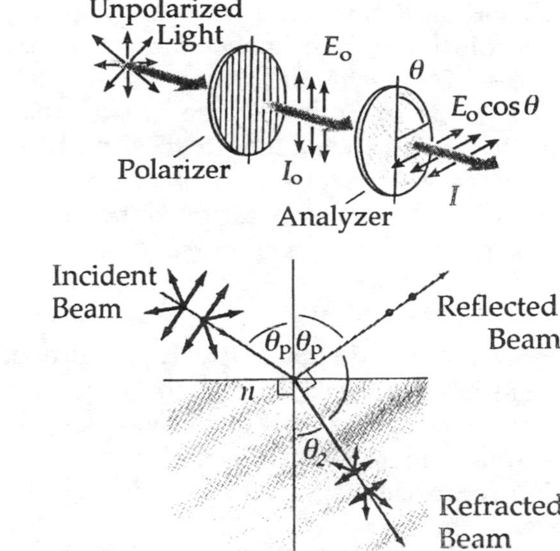

Figure 38.2 Only a fraction of the light is transmitted through the analyzer when θ lies between 0° and 180°.

Figure 38.3 Light reflected from a surface at the angle θ_p is completely polarized, when $\tan(\theta_p) = n$.

Polarization by reflection occurs when a light beam is reflected from a surface such that the reflected beam is perpendicular to the refracted beam. The *reflected component will be completely polarized* with its electric field vector parallel to the reflecting surface; and the *refracted (or transmitted) beam will be partially polarized.* This situation is illustrated in Figure 38.3.

EQUATIONS AND CONCEPTS

In single-slit diffraction, the total phase difference β between the first and last phasors will depend on the angle θ which determines the direction to an arbitrary point on the screen.

$$\beta = \frac{2\pi}{\lambda} a \sin\theta \tag{38.3}$$

The resultant phasor will be zero (the condition for destructive interference) when β equals an integer multiple of 2π. This leads to a general *condition for destructive interference* in terms of θ.

$$\sin\theta = m\frac{\lambda}{a} \tag{38.1}$$

$(m = \pm 1, \pm 2, \pm 3, \ldots)$

(Destructive interference)

The *intensity* at any point on the screen is given in terms of the intensity I_0 at $\theta = 0$ where β is given by Equation 38.3.

$$I_\theta = I_0 \left[\frac{\sin(\beta/2)}{\beta/2}\right]^2 \tag{38.4}$$

Rayleigh's criterion states the condition for the resolution of two images due to nearby sources. For a slit, the angular separation between the sources must be greater than the ratio of the wavelength to slit width. In the case of a *circular aperture*, the minimum angular separation depends on D, the diameter of the aperture (or lens).

$$\theta_{\min} = \frac{\lambda}{a} \quad \text{(slit)} \tag{38.6}$$

$$\theta_{\min} = 1.22\frac{\lambda}{D} \quad \text{(circular aperture)} \tag{38.7}$$

A grating of equally spaced parallel slits (separated by a distance d) will produce an interference pattern in which there is a series of maxima for each wavelength. Maxima due to wavelengths of different value comprise a spectral order denoted by order number m.

$$d\sin\theta = m\lambda \tag{38.8}$$

$$(m = 0, 1, 2, 3, \ldots)$$

The *resolving power* of a grating increases as the *number of lines illuminated* is increased and is proportional to the order in which the spectrum is observed.

$$R = \frac{\lambda}{\Delta\lambda} \tag{38.9}$$

$$R = Nm \tag{38.10}$$

When the order number m is substituted from Equation 38.8, $d\sin\theta = m\lambda$, into Equation 38.10, it becomes clear that *the resolution depends on the width of the grating* (Nd). It should also be noted that the angular width $\Delta\theta$ of a spectral line formed by a diffraction grating is inversely proportional to the width of the grating. This statement can be written as $\Delta\theta \propto 1/Nd$.

$$R = \frac{Nd}{\lambda}\sin\theta$$

In the zeroth order (central maximum) all wavelengths are indistinguishable. If in a particular order ($m > 0$) R = 10 000, the grating will produce a spectrum in which wavelengths differing in value by 1 part in 10 000 can be resolved.

Bragg's law gives the conditions for *constructive interference* of x-rays reflected from the parallel planes of a crystalline solid separated by a distance *d. The angle* θ *is the angle between the incident beam and the surface.*

$$2d\sin\theta = m\lambda \qquad (m = 1, 2, 3, \ldots) \qquad (38.11)$$

When polarized light is incident on a polarizing film, the fraction of the incident intensity transmitted through the film depends on the angle between the transmission axis of the polarizer and the electric field vector of the incident light. This is known as *Malus's law.*

$$I = I_0\cos^2\theta \qquad (38.12)$$

Brewster's law gives the polarizing angle for a particular surface. The *polarizing angle* θ_p is the angle of incidence for which the reflected beam is *completely polarized.*

$$n = \tan\theta_p \qquad (38.13)$$

REVIEW CHECKLIST

▷ Determine the positions of the maxima and minima in a single-slit diffraction pattern and calculate the intensities of the secondary maxima relative to the intensity of the central maximum. Determine the positions of the principal maxima in the interference pattern of a diffraction grating.

▷ Determine whether or not two sources under a given set of conditions are resolvable as defined by Rayleigh's criterion.

▷ Understand what is meant by the resolving power and the dispersion of a grating, and calculate the resolving power of a grating under specified conditions.

▷ Describe the technique of x-ray diffraction and make calculations of the lattice spacing using Bragg's law.

▷ Understand how the state of polarization of a light beam can be determined by use of a polarizer-analyzer combination. Describe qualitatively the polarization of light by selective absorption, reflection, scattering, and double refraction. Also, make appropriate calculations using Malus's law and Brewster's law.

SOLUTIONS TO SELECTED END-OF-CHAPTER PROBLEMS

7. A screen is placed 50 cm from a single slit, which is illuminated with 690-nm light. If the distance between the first and third minima in the diffraction pattern is 3.0 mm, what is the width of the slit?

Solution In the equation for single-slit diffraction minima at small angles,

$$\frac{y}{L} \cong \sin\theta = \frac{m\lambda}{a},$$

take differences between the first and third minima, to see that

$$\frac{\Delta y}{L} = \frac{\Delta m\lambda}{a} \quad \text{with } \Delta y = 3.0\times10^{-3}\text{ m, and } \Delta m = 3-1 = 2$$

The width of the slit is then

$$a = \frac{\lambda L\Delta m}{\Delta y} = \frac{(690\times10^{-9}\text{ m})(0.50\text{ m})(2)}{3.0\times10^{-3}\text{ m}} = 2.3\times10^{-4}\text{ m} \quad \Diamond$$

---

8. In Equation 38.4, let $\beta/2 \equiv \phi$ and show that $I = 0.5I_0$ when $\sin\phi = \phi/\sqrt{2}$.

Solution

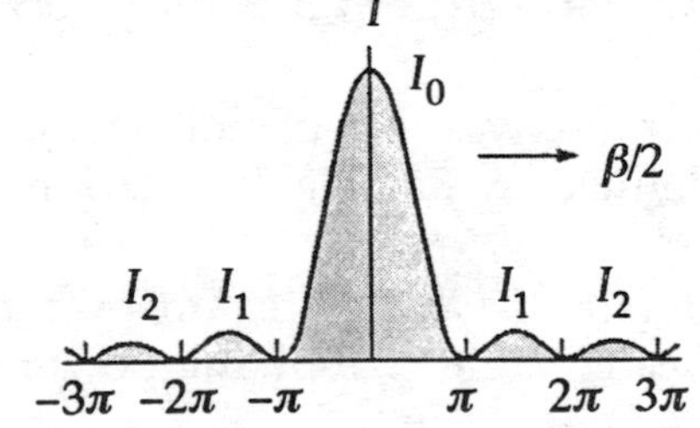

This problem, along with Problems 9 and 76, gives the half-width at half-maximum of the central peak in a single-slit diffraction pattern, which is one measure of how much a wave spreads when it goes through an opening. Equation 38.4 becomes

$$I = I_0\left(\frac{\sin\phi}{\phi}\right)^2 = 0.5I_0$$

Thus, $$\frac{\sin\phi}{\phi} = \frac{1}{\sqrt{2}} \quad \text{and} \quad \sin\phi = \frac{\phi}{\sqrt{2}} \quad \Diamond$$

---

11. A diffraction pattern is formed on a screen 120 cm away from a 0.40-mm-wide slit. Monochromatic 546.1-nm light is used. Calculate the fractional intensity I/I_0 at a point on the screen 4.1 mm from the center of the principle maximum.

Solution

$$\sin\theta \cong \frac{y}{L} = \frac{4.1\times10^{-3}\text{ m}}{1.2\text{ m}} = 3.417\times10^{-3}$$

$$\frac{\beta}{2} = \frac{\pi a \sin\theta}{\lambda} = \frac{\pi\left(4.0\times10^{-4}\text{ m}\right)\left(3.417\times10^{-3}\text{ m}\right)}{546.1\times10^{-9}\text{ m}} = 7.863\text{ rad}$$

$$\frac{I}{I_0} = \left[\frac{\sin\beta/2}{\beta/2}\right]^2 = \left[\frac{\sin(7.863\text{ rad})}{7.863\text{ rad}}\right]^2 = 1.6\times10^{-2} \quad \Diamond$$

---

13. A helium-neon laser emits light that has a wavelength of 632.8 nm. The circular aperture through which the beam emerges has a diameter of 0.50 cm. Estimate the diameter of the beam 10.0 km from the laser.

Solution Following Equation 38.7 for diffraction from a circular opening, the beam spreads into a cone of half-angle

$$\theta_m = 1.22\frac{\lambda}{D} = 1.22\frac{\left(632.8\times10^{-9}\text{ m}\right)}{(0.0050\text{ m})} = 1.54\times10^{-4}\text{ rad}$$

The radius of the beam ten kilometers away is, from the definition of radian measure,

$$r_{\text{beam}} = \theta_m\left(1.0\times10^4\text{ m}\right) = 1.544\text{ m}$$

and its diameter is

$$d_{\text{beam}} = 2r_{\text{beam}} = 3.1\text{ m} \quad \Diamond$$

---

19. At what distance could one theoretically distinguish two automobile headlights separated by 1.4 m? Assume a pupil diameter of 6.0 mm and yellow headlights ($\lambda = 580$ nm). The index of refraction in the eye is approximately 1.33.

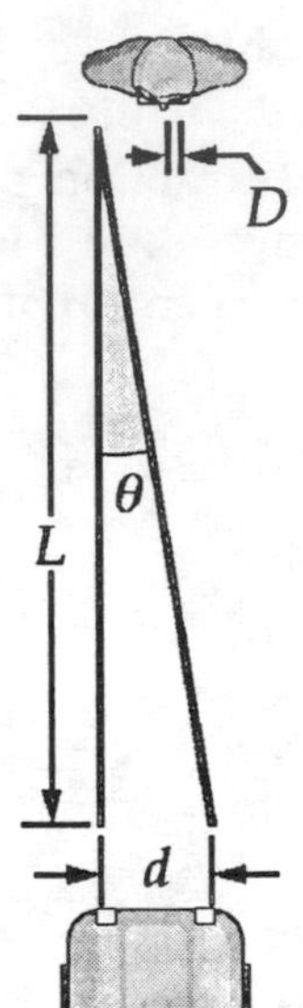

Solution Light from each source diffracts as it passes through the pupil of the eye. When barely resolved, the central maximum of one diffraction pattern on the retina will fall on the first minimum in the other diffraction pattern, the angle between them being:

$$\theta = \frac{d}{L} = 1.22\frac{\lambda}{D}$$

where $d = 1.4 \text{ m}$ = separation of the headlights

L = distance to the car

$\lambda = \frac{580}{1.33} \text{ nm}$ = wavelength of light in the eye

$D = 0.0060 \text{ m}$ = distance across an enlarged pupil

Substituting the given values, we have

$$L = \frac{Dd}{1.22\lambda} = \frac{(0.0060 \text{ m})(1.4 \text{ m})}{1.22\left(436 \times 10^{-9} \text{ m}\right)} = 16 \text{ km} \quad \Diamond$$

In real life, the distance to barely-resolved headlights will be much shorter because of what astronomers call seeing. The variable-density atmosphere refracts the light to change its direction, to shake the images, and at long distances, to make the headlights twinkle.

---

21. The Impressionist painter Georges Seurat created paintings with an enormous number of dots of pure pigment approximately 2.0 mm in diameter. The idea was to have colors such as red and green next to each other to form a scintillating canvas. Outside what distance would one be unable to discern individual dots on the canvas? (Assume $\lambda = 500$ nm within the eye and a pupil diameter of 4.0 mm.)

Solution By Rayleigh's criterion, two dots separated center-to-center by 2 mm would be seen to overlap when

$$\theta = \frac{d}{L} = 1.22\frac{\lambda}{D}, \quad \text{where} \quad d = 2.0 \text{ mm}, \quad \lambda = 500 \text{ nm}, \quad \text{and} \quad D = 4.0 \text{ mm}$$

$$L = \frac{Dd}{1.22\lambda} = \frac{\left(4.0 \times 10^{-3} \text{ m}\right)\left(2.0 \times 10^{-3} \text{ m}\right)}{1.22\left(500 \times 10^{-9} \text{ m}\right)} = 13 \text{ m} \quad \Diamond$$

---

27. A grating with 250 lines/mm is used with an incandescent light source. Assume the visible spectrum to range in wavelength from 400 to 700 nm. In how many orders can one see (a) the entire visible spectrum and (b) the short-wavelength region?

Solution The grating spacing is $d = 1.00 \text{ mm}/250 = 4.00 \times 10^{-6} \text{ m}$

(a) In each order of interference m, red light diffracts at a larger angle than the other colors with shorter wavelengths. We find the largest number m satisfying

$$d \sin\theta = m\lambda \quad \text{with} \quad \lambda = 700 \text{ nm};$$

With $\sin\theta$ having its largest value $\quad \left(4.00 \times 10^{-6} \text{ m}\right)\sin 90° = m\left(700 \times 10^{-9} \text{ m}\right)$

so that $\quad m = 5.71$

Thus red light cannot be seen in the 6th order, and the full visible spectrum appears in only five orders. ◊

(b) Now consider light at the boundary between violet and ultraviolet.

$d\sin\theta = m\lambda$ becomes $\left(4.00 \times 10^{-6} \text{ m}\right)\sin 90° - m\left(400 \times 10^{-9} \text{ m}\right)$, and $m = 10.0$ ◊

---

29. The full width of a 3.00-cm-wide grating is illuminated by a sodium discharge tube. The lines in the grating are uniformly spaced at 775 nm. Calculate the angular separation in the first-order spectrum between the two wavelengths forming the sodium doublet ($\lambda_1 = 589.0$ nm and $\lambda_2 = 589.6$ nm).

Solution $m\lambda = d\sin\theta$

$$\theta_2 = \sin^{-1}\left(\frac{m\lambda_2}{d}\right) = \sin^{-1}\left[\frac{(1)\left(5.896 \times 10^{-7} \text{ m}\right)}{7.75 \times 10^{-7} \text{ m}}\right] = 49.532°$$

$$\theta_1 = \sin^{-1}\left(\frac{m\lambda_1}{d}\right) = \sin^{-1}\left[\frac{(1)\left(5.890 \times 10^{-7} \text{ m}\right)}{7.75 \times 10^{-7} \text{ m}}\right] = 49.464°$$

$\Delta\theta = \theta_2 - \theta_1 = 0.0683°$ ◊

---

31. A source emits 531.62-nm and 531.81-nm light. (a) What minimum number of lines is required for a grating that resolves the two wavelengths in the first-order spectrum? (b) Determine the slit spacing for a grating 1.32 cm wide that has the required minimum number of lines.

Solution The resolving power of the diffraction grating is

(a) $$Nm = \frac{\lambda}{\Delta\lambda}$$

so $$N(1) = \frac{531.7\ \text{nm}}{0.19\ \text{nm}} = 2800\ \text{lines} \quad \Diamond$$

(b) $$\frac{1.32\times 10^{-2}\ \text{m}}{2800} = 4.72\ \mu\text{m} \quad \Diamond$$

---

35. White light is spread out into its spectral components by a diffraction grating. If the grating has 2000 lines per centimeter, at what angle does red light ($\lambda = 640$ nm) appear in first order?

Solution The grating spacing is

$$d = 1.00\times 10^{-2}\ \text{m}/2000$$

$$= 5.00\times 10^{-6}\ \text{m}$$

Therefore $$\sin\theta = \frac{m\lambda}{d} = \frac{1(640\times 10^{-9}\ \text{m})}{5.00\times 10^{-6}\ \text{m}} = 0.128$$

and $$\theta = 7.35° \quad \Diamond$$

---

45. If the interplanar spacing of NaCl is 0.281 nm, what is the predicted angle at which 0.140-nm x-rays are diffracted in a first-order maximum?

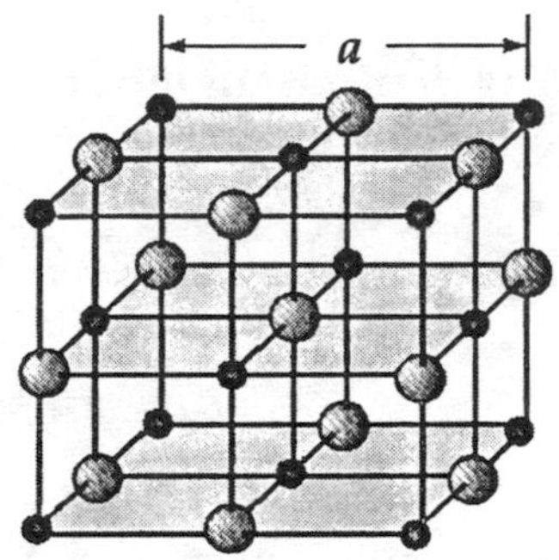

Figure 38.21

Solution The atomic planes in this crystal are shown in Figure 38.21. The diffraction they produce is described by the Bragg condition, that

$$2d\sin\theta = m\lambda$$

$$\sin\theta = \frac{m\lambda}{2d} = \frac{1(0.140\times10^{-9}\text{ m})}{2(0.281\times10^{-9}\text{ m})} = 0.249$$

$$\theta = 14.4° \quad \Diamond$$

---

53. Plane-polarized light is incident on a single polarizing disk with the direction of E_0 parallel to the direction of the transmission axis. Through what angle should the disk be rotated so that the intensity in the transmitted beam is reduced by a factor of (a) 3, (b) 5, (c) 10?

Solution We start from $\theta = 0$, in $I = I_0\cos^2\theta$.

(a) For $I = I_0/3$, $\cos\theta = \frac{1}{\sqrt{3}}$ and $\theta = 54.7° \quad \Diamond$

(b) For $I = I_0/5$, $\cos\theta = \frac{1}{\sqrt{5}}$ and $\theta = 63.4° \quad \Diamond$

(c) For $I = I_0/10$, $\cos\theta = \frac{1}{\sqrt{10}}$ and $\theta = 71.6° \quad \Diamond$

---

55. The critical angle for sapphire surrounded by air is 34.4°. Calculate the polarizing angle for sapphire.

Solution $n = \tan\theta_p$

$$\theta_p = \tan^{-1} n$$

and $\sin\theta_c = \frac{1}{n}$ $n = \frac{1}{\sin\theta_c}$

Therefore, $\theta_p = \tan^{-1}\left(\frac{1}{\sin\theta_c}\right) = \tan^{-1}\left(\frac{1}{\sin 34.4°}\right) = 60.5°$ ◊

---

56. For a particular transparent medium surrounded by air, show that the critical angle for internal reflection and the polarizing angle are related by cot θ_p = sin θ_c.

Solution $\sin\theta_c = \frac{1}{n}$ and $\tan\theta_p = n$

Thus, $\sin\theta_c = \frac{1}{\tan\theta_p}$ or $\cot\theta_p = \sin\theta_c$ ◊

---

57. If the polarizing angle for cubic zirconia (ZrO_2) is 65.6°, what is the index of refraction for this material?

Solution $n = \tan\theta_p = \tan(65.6°) = 2.20$ ◊

---

59. The hydrogen spectrum has a red line at 656 nm and a blue line at 434 nm. What is the angular separation between two spectral lines obtained with a diffraction grating that has 4500 lines/cm?

Solution The grating spacing is $d = (1.00 \times 10^{-2}\text{ m})/4500 = 2.22 \times 10^{-6}\text{ m}$

In the first-order spectrum, the angles of diffraction are given by $\sin\theta = \dfrac{\lambda}{d}$:

$$\sin\theta_1 = \frac{656 \times 10^{-9}\text{ m}}{2.22 \times 10^{-6}\text{ m}} = 0.295 \quad \text{so that} \quad \theta_1 = 17.17°$$

$$\sin\theta_2 = \frac{434 \times 10^{-9}\text{ m}}{2.22 \times 10^{-6}\text{ m}} = 0.195 \quad \text{so that} \quad \theta_2 = 11.26°$$

The angular separation is then $\theta_2 - \theta_1 = 5.91°$ ◊

In the second-order spectrum, we proceed similarly:

$$\Delta\theta = \sin^{-1}(2\lambda_1/d) - \sin^{-1}(2\lambda_2/d) = 13.2° \quad ◊$$

Again, in the third order,

$$\Delta\theta = \sin^{-1}(3\lambda_1/d) - \sin^{-1}(3\lambda_2/d) = 26.5° \quad ◊$$

Since the red line does not appear in the fourth-order spectrum, the answer is complete.

---

65. A diffraction grating of length 4.00 cm contains 6000 rulings over a width of 2.00 cm. (a) What is the resolving power of this grating in the first three orders? (b) If two monochromatic waves incident on this grating have a mean wavelength of 400 nm, what is their wavelength separation if they are just resolved in the third order?

Solution

(a) From Equation 38.10,

$$R = mN \quad \text{where} \quad N = \left(\frac{6000\text{ lines}}{2.00\text{ cm}}\right)(4.00\text{ cm}) = 12.0 \times 10^3\text{ lines}$$

In the 1st order, $R = (1)(12000 \text{ lines}) = 12000$ ◊

In the 2nd order, $R = (2)(12000 \text{ lines}) = 24000$ ◊

In the 3rd order, $R = (3)(12000 \text{ lines}) = 36000$ ◊

(b) From Equation 38.9, $R = \dfrac{\lambda}{\Delta\lambda}$

In the third order, $\Delta\lambda = \dfrac{\lambda}{R} = \dfrac{4.00 \times 10^{-7}\text{ m}}{3.60 \times 10^{4}} = 1.11 \times 10^{-11}\text{ m} = 0.0111\text{ nm}$ ◊

---

67. Light of wavelength 500 nm is incident normally on a diffraction grating. If the third-order maximum of the diffraction pattern is observed at 32°, (a) what is the number of rulings per centimeter for the grating? (b) Determine the total number of primary maxima that can be observed in this situation.

Solution

(a) Use Equation 38.8, $d\sin\theta = m\lambda$

$$d = \frac{m\lambda}{\sin\theta} = \frac{(3)(5.0 \times 10^{-7}\text{ m})}{\sin(32°)} = 2.83 \times 10^{-6}\text{ m}$$

Therefore, the number of lines per unit length $= \dfrac{1}{d} = 3.534 \times 10^{5}$ lines/m

or 3.53×10^{3} lines/cm ◊

(b) $\sin\theta = m\left(\dfrac{\lambda}{d}\right) = \dfrac{m(5 \times 10^{-7}\text{ m})}{2.83 \times 10^{-6}\text{ m}} = m(0.177)$

For $\sin\theta \le 1$, we require that $m(0.177) \le 1$ or $m \le 5.65$.

Therefore, the total number of maxima $= 2m + 1 = 11$ ◊

---

73. Suppose that the single slit in Figure 38.1b (of this study guide) is 6.0 cm wide and in front of a microwave source operating at 7.5 GHz. (a) Calculate the angle subtended by the first minimum in the diffraction pattern. (b) What is the relative intensity I/I_0 at $\theta = 15°$? (c) Consider the case when there are two such sources, separated laterally by 20 cm, behind the slit. What must the maximum distance between the plane of the sources and the slit be if the diffraction patterns are to be resolved? (In this case, the approximation $\sin\theta \approx \tan\theta$ is not valid because of the relatively small value of a/λ.)

Solution

(a) From Equation 38.1, $$\theta = \sin^{-1}\left(\frac{m\lambda}{a}\right)$$

In this case, $m = 1$; $$\lambda = \frac{c}{f} = \frac{(3.0\times10^8 \text{ m/s})}{(7.5\times10^9 \text{ s}^{-1})} = 4.0\times10^{-2} \text{ m} = 0.040 \text{ m}$$

so $$\theta = \sin^{-1}(0.666) = 41.8° \quad \Diamond$$

(b) From Equation 38.4, $$\frac{I}{I_0} = \left(\frac{\sin(\beta/2)}{\beta/2}\right)^2 \quad \text{where} \quad \beta = \frac{2\pi a \sin\theta}{\lambda}$$

When $\theta = 15°$, $\beta = 2.44$ rad and $\frac{I}{I_0} = 0.593 \quad \Diamond$

(c) $\sin\theta = \frac{\lambda}{a}$ so $\theta = 41.81°$

This is the *minimum* angle subtended by the two sources at the slit. Let α = the half angle between the sources, each a distance L' from the center line and a distance L from the slit plane. Next,

$$\sin\alpha = \frac{L'}{\sqrt{L'^2 + L^2}} = \sin\left(\frac{41.81}{2}\right) = 0.357$$

and $$L^2 = \frac{[1-(0.357)^2](0.10 \text{ m})^2}{(0.357)^2} \quad \text{and} \quad L = 0.262 \text{ m} \quad \Diamond$$

76. Another method to solve the equation $\phi = \sqrt{2}\sin\phi$ in Problem 9 is to use a calculator, guess a first value of ϕ, see if it fits, and continue to update your estimate until the equation balances. How many steps (iterations) did this take?

Solution We can list each trial as we try to home in on the solution to $\phi = \sqrt{2}\sin\phi$ by narrowing the range in which it must lie:

ϕ	$\sqrt{2}\sin\phi$	
1	1.19	bigger than ϕ
2	1.29	smaller than ϕ
1.5	1.41	smaller
1.4	1.394	
1.39	1.391	bigger
1.395	1.392	
1.392	1.3917	smaller
1.3915	1.39154	bigger
1.39152	1.39155	bigger
1.3916	1.391568	smaller
1.39158	1.391563	
1.39157	1.391560	
1.39156	1.391558	
1.391559	1.3915578	
1.391558	1.3915575	
1.391557	1.3915573	
1.3915574	1.3915574	

We get the answer to seven digits after 17 steps. ◊ Clever guessing, like just using the value of $\sqrt{2}\sin\phi$ as the next guess for ϕ, could reduce this to around 13 steps. Looking back at Problem 38.8, we see that the half-width at half-maximum of the central peak in a single-slit diffraction pattern is described by $1.39 = \phi = \beta/2 = \pi a \sin\theta/\lambda$ and $\sin\theta = 0.443\lambda/a$, so the full angular width at half-maximum is 2 Arcsin $(0.443\lambda/a)$.

Chapter 39

Relativity

Chapter 39

RELATIVITY

INTRODUCTION

Most of our everyday experiences and observations have to do with objects that move at speeds much less than that of light. Newtonian mechanics and early ideas on space and time were formulated to describe the motion of such objects. This formalism is very successful in describing a wide range of phenomena that occur at low speeds. It fails, however, when applied to particles whose speeds approach that of light. For example, it is possible to accelerate an electron to a speed of $0.99c$ (where c is the speed of light) by using a potential difference of several million volts. According to Newtonian mechanics, if the potential difference is increased by a factor of 4, the electron speed should jump to $1.98c$. However, experiments show that the speed of the electron always remains less than the speed of light, regardless of the size of the accelerating voltage. Thus, we can see that the validity of Newtonian mechanics is limited to low-velocity interactions.

In 1905, at the age of only 26, Einstein published his special theory of relativity. With this theory, experimental observations can be correctly predicted over the range of speeds from $v = 0$ to speeds approaching the speed of light. Newtonian mechanics, which was accepted for over 200 years, is in fact a special case of Einstein's theory. This chapter gives an introduction to the special theory of relativity, with emphasis on some of its consequences.

NOTES FROM SELECTED CHAPTER SECTIONS

39.1 The Principle of Newtonian Relativity

The special theory of relativity is based on two basic postulates:

- The laws of physics are the same in all *inertial* reference systems.
- The speed of light in vacuum is always measured to be 3.00×10^8 m/s, and the measured value is *independent* of the motion of the observer or of the motion of the source of light.

According to the principle of Newtonian relativity, the laws of mechanics are the same in all inertial frames of reference. Inertial frames of reference are those coordinate systems which are *at rest with respect to one another* or which move at constant velocity with respect to one another.

39.2 The Michelson-Morley Experiment

This experiment was designed to detect the velocity of the Earth with respect to the *hypothetical ether.* The outcome of the experiment was *negative,* contracting the ether hypothesis.

39.3 Einstein's Principle of Relativity

Einstein's special theory of relativity is based upon two postulates:

- The laws of physics are the same in every inertial frame of reference.
- The speed of light has the same value as measured by all observers, independent of the motion of the light source or observer.

The second postulate is consistent with the negative results of the Michelson-Morley experiment which failed to detect the presence of an ether and suggested that the speed of light is the same in all inertial frames.

39.4 Consequences of Special Relativity

Two events that are simultaneous in one reference frame are, in general, not simultaneous in another frame which is moving with respect to the first.

The *proper time* is always the time measured by a clock at rest in the frame of reference of the measurement. According to a stationary observer, a moving clock runs slower than an identical clock at rest. This effect is known as *time dilation.*

The *proper length* of an object is defined as the length of the object measured in the *reference frame in which the object is at rest.* The length of an object measured in a reference frame in which it is moving is always less than the proper length. This effect is known as *length contraction.* The contraction occurs only *along the direction of motion.*

39.6 Relativistic Momentum and the Relativistic Form of Newton's Laws

To account for *relativistic effects,* it is necessary to modify the definition of momentum to satisfy the following conditions:

- The relativistic momentum must be conserved in all collisions.
- The relativistic momentum must approach the classical value, mv, as the quantity v/c approaches zero.

EQUATIONS AND CONCEPTS

In order to explain the motion of particles moving at speeds approaching the speed of light, one must use the *Lorentz coordinate transformation equations.* These equations represent the transformation between any two inertial frames in *relative* motion with velocity v in the x direction.

$$\left.\begin{aligned} t' &= \gamma\left(t - \frac{v}{c^2}x\right) \\ x' &= \gamma(x - vt) \end{aligned}\right\} \quad \text{where} \quad \gamma \equiv \frac{1}{\sqrt{1-\frac{v^2}{c^2}}} \tag{39.9}$$

$$y' = y$$

$$z' = z$$

The *Lorentz velocity transformation equations* relate the observed velocity u' in the moving (S') frame to the measured velocity u in the S frame, and the relative velocity of S' with respect to S.

$$u'_x = \frac{u_x - v}{1 - \frac{u_x v}{c^2}} \tag{39.14}$$

$$u'_y = \frac{u_y}{\gamma\left(1 - \frac{u_x v}{c^2}\right)} \tag{39.15}$$

$$u'_z = \frac{u_z}{\gamma\left(1 - \frac{u_x v}{c^2}\right)}$$

Consider a single clock in the S' frame in which an observer in this frame measures the time interval of an event $\Delta t'$ (say the time it takes a light pulse to travel from a source to a mirror and back to the source). The time interval Δt for this event as measured by an observer in S is greater than $\Delta t'$, since the observer in S must make clock readings at two different positions. That is, *a moving clock appears to run slower than an identical clock at rest with respect to the observer.* This is known as time dilation.

$$\Delta t = \frac{\Delta t'}{\sqrt{1 - \frac{v^2}{c^2}}} = \gamma \Delta t' \tag{39.7}$$

If an object moves along the x axis with a speed v, and has a length L_p as measured in the moving frame, the length L as measured by a stationary observer is shorter than L_p by the factor $1/\gamma$. This is called *length contraction*, and the length L_p measured in the reference frame in which the object is at rest is called the *proper length*.

$$L = \frac{1}{\gamma}L_p = L_p\sqrt{1-\frac{v^2}{c^2}} \quad (39.8)$$

The fact that observers in the S and S' frames do not reach the same conclusions regarding such measurements can be understood by recognizing that simultaneity is not an absolute concept. That is, *Two events that are simultaneous in one reference frame are not simultaneous in another reference frame that is in motion relative to the first.*

The relativistic equation for the *momentum* of a particle of mass m moving with a speed u satisfies the conditions that (1) momentum is conserved in all collisions and (2) p approaches the classical expression as u approaches 0. That is, as $u \Rightarrow 0, p \Rightarrow mu$.

$$\mathbf{p} \equiv \frac{m\mathbf{u}}{\sqrt{1-\frac{u^2}{c^2}}} \quad (39.17)$$

In the relativistic equation for the *kinetic energy* of a particle of mass m moving with a speed u, the term mc^2 is called the *rest energy*.

$$K = \frac{mc^2}{\sqrt{1-\frac{u^2}{c^2}}} - mc^2 \quad (39.22)$$

The *total energy* E of a particle is the sum of the kinetic energy and the rest energy. This expression shows that *mass and energy are equivalent concepts*; that is, mass is a form of energy.

$$E = \frac{mc^2}{\sqrt{1-\frac{u^2}{c^2}}} \quad (39.24)$$

When the momentum or energy of a particle is known (rather than the speed), it is useful to have an expression relating the total energy and relativistic momentum.

$$E^2 = p^2c^2 + (mc^2)^2 \qquad (39.25)$$

There is an exact expression relating energy and momentum for particles which have zero mass (e.g. photons). These zero-mass particles always travel at the speed of light.

$$E = pc \qquad (39.26)$$

A convenient energy unit to use to express the energies of electrons and other subatomic particles is the electron volt (eV).

$$1 \text{ eV} = 1.60 \times 10^{-19} \text{ J}$$

REVIEW CHECKLIST

- ▷ State Einstein's two postulates of the special theory of relativity.
- ▷ Understand the Michelson-Morley experiment, its objectives, results, and the significance of its outcome.
- ▷ Understand the idea of simultaneity, and the fact that simultaneity is not an absolute concept. That is, two events which are simultaneous in one reference frame are not simultaneous when viewed from a second frame moving with respect to the first.
- ▷ Make calculations using the equations for time dilation, length contraction, and relativistic mass.
- ▷ State the correct relativistic expressions for the momentum, kinetic energy, and total energy of a particle. Make calculations using these equations.

SOLUTIONS TO SELECTED END-OF-CHAPTER PROBLEMS

1. In a laboratory frame of reference, an observer notes that Newton's second law is valid. Show that it is also valid for an observer moving at a constant speed relative to the laboratory frame.

Solution The first observer watches some object accelerate under applied forces. Call the instantaneous velocity of the object u_x. The second observer has constant velocity v relative to the first, which we assume to be small compared with light, and measures the object to have velocity

$$u'_x = u_x - v$$

The acceleration is $$\frac{du'_x}{dt} = \frac{du_x}{dt} - 0$$

This is the same as that measured by the first observer. In this nonrelativistic case, they measure also the same mass and forces; so the second observer also confirms that $\Sigma F = ma$. ◊

---

3. A 2000-kg car moving at 20 m/s collides with and sticks to a 1500-kg car at rest at a stop sign. Show that momentum is conserved in a reference frame moving at 10 m/s in the direction of the moving car.

Solution In the rest frame,

$$p_i = m_1 v_{1i} + m_2 v_{2i} = (2000\text{ kg})(20\text{ m/s}) + (1500\text{ kg})(0\text{ m/s}) = 4.0\times10^4\text{ kg}\cdot\text{m/s}$$

$$p_f = (m_1 + m_2)v_f = (2000\text{ kg} + 1500\text{ kg})v_f$$

Since $p_i = p_f$, $v_f = \dfrac{4.0\times10^4\text{ kg}\cdot\text{m/s}}{2000\text{ kg} + 1500\text{ kg}} = 11.429\text{ m/s}$

In the moving frame, these velocities are all reduced by +10 m/s.

$$v'_{1i} = v_{1i} - v' = 20\text{ m/s} - (+10\text{ m/s}) = 10\text{ m/s}$$

$$v'_{2i} = v_{2i} - v' = 0\text{ m/s} - (+10\text{ m/s}) = -10\text{ m/s}$$

$$v'_f = 11.429\text{ m/s} - (+10\text{ m/s}) = 1.429\text{ m/s}$$

Our initial momentum is then

$$p_i' = m_1 v_{1i}' + m_2 v_{2i}' = (2000 \text{ kg})(10 \text{ m/s}) + (1500 \text{ kg})(-10 \text{ m/s}) = 5000 \text{ kg} \cdot \text{m/s}$$

and our final momentum is

$$p_f' = (2000 \text{ kg} + 1500 \text{ kg})v_f' = (3500 \text{ kg})(1.429 \text{ m/s}) = 5000 \text{ kg} \cdot \text{m/s} \quad \Diamond$$

9A. A spaceship of proper length L_p takes t seconds to pass an Earth observer. Determine its speed as measured by the Earth observer.

Solution Here our answer will include both the case where the relativistic correction is negligible and the case where it is important:

$$L = L_p / \gamma$$

$$vt = L_p\left(1 - v^2/c^2\right)^{1/2}$$

$$v^2t^2 = L_p^2\left(1 - v^2/c^2\right)$$

$$v^2c^2 = L_p^2c^2/t^2 - v^2L_p^2/t^2$$

$$v = \frac{cL_p/t}{\sqrt{c^2 + L_p^2/t^2}} \quad \Diamond$$

However small t may become, v is less than c. On the other hand, if t is rather large, this reduces to the classical answer L_p/t.

Related Calculation Find the ship's speed if the proper length is 300 m, and it takes 0.75 μs to pass the Earth observer.

Solution Substituting the given values into our answer, we find:

$$v = \frac{\left(3.00 \times 10^8\right)(300 \text{ m})/\left(0.75 \times 10^{-6} \text{ s}\right)}{\sqrt{\left(3.00 \times 10^8\right)^2 + (300 \text{ m})^2/\left(0.75 \times 10^{-6} \text{ s}\right)^2}} = 2.4 \times 10^8 \text{ m/s} \quad \Diamond$$

---

12. The cosmic rays of highest energy are protons, having kinetic energy of 10^{13} MeV. (a) How long would it take a proton of this energy to travel across the Milky Way galaxy, of diameter 10^5 light-years, as measured in the proton's frame? (b) From the point of view of the proton, how many kilometers across is the galaxy?

Solution We first find the proton's speed in the frame of reference where the galaxy is stationary:

For a proton,

$$mc^2 = \left(1.67\times10^{-27}\text{ kg}\right)\left(2.9979\times10^8\text{ m/s}\right)^2\left(1\text{ eV}/1.6022\times10^{-19}\text{ kg}\cdot\text{m}^2/\text{s}^2\right) = 938\text{ MeV}$$

Now, $K = (\gamma - 1)(938\text{ MeV})$

$$10^{13}\text{ MeV} = (\gamma - 1)(938\text{ MeV})\quad\text{and}\quad \gamma = 1.07\times10^{10} = 1/\sqrt{1-v^2/c^2}$$

$$1 - v^2/c^2 = 8.80\times10^{-21}$$

$$v = c\sqrt{1 - 8.80\times10^{-21}}$$

This speed is nearly as large as the speed of light. In the galaxy frame the traversal time is

$$\Delta t = x/v = 10^5\text{ light years}/c = 10^5\text{ years}$$

(a) This is dilated from the proper time measured in the proton's frame:

$$\Delta t = \gamma\Delta t'$$

$$\Delta t' = \Delta t/\gamma = 10^5\text{ y}/1.07\times10^{10} = 9.38\times10^{-6}\text{ years} = 296\text{ s}\quad\Diamond$$

(b) The proton sees the galaxy moving across the proton, at speed nearly equal to c, in 296 s:

$$\Delta L' = \left(3.00\times10^8\right)(296\text{ s}) = 8.88\times10^7\text{ km}\quad\Diamond$$

The observers on the proton and on the galaxy agree on the value v of their speed of relative motion and on the value of γ.

---

17. Two jets of material from the center of a radio galaxy fly away in opposite directions. Both jets move at 0.75*c* relative to the galaxy. Determine the speed of one jet relative to the other.

Solution Arbitrarily take the galaxy as the unmoving frame, the jet moving toward the right as the object, and the jet moving toward the left as the "moving" frame.

$$u_x = 0.75\ c \qquad v = -0.75\ c$$

Thus,
$$u_x' = \frac{u_x - v}{1 - \frac{u_x v}{c^2}} = \frac{0.75c - (-0.75c)}{1 - \frac{(0.75c)(-0.75c)}{c^2}} = \frac{1.5c}{1 + 0.75^2}$$

$$u_x' = 0.960c \quad \lozenge$$

---

19. A cube of steel has a volume of 1.0 cm³ and a mass of 8.0 g when at rest on the Earth. If this cube is now given a speed $v = 0.90\ c$, what is its density as measured by a stationary observer? Note that relativistic density is $m/V = E/c^2V$.

Solution The measured volume will now be

$$V = (1\text{ cm})(1\text{ cm})\left(\sqrt{1 - v^2/c^2}\text{ cm}\right) = 0.436\text{ cm}^3$$

The energy is
$$E = \gamma mc^2 = mc^2 / \sqrt{1 - v^2/c^2} = 2.29mc^2$$

The density measured by the stationary observer is

$$\frac{E}{c^2 V} = \frac{2.29\ mc^2}{c^2\left(0.436\text{ cm}^3\right)} = 5.26\left(\frac{8\text{ g}}{1\text{ cm}^3}\right) = 42.1\text{ g}/\text{cm}^3 \quad \lozenge$$

---

21. Find the momentum of a proton in MeV/c units if its total energy is twice its rest energy.

Solution For a proton,

$$mc^2 = \left(1.67\times10^{-27}\text{ kg}\right)\left(2.9979\times10^{8}\text{ m/s}\right)^2\left(1\text{ eV}/1.6022\times10^{-19}\text{ kg}\cdot\text{m}^2/\text{s}^2\right)$$

$$= 938\text{ MeV}$$

The total energy $\gamma mc^2 = 2mc^2$, so $\gamma = 2$

Since $\dfrac{1}{\sqrt{1-v^2/c^2}} = 2,$ $v/c = 0.866$

Therefore, the momentum is $p = \gamma mv = \dfrac{\left(\gamma mc^2\right)}{c}(v/c)$

$$p = \frac{2(938\text{ MeV})}{c}(0.866) = 1620\text{ MeV}/c \quad \Diamond$$

---

23. Show that the energy-momentum relationship $E^2 = p^2c^2 + (mc^2)^2$ follows from the expressions $E = \gamma mc^2$ and $p = \gamma mu$.

Solution $E = \gamma mc^2$ $p = \gamma mu$

$E^2 = (\gamma mc^2)^2$ $p^2 = (\gamma mu)^2$

Substituting terms,

$$E^2 - p^2c^2 = (\gamma mc^2)^2 - (\gamma mu)^2c^2 = \gamma^2\left[\left(mc^2\right)^2 - \left(mc^2\right)u^2\right]$$

But that is just $E^2 - p^2c^2 = \left(mc^2\right)^2\left(1-\dfrac{u^2}{c^2}\right)\left(1-\dfrac{u^2}{c^2}\right)^{-1}$

Thus, $E^2 - p^2c^2 = \left(mc^2\right)^2 \quad \Diamond$

---

25. A proton moves at $0.95c$. Calculate its (a) rest energy, (b) total energy, and (c) kinetic energy.

Solution At $v = 0.95c$, $\gamma = 3.2$

(a) $E_0 = m_0c^2 = (1.67\times10^{-27}\text{ kg})(3.00\times10^8\text{ m/s})^2 = 1.50\times10^{-10}\text{ J} = 939\text{ MeV}$ ◊

(b) $E = \gamma m_0c^2 = \gamma E_0 = (3.2)(939\text{ MeV}) = 3000\text{ MeV}$ ◊

(c) $K = E - E_0 = 3000\text{ MeV} - 939\text{ MeV} = 2100\text{ MeV}$ ◊

---

33. A pion at rest ($m_\pi = 270m_e$) decays to a muon ($m_\mu = 206m_e$) and an antineutrino ($m_\nu = 0$): $\pi^- \to \mu^- + \bar{\nu}$. Find the kinetic energy of the muon and the antineutrino in electron volts. (*Hint*: Relativistic momentum is conserved.)

Solution By conservation of energy, $m_\pi c^2 = \gamma m_\mu c^2 + p_\nu c$

By conservation of momentum, $p_\nu = -p_\mu = -\gamma m_\mu v$

Substituting, $$138\text{ MeV} = \left(\gamma + \frac{\gamma v}{c}\right)(105\text{ MeV}) = \frac{1+v/c}{\sqrt{1-(v/c)^2}}(105\text{ MeV})$$

Solving for v/c, $\dfrac{v}{c} = 0.266$ and $\gamma = \dfrac{1}{\sqrt{1-v^2/c^2}} = 1.0374$

and $$KE_\mu = (0.0374)(105) = 3.95\text{ MeV} \quad ◊$$

$$KE_{\bar{\nu}} = 138 - (105 + 4) = 29.1\text{ MeV} \quad ◊$$

---

37. The power output of the Sun is 3.8×10^{26} W. How much rest mass is converted to kinetic energy in the Sun each second?

Solution From $E = mc^2$, we have

$$m = \frac{E}{c^2} = \frac{3.8\times10^{26}\text{ J}}{(3.00\times10^8\text{ m/s})^2} = 4.2\times10^9\text{ kg} \quad ◊$$

---

41. The net nuclear reaction inside the Sun is $4p \rightarrow {}^4He + \Delta E$. If the rest mass of each proton is 938.2 MeV and the rest mass of the ^4He nucleus is 3727 MeV, calculate the percentage of the starting mass that is released as energy.

Solution The original rest energy is $E = 4(938.2\text{ MeV}) = 3752.8\text{ MeV}$

The energy given off is $\Delta E = (3752.8 - 3727)\text{ MeV} = 25.8\text{ MeV}$

The fractional energy released is $\frac{\Delta E}{E} = \frac{25.8\text{ MeV}}{3753\text{ MeV}} \times 100\% = 0.687\%$ ◊

---

47. Spaceship I, which contains students taking a physics exam, approaches Earth with a speed of 0.60c (relative to Earth), while spaceship II, which contains professors proctoring the exam, moves at 0.28c (relative to Earth) directly toward the students. If the professors stop the exam after 50 min have passed on their clock, how long does the exam last as measured by (a) the students and (b) an observer on Earth?

Solution Suppose that in the Earth frame, the students are moving to the right ($u_x = 0.60c$), and the professors are moving to the left, at a speed of ($v = -0.28c$).

In the professor's frame,
the Earth moves to the right at $u_e = 0.28c$,

and the students move to the right at $u_x' = \frac{u_x - v}{1 - u_x v/c^2} = \frac{0.60c - (-0.28c)}{1 - (0.60c)(-0.28c)/c^2} = 0.753c$

The professors measure 50 minutes on a clock at rest in their frame: they measure proper time and everyone else sees longer, dilated time intervals.

(a) For the students, $\Delta t = \gamma \Delta t' = \frac{50\text{ min}}{\sqrt{1 - (0.753)^2}} = 76\text{ min}$ ◊

(b) On Earth, $\Delta t = \gamma \Delta t' = \frac{50\text{ min}}{\sqrt{1 - (0.28)^2}} = 52\text{ min}$ ◊

---

53. Two rockets are on a collision course. They are moving at 0.800*c* and 0.600*c* and are initially 2.52×10^{12} m apart as measured by Liz, the Earth observer in Figure P39.53. Both rockets are 50.0 m in length as measured by Liz.
(a) What are their respective proper lengths? (b) What is the length of each rocket as measured by an observer in the other rocket? (c) According to Liz, how long before the rockets collide? (d) According to rocket 1, how long before they collide? (e) According to rocket 2, how long before they collide? (f) If both rocket crews are capable of total evacuation within 90 minutes (their own time), will there be any casualties?

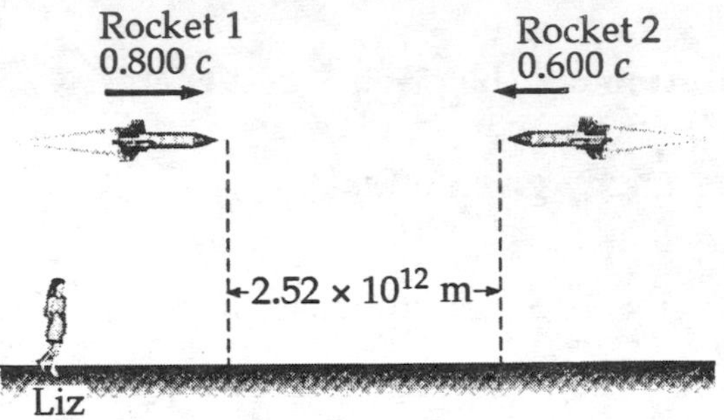

Figure P39.53

Solution

(a) $L=\frac{L'}{\gamma}$ so $L'=\gamma L$ is the proper length where L is the measured length.

Relative to Earth, $v_1=0.800c$ gives $\gamma_1=1.67$, and $v_2=-0.600c$ gives $\gamma_2=1.25$.

Since the measured length is L = 50.0 m for each rocket, their respective proper lengths are

$$L_1'=1.67(50.0\text{ m})=83.3\text{ m} \quad \Diamond \qquad \text{and} \qquad L_2'=1.25(50.0\text{ m})=62.5\text{ m} \quad \Diamond$$

(b) Compute the speed of one rocket relative to the other by considering an observer in rocket #1. He sees the Earth moving with velocity $v=-0.800c$. The velocity addition equation then gives the velocity of rocket #2 (relative to rocket #1) as

$$u_{12}=\frac{v+u'}{1+\frac{vu'}{c^2}}=\frac{(-0.800c)+(-0.600c)}{1+\frac{(-0.600c)(-0.800c)}{c^2}}=\frac{-1.400c}{1.48}=-0.946c \qquad \text{(\#2 rel to \#1)}$$

Similarly, the velocity of #1 relative to #2 is 0.946*c*. The γ factor for the relative motion between rockets is $\gamma_{12}=3.08$.

The contracted length of rocket #2 as measured by an observer in #1 is

$$L_2=\frac{L_2'}{\gamma_{12}}=\frac{62.5\text{ m}}{3.08}=20.3\text{ m} \quad \Diamond$$

Similarly, the contracted length of #1 as measured by an observer in #2 is

$$L_1 = \frac{L_1'}{\gamma_{12}} = \frac{83.3\text{ m}}{3.08} = 27.0\text{ m} \quad \Diamond$$

(c) Liz sees the initial 2.52×10^{12} m-gap between the rockets decreasing at a rate of $1.400c$. Thus, according to her, the time before collision is

$$t = \frac{2.52 \times 10^{12}\text{ m}}{1.40(3.00 \times 10^{8}\text{ m/s})} = 6000\text{ s} = 1.67\text{ h} \quad \Diamond$$

(d) Liz, at rest in S, measures the proper length of the initial gap between the rockets. The contracted initial gap as measured by an observer in rocket #1 is

$$L_1 = \frac{L'}{\gamma_1} = \frac{2.52 \times 10^{12}\text{ m}}{1.67} = 1.51 \times 10^{12}\text{ m}$$

The time before collision according to this observer is

$$t_1 = \frac{L_1}{u_{12}} = \frac{1.51 \times 10^{12}\text{ m}}{0.946(3.00 \times 10^{8}\text{ m/s})} = 5328\text{ s} = 1.48\text{ h} \quad \Diamond$$

(e) The contracted initial gap between rockets as measured by an observer in rocket #2 is

$$L_2 = \frac{L}{\gamma_2} = \frac{2.52 \times 10^{12}\text{ m}}{1.25} = 2.02 \times 10^{12}\text{ m}$$

The time before collision according to this observer is

$$t_2 = \frac{L_2}{u_{12}} = \frac{2.02 \times 10^{12}\text{ m}}{0.946(3.00 \times 10^{8}\text{ m/s})} = 7104\text{ s} = 1.97\text{ h} \quad \Diamond$$

(f) Since it requires 90 min (1.50 h) on their own clock for each crew to evacuate, the crew of rocket #2 makes it, but the crew of rocket #1 does not. ◊

---

55. A particle having charge q moves at speed v along a straight line in a uniform electric field **E**. If the motion and the electric field are both in the x direction, (a) show that the acceleration of the particle in the x direction is

$$a = \frac{dv}{dt} = \frac{qE}{m}\left(1 - \frac{v^2}{c^2}\right)^{3/2}$$

(b) Discuss the significance of the fact that the acceleration depends on the speed. (c) If the particle starts from rest at $x = 0$ at $t = 0$, how would you find its speed and position after a time t has elapsed.

Solution

(a) At any speed, the *momentum* of the particle is given by $p = \gamma m v = \dfrac{mv}{\sqrt{1 - \dfrac{v^2}{c^2}}}$.

Since $F = qE = \dfrac{dp}{dt}$, $\qquad qE = \dfrac{d}{dt}\left[mv\left(1 - \dfrac{v^2}{c^2}\right)^{-1/2}\right]$

$$qE = m\left(1 - \frac{v^2}{c^2}\right)^{-1/2}\frac{dv}{dt} + \tfrac{1}{2}mv\left(1 - \frac{v^2}{c^2}\right)^{-3/2}\left(\frac{2v}{c^2}\right)\frac{dv}{dt}$$

Simplifying, we find that $\qquad \dfrac{qE}{m} = \dfrac{dv}{dt}\left(1 - \dfrac{v^2}{c^2}\right)^{-3/2}$

$$a = \frac{dv}{dt} = \frac{qE}{m}\left(1 - \frac{v^2}{c^2}\right)^{3/2} \quad \Diamond$$

(b) As $v \to c$, we see that $a \to 0$. The particle thus never attains the speed of light.

(c) $\displaystyle\int_0^v \left(1 - \frac{v^2}{c^2}\right)^{-3/2} dv = \int_0^t \frac{qE}{m}\, dt \qquad v = \frac{qEct}{\sqrt{m^2c^2 + q^2E^2t^2}} = \frac{dx}{dt}$

$$x = \int_0^x dx = qEc\int_0^t \frac{t\,dt}{\sqrt{m^2c^2 + q^2E^2t^2}} = \frac{c}{qE}\left(\sqrt{m^2c^2 + q^2E^2t^2} - mc\right) \quad \Diamond$$

Chapter 40

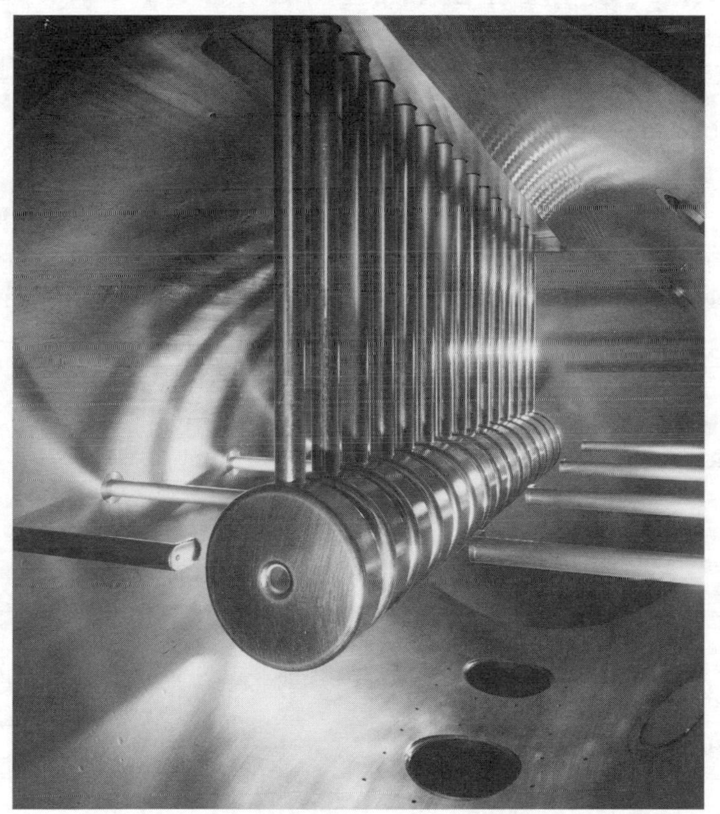

Introduction to Quantum Physics

Chapter 40

INTRODUCTION TO QUANTUM PHYSICS

INTRODUCTION

In the previous chapter, we noted that Newtonian mechanics must be replaced by Einstein's special theory of relativity when dealing with particles whose speeds are comparable to the speed of light. Although many problems were indeed resolved by the theory of relativity in the early part of the 20th century, many experimental and theoretical problems remained unsolved. Attempts to explain the behavior of matter on the atomic level with the laws of classical physics were totally unsuccessful. Various phenomena, such as blackbody radiation, the photoelectric effect, and the emission of sharp spectra lines by atoms in a gas discharge tube, could not be understood within the framework of classical physics. We shall describe these phenomena because of their importance in subsequent developments.

Another revolution took place in physics between 1900 and 1930. This was the era of a new and more general formulation called *quantum mechanics.* This new approach was highly successful in explaining the behavior of atoms, molecules, and nuclei. As with relativity, the quantum theory requires a modification of our ideas concerning the physical world.

An extensive study of quantum theory is certainly beyond the scope of this book. This chapter is simply an introduction to the underlying ideas of quantum theory and the wave-particle nature of matter. We also discuss some simple applications of quantum theory, including the photoelectric effect, the Compton effect, and x-rays.

NOTES FROM SELECTED CHAPTER SECTIONS

40.1 Blackbody Radiation and Planck's Hypothesis

A *black body* is an ideal body that absorbs all radiation incident on it. Any body at some temperature T emits thermal radiation which is characterized by the properties of the body and its temperature. The spectral distribution of blackbody radiation at various temperatures is sketched in the figure. As the temperature increases, the intensity of the radiation (area under the curve) increases, while the peak of the distribution shifts to lower wavelengths.

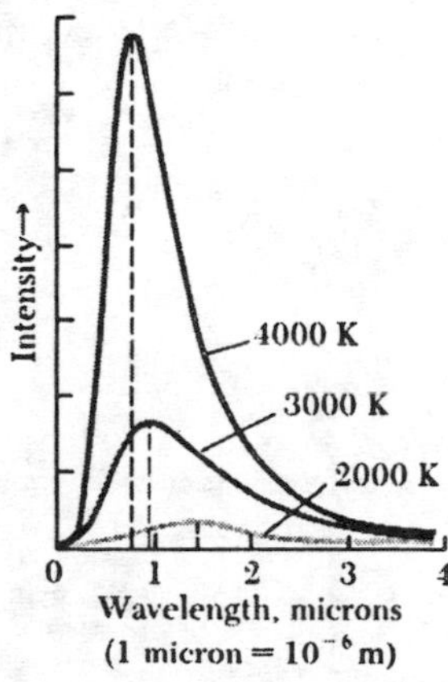

Classical theories failed to explain blackbody radiation. An empirical formula, proposed by Max Planck, is consistent with this distribution at all wavelengths. Planck made two basic assumptions in the development of this result:

- The oscillators emitting the radiation could only have *discrete energies* given by $E_n = nhf$, where n is a quantum number ($n = 1, 2, 3, \ldots$), f is the oscillator frequency, and h is Planck's constant.

- These oscillators can emit or absorb energy in discrete units called quanta (or photons), where the energy of a light quantum obeys the relation $E = hf$.

Subsequent developments showed that the quantum concept was necessary in order to explain several phenomena at the atomic level, including the photoelectric effect, the Compton effect, and atomic spectra.

40.2 The Photoelectric Effect

When light is incident on certain metallic surfaces, electrons can be emitted from the surfaces. This is called the photoelectric effect, discovered by Hertz. One cannot explain many features of the photoelectric effect using classical concepts. In 1905, Einstein provided a successful explanation of the photoelectric effect by extending Planck's quantum concept to include electromagnetic fields.

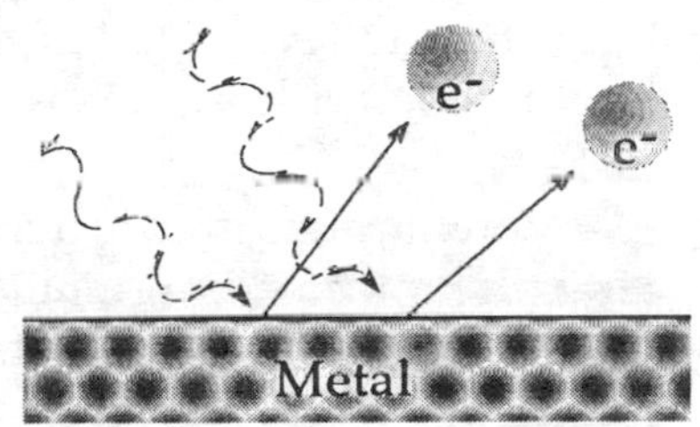

Several features of the photoelectric effect could not be explained with classical physics or with the wave theory of light. However, each of these features can be explained and understood on the basis of the photon theory of light. These observations and their explanations include:

- No electrons are emitted if the incident light frequency falls below some cutoff frequency, f_c, which is characteristic of the material being illuminated. For example, in the case of sodium, $f_c = 5.50 \times 10^{14}$ Hz. This is inconsistent with the wave theory, which predicts that the photoelectric effect should occur at any frequency, provided the light intensity is high enough.

The photoelectric effect is not observed below a certain cutoff frequency, because the energy of the photon must be greater than or equal to ϕ, the energy required to lift the electron out of the metal. If the energy of the incoming photon is not equal to or greater than ϕ, the electrons will never be ejected from the surface, regardless of the intensity of the light.

- If the light frequency exceeds the cutoff frequency, a photoelectric effect is observed and the number of photoelectrons emitted is proportional to the light intensity. However, the maximum kinetic energy of the photoelectrons is independent of light intensity, a fact that cannot be explained by the concepts of classical physics.

  That K_{max} is independent of the light intensity can be understood with the following argument: If the light intensity is doubled, the number of photons is doubled, which doubles the number of photoelectrons emitted. However, their kinetic energy, which equals $hf - \phi$, depends only on the light frequency and the work function, not on the light intensity.

- The maximum kinetic energy of the photoelectrons increases with increasing light frequency. That K_{max} increases with increasing frequency is easily understood with Equation 40.8.

- Electrons are emitted from the surface almost instantaneously (less than 10^{-9} s after the surface is illuminated), even at low light intensities. Classically, one would expect that the electrons would require some time to absorb the incident radiation before they acquire enough kinetic energy to escape from the metal.

  Finally, the almost instantaneous electron emission is consistent with the particle theory of light, in which the incident energy appears in small packets and there is a one-to-one interaction between photons and electrons. This is in contrast to having the energy of the photons distributed uniformly over a large area.

40.4 The Compton Effect

The Compton effect involves the scattering of an x-ray by an electron. The scattered x-ray undergoes a *change* in wavelength $\Delta\lambda$, called the Compton shift, which cannot be explained using classical concepts. By treating the x-ray as a photon (the quantum concept), the scattering process between the photon and electron predicts a shift in photon (x-ray) wavelength given by Equation 40.10, where θ is the angle between the incident and scattered x-ray and m is the mass of the electron. The formula is in excellent agreement with experimental results.

The figure to the right represents the data for Compton scattering of x-rays from graphite at $\theta = 90.0°$. In this case, the Compton shift is $\Delta\lambda = 0.0236$ Å and λ_0 is the wavelength of the incident x-ray beam.

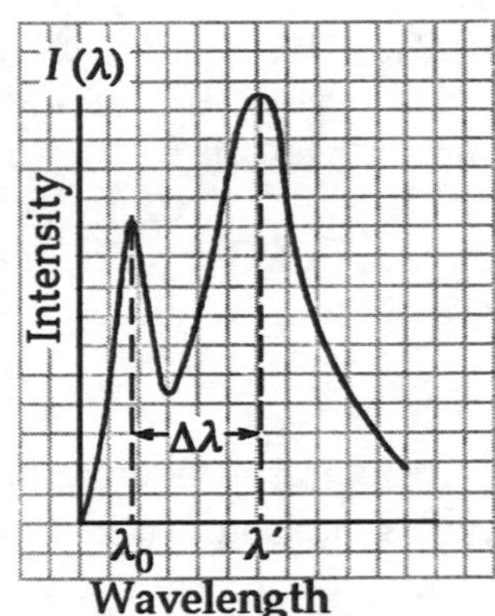

40.5 Atomic Spectra
40.6 Bohr's Quantum Model of the Atom

It is important to understand the behavior of the hydrogen atom as an atomic system for the following reasons:

- Much of what is learned about the hydrogen atom with its single electron can be extended to such single-electron ions as He^+ and Li^{2+}, which are hydrogen-like in their atomic structure.

- The hydrogen atom is an ideal system for performing precise tests of theory against experiment and for improving our overall understanding of atomic structure.

- The quantum numbers used to characterize the allowed states of hydrogen can be used to describe the allowed states of more complex atoms. This enables us to understand the periodic table of the elements, which is one of the greatest triumphs of quantum mechanics.

- The basic ideas about atomic structure must be well understood before we attempt to deal with the complexities of molecular structures and the electronic structure of solids.

The basic assumptions of the Bohr theory as it applies to the hydrogen atom are:

- The electron moves in circular orbits around the proton under the influence of the Coulomb force of attraction. In this case, the Coulomb force is the centripetal force.

- Only certain electron orbits are stable. These are orbits in which the hydrogen atom does not emit energy in the form of radiation. Hence, the total energy of the atom remains constant, and classical mechanics can be used to describe the electron's motion.

- Radiation is emitted by the hydrogen atom when the electron "jumps" from a more energetic initial state to a lower state. The "jump" cannot be visualized or treated classically. In particular, the frequency, f, of the radiation emitted in the jump is related to the change in the atom's energy and is *independent of the frequency of the electron's orbital motion.* The frequency of the emitted radiation is described by

$$E_i - E_f = hf$$

where E_i is the energy of the initial state, E_f is the energy of the final state, h is Planck's constant, and $E_i > E_f$.

- The size of the allowed electron orbits is determined by a condition imposed on the electron's orbital angular momentum: The allowed orbits are those for which the electron's orbital angular momentum about the nucleus is an integral multiple of $\hbar = h/2\pi$.

$$mvr = n\hbar \quad \text{where} \quad n = 1, 2, 3, \ldots.$$

The lowest stationary state of an electron is called the ground state. The minimum energy required to ionize an atom (remove an electron in the ground state from the influence of the proton) is called the ionization energy.

According to the correspondence principle, quantum mechanics is in agreement with classical mechanics when the quantum numbers are very large.

EQUATIONS AND CONCEPTS

The Wien displacement law properly describes the distribution of wavelengths in the energy spectrum emitted by a blackbody radiator.

$$\lambda_{max} T = 0.2898 \times 10^{-2}\ \text{m} \cdot \text{K} \qquad (40.1)$$

A *black body* is an ideal body that absorbs all radiation incident on it. Any body at some temperature T emits thermal radiation which is characterized by the properties of the body and its temperature. As the temperature of a blackbody increases, the intensity of the radiation increases, while the peak of the distribution shifts to shorter wavelengths.

Vibrating molecules are characterized by discrete energy levels called quantum states. The positive integer n is called a quantum number.

$$E_n = nhf \quad (40.5)$$

$$h = 6.626 \times 10^{-34}\ \text{J}\cdot\text{s} \quad (40.4)$$

Molecules emit or absorb energy in discrete units of light energy called quanta. The energy of a quantum or photon corresponds to the energy difference between adjacent quantum states.

$$E = hf \quad (40.6)$$

When light is incident on certain metallic surfaces, electrons can be emitted from the surfaces. This is the photoelectric effect, and was discovered by Hertz. One cannot explain many of the features of the photoelectric effect using classical concepts. In 1905, Einstein provided a successful explanation of the photoelectric effect by extending Planck's quantum concept to include electromagnetic fields. In his model, Einstein assumed that light consists of a stream of particles called *photons* whose energy is given by $E = hf$, where h is Planck's constant and f is their frequency.

The maximum kinetic energy of the ejected photoelectron also depends on the work function of the metal, ϕ, which is typically a few eV. This model is in excellent agreement with experimental results, including the prediction of a cutoff (or threshold) wavelength above which no photoelectric effect is observed.

$$K_{\text{max}} = hf - \phi \quad (40.8)$$

$$\lambda_c = \frac{hc}{\phi} \quad (40.9)$$

The Compton effect involves the scattering of an x-ray by an electron. The scattered x-ray undergoes a change in wavelength called the Compton shift, which cannot be explained using classical concepts.

$$\lambda' - \lambda_0 = \frac{h}{mc}(1 - \cos\theta) \quad (40.10)$$

By treating the x-ray as a photon (the quantum concept), the scattering process between the photon and electron predicts a shift in photon (x-ray) wavelength, where θ is the angle between the incident and scattered x-ray and m is the mass of the electron. The formula is in excellent agreement with experimental results.

The quantity $h/(mc) = 0.00243$ nm is called the Compton wavelength.

When an atomic electron undergoes a transition from one allowed orbit to another, the frequency of the emitted photon is proportional to the difference in energies of the initial and final states.

$$E_i - E_f = hf \qquad (40.18)$$

The Bohr model of the atom assumed that the electron orbited the nucleus in a circular path under the influence of the Coulomb force of attraction. Also, the angular momentum of the electron about the nucleus must be quantized in units of $nh/(2\pi)$.

$$mvr = n\hbar \qquad (40.19)$$

where $n = 1, 2, 3, \ldots$

While in one of the allowed orbits or stationary states (determined by quantization of the orbital angular momentum), the electron does not radiate energy.

The total energy of the hydrogen atom ($KE + PE$) depends on the radius of the allowed orbit of the electron.

$$E = -\frac{k_e e^2}{2r} \qquad (40.22)$$

The electron can exist only in certain allowed orbits, the radii of which can be expressed in terms of the Bohr radius, a_0.

$$r_n = \frac{n^2\hbar^2}{mk_e e^2} \qquad (40.23)$$

where $n = 1, 2, 3, \ldots$

The *Bohr radius* corresponds to $n = 1$.

$$a_0 = \frac{\hbar^2}{mk_e e^2} = 0.0529 \text{ nm} \qquad (40.24)$$

Any orbit radius in the hydrogen atom can then be expressed in terms of the Bohr radius.

$$r_n = n^2 a_0 = n^2(0.0529 \text{ nm}) \qquad (40.25)$$

When numerical values of the constants are used, the energy level values can be expressed in units of electron volts (eV).

$$E_n = -\frac{13.6}{n^2} \text{ eV} \qquad (40.27)$$

where $n = 1, 2, 3, \ldots$

The lowest energy state or ground state corresponds to the principal quantum number $n = 1$. The energy level approaches $E = 0$ as r approaches infinity. This is the ionization energy for the atom.

A photon of frequency, f, (and wavelength λ) is emitted when an electron undergoes a transition from an initial energy level to a final lower level.

$$f = \frac{E_i - E_f}{h} \qquad (40.28)$$

$$\frac{1}{\lambda} = R_H\left(\frac{1}{n_f^2} - \frac{1}{n_i^2}\right) \qquad (40.30)$$

where $R_H = \dfrac{k_e e^2}{2a_0 h}$

The lines observed in the hydrogen spectrum can be arranged into series corresponding to assigned values of principal quantum numbers of the initial and final states.

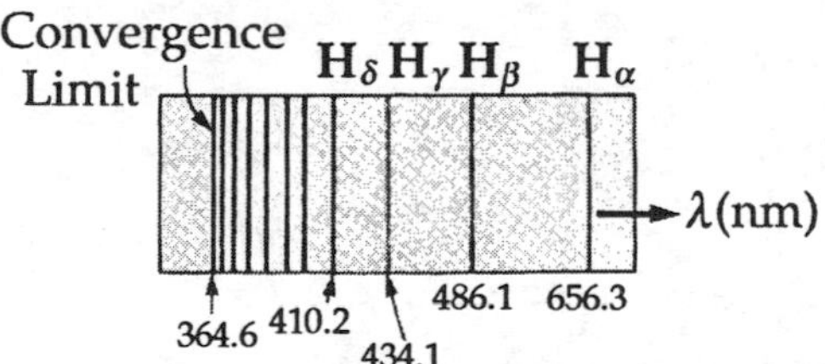

For the Lyman series,
$n_f = 1$ and $n_i = 2, 3, 4, \ldots$

For the Balmer series, as illustrated,
$n_f = 2$ and $n_i = 3, 4, 5, \ldots$

For the Paschen series,
$n_f = 3$ and $n_i = 4, 5, 6, \ldots$

For the Brackett series,
$n_f = 4$ and $n_i = 5, 6, 7, \ldots$

In the case of very large values of the principal quantum number, the energy differences between adjacent levels approach zero and essentially a continuous range (as opposed to a quantized set) of energy values of the emitted photon is possible. In this limit of large quantum numbers, the classical model is reasonable.

REVIEW CHECKLIST

▷ Describe the formula for blackbody radiation proposed by Planck, and the assumption made in deriving this formula.

▷ Describe the Einstein model for the photoelectric effect, and the predictions of the fundamental photoelectric effect equation for the maximum kinetic energy of photoelectrons. Recognize that Einstein's model of the photoelectric effect involves the photon concept ($E = hf$), and the fact that the basic features of the photoelectric effect are consistent with this model.

▷ Describe the Compton effect (the scattering of x-rays by electrons) and be able to use the formula for the Compton shift. Recognize that the Compton effect can only be explained using the photon concept.

▷ State the basic postulates of the Bohr model of the hydrogen atom.

▷ Sketch an energy level diagram for hydrogen (include assignment of values of the principle quantum number, n), show transitions corresponding to spectral lines in the several known series, and make calculations of the wavelength values.

SOLUTIONS TO SELECTED END-OF-CHAPTER PROBLEMS

1. Calculate the energy, in electron volts, of a photon whose frequency is (a) 6.2×10^{14} Hz, (b) 3.1 GHz, (c) 46 MHz. (d) Determine the corresponding wavelengths for these photons.

Solution $E = hf$

(a) $E = (6.63 \times 10^{-34}\ \text{J}\cdot\text{s})(6.2 \times 10^{14}\ \text{Hz}) = 4.11 \times 10^{-19}\ \text{J} = 2.57\ \text{eV}$ ◊

(b) $E = (6.63 \times 10^{-34}\ \text{J}\cdot\text{s})(3.1 \times 10^{9}\ \text{Hz}) = 2.06 \times 10^{-24}\ \text{J} = 12.8\ \mu\text{eV}$ ◊

(c) $E = (6.63 \times 10^{-34}\ \text{J}\cdot\text{s})(46 \times 10^{6}\ \text{Hz}) = 3.05 \times 10^{-26}\ \text{J} = 1.91 \times 10^{-7}\ \text{eV}$ ◊

(d) $\lambda_a = c/f = (3.00 \times 10^{8}\ \text{m/s})/(6.2 \times 10^{14}\ \text{s}^{-1}) = 4.84 \times 10^{-7}\ \text{m} = 484\ \text{nm}$ ◊ (blue light)

$\lambda_b = (3.00 \times 10^{8}\ \text{m/s})/(3.1 \times 10^{9}\ \text{s}^{-1}) = 0.0967\ \text{m} = 9.67\ \text{cm}$ ◊ (Microwave)

$\lambda_c = (3.00 \times 10^{8}\ \text{m/s})/(46 \times 10^{6}\ \text{s}^{-1}) = 6.52\ \text{m}$ ◊ (A radio wave in the public LO band)

3. An FM radio transmitter has a power output of 150 kW and operates at a frequency of 99.7 MHz. How many photons per second does the transmitter emit?

Solution Each photon has an energy

$$E = hf = \left(6.63\times10^{-34}\ \text{J}\cdot\text{s}\right)\left(99.7\times10^{6}\ \text{s}^{-1}\right) = 6.61\times10^{-26}\ \text{J}$$

The number of photons per second is the power divided by the energy per photon:

$$R = \frac{P}{E} = \frac{150\times10^{3}\ \text{J/s}}{6.61\times10^{-26}\ \text{J}} = 2.27\times10^{30}\ \text{photons/s} \quad \lozenge$$

---

11. The human eye is most sensitive to 560-nm light. What temperature black body would radiate most intensely at this wavelength?

Solution We use Wien's law: $\lambda_{\text{max}}T = 0.290\times10^{-2}\ \text{m}\cdot\text{K}$

Thus, $T = \dfrac{2.90\ \text{mm}\cdot\text{K}}{560\times10^{-6}\ \text{mm}} = 5180\ \text{K} \quad \lozenge$

Related Information: This is close to the temperature of the surface of the Sun (which is, in turn, a pretty good black body). Living things on Earth evolved having sensitivity to electromagnetic waves around this wavelength because there is such a lot of it bouncing around, carrying information.)

---

13. Show that at short wavelengths or low temperatures, Planck's radiation law (Eq. 40.3) predicts an exponential decrease in $I(\lambda, T)$ given by Wien's radiation law:

$$I(\lambda,T) = \frac{2\pi hc^2}{\lambda^5}e^{-hc/\lambda k_BT}$$

Solution

Planck's equation is $I(\lambda,T) = \dfrac{2\pi hc^2}{\lambda^5\left(e^{hc/\lambda k_BT}-1\right)}$

When λ or T are small, the term $hc/\lambda k_B T$ grows larger than 1, and the exponential term $e^{hc/\lambda k_B T}$ grows much larger than 1.

As the exponent increases in importance, the '−1' term decreases in importance, until it can be left out entirely. The expression can then be written,

$$I(\lambda,T)=\frac{2\pi hc^2}{\lambda^5}e^{-hc/\lambda k_B t} \quad \lozenge$$

---

16. Molybdenum has a work function of 4.2 eV. (a) Find the cutoff wavelength and threshold frequency for the photoelectric effect. (b) Calculate the stopping potential if the incident light has a wavelength of 180 nm.

Solution We write Einstein's equation as $eV_s = hf - \phi$.

(a) The cutoff wavelength and threshold frequency describe light barely able to produce photoelectrons, with zero kinetic energy.

$$0 = hf_c - \phi = \frac{hc}{\lambda_c} - \phi$$

$$\lambda_c = \frac{hc}{\phi} = \frac{(6.63\times10^{-34}\ \text{J}\cdot\text{s})(3.00\times10^{8}\ \text{m/s})}{(4.2\ \text{eV})(1.602\times10^{-19}\ \text{J/eV})} = 296\ \text{nm} \quad \lozenge$$

$$f = \frac{c}{\lambda} = \frac{3.00\times10^{8}\ \text{m/s}}{296\times10^{-9}\ \text{m}} = 1.01\times10^{15}\ \text{Hz} \quad \lozenge$$

(b) $\dfrac{hc}{\lambda} = \phi + eV_s$

$$\frac{(6.63\times10^{-34}\ \text{J}\cdot\text{s})(3.00\times10^{8}\ \text{m/s})}{180\times10^{-9}\ \text{m}} = (4.2\ \text{eV})(1.60\times10^{-19}\ \text{J/eV}) + (1.60\times10^{-19}\ \text{J/eV})V_s$$

Thus, $V_s = 2.7\ \text{V}$ $\lozenge$

---

19. Two light sources are used in a photoelectric experiment to determine the work function for a particular metal surface. When green light from a mercury lamp (λ = 546.1 nm) is used, a retarding potential of 1.70 V reduces the photocurrent to zero. (a) Based on this measurement, what is the work function for this metal? (b) What stopping potential would be observed when using the yellow light from a helium discharge tube (λ = 587.5 nm)?

Solution First calculate the energy of the photon: $hf = \frac{hc}{\lambda} = 2.27 \text{ eV}$

(a) Next, note that 1 eV is required to raise an electron through 1 V potential, and by Equation 40.7, $K_{max} = 1.70 \text{ eV}$

Finally, by the definition of the work function, $K_{max} = hf - \phi$

$$\phi = 2.27 \text{ eV} - (1.70 \text{ eV}) = 0.571 \text{ eV} \quad \Diamond$$

(b) In this case, $hf = \frac{hc}{\lambda} = 2.11 \text{ eV}$

Using the answer in (a), $K_{max} = hf - \phi$, and

$$K_{max} = 2.11 \text{ eV} - 0.57 \text{ eV} = 1.54 \text{ eV, and } V_s = 1.54 \text{ V.} \quad \Diamond$$

---

21. Lithium, beryllium, and mercury have work functions of 2.3 eV, 3.9 eV, and 4.5 eV, respectively. If 400-nm light is incident on each of these metals, determine (a) which metals exhibit the photoelectric effect and (b) the maximum kinetic energy for the photoelectron in each case.

Solution The energy of a 400-nm wavelength photon is

$$E = h\frac{c}{\lambda} = 3.11 \text{ eV}$$

(a) In order for the photoelectric effect to occur, the energy of the photon must be greater than the work function. Thus, the effect will occur only in lithium. ◊

(b) For lithium, $K_{max} = hf - \phi = 3.11 \text{ eV} - 2.30 \text{ eV} = 0.81 \text{ eV}$ ◊

---

31. After a 0.80-nm x-ray photon scatters from a free electron, the electron recoils at 1.4×10^6 m/s. (a) What was the Compton shift in the photon's wavelength? (b) Through what angle was the photon scattered?

Solution $KE = \frac{1}{2}mv^2 = \frac{1}{2}(9.11\times10^{-31}\text{ kg})(1.40\times10^6\text{ m/s})^2 = 8.93\times10^{-19}\text{ J}$

but $KE = E_\gamma - E'_\gamma = \frac{hc}{\lambda} - \frac{hc}{\lambda'} = hc\left(\frac{\lambda'-\lambda}{\lambda'\lambda}\right) \approx hc\left(\frac{\Delta\lambda}{\lambda^2}\right)$

(a) Thus, $\Delta\lambda = \frac{\lambda^2(KE)}{hc} = \frac{(8.00\times10^{-10}\text{ m})^2(8.93\times10^{-19}\text{ J})}{(6.63\times10^{-34}\text{ J}\cdot\text{s})(3.00\times10^8\text{ m/s})} = 2.87\times10^{-3}\text{ nm}$ ◊

(b) From the Compton equation, $\Delta\lambda = \lambda_C(1-\cos\theta)$.

Thus, $\cos\theta = \left(1 - \frac{\Delta\lambda}{\lambda_C}\right) = 1 - \frac{2.87\times10^{-3}\text{ nm}}{2.43\times10^{-3}\text{ nm}} = -0.1811$

and $\theta = 100°$ ◊

---

35. A 0.0016-nm photon scatters from a free electron. For what (photon) scattering angle do the recoiling electron and the scattered photon have the same kinetic energy?

Solution The energy of the incoming photon is

$$E_i = \frac{hc}{\lambda} = \frac{(6.63\times10^{-34}\text{ J}\cdot\text{s})(3.00\times10^8\text{ m/s})}{0.0016\times10^{-9}\text{ m}} = 1.24\times10^{-13}\text{ J}$$

Since the outgoing photon and the electron each have half of this energy in kinetic form,

$$E' = 6.21\times10^{-14}\text{ J} \quad \text{and} \quad \lambda' = \frac{hc}{E'} = 3.2\times10^{-12}\text{ m}$$

The shift in wavelength is $\Delta\lambda = \lambda' - \lambda = 1.6\times10^{-12}\text{ m}$

But by Equation 40.10, $\Delta\lambda = \lambda_c(1-\cos\theta)$

so $\cos\theta = 1 - \frac{\Delta\lambda}{\lambda_c} = 1 - \frac{1.6\times10^{-12}\text{ m}}{0.00243\times10^{-9}\text{ m}} = -0.342$ and $\theta = 70°$ ◊

---

36. Show that the wavelengths for the Balmer series satisfy the equation

$$\lambda = \frac{364.5n^2}{n^2 - 4}\ \text{nm}$$

where $n = 3, 4, 5, \ldots.$

Solution Start with Balmer's equation, $\frac{1}{\lambda} = R_H\left(\frac{1}{2^2} - \frac{1}{n^2}\right)$ or

$$\lambda = \frac{(4n^2/R_H)}{(n^2 - 4)}$$

Substituting $R_H = 1.0973732 \times 10^7\ \text{m}^{-1}$, we obtain

$$\lambda = \frac{(3.645 \times 10^{-7}\ \text{m})n^2}{n^2 - 4} = \frac{364.5n^2}{n^2 - 4}\ \text{nm}, \quad \text{where } n = 3, 4, 5, \ldots \ \lozenge$$

---

39. (a) What value of n is associated with the 94.96-nm line in the Lyman hydrogen series? (b) Could this wavelength be associated with the Paschen or Brackett series?

Solution Our equation is $\frac{1}{\lambda} = R_H\left(\frac{1}{n_f^2} - \frac{1}{n_i^2}\right)$

where $R_H = 1.097 \times 10^7\ \text{m}^{-1}$

and for the Lyman series, $n_f = 1$, and $n_i = 2, 3, 4, \ldots$

Substituting the given values, $\frac{1}{94.96 \times 10^{-9}\ \text{m}} = (1.097 \times 10^{-9}\ \text{m}^{-1})\left(1 - \frac{1}{n_i^2}\right)$

Solving for n_i, $n_i = 5$ $\lozenge$

(b) By Figure 40.18, spectral lines in the Balmer, Paschen, and Brackett series all have much longer wavelengths, since much smaller energy losses put the atom into energy levels 2, 3, or 4. $\lozenge$

---

43. Construct an energy level diagram for the He^+ ion, for which Z = 2. (b) What is the ionization energy for He^+?

Solution From Equation 40.32,

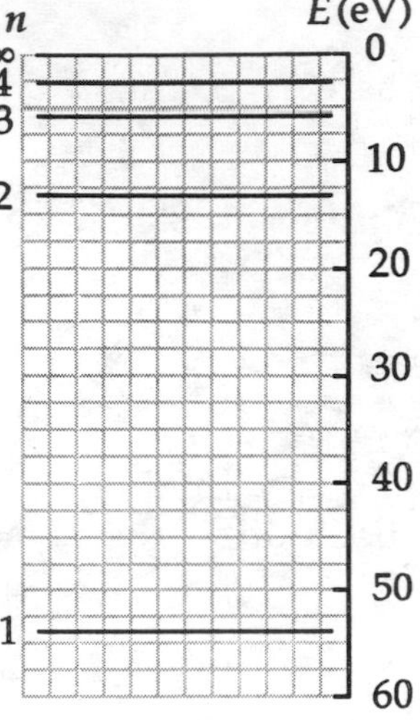

$$E_n = -\frac{k_e e^2 Z^2}{2a_0 n^2} = \frac{(-13.6\text{ eV})(2)^2}{n^2}$$

For the lowest energy level, $n = 1$ and $E_1 = -54.4$ eV

For the first excited state, $n = 2$ and $E_2 = -13.6$ eV

(a) The diagram is shown on the right. ◊

(b) The ionization energy is the energy required to lift the electron from the ground state to the point of breaking free from the nucleus, or from $n = 1$ to $n = \infty$, from energy -54.4 eV to 0 eV. The answer then, is 54.4 eV. ◊

---

47. A photon is emitted as a hydrogen atom undergoes a transition from the $n = 6$ state to the $n = 2$ state. Calculate (a) the energy, (b) the wavelength, and (c) the frequency of the emitted photon.

Solution By conservation of energy, the energy of the photon is equal to the energy lost in the transition.

(a) From Equation 40.27, $E_{6-2} = (-13.6\text{ eV})\left(\frac{1}{6^2} - \frac{1}{2^2}\right) = 3.02\text{ eV}$ ◊

(b) $\frac{1}{\lambda} = R_H\left(\frac{1}{n_f^2} - \frac{1}{n_i^2}\right) = \left(1.097 \times 10^7\text{ m}^{-1}\right)\left(\frac{1}{4} - \frac{1}{36}\right)$

$\lambda = 410$ nm ◊ (This is the deep violet spectral line farthest to the left in the hydrogen spectrum of the text's figure 40.14)

(c) $f = c/\lambda = (3.00 \times 10^8\text{ m/s})/(410 \times 10^{-9}\text{ m}) = 7.32 \times 10^{14}$ Hz ◊

---

50. An electron is in the nth Bohr orbit of the hydrogen atom. (a) Show that the period of the electron is $T = \tau_0 n^3$, and determine the numerical value of τ_0. (b) On the average, an electron remains in the $n = 2$ orbit for about 10 μs before it jumps down to the $n = 1$ (ground-state) orbit. How many revolutions does the electron make before it jumps to the ground state? (c) If one revolution of the electron is defined as an "electron year" (analogous to an Earth year being one revolution of the Earth around the Sun), does the electron in the $n = 2$ orbit "live" very long? Explain. (d) How does the above calculation support the "electron cloud" concept?

Solution

(a) The time for one complete orbit is $T = \dfrac{2\pi r}{v}$

From Bohr's quantization postulate, $L = mvr = n\hbar$, we see that $v = \dfrac{n\hbar}{mr}$

Thus, the orbital period becomes $T = \dfrac{2\pi m r^2}{n\hbar} = \dfrac{2\pi m (a_0 n^2)^2}{n\hbar} = \dfrac{2\pi m a_0^2}{\hbar} n^3$

or $\quad T = \tau_0 n^3 \quad$ where $\quad \tau_0 = \dfrac{2\pi m a_0^2}{\hbar} = \dfrac{2\pi(9.11\times 10^{-31}\text{ kg})(0.0529\times 10^{-9}\text{ m})^2}{(1.055\times 10^{-34}\text{ J}\cdot\text{s})}$

$$\tau_0 = 1.52\times 10^{-16}\text{ s} \quad \Diamond$$

(b) With $n = 2$, we have $\quad T = 8\tau_0 = 8(1.52\times 10^{-16}\text{ s}) = 1.21\times 10^{-15}\text{ s}$

Thus, if the electron stays in the $n = 2$ state for 10^{-8} s, it will make

$$\frac{10^{-8}\text{ s}}{1.21\times 10^{-15}\text{ s/rev}} = 8.23\times 10^6 \text{ revolutions of the nucleus} \quad \Diamond$$

(c) Yes, for 8.23×10^6 "electron years" $\Diamond$

(d) With that many revolutions, it appears like the electron (or at least some part of it) is present everywhere in the $n = 2$ orbit. $\Diamond$

53. Positronium is a hydrogen-like atom consisting of a positron (a positively charged electron) and an electron revolving around each other. Using the Bohr model, find the allowed radii (relative to the center of mass of the two particles) and the allowed energies of the system.

Solution

Let r represent the distance between the electron and the positron. The two move with equal-size and opposite velocities in the same circle, radius $r/2$, around their center of mass. The total angular momentum is quantized according to

$$\frac{mvr}{2}+\frac{mvr}{2}=n\hbar, \qquad \text{where} \quad n=1, 2, 3, \ldots$$

For each particle, $\Sigma F = ma$ reads $\dfrac{k_e e^2}{r^2}=\dfrac{mv^2}{r/2}$

We can eliminate $v=\dfrac{n\hbar}{mr}$ to find $\dfrac{k_e e^2}{r}=\dfrac{2mn^2\hbar}{m^2r^2}$

and $r=\dfrac{2n^2\hbar^2}{mk_e e^2}=2a_0n^2=\left(1.06\times10^{-10}\text{ m}\right)n^2$ for the allowed separation distances. ◊

The energy can be calculated from: $E=\frac{1}{2}mv^2+\frac{1}{2}mv^2-\dfrac{k_e e^2}{r}$

Since $mv^2=\dfrac{k_e e^2}{2r}$, $\qquad E=\dfrac{k_e e^2}{2r}-\dfrac{k_e e^2}{r}=-\dfrac{k_e e^2}{2r}=\dfrac{k_e e^2}{4a_0n^2}$

This is one-half the energy of hydrogen, or $-(6.80\text{ eV})/n^2$. ◊

55. An electron in chromium moves from the $n = 2$ state to the $n = 1$ state without emitting a photon. Instead, the excess energy is transferred to an outer electron (one in the $n = 4$ state), which is then ejected by the atom. [This is called an Auger (pronounced 'ohjay') process, and the ejected electron is referred to as an Auger electron.] Use the Bohr theory to estimate the kinetic energy of the Auger electron.

Solution

We ignore the shielding effect of all the other electrons on the electrons that change state. This will not be a good quantitative estimate. However, from Equation 40.32, the inner electron starts with an energy of

$$E_2 = -\frac{k_e e^2 Z^2}{2a_0 n^2} = -(13.6\text{ eV})\frac{24^2}{2^2} = -1.958\text{ keV}$$

and drops to a level of $\quad E_1 = -(13.6\text{ eV})\frac{24^2}{1^2} = -7.83\text{ keV}$

for a net energy change of $\quad \Delta E = -5.88\text{ keV}.$

This energy goes to the outer electron, which starts with a shell energy of

$$E_4 = -(13.6\text{ eV})\frac{24^2}{4^2} = -0.490\text{ keV}.$$

In its free state, the electron's shell energy $E_\infty = 0$.

Thus it ends with a net kinetic energy of

$$K = E_4 - E_\infty + E_1 - E_2 = 5.39\text{ keV} \quad \lozenge$$

60. **Use Bohr's model of the hydrogen atom to show that when the electron moves from the state n to the state $n-1$, the frequency of the emitted light is**

$$f=\frac{2\pi^2 mk_e{}^2e^4}{h^3}\left[\frac{2n-1}{(n-1)^2n^2}\right]$$

Show that as $n\to\infty$, this expression varies as $1/n^3$ and reduces to the classical frequency the atom is expected to emit. (*Hint*: To calculate the classical frequency, note that the frequency of revolution is $v/2\pi r$, where r is given by Equation 40.23.)

Solution Combining Equations 40.22 and 40.23, $E_n=-\dfrac{mk_e{}^2e^4}{2n^2\hbar^2}$, with $\hbar=h/2\pi$.

Thus, in the Bohr model, $$hf=\Delta E=\frac{4\pi^2 mk_e{}^2e^4}{2h^2}\left(\frac{1}{(n-1)^2}-\frac{1}{n^2}\right)$$

which reduces to $$f=\frac{2\pi^2 mk_e{}^2e^4}{h^3}\left(\frac{2n-1}{(n-1)^2n^2}\right)\quad\Diamond$$

As $n\to\infty$, the '-1' terms lose importance and drop out, leaving the right-hand term equal to

$$\lim_{n\to\infty}\left(\frac{2n-1}{(n-1)^2n^2}\right)=\frac{2n}{n^4}=\frac{2}{n^3}\quad\Diamond$$

Thus the frequency becomes $$f=\frac{4\pi^2 mk_e{}^2e^4}{h^3}\left(\frac{1}{n^3}\right)$$

But in the Bohr model, $$v^2=\frac{k_e e^2}{mr}\quad\text{and}\quad r=\frac{n^2h^2}{4\pi^2 mk_e e^2},$$

and the classical frequency is $$f=\frac{v}{2\pi r}=\frac{1}{2\pi r^{3/2}}\sqrt{\frac{k_e e^2}{m}}=\frac{1}{2\pi}\left(\sqrt{\frac{k_e e^2}{m}}\right)\left(\frac{4\pi^2 mk_e e^2}{n^2h^2}\right)$$

But, solving, we find this is identical to $f=\dfrac{4\pi^2 mk_e{}^2e^4}{h^3}\left(\dfrac{1}{n^3}\right)\ \Diamond$

---

61. A muon (problem 59) is captured by a deuteron to form a muonic atom. (a) Find the energy of the ground state and the first excited state. (b) What is the wavelength of the photon emitted when the atom makes a transition from the first excited state to the ground state?

Solution

(a) First, we find the "Bohr radius" of the atom, noting that we must use the reduced mass μ of the atom in Equation 40.23, as it is defined in problem 3 of chapter 42.

$$\mu = \frac{m_1 m_2}{m_1 + m_2} = \frac{(207\ m_e)(3680\ m_e)}{(207 + 3680)\ m_e} = 196\ m_e$$

Thus,

$$R_0 = \frac{\hbar^2}{\mu k_e e^2} = \frac{\hbar^2}{(196\ m_e)k_e e^2} = \frac{a_0}{196}$$

But a_0 is the standard Bohr radius. Substituting R_0 into Equation 40.32, with $Z = 1$,

$$E_n = -\frac{k_e e^2}{2R_0}\left(\frac{1}{n^2}\right) = -196\left(\frac{k_e e^2}{2a_0}\right)\left(\frac{1}{n^2}\right) = -196(13.6\ \text{eV})\left(\frac{1}{n^2}\right) = -(2.67\ \text{keV})\left(\frac{1}{n^2}\right)$$

Thus,

$$E_1 = -2.67\ \text{keV}$$

and

$$E_2 = -0.667\ \text{keV}. \quad \Diamond$$

(b) Since $\Delta E = E_2 - E_1 = 2.00\ \text{keV}$, and $\Delta E = hc/\lambda$,

$$\lambda = \frac{hc}{\Delta E} = \frac{1240\ \text{eV}\cdot\text{nm}}{2.00 \times 10^3\ \text{eV}} = 0.620\ \text{nm} \quad \Diamond$$

64. The total power per unit area radiated by a black body at a temperature T is the area under the $I(\lambda, T)$ versus λ curve, as in Figure 40.2. (a) Show that this power per unit area is

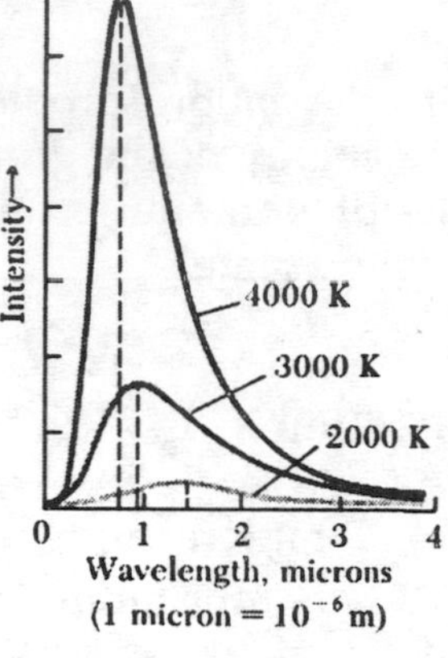

$$\int_0^\infty I(\lambda,T)d\lambda = \sigma T^4$$

where $I(\lambda, T)$ is given by Planck's radiation law and σ is a constant independent of T. This result is known as the Stefan-Boltzman law. To carry out the integration, you should make the change of variable $x = hc/\lambda k_B T$ and use the fact that

$$\int_0^\infty \left(\frac{x^3 dx}{e^x - 1}\right) = \frac{\pi^4}{15}$$

(b) Show that the Stefan-Boltzman constant σ has the value

$$\sigma = \frac{2\pi^5 k_B{}^4}{15c^2h^3} = 5.7\times10^{-8}\ \text{W/m}^2\cdot\text{K}^4.$$

Solution In order to make the suggested substitution, we find λ and $d\lambda$:

$$x = \frac{hc}{\lambda k_B T} \qquad \lambda = \frac{hc}{x k_B T} \qquad d\lambda = -\frac{hc\, dx}{x^2 k_B T}$$

We also note that the limits of integration change from $\lambda = (0,\infty)$ to $x = (\infty, 0)$. Substituting these variables into the integral, the intensity of the blackbody radiation is:

$$\int_0^\infty I(\lambda,T)d\lambda = \int_0^\infty \frac{2\pi hc^2}{\lambda^5\left(e^{hc/\lambda k_B T} - 1\right)}d\lambda = \int_\infty^0 -\left(\frac{2\pi hc^2}{e^x - 1}\right)\left(\frac{x^5 k_B{}^5 T^5}{h^5 c^5}\right)\left(\frac{hc\, dx}{x^2 k_B T}\right)$$

Pulling out constants, $$\int_0^\infty I(\lambda,T)d\lambda = \frac{2\pi k_B{}^4 T^4}{h^3c^2}\int_\infty^0 -\frac{x^3}{e^x - 1}dx = \frac{2\pi k_B{}^4 T^4}{h^3c^2}\int_0^\infty \frac{x^3}{e^x - 1}dx$$

But the integral term is just $\pi^4/15$, so $$\int_0^\infty I(\lambda,T)d\lambda = \frac{2\pi^5 k_B{}^4 T^4}{15h^3c^2} = \sigma T^4 \quad \Diamond$$

with $$\sigma = \frac{2\pi^5 k_B{}^4}{15c^2h^3} = 5.7\times10^{-8}\ \text{W/m}^2\cdot\text{K}^4 \quad \Diamond$$

67. Show that the ratio of the Compton wavelength λ_c to the de Broglie wavelength $\lambda = h/p$ for a relativistic electron is

$$\frac{\lambda_c}{\lambda} = \left[\left(\frac{E}{mc^2}\right)^2 - 1\right]^{1/2}$$

where E is the total energy of the electron and m is its mass.

Solution

From the definition of the Compton wavelength, $\lambda_c = h/mc$

Taking the ratio of the Compton wavelength to the de Broglie wavelength,

$$\frac{\lambda_c^2}{\lambda^2} = \frac{p^2}{(mc)^2}$$

But from Equation 39.25, the relativistic momentum is

$$p^2 = \frac{E^2 - m^2c^4}{c^2}$$

Substituting and simplifying,

$$\frac{\lambda_c^2}{\lambda^2} = \frac{\left(E^2 - m^2c^4\right)}{\left(mc^2\right)^2}$$

Finally,

$$\frac{\lambda_c^2}{\lambda^2} = \left(\frac{E}{mc^2}\right)^2 - 1 \quad \text{and} \quad \frac{\lambda_c}{\lambda} = \sqrt{\left(\frac{E}{mc^2}\right)^2 - 1} \quad \Diamond$$

70. The table below shows data obtained in a photoelectric experiment. (a) Using these data, make a graph similar to Figure 40.8 that plots as a straight line. From the graph, determine (b) an experimental value for Planck's constant (in joule-seconds) and (c) the work function (in electron volts) for the surface. (Two significant figures for each answer are sufficient.)

Wavelength (nm)	Maximum Kinetic Energy of Photoelectrons (eV)
588	0.67
505	0.98
445	1.35
399	1.63

Solution We first must convert the wavelengths to frequency.

Using $f = c/\lambda$,

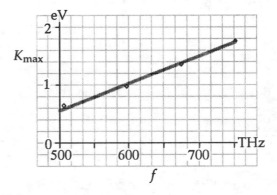

$$f = \frac{3\times10^{8}\ \text{m/s}}{588\times10^{-9}\ \text{m}} = 5.10\times10^{14}\ \text{Hz} = 510\ \text{THz}$$

Thus the corresponding frequencies are

{510 THz, 594 THz, 674 THz, and 752 THz}

(a) The graph is as shown to the right. ◊

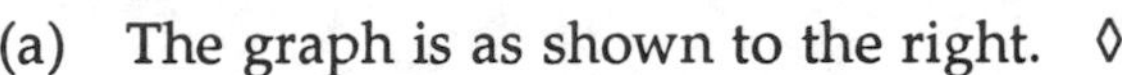

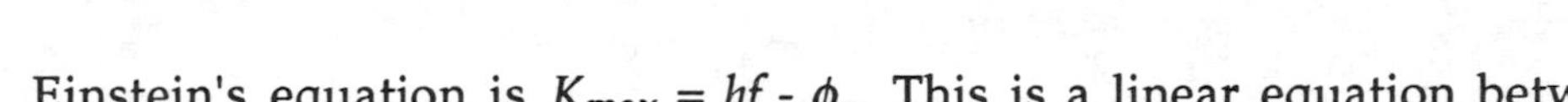

Einstein's equation is $K_{max} = hf - \phi$. This is a linear equation between K_{max} and f, with a slope of h and a y intercept of $(-\phi)$.

(b) The slope of the graph can be estimated as $4.0\times10^{-15}\ \text{eV}\cdot\text{s} = 6.4\times10^{-34}\ \text{J}\cdot\text{s} \pm 10\%$ ◊

(c) The y intercept can be found by extending the graph to the left, to $f = 0$:

$$\phi = -K_{max}(0) = -(-1.4) = 1.4\ \text{eV} \pm 10\% \quad ◊$$

Both of these answers are reasonable.

Chapter 41

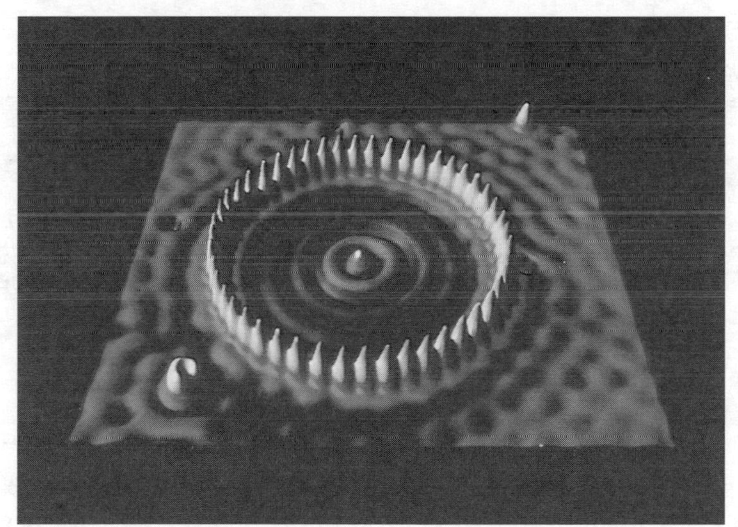

Quantum Mechanics

Chapter 41

QUANTUM MECHANICS

INTRODUCTION

In this chapter, we introduce quantum mechanics, a theory for explaining atomic structure. This scheme, developed from 1925 to 1926 by Schrödinger, Heisenberg, and others, addresses the limitations of the Bohr model and lets us understand a host of phenomena involving atoms, molecules, nuclei, and solids. We begin by describing wave-particle duality, wherein a particle can be viewed as having wave-like properties, and its wavelength can be calculated if its momentum is known. Next, we describe some of the basic features of the formalism of quantum mechanics and their application to simple, one-dimensional systems. For example, we shall treat the problem of a particle confined to a potential well having infinitely high barriers.

NOTES FROM SELECTED CHAPTER SECTIONS

41.1 Photons and Electromagnetic Waves

The results of some experiments are better described on the basis of the photon model of light; other experimental outcomes are better described in terms of the wave model. The photon theory and the wave theory complement each other—light exhibits both wave and photon characteristics.

41.2 The Wave Properties of Particles

De Broglie postulated that a particle in motion has wave properties and a corresponding wavelength inversely proportional to the particle's momentum.

41.4 The Uncertainty Principle

Quantum theory predicts that it is fundamentally impossible to make simultaneous measurements of a particle's position and speed with infinite accuracy.

41.5 Introduction to Quantum Mechanics

The wave function is a complex valued quantity, the absolute square of which gives the probability of finding a particle at a given point at some instant; and the wave function contains all the information that can be known about the particle.

The *Schrödinger equation* describes the manner in which matter waves change in time and space.

The probability of finding a certain value for a quantity (position, energy) is called the *expectation value* of the quantity.

41.6 A Particle in a Box

A particle confined to a box and represented by a well-defined de Broglie wave function is represented by a sinusoidal wave. The allowed states of the system are called *stationary states* since they represent *standing waves.*

The minimum energy which the particle can have is called the *zero-point energy.*

41.7 The Schrödinger Equation

The basic problem in wave mechanics is to determine a solution to the Schrödinger equation. The solution will provide the allowed wave functions and energy levels of the system.

41.9 Tunneling Through a Barrier

When a particle is incident onto a barrier, the height of which is greater than the energy of the particle, there is a finite probability that the particle will penetrate the barrier. In this process, called *tunneling,* part of the incident wave is transmitted and part is reflected.

EQUATIONS AND CONCEPTS

According to the de Broglie hypothesis, a material particle should have an associated wavelength λ which depends on its momentum.

$$\lambda = \frac{h}{mv} \tag{41.1}$$

Matter waves can be represented by a wave function, ψ, which, in general, depends on position and time. In the expression for the part of the wave function which depends only on position, k is the wave number.

$$\psi(x) = A\sin\left(\frac{2\pi x}{\lambda}\right) \qquad (41.5)$$

The normalization condition is a statement of the requirement that the particle exists at some point (along the x axis in the one-dimensional case) at all times.

$$\int_{-\infty}^{\infty} |\psi|^2\, dx = 1 \qquad (41.6)$$

Although it is not possible to specify the position of a particle exactly, it is possible to calculate the probability P_{ab} of finding the particle within an interval $a \le x \le b$.

$$P_{ab} = \int_a^b |\psi|^2\, dx \qquad (41.7)$$

The probability of measuring a certain value for the position of a particle is called the expectation value of the coordinate x.

$$\langle x \rangle \equiv \int_{-\infty}^{\infty} x|\psi|^2\, dx \qquad (41.8)$$

The uncertainty principle states that if a measurement of position is made with a precision Δx and a *simultaneous* measurement of momentum is made with a precision Δp, then the product of the two uncertainties can never be smaller than a number of the order of $\hbar$.

$$\Delta x \Delta p_x \ge \frac{\hbar}{2} \qquad (41.3)$$

The allowed wave functions for a particle in a rigid box of width L are sinusoidal.

$$\psi(x) = A\sin\left(\frac{n\pi x}{L}\right) \qquad (41.10)$$

$$n = 1, 2, 3, \ldots$$

The energy of a particle in a box is quantized and the least energy which the particle can have is called the zero-point energy.

$$E_n = \left(\frac{h^2}{8mL^2}\right)n^2 \qquad (41.11)$$

$$n = 1, 2, 3, \ldots$$

The time independent Schrödinger equation for a bound system (total energy E is constant) allows in principle the determination of the wave functions and energies of the allowed states if the potential energy function is known.

$$\frac{\partial^2 \psi}{\partial x^2} = -\frac{2m}{\hbar^2}(E - U)\psi \qquad (41.15)$$

The Schrödinger equation can be written for a harmonic oscillator of total energy, E.

$$\frac{d^2 \psi}{dx^2} = -\left[\left(\frac{2mE}{\hbar^2}\right) - \left(\frac{m\omega}{\hbar}\right)^2 x^2\right]\psi \qquad (41.22)$$

The energy levels of the harmonic oscillator are quantized.

$$E_n = \left(n + \tfrac{1}{2}\right)\hbar\omega$$

$$n = 0, 1, 2, \ldots$$

REVIEW CHECKLIST

▷ Discuss the wave properties of particles, the de Broglie wavelength concept, and the dual nature of both matter and light.

▷ Describe the concept of wave function for the representation of matter waves and state in equation form the normalization condition and expectation value of the coordinate.

▷ Discuss the manner in which the uncertainty principle makes possible a better understanding of the dual wave-particle nature of light and matter.

▷ State the time-independent form of the Schrödinger equation for a bound system of total energy, E, and discuss the required conditions on the wave function.

▷ Describe the allowed wave functions and energy levels for a particle in a one-dimensional box.

SOLUTIONS TO SELECTED END-OF-CHAPTER PROBLEMS

2. Calculate the de Broglie wavelength for an electron that has kinetic energy (a) 50 eV and (b) 50 keV.

Solution

(a) We can find the electron speed first, and then the wavelength.

$$K = \tfrac{1}{2}mv^2 = (50\text{ eV})(1.6\times10^{-19}\text{ C}/e^-)(1\text{ J}/\text{C}\cdot\text{V})$$

$$v = \sqrt{2\frac{(80\times10^{-19}\text{ J})}{(9.11\times10^{-31}\text{ kg})}} = 4.19\times10^6\text{ m/s}$$

$$\lambda = \frac{h}{p} = \frac{h}{mv}$$

$$\lambda = \frac{6.63\times10^{-34}\text{ J}\cdot\text{s}}{(9.11\times10^{-31}\text{ kg})(4.19\times10^6\text{ m/s})}$$

$$\lambda = 0.17\text{ nm} \quad \Diamond$$

(b) Or we can skip computing the speed by using

$$K = \frac{mv^2}{2} = \frac{p^2}{2m} = 50{,}000(1.6\times10^{-19}\text{ J})$$

$$p = 1.20\times10^{-22}\text{ kg}\cdot\text{m/s}$$

$$\lambda = \frac{h}{p} = 5.5\times10^{-12}\text{ m} \quad \Diamond$$

Make sure you can also find the wavelength of a photon when given its energy. Check that a 50-eV photon of ultraviolet light has $\lambda = \frac{hc}{E} = 24.9\text{ nm}$.

10. For an electron to be confined to a nucleus, its de Broglie wavelength would have to be less than 10^{-14} m. (a) What would be the kinetic energy of an electron confined to this region? (b) On the basis of this result, would you expect to find an electron in a nucleus? Explain.

Solution

(a) In this problem, the electron must be treated relativistically. The momentum of the electron is not equal to mv; but it is still

$$p = \frac{h}{\lambda} = \frac{6.626 \times 10^{-34}\ \text{J} \cdot \text{s}}{1 \times 10^{-14}\ \text{m}} = 6.626 \times 10^{-20}\ \text{kg} \cdot \text{m/s}$$

The energy of the electron is

$$E - \left(p^2c^2 + m^2c^4\right)^{1/2}$$

$$E = \sqrt{(6.626 \times 10^{-20})^2(3.00 \times 10^8)^2 + (0.511 \times 10^6)^2(1.6 \times 10^{-19})^2}$$

$$E = 1.99 \times 10^{-11}\ \text{J} = 1.24 \times 10^8\ \text{eV}$$

so that $$K = E - m_e c^2 \approx 124\ \text{MeV} \quad \Diamond$$

(b) The kinetic energy is too large to expect that the electron could be confined to a region the size of the nucleus. The electric potential energy of an electron distant by 10^{-14} m from a proton is

$$U = qV = -e(k_e e / r)$$

$$U = -e(8.99 \times 10^9\ \text{N} \cdot \text{m}^2/\text{C}^2)(1.6 \times 10^{-19}\ \text{C})/(10^{-14}\ \text{m})$$

$$U = -0.144\ \text{MeV} \quad \Diamond$$

The kinetic energy of +124 MeV would make the electron immediately escape the proton's attraction.

12. Robert Hofstadter won the 1961 Nobel Prize in physics for his pioneering work in scattering 20-GeV electrons from nuclei. (a) What is the γ-factor for a 20-GeV electron, where $\gamma = \left(1 - v^2/c^2\right)^{-1/2}$? What is the momentum of the electron in $\text{kg}\cdot\text{m/s}$? (b) What is the wavelength of a 20-GeV electron and how does it compare with the size of a nucleus?

Solution

(a) From $E = \gamma m_e c^2$, $$\gamma = \frac{20 \times 10^3 \text{ MeV}}{0.511 \text{ MeV}} = 39{,}139 \quad \lozenge$$

For these extreme-relativistic electrons, with $m_e c^2 << pc$, $E^2 = p^2c^2 + m^2c^4$ simplifies to $E^2 \cong p^2c^2$.

$$p = \frac{E}{c} = \frac{(2.0 \times 10^4 \text{ MeV})(1.6 \times 10^{-13} \text{ J/MeV})}{3.0 \times 10^8 \text{ m/s}}$$

$$p = 1.07 \times 10^{-17} \text{ kg}\cdot\text{m/s} \quad \lozenge$$

(b) $$\lambda = \frac{h}{p} = \frac{6.626 \times 10^{-34} \text{ J}\cdot\text{s}}{1.07 \times 10^{-17} \text{ kg}\cdot\text{m/s}} \cong 6.2 \times 10^{-17} \text{ m} \quad \lozenge$$

Since the size of a nucleus is on the order of 10^{-14} m, the 20-GeV electrons would be small enough to go through the nucleus.

---

16. Neutrons traveling at 0.40 m/s are directed through a double slit having a 1.0-mm separation. An array of detectors is placed 10 m from the slit. (a) What is the de Broglie wavelength of the neutrons? (b) How far off axis is the first zero-intensity point on the detector array? (c) Can we say which slit the neutron passed through? Explain.

Solution

(a) $$\lambda = \frac{h}{mv} = \frac{6.63 \times 10^{-34}}{(1.67 \times 10^{-27})(0.40)} = 9.9 \times 10^{-7} \text{ m} \quad \lozenge$$

(b) The condition for destructive interference in a multiple-slit experiment is $d\sin\theta = (m+\frac{1}{2})\lambda$ with $m = 0$ for the first minimum. Then,

$$\theta = \sin^{-1}\left(\frac{\lambda}{2d}\right) = 0.0284^\circ \qquad \frac{y}{L} = \tan\theta$$

$$y = L\tan\theta = (10\text{ m})(\tan 0.0284^\circ) = 5.0\text{ mm} \quad \lozenge$$

(c) We cannot say the neutron passed through one slit. We can only say it passed through the pair of slits, as a water wave does to produce an interference pattern.

---

17. The resolving power of a microscope is proportional to the wavelength used. If one wished to use a microscope to "see" an atom, a resolution of approximately 10^{-11} m would have to be obtained. (a) If electrons are used (electron microscope), what minimum kinetic energy is required for the electrons? (b) If photons are used, what minimum photon energy is needed to obtain the required resolution?

Solution

(a) Since the de Broglie wavelength is $\lambda = \frac{h}{p}$,

$$p_e = \frac{h}{\lambda} = \frac{6.626\times10^{-34}\text{ J}\cdot\text{s}}{1.00\times10^{-11}\text{ m}} = 6.626\times10^{-23}\text{ kg}\cdot\text{m/s}$$

$$K_e = \frac{p_e^{\,2}}{2m} = \frac{(6.626\times10^{-23}\text{ kg}\cdot\text{m/s})^2}{2(9.11\times10^{-31}\text{ kg})} = 2.41\times10^{-15}\text{ J} = 15.1\text{ keV} \quad \lozenge$$

(b) For photons: $\lambda = \frac{c}{f} = \frac{hc}{hf}$

$$E = hf = \frac{hc}{\lambda} = \frac{(6.626\times10^{-34}\text{ J}\cdot\text{s})(3\times10^{8}\text{ m/s})}{1\times10^{-11}\text{ m}} = 1.99\times10^{-15}\text{ J} = 124\text{ keV} \quad \lozenge$$

For the photon, this wavelength $\lambda = 0.1$ Å is in the x-ray range of the electromagnetic spectrum.

---

22. An electron (m = 9.11 × 10^{-31} kg) and a bullet (m = 0.020 kg) each have a speed of 500 m/s, accurate to within 0.010%. Within what limits could we determine the position of the objects?

Solution For the electron, the uncertainty in momentum is

$$\Delta p = m\Delta v = (9.11 \times 10^{-31}\ \text{kg})(500\ \text{m/s})(10^{-4})$$

$$\Delta p = 4.56 \times 10^{-32}\ \text{kg} \cdot \text{m/s}$$

The minimum uncertainty in position is then

$$\Delta x = \frac{h}{4\pi\Delta p} = \frac{6.63 \times 10^{-34}\ \text{J} \cdot \text{s}}{4\pi(4.56 \times 10^{-32}\ \text{kg} \cdot \text{m/s})} = 1.16\ \text{mm} \quad \Diamond$$

For the bullet,

$$\Delta p = m\Delta v = (0.020\ \text{kg})(500\ \text{m / s})(10^{-4}) = 10^{-3}\ \text{kg} \cdot \text{m / s}$$

$$\Delta x = \frac{h}{4\pi\,\Delta p} = 5.3 \times 10^{-32}\ \text{m} \quad \Diamond$$

Quantum mechanics describes all objects, but the quantum fuzziness in position is unobservably small for the bullet and large for the small-mass electron.

---

24. A free electron has a wave function $\psi(x) = A\sin\left(5.00 \times 10^{10}\, x\right)$, where x is in meters. Find (a) the de Broglie wavelength, (b) the momentum, and (c) the energy in electron volts.

Solution The wave function A sin(5.00 × 10^{10} x) will go through one full cycle between x_1 = 0 and (5.00 × 10^{10})x_2 = 2π. The wavelength is then

(a) $$\lambda = x_2 - x_1 = (2\pi\,\text{m}) / 5 \times 10^{10} = 1.26 \times 10^{-10}\ \text{m} \quad \Diamond$$

To say the same thing, we can inspect A sin(5 × 10^{10} x/m) to see that the wave number is k = 5 × 10^{10}/m.

(b) Since $\lambda = \frac{h}{p}$,

The momentum $$p = \frac{h}{\lambda} = \frac{6.63 \times 10^{-34} \text{ J} \cdot \text{s}}{1.26 \times 10^{-10} \text{ m}} = 5.28 \times 10^{-24} \text{ kg} \cdot \text{m/s} \quad \lozenge$$

(c) The electron's kinetic energy is

$$K = \tfrac{1}{2}mv^2 = \frac{p^2}{2m} = \frac{(5.28 \times 10^{-24} \text{ kg} \cdot \text{m/s})^2}{2(9.11 \times 10^{-31} \text{ kg})}\left(\frac{1 \text{ eV}}{1.6 \times 10^{-19} \text{ J}}\right) = 95.5 \text{ eV} \quad \lozenge$$

Its relativistic total energy is 511 keV + 95.5 eV.

---

27. Use the particle-in-a-box model to calculate the first three energy levels of a neutron trapped in a nucleus of diameter 2.00 × 10^{-5} nm. Are the energy-level differences realistic?

Solution

$$E_n = \frac{h^2}{8mL^2}n^2$$

$$E_1 = \frac{(6.626 \times 10^{-34} \text{ J} \cdot \text{s})^2}{8(1.67 \times 10^{-27} \text{ kg})(2 \times 10^{-14} \text{ m})^2}(1)^2 = 8.21 \times 10^{-14} \text{ J} = 0.513 \text{ MeV} \quad \lozenge$$

$$E_2 = 4E_1 = 2.05 \text{ MeV} \quad \lozenge$$

$$E_3 = 9E_1 = 4.62 \text{ MeV} \quad \lozenge$$

Yes, the energy differences are of the order of 1 MeV, which is a typical energy for a γ-ray photon, as emitted by a nucleus in an excited state.

---

33. An electron is contained in a one-dimensional box of width 0.100 nm. (a) Draw an energy-level diagram for the electron for levels up to $n = 4$. (b) Find the wavelengths of all photons that can be emitted by the electron in making transitions that will eventually get it from the $n = 4$ state to the $n = 1$ state.

Solution

(a) We can base the solution on pictures, paralleling our treatment of standing mechanical waves in Chapter 18.

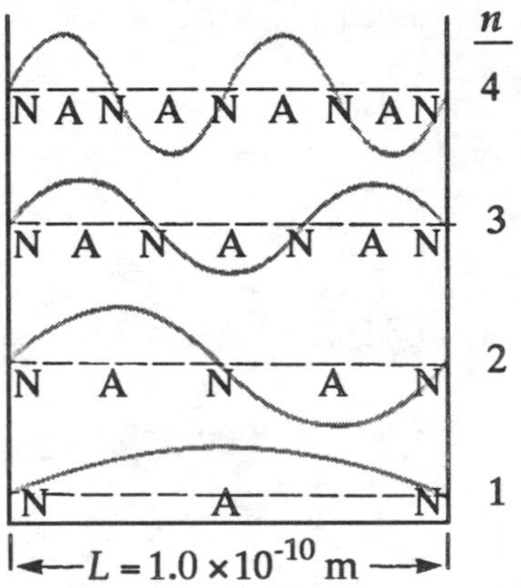

n
N A N A N A N A N 4
N A N A N A N 3
N A N A N 2
N A N 1
$L = 1.0 \times 10^{-10}$ m

In state 1,

$$d_{N \text{ to } N} = 10^{-10} \text{ m} = \frac{\lambda}{2}$$

$$\lambda = 2 \times 10^{-10} \text{ m} = \frac{h}{p}$$

$$p = \frac{h}{\lambda} = 3.32 \times 10^{-24} \text{ kg} \cdot \text{m/s}$$

$$K_1 = \frac{p^2}{2m} = \frac{(3.32 \times 10^{-24} \text{ kg} \cdot \text{m/s})^2}{2(9.11 \times 10^{-31} \text{ kg})} = 6.03 \times 10^{-18} \text{ J} = 37.7 \text{ eV}$$

In state 2, similarly,

$$d_{N \text{ to } N} = 0.5 \times 10^{-10} \text{ m}$$

$$\lambda = 1 \times 10^{-10} \text{ m}$$

$$p = 6.63 \times 10^{-24} \text{ kg} \cdot \text{m/s}$$

$$K_2 = 151 \text{ eV}$$

In state 3, the wavelength is three times smaller than in the ground state, so the momentum is three times larger and the kinetic energy nine times larger:

$$K_3 = 339 \text{ eV}$$

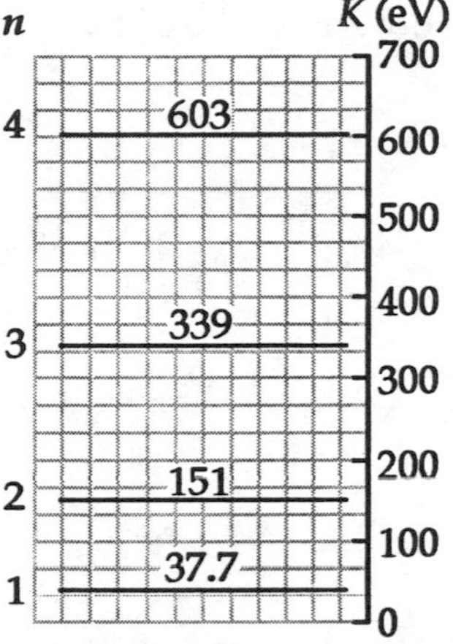

In state 4,

$$K_4 = p^2/2m = h^2/2m\lambda^2$$

$$K_4 = h^2/8m\, d_{NN}^2 = h^2/8m(L/4)^2 = 16K_1 = 603 \text{ eV}$$

The energy-level diagram is shown at the right.

(b) When the charged, massive electron inside the box makes a downward transition between energy level 4 and level 3, a chargeless, massless photon comes out of the box, carrying energy 603 eV – 339 eV = 264 eV. Its wavelength is

$$\lambda = \frac{c}{f} = \frac{hc}{E} = \frac{(6.63\times10^{-34} \text{ J}\cdot\text{s})(3.00\times10^{8} \text{ m/s})}{(264 \text{ eV})\left(1.6\times10^{-19} \text{ J/eV}\right)} = 4.71 \text{ nm} \quad \Diamond$$

We consider four electron standing-wave states, so there are six downward tumbles among them, six traveling-wave photons that can come out. If the electron drops $4 \to 2$, its change in energy is 151 – 603 eV = –452 eV.

The photon radiated has	energy	$E = +452$ eV,
	frequency	$f = E/h$,
	and wavelength	$\lambda = c/f = hc/E = 2.75$ nm. ◊

In the transition $4 \to 1$, a spectral line of wavelength

$$\lambda = \frac{hc}{E} = \frac{(6.63\times10^{-34} \text{ J}\cdot\text{s})(3\times10^{8} \text{ m/s})}{(603-37.7 \text{ eV})\left(1.6\times10^{-19} \text{ J/eV}\right)} = 2.20 \text{ nm} \text{ is emitted.} \quad \Diamond$$

Similarly, when the electron tumbles $3 \to 2$, it puts out light with $\lambda = 6.60$ nm. ◊

In the transitions $3 \to 1$ and $2 \to 1$, the radiated light has $\lambda = 4.12$ nm and $\lambda = 11.0$ nm respectively. ◊

41. Show that the time-dependent wave function

$$\psi = Ae^{i(kx-\omega t)}$$

is a solution to both the wave equation (Eq. 41.12) and the Schrödinger equation $(k = 2\pi/\lambda)$ (Eq. 41.15).

Solution From $\psi = A\, e^{i(kx - \omega t)}$, we evaluate

$$\frac{\partial\psi}{\partial x} = ikAe^{i(kx-\omega t)} \qquad \frac{\partial^2\psi}{\partial x^2} = -k^2Ae^{i(kx-\omega t)}$$

$$\frac{\partial\psi}{\partial t} = -i\omega Ae^{i(kx-\omega t)} \qquad \frac{\partial^2\psi}{\partial t^2} = -\omega^2Ae^{i(kx-\omega t)}$$

We test for having a solution to $\frac{\partial^2\psi}{\partial x^2} = \frac{1}{v^2}\frac{\partial^2\psi}{\partial t^2}$ by substituting.

We have a solution if $-k^2Ae^{i(kx-\omega t)} = -(\omega^2 A / v^2)e^{i(kx-\omega t)}$.

Both sides depend on x and t in the same way, so we have a solution if $k^2 = \omega^2/v^2$.

But this is true since $k = \frac{2\pi}{\lambda} = \frac{2\pi f}{v} = \frac{\omega}{v}$

We test for having a solution to $\frac{\partial^2\psi}{\partial x^2} = -\frac{2m}{\hbar^2}(E-U)\psi$ by substituting.

We have a solution if $-k^2Ae^{i(kx-\omega t)} = -\frac{2m}{\hbar^2}(K)Ae^{i(kx-\omega t)}$

Both sides depend on x and t in the same way, so we have a solution if $k^2 = \frac{2m}{\hbar^2}K$.

But this is true for a nonrelativistic particle with mass, since

$$\frac{2mK}{\hbar^2} = \frac{2m\left(\frac{1}{2}mv^2\right)}{(h/2\pi)^2} = \frac{4\pi^2m^2v^2}{h^2} = \left(\frac{2\pi p}{h}\right)^2 = \left(\frac{2\pi}{\lambda}\right)^2 = k^2$$

43. Suppose a particle is trapped in its ground state in a box that has infinitely high walls (Fig. 41.13a). Now suppose the left-hand wall is suddenly lowered to a finite height. (a) Qualitatively sketch the wave function for the particle a short time later. (b) If the box has a width L, what is the wavelength of the wave that penetrates the barrier?

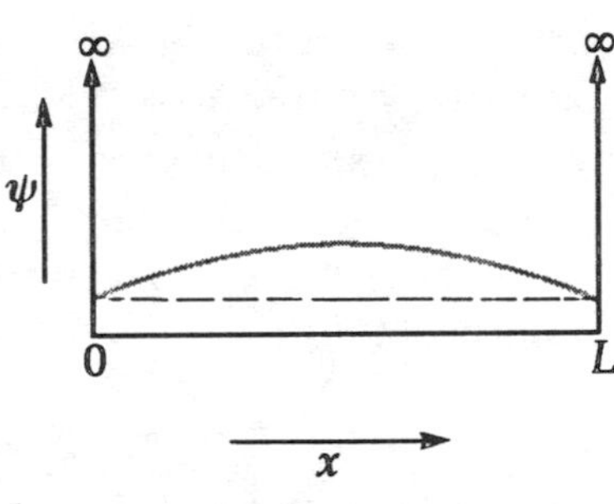

Figure 41.13a

Solution Before the wall is lowered, the wave function looks the wave in the figure to the right.

(a) If the left-hand wall is shrunk to finite height and finite thickness, the wave function will look much the same in region I. It will show exponential decay in region II and a small-amplitude traveling wave in region III as in the second figure, below. ◊

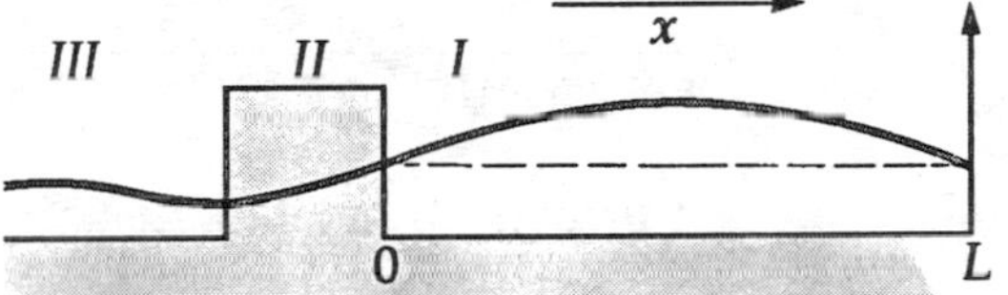

(b) Nothing changes the particle's energy; so it has the same momentum and wavelength as before, namely $2d_{NN} = 2L$. ◊

---

45. A 5.0-eV electron is incident on a barrier 0.20 nm thick and 10 eV high (Fig. P41.45). What is the probability that the electron (a) will tunnel through the barrier and (b) will be reflected?

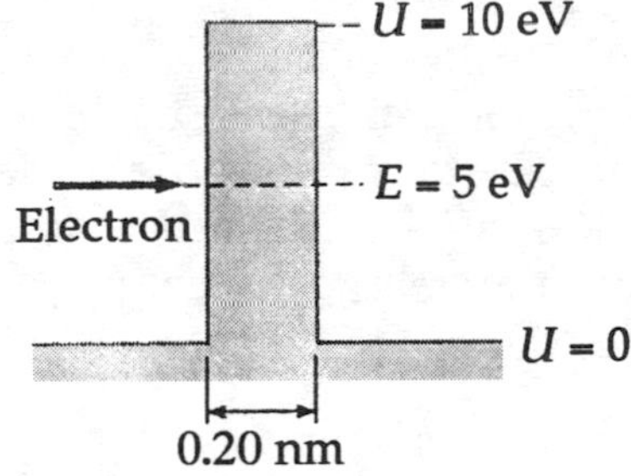

Figure P41.45

Solution We apply Equation 41.20 with transmission coefficient:

$$K = \frac{\sqrt{2m(U-E)}}{\hbar}$$

Thus,
$$K = \frac{\sqrt{2(9.11\times10^{-31}\text{ kg})(10\text{ eV} - 5\text{ eV})(1.6\times10^{-19}\text{ J/eV})}}{6.63\times10^{-34}\text{ J}\cdot\text{s}/2\pi} = 1.14\times10^{10}\text{ /m}$$

(a) The probability of transmission is

$$T \cong e^{-2KL} = e^{-2(1.14\times10^{10}\text{ m}^{-1})(2.0\times10^{-10}\text{ m})} = e^{-4.58} = 0.010 \quad ◊$$

(b) If the electron does not tunnel, it is reflected, with probability $1 - 0.010 = 0.990$ ◊

---

49. The nuclear potential energy that binds protons and neutrons in a nucleus is often approximated by a square well. Imagine a proton confined in an infinitely high square well of width 1.0×10^{-5} nm, a typical nuclear diameter. Calculate the wavelength and energy associated with the photon emitted when the proton moves from the $n = 2$ state to the ground state. In what region of the electromagnetic spectrum does this wavelength belong?

Solution In level 1, the node-to-node distance of the standing wave is 10^{-14} m. The wavelength is 2.0×10^{-14} m $= h/p$. The proton's kinetic energy is

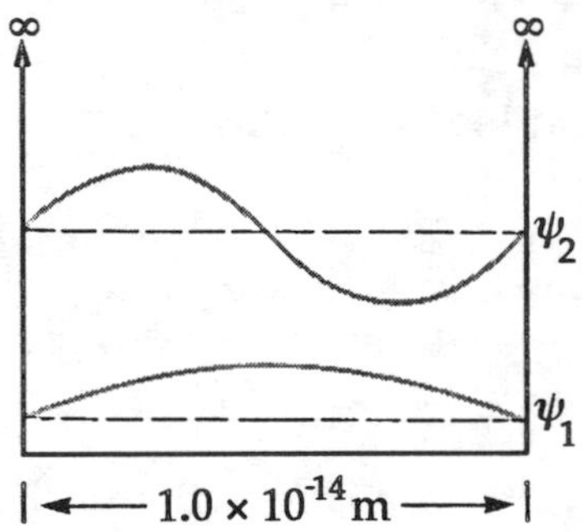

$$K = \frac{1}{2}mv^2 = \frac{p^2}{2m} = \frac{h^2}{2m\lambda^2}$$

$$= \frac{(6.63 \times 10^{-34}\ \mathrm{J \cdot s})^2}{2(1.67 \times 10^{-27}\ \mathrm{kg})(2 \times 10^{-14}\ \mathrm{m})^2}$$

$$= 3.29 \times 10^{-13}\ \mathrm{J} = 2.06\ \mathrm{MeV}$$

In the first excited state, level 2, the node-to-node distance is two times smaller than in state 1. The momentum is two times larger and the energy is four times larger: $K = 8.23$ MeV.

The proton has mass, has charge, moves slowly compared to light in a standing-wave state, and stays inside the nucleus. When it falls from level 2 to level 1, its energy change is 2.06 MeV – 8.23 MeV = –6.17 MeV. Then +6.17 MeV ◊ is the energy of the photon with no mass and no charge, moving at the speed of light in a traveling wave state, that comes out of the nucleus. Its frequency is

$$f = \frac{E}{h} = \frac{6.17 \times 10^6 \times 1.6 \times 10^{-19}\ \mathrm{J}}{6.63 \times 10^{-34}\ \mathrm{J \cdot s}} = 1.49 \times 10^{21}\ \mathrm{Hz}$$

and its wavelength is

$$\lambda = \frac{c}{f} = \frac{(3 \times 10^8\ \mathrm{m/s})}{1.49 \times 10^{21}\ /\mathrm{s}} = 2.02 \times 10^{-13}\ \mathrm{m} \quad \lozenge$$

making it a gamma ray. ◊

57. For a particle described by a wave function $\psi(x)$, the expectation value of a physical quantity $f(x)$ associated with the particle is defined by

$$\langle f(x)\rangle \equiv \int_{-\infty}^{\infty} f(x)|\psi|^2\,dx$$

For a particle in a one-dimensional box extending from $x = 0$ to $x = L$, show that

$$\langle x^2\rangle = \frac{L^2}{3} - \frac{L^2}{2n^2\pi^2}$$

Solution $\quad <x^2> \; = \int_{-\infty}^{\infty} x^2|\psi|^2\,dx$

$$\psi_n(x) = A\sin\left(\frac{n\pi x}{L}\right) \quad \text{where} \quad A = \sqrt{\frac{2}{L}} \text{ is the normalization factor}$$

$$<x^2> \; = \left(\frac{2}{L}\right)\int_0^L x^2\sin^2\left(\frac{n\pi x}{L}\right)dx = \frac{L^2}{3} - \frac{L^2}{2n^2\pi^2} \quad \Diamond$$

---

59. Particles incident from the left are confronted with a step potential energy shown in Figure P41.59. The step has a height U_0, and the particles have energy $E > U_0$. Classically, all the particles would pass into the region of higher potential energy at the right. However, according to quantum mechanics, a fraction of the particles are reflected at the barrier. The reflection coefficient R for this case is

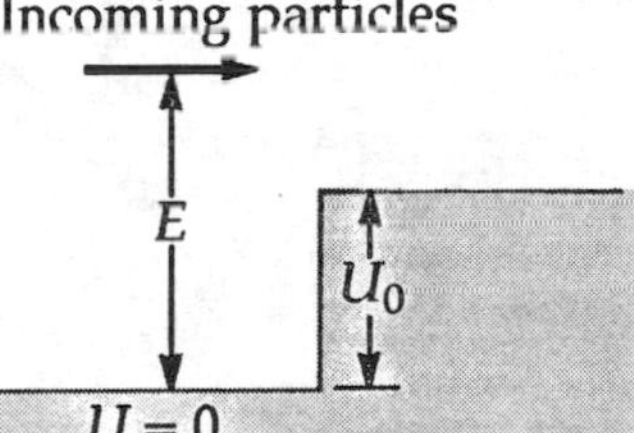

Figure P41.59

$$R = \frac{(k_1 - k_2)^2}{(k_1 + k_2)^2}$$

where $k_1 = 2\pi/\lambda_1$ and $k_2 = 2\pi/\lambda_2$ are the angular wave numbers for the incident and transmitted particles, respectively. If $E = 2U_0$, what fraction of the incident particles are reflected? (This situation is analogous to the partial reflection and transmission of light striking an interface between two different media.)

Solution $\quad R = \dfrac{(k_1 - k_2)^2}{(k_1 + k_2)^2} = \dfrac{(1 - k_2/k_1)^2}{(1 + k_2/k_1)^2} \quad$ and $\quad \dfrac{\hbar^2 k^2}{2m} = E - U$ for constant U

In this case, since $U = 0$ to the left of the barrier and $U = U_0$ to the right,

$$\frac{\hbar^2 k_1^2}{2m} = E \quad (1) \qquad \text{and} \qquad \frac{\hbar^2 k_2^2}{2m} = E - U_0 \quad (2)$$

Dividing (2) by (1) gives $\frac{k_2^2}{k_1^2} = 1 - \frac{U_0}{E} = 1 - \frac{1}{2} = \frac{1}{2}$ so $\frac{k_2}{k_1} = \frac{1}{\sqrt{2}}$

Therefore, $$R = \frac{(1 - 1/\sqrt{2})^2}{(1 + 1/\sqrt{2})^2} = \left[\frac{(\sqrt{2} - 1)}{(\sqrt{2} + 1)}\right]^2 = 0.0294 \quad \Diamond$$

---

60. A particle has a wave function

$$\psi(x) = \begin{cases} \sqrt{\frac{2}{a}}\, e^{-x/a} & \text{for } x > 0 \\ 0 & \text{for } x < 0 \end{cases}$$

(a) Find and sketch the probability density. (b) Find the probability that the particle will be at any point where $x < 0$. (c) Show that ψ is normalized and then find the probability that the particle will be found between $x = 0$ and $x = a$.

Solution

(a) $$|\psi|^2 = \begin{cases} \frac{2}{a} e^{-2x/a} & \text{for } x > 0 \\ 0 & \text{for } x < 0 \end{cases} \quad \Diamond$$

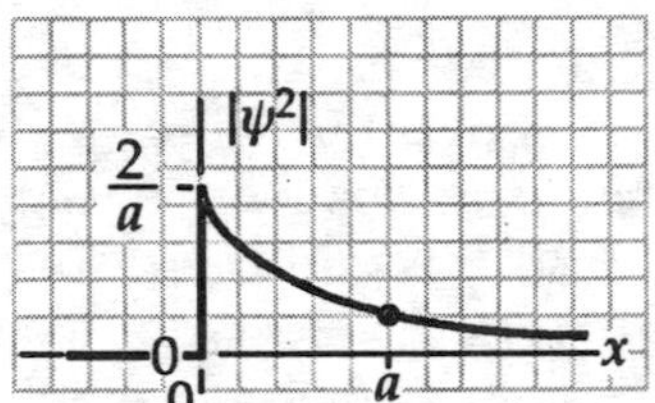

(b) The particle has zero probability of being at any point where $x < 0$. $\Diamond$

(c) For normalization, $\int_{\text{all } x} |\psi|^2\, dx = 1$

$$0 + \int_0^\infty \frac{2}{a} e^{-2x/a}\, dx = -\int_0^\infty e^{-2x/a}(-2\,dx/a) = -e^{-2x/a}\Big|_0^\infty = -[0 - 1] = 1$$

Thus, $\sqrt{2/a}$ was the right normalization coefficient for ψ in the first place. Probability of finding the particle in $0 < x < a$ is

$$\int_0^a \frac{2}{a} e^{-2x/a}\, dx = -e^{-2x/a}\Big|_0^a = -[e^{-2} - 1] = 0.865 \quad \Diamond$$

---

61. An electron of momentum p is at a distance r from a stationary proton. The electron has kinetic energy $K = p^2/2m$, potential energy $U = -k_e e^2/r$, and total energy $E = K + U$. If the electron is bound to the proton to form a hydrogen atom, its average position is at the proton, but the uncertainty in its position is approximately equal to the radius r of its orbit. The electron's average momentum is zero, but the uncertainty in its momentum is given by the uncertainty principle. Treating the atom as a one-dimensional system, (a) estimate the uncertainty in the electron's momentum in terms of r. (b) Estimate the electron's kinetic, potential, and total energies in terms of r. (c) The actual value of r is the one that *minimizes the total energy,* resulting in a stable atom. Find the value of r and the resulting total energy. Compare your answer with the predictions of the Bohr theory.

Solution

(a) $\Delta x \Delta p \geq \dfrac{\hbar}{2}$ so if $\Delta x = r$, $\quad \Delta p \geq \dfrac{\hbar}{2r}$ ◊

(b) Suppose arbitrarily that $p = \dfrac{\hbar}{r}$

$$K = \frac{p^2}{2m} = \frac{\hbar^2}{2mr^2} \quad ◊$$

$$U = -\frac{k_e e^2}{r} \quad ◊$$

$$E = \frac{\hbar^2}{2mr^2} - \frac{k_e e^2}{r} \quad ◊$$

(c) To minimize E, $\quad \dfrac{dE}{dr} = -\dfrac{\hbar^2}{mr^3} + \dfrac{k_e e^2}{r^2} = 0$ or $\quad r = \dfrac{\hbar^2}{mk_e e^2} =$ Bohr radius

Then, $$E = \frac{\hbar^2}{2m}\left(\frac{mk_e e^2}{\hbar^2}\right) - k_e e^2\left(\frac{mk_e e^2}{\hbar^2}\right) = -\frac{mk_e^2 e^4}{2\hbar^2} = -13.6 \text{ eV} \quad ◊$$

This is the same value as that predicted by the Bohr theory for the ground state of hydrogen.

65. An electron is represented by the time-independent wave function

$$\psi(x) = \begin{cases} Ae^{-\alpha x} & \text{for } x > 0 \\ Ae^{+\alpha x} & \text{for } x < 0 \end{cases}$$

(a) Sketch the wave function as a function of x. (b) Sketch the probability that the electron is found between x and $x + dx$. (c) Why do you suppose this is a physically reasonable wave function? (d) Normalize the wave function. (e) Determine the probability of finding the electron somewhere in the range

$$x_1 = -\frac{1}{2\alpha} \quad \text{to} \quad x_2 = \frac{1}{2\alpha}$$

Solution

(a)

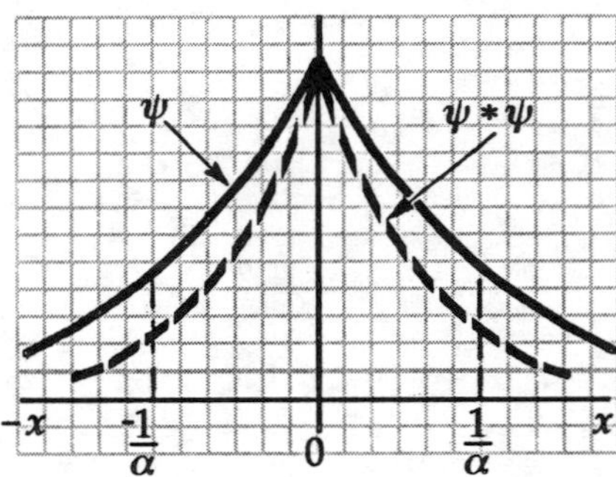

(b) $P(x \to x + dx) = |\psi|^2\, dx = A^2 e^{-2\alpha x}\, dx,$

for $x > 0$

$P(x \to x + dx) = |\psi|^2\, dx = A^2 e^{+2\alpha x}\, dx,$

for $x < 0$

(c) ψ is continuous; $\psi \to 0$ as $x \to \infty$; mimics an electron bound at $x = 0$. . . .

(d) As ψ is symmetric, $\int_{-\infty}^{\infty} \psi^* \psi\, dx = 2\int_0^{\infty} \psi^* \psi\, dx = 1$

or $2A^2 \int_0^{\infty} e^{-2\alpha x}\, dx = 1$ or $\left[\frac{2A^2}{(-2\alpha)}\right]\left[e^{-\infty} - e^0\right] = 1$

This gives $A = \sqrt{\alpha}$

(e) $P_{(-1/2\alpha)\to(1/2\alpha)} = 2\int_{x=0}^{1/2\alpha} (\sqrt{\alpha})^2 e^{-2\alpha x}\, dx = \left[\frac{2\alpha}{(-2\alpha)}\right]\left[e^{-2\alpha/2\alpha} - 1\right] = \left[1 - e^{-1}\right] = 0.632$ ◊

Chapter 42

Atomic Physics

Chapter 42

ATOMIC PHYSICS

INTRODUCTION

In Chapter 41, we introduced some of the basic concepts and techniques used in quantum mechanics, along with their applications to various simple systems. This chapter deals with the application of quantum mechanics to the real world of atomic structure.

A large portion of this chapter is an application of quantum mechanics to the study of the hydrogen atom. Understanding the hydrogen atom, the simplest atomic system, is especially important for several reasons:

- Much of what is learned about the hydrogen atom with its single electron can be extended to such single-electron ions as He^+ and Li^{2+}.
- The hydrogen atom is an ideal system for performing precise tests of theory against experiment and for improving our overall understanding of atomic structure.
- The quantum numbers used to characterize the allowed states of hydrogen can be used to describe the allowed states of more complex atoms. This enables us to understand the periodic table of the elements, one of the greatest triumphs of quantum mechanics.
- The basic ideas about atomic structure must be well understood before we attempt to deal with the complexities of molecular structures and the electronic structure of solids.

NOTES FROM SELECTED CHAPTER SECTIONS

42.1 Early Models of the Atom

One of the first indications that there was a need for modification of the Bohr theory became apparent when improved spectroscopic techniques were used to examine the spectral lines of hydrogen. It was found that many of the lines in the Balmer and other series were not single lines at all. Instead, each was a group of lines spaced very close together. An additional difficulty arose when it was observed that, in some situations, certain single spectral lines were split into three closely spaced lines when the atoms were placed in a strong magnetic field.

42.2 The Hydrogen Atom Revisited

In the three-dimensional problem of the hydrogen atom, three quantum numbers are required for each stationary state, corresponding to the three independent degrees of freedom for the electron.

The three quantum numbers which emerge from the theory are represented by the symbols n, ℓ, and m_ℓ. The quantum number n is called the *principal quantum number*; ℓ is called the *orbital quantum number*; and m_ℓ is called the *orbital magnetic quantum number.*

There are certain important relationships between these quantum numbers, as well as certain restrictions on their values. These restrictions are:

The values of n can range from 1 to ∞.
The values of ℓ can range from 0 to $n - 1$.
The values of m_ℓ can range from $-\ell$ to ℓ.

For historical reasons, *all states with the same principal quantum number are said to form a shell.* These shells are identified by the letters K, L, M, . . . , which designate the states for which $n = 1, 2, 3, \ldots$. Likewise, *the states having the same values of n and ℓ are said to form a subshell.* The letters $s, p, d, f, g, h, \ldots$ are used to designate the states for which $\ell = 0, 1, 2, 3, \ldots$.

42.3 The Spin Magnetic Quantum Number

The *spin magnetic quantum number* m_s accounts for the two closely spaced energy states corresponding to the two possible orientations of the electron spin.

42.4 The Wave Functions for Hydrogen

Since the potential energy of the electron in the hydrogen atom depends only on the radial distance r, the wave functions that describe the s states are *spherically symmetric.*

42.5 The "Other" Quantum Numbers

The allowed energy values for the electron in a hydrogen atom are determined by the value of the *principal quantum number, n.* There are three "other" quantum numbers which also serve to characterize the wave function of the hydrogen electron. The *orbital quantum number, ℓ,* restricts the orbital angular momentum to discrete values. The *magnetic orbital quantum number, m_ℓ,* specifies the allowed directions of the angular momentum with respect to an external magnetic field. The *spin magnetic quantum number*, m_s, restricts the spin angular momentum of the hydrogen electron to have only two possible orientations in space. The spin contribution to the magnetic moment is twice the contribution of the orbital motion.

42.6 The Exclusion Principle and the Periodic Table

The *exclusion principal* states that no two electrons can exist in identical quantum states. This means that no two electrons in a given atom can be characterized by the same set of quantum numbers at the same time.

Hund's rule states that when an atom has orbitals of equal energy, the order in which they are filled by electrons is such that a maximum number of electrons will have unpaired spins.

42.7 Atomic Spectra: Visible and X-Ray

An atom will emit electromagnetic radiation if an electron in an excited state makes a transition to a lower energy state. The set of wavelengths observed for a species by such processes is called an *emission spectrum*. Likewise, atoms with electrons in the ground-state configuration can also absorb electromagnetic radiation at specific wavelengths, giving rise to an *absorption spectrum*. Such spectra can be used to identify the elements in the gases.

Since the orbital angular momentum of an atom changes when a photon is emitted or absorbed (that is, as a result of a transition) and since angular momentum must be conserved, we conclude that *the photon involved in the process must carry angular momentum.*

X-rays are emitted by atoms when an electron undergoes a transition from an outer shell into an electron vacancy in one of the inner shells. Transitions into a vacant state in the K shell give rise to the K series of spectral lines; in a similar way, transitions into a vacant state in the L shell create the L series of lines, and so on. The x-ray spectrum of a metal target consists of a set of sharp characteristic lines superimposed on a broad, continuous spectrum.

42.8 Atomic Transitions

Stimulated absorption occurs when light with photons of an energy which matches the energy separation between two atomic energy levels is absorbed by an atom.

An atom in an excited state has a certain probability of returning to its original energy state. This process is called *spontaneous emission*.

When a photon with an energy equal to the excitation energy of an excited atom is incident on the atom, it can increase the probability of de-excitation. This is called *stimulated emission*, and results in a second photon of energy equal to that of the incident photon.

42.9 Lasers and Holography

The following three conditions must be satisfied in order to achieve laser action:

- The system must be in a state of population inversion (that is, more atoms in an excited state than in a ground state).
- The excited state of the system must be a *metastable state*, which means its lifetime must be long compared with the usually short lifetimes of excited states. When such is the case, stimulated emission will occur before spontaneous emission.
- The emitted photons must be confined in the system long enough to allow them to stimulate further emission from other excited atoms. This is achieved by the use of reflecting mirrors at the ends of the system. One end is made totally reflecting, and the other is slightly transparent to allow the laser beam to escape.

EQUATIONS AND CONCEPTS

The *potential energy* of the hydrogen atom depends on the radius of the allowed orbit of the electron.

$$U(r) = -\frac{k_e e^2}{r} \qquad (42.1)$$

When numerical values of the constants are used, the energy level values for hydrogen can be expressed in units of electron volts (eV).

$$E_n = -\frac{13.6}{n^2}\ \text{eV} \qquad (42.2)$$

The lowest energy state or ground state corresponds to the principal quantum number $n = 1$. The energy level approaches $E = 0$ as n approaches infinity. This is the *ionization energy* for the atom.

The simplest wave function for hydrogen is the one which describes the 1s state and depends only on the radial distance r. The parameter a_0 is the Bohr radius.

$$\psi_{1s}(r) = \frac{1}{\sqrt{\pi a_0^{\,3}}} e^{-r/a_0} \qquad (42.3)$$

$$a_0 = 0.0529 \text{ nm} \qquad (42.4)$$

The radial probability density for the 1s state of hydrogen is defined as the probability of finding the electron in a spherical shell of radius r and thickness dr.

$$P_{1s}(r) = \left(\frac{4r^2}{a_0^{\,3}}\right) e^{-2r/a_0} \qquad (42.7)$$

The value of the quantum number ℓ determines the magnitude of the electron's *orbital angular momentum, L.*

$$L = \sqrt{\ell(\ell+1)}\hbar \qquad (42.10)$$

$$(\ell = 0, 1, 2, 3, \ldots, n-1)$$

When an atom is placed in an external magnetic field, the projection (or component) of the orbital angular momentum L_z along the direction of the magnetic field is quantized.

$$L_z = m_\ell \hbar \qquad (42.11)$$

$$(m_\ell = -\ell,\ -\ell+1,\ -\ell+2, \ldots \ell)$$

If an orbiting electron is placed in a weak magnetic field directed along the z axis, the projection of the angular momentum vector of the electron (**L**) along the z axis can have only discrete values.

$$\cos\theta = \frac{m_\ell}{\sqrt{\ell(\ell+1)}} \qquad (42.12)$$

In addition to orbital angular momentum, the electron has an intrinsic angular momentum or *spin angular momentum* S, as if due to spinning on its axis. This spin angular momentum is described by a single quantum number s whose value can only be $\frac{1}{2}$.

$$S = \sqrt{s(s+1)}\hbar \qquad (42.13)$$

$$S = \frac{\sqrt{3}}{2}\hbar$$

The spin angular momentum is space quantized with respect to the direction of an external magnetic field (along the z direction). The spin magnetic quantum number m_s can have values of $\pm\frac{1}{2}$.

$$S_z = m_s\hbar = \pm\frac{1}{2}\hbar \qquad (42.14)$$

The *spin magnetic moment* μ_s is related to the spin angular momentum.

$$\mu_s = \left(-\frac{e}{m}\right)\mathbf{S} \qquad (42.15)$$

The shielding effect of the nuclear charge by inner-core electrons must be taken into account when calculating the allowed energy levels of multielectron atoms. The atomic number is replaced by an effective atomic number, Z_{eff}, which depends on the values of n and ℓ.

$$E_n = -\frac{13.6 Z_{eff}{}^2}{n^2}\text{eV} \qquad (42.19)$$

Z_{eff} is the atomic number, modified by the number of shielding electrons: for K-shell electrons, $Z_{eff} = Z - 1$; For an M shell electron, $Z_{eff} = Z - 9$. (See Example 42.7 in the text.)

$$Z_{eff} = Z - n_s$$

where

n_s = the number of shielding electrons

SUGGESTIONS, SKILLS, AND STRATEGIES

After reading the chapter in your text, review the significance and set of allowed values for each of the quantum numbers used to describe the various electronic states of electrons in an atom.

In addition to the principal quantum number n (which can range from 1 to ∞), other quantum numbers are necessary to completely specify the possible energy levels in the hydrogen atom and also in more complex atoms.

All energy states with the same principal quantum number, n, form a shell. These shells are identified by the spectroscopic notation K, L, M, . . . corresponding to $n = 1, 2, 3, \ldots$.

The orbital quantum number ℓ [which can range from 0 to $(n - 1)$], determines the allowed value of orbital angular momentum. All energy states having the same values of n and ℓ form a subshell. The letter designations $s, p, d, f, \ldots$ correspond to values of $\ell = 0, 1, 2, 3, \ldots$.

The magnetic orbital quantum number m_ℓ (which can range from $-\ell$ to ℓ) determines the possible orientations of the electron's orbital angular momentum vector in the presence of an external magnetic field.

The spin magnetic quantum number m_s, can have only two values, $m_s = -\frac{1}{2}$ and $m_s = \frac{1}{2}$, which in turn correspond to the two possible directions of the electron's intrinsic spin.

REVIEW CHECKLIST

▷ Understand the significance of the wave function and the associated radial probability density for the ground state of hydrogen.

▷ For each of the quantum numbers, n, ℓ (the orbital quantum number), m_ℓ (the orbital magnetic quantum number), and m_s (the spin magnetic quantum number):

- qualitatively describe what each implies concerning atomic structure;
- state the allowed values which may be assigned to each, and the number of allowed states which may exist in a particular atom corresponding to each quantum number.

▷ Associate the customary shell and subshell spectroscopic notations with allowed combinations of quantum numbers n and ℓ. Calculate the possible values of the orbital angular momentum, L, corresponding to a given value of the principal quantum number.

▷ Describe how allowed values of the magnetic orbital quantum number, m_ℓ, may lead to a restriction on the orientation of the orbital angular momentum vector in an external magnetic field. Find the allowed values for L_z (the component of the angular momentum along the direction of an external magnetic field) for a given value of L.

▷ State the Pauli exclusion principal and describe its relevance to the periodic table of the elements. Show how the exclusion principal leads to the known electronic ground state configuration of the light elements.

SOLUTIONS TO SELECTED END-OF-CHAPTER PROBLEMS

3. A general expression for the energy levels of one-electron atoms and ions is

$$E_n = -\left(\frac{\mu k_e^{\,2} q_1 q_2}{2\hbar^2}\right)\frac{1}{n^2}$$

where k_e is the Coulomb constant, q_1 and q_2 are the charges of the two particles, and μ is the reduced mass, given by $\mu = m_1 m_2 / (m_1 + m_2)$. In Example 40.7, we found that the wavelength for the $n = 3$ to $n = 2$ transition of the hydrogen atom is 656.3 nm (visible red light). What are the wavelengths for this same transition in (a) positronium, which consists of an electron and a positron, and (b) singly ionized helium? (*Note*: A positron is a positively charged electron.)

Solution For hydrogen, $\mu = \dfrac{m_{\text{proton}} m_e}{m_{\text{proton}} + m_e} \cong m_e$

The photon energy is $E_3 - E_2$ and its wavelength is $\lambda = \dfrac{c}{f} = \dfrac{hc}{E_3 - E_2} = 656.3\text{ nm}$

(a) For positronium, $\mu = \dfrac{m_e m_e}{m_e + m_e} = \dfrac{m_e}{2}$,

so the energy of each level is one half as large as in hydrogen (protonium). The photon energy is inversely proportional to its energy, so $\lambda_{32} = 2(656\text{ nm}) = 1312\text{ nm}$ ◊, which is in the infrared region.

(b) For He$^+$, $\mu \approx m_e$, $q_1 = e$, and $q_2 = 2e$, so each energy is $2^2 = 4$ times larger than hydrogen. Then,

$$\lambda_{32} = \left(\frac{656}{4}\right) \text{nm} = 164 \text{ nm} \quad \Diamond$$

which is in the ultraviolet region.

---

7. The wave function for an electron in the $2p$ state of hydrogen is

$$\psi_{2p} = \frac{1}{\sqrt{3}(2a_0)^{3/2}} \frac{r}{a_0} e^{-r/2a_0}$$

What is the most likely distance from the nucleus to find an electron in the $2p$ state? (See Fig. 42.8.)

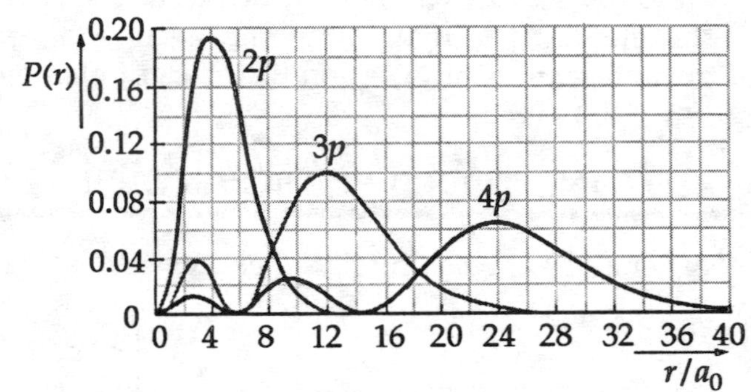

Figure 42.8

Solution The radial probability density function is given by

$$P = 4\pi r^2 |\psi|^2 = 4\pi r^2 \left(\frac{r^2}{24 a_0^5}\right) e^{-r/a_0}$$

The most likely position for the electron corresponds to the point at which $\frac{dP}{dr} = 0$ (where P reaches its *maximum* value).

$$\frac{dP}{dr} = \frac{4\pi}{24 a_0^5}\left[4r^3 e^{-r/a_0} + r^4\left(-\frac{1}{a_0}\right)e^{-r/a_0}\right] = 0$$

Solving for r gives $$r = 4a_0 \quad \Diamond$$

---

8. Show that the 1*s* wave function for an electron in hydrogen

$$\psi(r) = \frac{1}{\sqrt{\pi a_0^{\,3}}} e^{-r/a_0}$$

satisfies the radially symmetric Schrödinger equation, $-\frac{\hbar^2}{2m}\left(\frac{d^2\psi}{dr^2} + \frac{2}{r}\frac{d\psi}{dr}\right) - \frac{k_e e^2}{r}\psi = E\psi$

Solution $$\frac{d\psi}{dr} = \frac{1}{\sqrt{\pi a_0^{\,3}}}\frac{d}{dr}\left(e^{-r/a_0}\right) = -\frac{1}{\sqrt{\pi a_0^{\,3}}}\left(\frac{1}{a_0}\right)e^{-r/a_0}$$

Hence, $$\frac{2}{r}\frac{d\psi}{dr} = -\frac{2}{\sqrt{\pi a_0^{\,5}}}\frac{1}{r}e^{-r/a_0} = -\frac{2}{a_0}\frac{\psi}{r} \qquad (1)$$

Likewise, $$\frac{d^2\psi}{dr^2} = -\frac{1}{\sqrt{\pi a_0^{\,5}}}\frac{d}{dr}e^{-r/a_0} = \frac{1}{\sqrt{\pi a_0^{\,7}}}e^{-r/a_0} = \frac{1}{a_0^{\,2}}\psi \qquad (2)$$

Substituting (1) and (2) into the Schrödinger equation gives

$$-\frac{\hbar^2}{2m}\left(\frac{1}{a_0^{\,2}} - \frac{2}{a_0 r}\right)\psi - \frac{k_e e^2}{r}\psi = E\psi \qquad (3)$$

From Equation 42.4, we have $a_0 = \frac{\hbar^2}{mk_e e^2}$ which, when substituted into (3) [where the second and third terms cancel each other], gives the ground state energy of hydrogen:

$$E = -\frac{k_e e^2}{2a_0}$$

---

9. If a muon (a negatively charged particle having a mass 206 times the electron's mass) is captured by a lead nucleus, $Z = 82$, the resulting system behaves like a one-electron atom. (a) What is the "Bohr radius" for a muon captured by a lead nucleus? (*Hint*: Use Equation 42.4.) (b) Using Equation 42.2 with e replaced by Ze, calculate the ground-state energy of a muon captured by a lead nucleus. (c) What is the transition energy of a muon descending from the $n = 2$ to the $n = 1$ level in a muonic lead atom?

Solution

(a) We expand the treatment of the Bohr atom in Section 40.6 to particle of charge $-e$ and mass m orbiting a much more massive nucleus of charge Ze. Angular momentum is quantized according to $mvr = n\hbar$.

$$n = 1, 2, 3, \ldots \quad \text{and} \quad \Sigma F = ma \quad \text{means} \quad \frac{k_e Ze^2}{r^2} = \frac{mv^2}{r}$$

We eliminate v with $v = \dfrac{n\hbar}{mr}$;

$$\frac{k_e Ze^2}{r^2} = \frac{mn^2\hbar^2}{m^2 r^3} \qquad r = \frac{n^2\hbar^2}{mk_e Ze^2}$$

So the generalized Bohr radius is $\dfrac{\hbar^2}{mk_e Ze^2}$

With $m = 206\ m_e$ and Z = 82, the radius for muonic lead is $a_\mu = \dfrac{a_0}{206 \times 82} = 3.13\ \text{fm}$ ◊

(b) The energy is

$$\tfrac{1}{2}mv^2 - \frac{k_e Ze^2}{r} = \frac{\tfrac{1}{2}k_e Ze^2}{r} - \frac{k_e Ze^2}{r} = -\frac{k_e Ze^2}{2r} = -\frac{k_e Ze^2}{2\left(\dfrac{n^2\hbar^2}{mk_e Ze^2}\right)} = -\frac{mk_e{}^2 Z^2 e^4}{2\hbar^2 n^2}$$

For the ground-state energy, we take $n = 1$:

$$E_{\mu 1} = -\frac{mk_e{}^2 Z^2 e^4}{2\hbar^2} = -206 \times 82^2 \times 13.6\ \text{eV} = -18.8\ \text{MeV} \quad ◊$$

(c) $E_{\mu 2} = \dfrac{-18.8}{4} = -4.7\ \text{MeV}$

$\Delta E_{\mu 2 \to \mu 1} = 14.1\ \text{MeV}$ ◊

13. How many sets of quantum numbers are possible for an electron for which (a) $n = 1$, (b) $n = 2$, (c) $n = 3$, (d) $n = 4$, and (e) $n = 5$? Check your results to show that they agree with the general rule that the number of sets of quantum numbers is equal to $2n^2$.

Solution

(a) $n = 1$: For $n = 1$, $\ell = 0$, $m_\ell = 0$, $m_s = \pm\frac{1}{2}$, $\rightarrow$ 2 sets

n	ℓ	m_ℓ	m_s
1	0	0	$-\frac{1}{2}$
1	0	0	$+\frac{1}{2}$

$2n^2 = 1(1)^2 = 2$ ◊

(b) For $n = 2$, we have

n	ℓ	m_ℓ	m_s
2	0	0	$\pm\frac{1}{2}$
2	1	−1	$\pm\frac{1}{2}$
2	1	0	$\pm\frac{1}{2}$
2	1	1	$\pm\frac{1}{2}$

Yields 8 sets; $2n^2 = 2(2)^2 = 8$ ◊

Note that the number is twice the number of m_ℓ values. Also, for each ℓ there are $(2\ell + 1)\, m_\ell$ values. Finally, ℓ can take on values ranging from 0 to $n - 1$. So the general expression is

$$s = \sum_{0}^{n-1} 2(2\ell + 1)$$

The series is an arithmetic progression 2 + 6 + 10 + 14, the sum of which is

$$s = \frac{n}{2}[2a + (n-1)d] \quad \text{where } a = 2,\ d = 4$$

$$s = \frac{n}{2}[4 + (n-1)4] = 2n^2$$

(c) $n = 3$: $2(1) + 2(3) + 2(5) = 2 + 6 + 10 = 18$

$2n^2 = 2(3)^2 = 18$ ◊

(d) $n = 4$: $2(1) + 2(3) + 2(5) + 2(7) = 32$

$2n^2 = 2(4)^2 = 32$ ◊

(e) $n = 5$: $32 + 2(9) = 32 + 18 = 50$

$2n^2 = 2(5)^2 = 50$ ◊

---

15. The ρ-meson has a charge of $-e$, a spin quantum number of 1, and a mass of 1507 times that of the electron. If the electrons in atoms were replaced by ρ-mesons, list the possible sets of quantum numbers for ρ-mesons in the 3*d* subshell.

Solution The 3*d* subshell has $\ell = 2$, and $n = 3$. Also, we have $s = 1$. Therefore, we can have $n = 3$, $\ell = 2$, $m_\ell = -2, -1, 0, 1, 2$, $s = 1$, and $m_s = -1, 0, 1$, leading to the following table:

n	ℓ	m_ℓ	s	m_s
3	2	-2	1	-1
3	2	-2	1	0
3	2	-2	1	+1
3	2	-1	1	-1
3	2	-1	1	0
3	2	-1	1	+1
3	2	0	1	-1
3	2	0	1	0
3	2	0	1	+1
3	2	+1	1	-1
3	2	+1	1	0
3	2	+1	1	+1
3	2	+2	1	-1
3	2	+2	1	0
3	2	+2	1	+1

---

26. (a) Scanning through Table 42.4 in order of increasing atomic number, note that the electrons fill the subshells in such a way that those subshells with the lowest values of $n + \ell$ are filled first. If two subshells have the same value of $n + \ell$, the one with the lower value of n is filled first. Using these two rules, write the order in which the subshells are filled through $n = 7$. (b) Predict the chemical valence for the elements that have atomic numbers 15, 47, and 86, and compare your predictions with the actual valences.

Solution

(a)

$n + \ell$	1	2	3	4	5	6	7
subshell	$1s$	$2s$	$2p, 3s$	$3p, 4s$	$3d, 4p, 5s$	$4d, 5p, 6s$	$4f, 5d, 6p, 7s$

(b) $Z = 15$: Filled subshells: $1s, 2s, 2p, 3s$ (12 electrons)

Valence subshell: 3 electrons in $3p$ subshell

Prediction: Valence = +3 or −5

Element is phosphorus = Valence +3 or −5

$Z = 47$: Filled subshells: $1s, 2s, 2p, 3s, 3p, 4s, 3d, 4p, 5s$ (38 electrons)

Outer subshell: 9 electrons in $4d$ subshell

Prediction: Valence = −1

Element is silver, valence +1 (Prediction fails)

$z = 86$: Filled shells: $1s, 2s, 2p, 3s, 3p, 4s, 3d, 4p, 5s, 4d, 5p, 6s, 4f, 5d, 6p$

Outer subshell is full—predict inert gas.
Element is radon, inert.

32. Use the method illustrated in Example 42.7 to calculate the wavelength of the x-ray emitted from a molybdenum target ($Z = 42$) when an electron moves from the L shell ($n = 2$) to the K shell ($n = 1$).

Solution Following Example 42.7, we suppose the electron is originally in the L shell with just one other electron in the K shell between it and the nucleus, so it moves in a field of effective charge $(42 - 1)e$. Its energy is then $E_L = -(42-1)^2\ 13.6\ \text{eV}/4$. In its final state we estimate the screened charge holding it in orbit as again $(42 - 1)e$, so its energy is $E_K = -(42-1)^2\ 13.6$ eV. The photon energy emitted is the difference.

$$E_\gamma = \tfrac{3}{4}(42-1)^2(13.6\ \text{eV}) = 1.71\times10^4\ \text{eV} = 2.74\times10^{-15}\ \text{J}$$

Then
$$f = \frac{E}{h} = 4.14\times10^{18}\ \text{Hz}$$

$$\lambda = \frac{c}{f} = 0.725\ \text{Å}\quad\Diamond$$

---

35. A ruby laser delivers a 10-ns pulse of 1.0 MW average power. If the photons have a wavelength of 694.3 nm, how many are contained in the pulse?

Solution The energy of the pulse is

$$E = (1.0\times10^6\ \text{W})(1.0\times10^{-8}\ \text{s}^{-1}) = 1.0\times10^{-2}\ \text{J}$$

The energy of each photon in the pulse is

$$E_\gamma = hf = \frac{hc}{\lambda} = \frac{(6.626\times10^{-34})(3\times10^8)}{694.3\times10^{-9}}\ \text{J} = 2.86\times10^{-19}\ \text{J}$$

So
$$N = \frac{E}{E_\gamma} = \frac{1.0\times10^{-2}}{2.86\times10^{-19}} = 3.5\times10^{16}\ \text{photons}\quad\Diamond$$

---

37. In the technique known as electron spin resonance (ESR), a sample containing unpaired electrons is placed in a magnetic field. Consider the simplest situation, that in which there is only one electron and therefore only two possible energy states, corresponding to $m_s = \pm\frac{1}{2}$. In ESR, the absorption of a photon causes the electron's spin magnetic moment to flip from a lower energy state to a higher energy state. (The lower energy state corresponds to the case where the magnetic moment μ_s is aligned with the magnetic field, and the higher energy state corresponds to the case where μ_s is aligned against the field.) What is the photon frequency required to excite an ESR transition in a 0.35-T magnetic field?

Solution

As in Section 29.4, the magnetic moment feels torque $\tau = \mu \times \mathbf{B}$ in an external field. In turning it from alignment with the field to the opposite direction, the field does work according to Equation 10.21,

$$W = \int dW = \int_0^{180^\circ} \tau\, d\theta = \int_0^{\pi} \mu B \sin\theta\, d\theta$$

$$W = -\mu B \cos\theta \Big|_0^{\pi} = 2\mu B$$

The photon must carry this much energy to make the electron flip:

$$\Delta E = 2\mu_B B = hf$$

$$2(9.27 \times 10^{-24})(0.35) = (6.63 \times 10^{-34})f$$

so

$$f = 9.79 \times 10^9 \text{ Hz} \quad \Diamond$$

---

39. A dimensionless number that often appears in atomic physics is the fine-structure constant $\alpha = k_e e^2 / \hbar c$, where k_e is the Coulomb constant. (a) Obtain a numerical value for $1/\alpha$. (b) In scattering experiments, the electron size is taken to be the classical electron radius, $r_e = k_e e^2 / m_e c^2$. In terms of α, what is the ratio of the Compton wavelength (Section 40.4), $\lambda_C = h/m_e c$, to the classical electron radius? (c) In terms of α, what is the ratio of the Bohr radius, a_0, to the Compton wavelength? (d) In terms of α, what is the ratio of the Rydberg wavelength, $1/R_H$, to the Bohr radius (Section 40.6)?

Solution

(a) $$\frac{1}{\alpha}=\frac{\hbar c}{k_e e^2}=\frac{(1.05457\times10^{-34}\ \text{J}\cdot\text{s})(2.997925\times10^{8}\ \text{m/s})}{(8.9875\times10^{9}\ \text{N}\cdot\text{m}^2/\text{C}^2)(1.60219\times10^{-19}\ \text{C})^2}=137.034\quad\Diamond$$

(b) $$\frac{\lambda_c}{r_e}=\frac{h/mc}{k_e e^2/mc^2}=\frac{hc}{k_e e^2}=\frac{2\pi}{\alpha}\quad\Diamond$$

(c) $$\frac{a_0}{\lambda_c}=\frac{\hbar^2/mk_e e^2}{h/mc}=\left(\frac{1}{2\pi}\right)\frac{\hbar c}{k_e e^2}=\frac{1}{2\pi\alpha}\quad\Diamond$$

(d) $$\frac{1}{R_H a_0}=\left(\frac{4\pi c\hbar^3}{mk_e{}^2e^4}\right)\left(\frac{mk_e e^2}{\hbar^2}\right)=4\pi\left(\frac{\hbar c}{k_e e^2}\right)=\frac{4\pi}{\alpha}\quad\Diamond$$

---

40. Show that the average value of r for the 1s state of hydrogen has the value $3a_0/2$. (*Hint:* Use Eq. 42.7.)

Solution The average value (expectation value) of r is $\quad \langle r\rangle = \int_0^\infty rP_{1s}(r)dr$

where $$P_{1s}(r)=\left(\frac{4r^2}{a_0{}^3}\right)e^{-2r/a_0}$$

$$\langle r\rangle=\frac{4}{a_0{}^3}\int_0^\infty r^3e^{-2r/a_0}dr$$

Letting $x=\frac{2r}{a_0}$, we find $$\langle r\rangle=\frac{a_0}{4}\int_0^\infty x^3e^{-x}dx$$

Integrating by parts (or using a table of integrals) gives $\quad \langle r\rangle=\frac{3}{2}a_0\quad\Diamond$

---

41. Suppose a hydrogen atom is in the 2*s* state. Taking $r = a_0$, calculate values for (a) $\psi_{2s}(a_0)$, (b) $|\psi_{2s}(a_0)|^2$, and (c) $P_{2s}(a_0)$. (*Hint:* Use Eq. 42.8.)

Solution The wave function for the 2*s* state is given by Equation 42.8:

$$\psi_{2s}(r) = \frac{1}{4\sqrt{2\pi}}\left(\frac{1}{a_0}\right)^{3/2}\left[2 - \frac{r}{a_0}\right]e^{-r/2a_0}$$

(a) Taking $r = a_0 = 0.529 \times 10^{-10}$ m, we find

$$\psi_{2s}(a_0) = \frac{1}{4\sqrt{2\pi}}\left(\frac{1}{0.529 \times 10^{-10}\text{ m}}\right)^{3/2}(2-1)e^{-1/2} = 1.57 \times 10^{14}\text{ m}^{-3/2} \quad \Diamond$$

(b) $|\psi_{2s}(a_0)|^2 = \left(1.57 \times 10^{14}\text{ m}^{-3/2}\right)^2 = 2.47 \times 10^{28}\text{ m}^{-3} \quad \Diamond$

(c) Using Equation 42.6 and the results to (c) gives

$$P_{2s}(a_0) = 4\pi a_0^{\,2}|\psi_{2s}(a_0)|^2 = 8.69 \times 10^8\text{ m}^{-1} \quad \Diamond$$

---

47. Consider a hydrogen atom in its ground state. (a) Treating orbiting the electron as an effective current loop of radius a_0, derive an expression for the magnetic field at the nucleus. (*Hint:* Use the Bohr theory of hydrogen and see Example 30.3.) (b) Find a numerical value for the magnetic field at the nucleus in this situation.

Solution

(a) The magnetic field at the center of a circular current loop of radius r is $B = \mu_0 I/2r$ (see Example 30.3). An electron moving in a circular orbit with frequency f corresponds to an *equivalent* current I given by

$$I = ef = \frac{e\omega}{2\pi}$$

From the Bohr theory, $mvr = n\hbar$, hence $\quad \omega = \frac{v}{r} = \frac{n\hbar}{mr^2}$

For the ground state, $n = 1$, and $r = a_0$, so

$$\omega = \frac{\hbar}{m a_0^2} \quad \text{and} \quad I = \frac{e\hbar}{2\pi m a_0^2}$$

Thus, $$B = \frac{\mu_0 I}{2a_0} = \frac{\mu_0 e\hbar}{4\pi m a_0^3} \quad \Diamond$$

(b) Substituting numerical values into the above expression gives

$$B = \frac{(4\pi \times 10^{-7}\ \text{N/A}^2)(1.60 \times 10^{-19}\ \text{C})(1.055 \times 10^{-34}\ \text{J}\cdot\text{s})}{4\pi(9.11 \times 10^{-31}\ \text{kg})(0.529 \times 10^{-10}\ \text{m})^3} = 12.5\ \text{T} \quad \Diamond$$

---

49. (a) Calculate the most probable position for an electron in the 2*s* state of hydrogen. (*Hint:* Let $x = r/a_0$, find an equation for x, and show that $x = 5.236$ is a solution to this equation.) (b) Show that the wave function given by Equation 42.8 is normalized.

Solution We use $\psi_{2s}(r) = \frac{1}{4}\left(2\pi a_0^3\right)^{-1/2}\left(2 - \frac{r}{a_0}\right)e^{-r/2a_0}$

and by Equation 42.6, the radial probability distribution function is

$$P(r) = 4\pi r^2 \psi^2 = \frac{1}{8}\left(\frac{r^2}{a_0^3}\right)\left(2 - \frac{r}{a_0}\right)^2 e^{-r/a_0}$$

Its extremes are given by

(a) $$\frac{dP(r)}{dr} = \frac{1}{8}\left[\frac{2r}{a_0^3}\left(2 - \frac{r}{a_0}\right)^2 - \frac{2r^2}{a_0^3}\left(\frac{1}{a_0}\right)\left(2 - \frac{r}{a_0}\right) - \frac{r^2}{a_0^3}\left(2 - \frac{r}{a_0}\right)^2\left(\frac{1}{a_0}\right)\right]e^{-r/a_0} = 0$$

or $$\frac{1}{8}\left(\frac{r}{a_0^3}\right)\left(2 - \frac{r}{a_0}\right)\left[2\left(2 - \frac{r}{a_0}\right) - \frac{2r}{a_0} - \frac{r}{a_0}\left(2 - \frac{r}{a_0}\right)\right]e^{-r/a_0} = 0$$

Therefore, $[\,.....\,] = 4 - \frac{6r}{a_0} + \left(\frac{r}{a_0}\right)^2 = 0$ which has solutions $r = \left(3 \pm \sqrt{5}\right)a_0$

$\left[\text{The roots of } \frac{dP}{dr} = 0 \text{ at } r = 0,\ r = 2a_0, \text{ and } r = \infty \text{ are minima } (\psi = 0).\right]$

We substitute the two roots into $P(r)$:

When $r = \left(3 - \sqrt{5}\right)a_0 = 0.764a_0$, then $P(r) = \frac{0.0519}{a_0}$

When $r = \left(3 + \sqrt{5}\right)a_0 = 5.24a_0$, then $P(r) = \frac{0.191}{a_0}$

Therefore, the most probable value of r is $\left(3 + \sqrt{5}\right)a_0 = 5.24a_0$ ◊

(b) $\int_0^\infty P(r)dr = \int_0^\infty \frac{1}{8}\left(\frac{r^2}{a_0{}^3}\right)\left(2 - \frac{r}{a_0}\right)^2 e^{-r/a_0}\,dr$

Let $u = \frac{r}{a_0}$, $dr = a_0 du$, then

$$\int_0^\infty P(r)dr = \int_0^\infty \tfrac{1}{8}u^2(4 - 4u + u^2)e^{-u}\,du$$

$$\int_0^\infty P(r)dr = \int_0^\infty \tfrac{1}{8}(u^4 - 4u^3 + 4u^2)e^{-u}\,du$$

Use a table of integrals or integrate by parts repeatedly to find

$$\int_0^\infty P(r)dr = -\tfrac{1}{8}(u^4 + 4u^2 + 8u + 8)e^{-u}\Big]_0^\infty = 1 \text{ as desired}$$ ◊

51. For hydrogen in the 1s state, what is the probability of finding the electron farther than $2.50a_0$ from the nucleus?

Solution The radial probability distribution function is $P(r) = 4\pi r^2|\psi|^2$

With $\psi_{1s} = \left(\pi a_0^3\right)^{-1/2} e^{-r/a_0}$, it is $P(r) = 4r^2 a_0^{-3} e^{-2r/a_0}$

The required probability is then $P = \int_{2.5a_0}^{\infty} P(r)dr = \int_{2.5a_0}^{\infty} \frac{4r^2}{a_0^3} e^{-2r/a_0}\, dr$

Let $z = \frac{2r}{a_0}$, $dz = \frac{2dr}{a_0}$: $\quad P = \frac{1}{2}\int_5^{\infty} z^2 e^{-z}\, dz$

$$P = -\tfrac{1}{2}(z^2 + 2z + 2)e^{-z}\Big|_5^{\infty}$$

$$P = -\tfrac{1}{2}[0] + \tfrac{1}{2}(25 + 10 + 2)e^{-5} = \left(\frac{37}{2}\right)(0.00674)$$

$$P = 0.125 \quad \Diamond$$

---

52. According to classical physics, an accelerated charge e radiates at a rate

$$\frac{dE}{dt} = -\frac{1}{6\pi\epsilon_0}\frac{e^2a^2}{c^3}$$

(a) Show that an electron in a classical hydrogen atom (see Fig. 42.3) spirals into the nucleus at a rate

$$\frac{dr}{dt} = -\frac{e^4}{12\pi^2\epsilon_0^2 r^2 m^2 c^3}$$

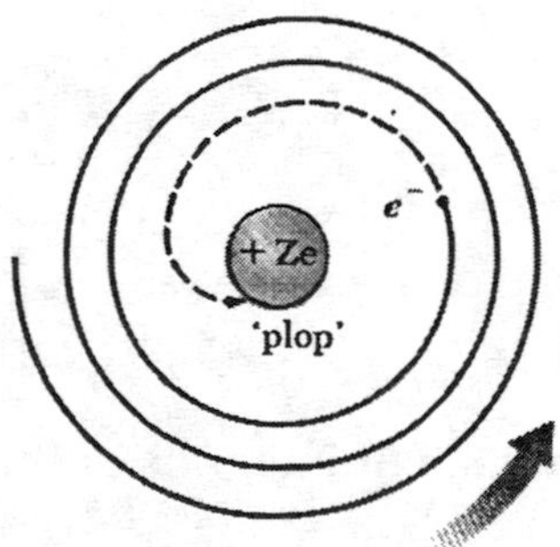

Figure 42.3

(b) Find the time it takes the electron to reach $r = 0$, starting from $r_0 = 2\times10^{-10}$ m.

Solution According to a classical model, the electron moving in a circular orbit about the proton in the hydrogen atom experiences a force $k_e e^2/r^2$; and from Newton's second law, $F = ma$, its acceleration is $k_e e^2/mr^2$.

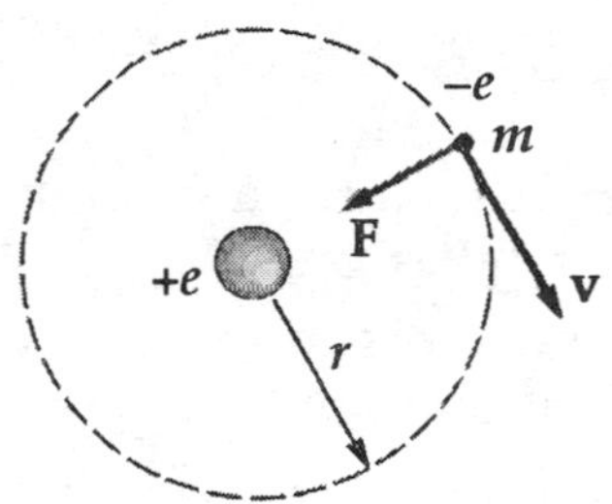

(a) Using the fact that the Coulomb constant $k_e = \dfrac{1}{4\pi \epsilon_0}$,

$$a = \frac{v^2}{r} = \frac{k_e e^2}{mr^2} = \frac{e^2}{4\pi \epsilon_0 mr^2} \qquad (1)$$

From the Bohr model of the atom (Section 40.6), we can write the total energy of the atom as

$$E = -\frac{k_e e^2}{2r} = -\frac{e^2}{8\pi \epsilon_0 r}$$

$$\frac{dE}{dt} = \frac{e^2}{8\pi \epsilon_0 r^2}\frac{dr}{dt} = -\frac{1}{6\pi \epsilon_0}\frac{e^2 a^2}{c^3} \qquad (2)$$

Substituting (1) into (2) for a, and solving for $\dfrac{dr}{dt}$, and simplifying gives

$$\frac{dr}{dt} = -\frac{4r^2}{3c^3}\left(\frac{e^2}{4\pi \epsilon_0 mr^2}\right)^2 = -\frac{e^4}{12\pi^2 \epsilon_0^2 r^2 m^2 c^3}$$

(b) We can express $\dfrac{dr}{dt}$ in the simpler form: $\dfrac{dr}{dt} = -\dfrac{A}{r^2} = -\dfrac{3.15\times10^{-21}}{r^2}$

Thus, $$-\int_{2\times10^{-10}\text{ m}}^{0} r^2 dr = 3.15\times10^{-21}\int_0^T dt$$

and $$T = \left(3.17\times10^{20}\right)\frac{r^3}{3}\Bigg]_0^{2\times10^{-10}\text{ m}} = 8.46\times10^{-10}\text{ s} = 0.846\text{ ns} \quad \Diamond$$

We know that atoms 'last' much longer than 0.8 ns; thus, classical physics does not hold (fortunately) for atomic systems.

53. Light from a certain He-Ne laser has a power output of 1.0 mW and a cross-sectional area of 10 mm^2. The entire beam is incident on a metal target that requires 1.5 eV to remove an electron from its surface. (a) Perform a classical calculation to determine how long it takes one atom in the metal to absorb 1.5 eV from the incident beam. (*Hint:* Assume the surface area of an atom is 1.0×10^{-20} m^2, and first calculate the energy incident on each atom per second.) (b) Compare the (wrong) answer obtained in part (a) to the actual response time for photoelectric emission $\left(\approx 10^{-9}\text{ s}\right)$, and discuss the reasons for the large discrepancy.

Solution

(a) The intensity of the beam is $I = \frac{P}{A} = \frac{10^{-3}\text{ W}}{10 \times 10^{-6}\text{ m}^2} = 100\text{ W / m}^2$

The power incident on our one atom is $P_a = IA = (100\text{ W/m}^2)10^{-20}\text{ m}^2 = 10^{-18}\text{ W}$

To absorb 1.5 eV = 2.4×10^{-19} J, the atom must be in the beam for time

$$t = \frac{E}{P} = \frac{2.4 \times 10^{-19}\text{ J}}{10^{-18}\text{ J/s}} = 0.24\text{ s} \quad \lozenge$$

(b) The classical answer is too large by a factor of about a billion because the beam does not carry smeared-out energy but energy in photon lumps. The atoms do not all have to wait together for energy to accumulate, like school teachers for retirement. In the first nanosecond of absorption, some atom will get hit with a photon and release an electron. The photocurrent begins right away.

Chapter 43

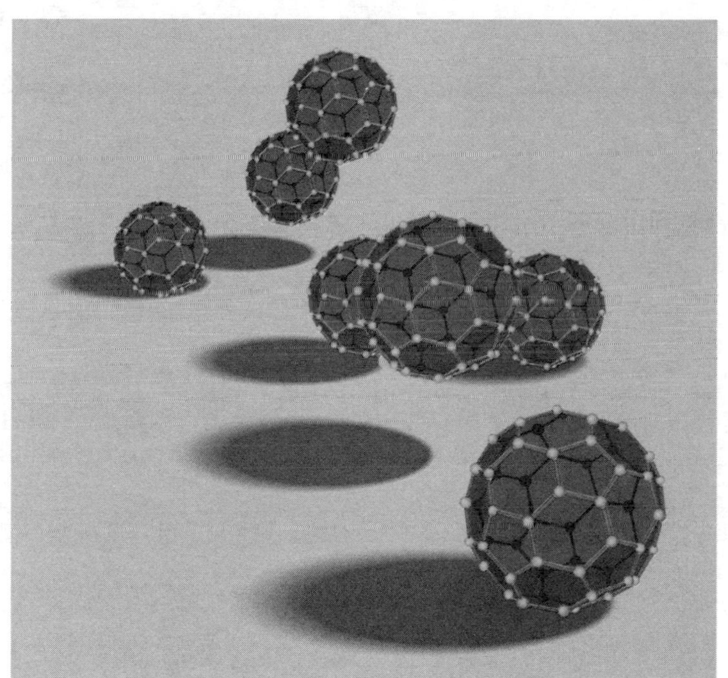

Molecules and Solids

Chapter 43

MOLECULES AND SOLIDS

INTRODUCTION

In this chapter, we study the aggregates of atoms known as molecules. First we describe the bonding mechanisms in molecules, the various modes of molecular excitation, and the radiation emitted or absorbed by molecules. We then take the next logical step and show how molecules combine to form solids. Then, by examining their electronic distributions, we explain the differences between insulating, metallic, and semiconducting crystals. The chapter concludes with discussions of semiconducting junctions, the basis for the operation of several semiconductor devices.

NOTES FROM SELECTED CHAPTER SECTIONS

43.1 Molecular Bonds

Two atoms combine to form a molecule because of a net attractive force between them when their separation is greater than their equilibrium separation in the molecule. Furthermore, the total energy of the stable bound molecule is *less* than the total energy of the separated atoms.

The potential energy for large atomic separations is negative, corresponding to a net attractive force. At the equilibrium separation, the attractive and repulsive forces just balance and the potential energy has its minimum value.

Ionic bonds are due to the Coulomb attraction between oppositely charged ions.

The energy released when an atom takes on an electron is called the *electron affinity.*

The *dissociation energy* is the energy required to separate a molecule into neutral atoms.

A *covalent bond* between two atoms results from a sharing of electrons from one or both atoms that combine to form the molecule.

In some cases, hydrogen forms a weak chemical bond called a *hydrogen bond.* Parts of negative ions are bound by the presence of the positive hydrogen ion between them.

The weak electrostatic attractions between molecules are called *van der Waals forces.* There are three types of van der Waals forces:

- The first type, called the *dipole-dipole force,* is an interaction between two molecules each having a permanent electric dipole moment. In effect, one molecule interacts with the electric field produced by another molecule.

- The second type of van der Waals force is a *dipole-induced force* in which a polar molecule having a permanent electric dipole moment *induces* a dipole moment in a nonpolar molecule.

- The third type of van der Waals force is called the *dispersion force.* The dispersion force is an attractive force that occurs between two nonpolar molecules. In this case, the interaction results from the fact that although the average dipole moment of a nonpolar molecule is zero, the average of the square of the dipole moment is nonzero because of charge fluctuations. Consequently, two nonpolar molecules near each other tend to be correlated so as to produce an attractive force.

43.2 The Energy and Spectra of Molecules

The energy of a molecule in the gaseous phase can be divided into four categories: (1) electronic energy, due to the mutual interactions of the molecule's electrons and nuclei; (2) translational energy, due to the motion of the molecule's center of mass through space; (3) rotational energy, due to the rotation of the molecule about its center of mass; and (4) vibrational energy, due to the vibration of the molecule's constituent atoms.

43.3 Bonding in Solids

Ionic crystals have the following general properties:

- They form relatively stable and hard crystals.
- They are poor electrical conductors because they contain no free electrons.
- They have high vaporization temperatures.
- They are transparent to visible radiation but absorb strongly in the infrared region.
- They are usually quite soluble in polar liquids such as water. The water molecule, which has a permanent electric dipole moment, exerts an attractive force on the charged ions, which breaks the ionic bonds and dissolves the solid.

The *covalent bond* is very strong and comparable to the ionic bond. In general, covalently bonded solids are very hard, have large bond energies, high melting points, are good insulators, and are transparent to visible light.

Metallic bonds are generally weaker than ionic or covalent bonds. The valence electrons in a metal are relatively free to move throughout the material. There are a large number of such mobile electrons in a metal, typically one or two per atom. The metal structure can be viewed as a "sea" or "gas" of nearly free electrons surrounded by a lattice of positive ions (Fig. 43.14). The binding mechanism in a metal is the attractive force between the positive ions and the electron gas.

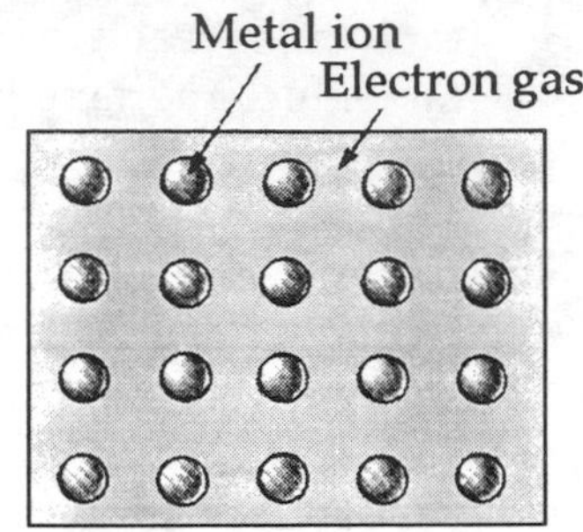

Figure 43.14

43.4 Band Theory of Solids

One can best understand the properties of metals, insulators, and semiconductors in terms of the *band theory of solids.* The valence band of a metal such as sodium is half filled. Therefore, there are many electrons free to move throughout the metal and contribute to the conduction current. In an insulator at T = 0 K, the valence band is completely filled with electrons, while the conduction band is empty. The region between the valence band and the conduction band is called the *energy gap* of the material.

43.5 Free-Electron Theory of Metals

In the free-electron theory of metals, the valence electrons fill the quantized levels in accordance with the Pauli exclusion principle. Only those electrons having energies near the *Fermi energy* can contribute to the electrical conductivity of the metal.

43.6 Electrical Conduction in Metals, Insulators, and Semiconductors

If an electric field is applied to metal, electrons having energies near the Fermi energy require only a small amount of additional energy from the applied field to reach nearby empty energy states. Thus electrons are free to move with only a small applied field in a metal because there are many empty states available for occupancy close to the occupied energy states.

Although an insulator has many vacant states in its conduction band that can accept electrons, there are so few electrons actually occupying conduction band states that the overall contribution to electrical conductivity is very small, resulting in a high resistivity for insulators.

Semiconductors have a small energy gap. At $T = 0$ K, all electrons in a pure semiconductor are in the valence band and there are no electrons in the conduction band. Thus, semiconductors are *poor* conductors at low temperatures. Thermal excitation across the narrow gap is more probable at higher temperature and the conductivity of semiconductors increases rapidly with temperature.

43.7 Semiconductor Devices

The band structures and electrical properties of a semiconductor can be modified by adding donor atoms with five valence electrons (such as arsenic), or by adding acceptor atoms with three valence electrons (such as indium). A semiconductor *doped* with donor impurity atoms is called an *n-type semiconductor,* while one doped with acceptor impurity atoms is called *p-type.*

The junction transistor consists of a semiconducting material with a very narrow n region sandwiched between two p regions. This configuration is called the *pnp transistor.* Another configuration is the *npn transistor,* which consists of a p region sandwiched between two n regions.

EQUATIONS AND CONCEPTS

The rotational energy of a molecule is quantized and depends on the value of the moment of inertia.

$$E_{\text{rot}} = \frac{\hbar^2}{2I} J(J+1) \qquad (43.6)$$

$$J = 0, 1, 2, \ldots$$

For a diatomic molecule, the moment of inertia I can be written in terms of the reduced mass, μ.

$$I = \mu r^2 \qquad (43.3)$$

$$\mu = \frac{m_1 m_2}{m_1 + m_2} \qquad (43.4)$$

The vibrational energy for a diatomic molecule is quantized and is characterized by the vibrational quantum number, v. The selection rule for allowed vibrational transitions is given by $\Delta v \pm 1$. The energy difference between successive vibrational levels is hf when f is the frequency of vibration.

$$E_{\text{vib}} = \left(v + \tfrac{1}{2}\right)\frac{h}{2\pi}\sqrt{\frac{k}{\mu}} \qquad (43.10)$$

$$v = 0, 1, 2, \ldots$$

$$\Delta E_{\text{vib}} = hf \qquad (43.11)$$

The ionic cohesive energy U_0 of a solid represents the energy necessary to separate the solid into a collection of positive and negative ions. In this expression, r_0 is the equilibrium ion separation and α, the Madelung constant, has a value which is characteristic of a specific crystal structure. The parameter m is a small integer.

$$U_0 = -\alpha k_e \frac{e^2}{r_0}\left(1 - \frac{1}{m}\right) \qquad (43.17)$$

The Fermi-Dirac distribution function $f(E)$ gives the probability of finding an electron in a particular energy state, E.

$$f(E) = \frac{1}{e^{(E-E_F)/k_B T} + 1} \qquad (43.18)$$

In the Fermi-Dirac distribution function, E_F is called the Fermi energy and is a function of the total number of electrons per unit volume, n.

$$E_F = \frac{h^2}{2m}\left(\frac{3n}{8\pi}\right)^{2/3} \qquad (43.25)$$

In thermal equilibrium, the number of electrons per unit volume with energy between E and $E + dE$ is the product of the probability of finding an electron in a particular state and the density of states.

$$n = \int_0^\infty N(E)\, dE \qquad (43.23)$$

$$= C\int_0^\infty \frac{E^{1/2}\, dE}{e^{(E-E_F)/k_B T} + 1}$$

$$\text{where } C = \frac{8\sqrt{2}\pi m^{3/2}}{h^3}$$

REVIEW CHECKLIST

▷ Understand the essential bonding mechanisms involved in ionic, covalent, hydrogen, and van der Waals bonding.

▷ Describe in terms of appropriate quantum numbers the allowed energy levels associated with rotational and vibrational motions of molecules. Use the selection rules to determine the separation between adjacent energy levels.

▷ Discuss the free-electron theory of metals including the significance of the Fermi-Dirac distribution function, Fermi energy, and particle distribution by energy interval.

▷ Use the band theory of solids as a basis for a qualitative discussion of the mechanisms for conduction in metals, insulators, and semiconductors.

▷ Describe a *p-n* junction and the diffusion of electrons and holes through the junction. Discuss the fabrication and function of a junction diode and junction transistor.

SOLUTIONS TO SELECTED END-OF-CHAPTER PROBLEMS

1. A K^+ ion and a Cl^- ion are 5.0×10^{-10} m from each other. Assuming the two ions act like point charges, determine (a) the force of attraction between them, and (b) the potential energy of attraction in electron volts.

Solution

(a) $$F = \frac{k_e|q_1||q_2|}{r^2} = \frac{8.99\times10^9\ \text{N}\cdot\text{m}^2\ (1.6\times10^{-19}\ \text{C})^2}{\text{C}^2(5\times10^{-10}\ \text{m})^2} = 9.21\times10^{-10}\ \text{N} \quad \Diamond$$

(b) $$U = \frac{k_e q_1 q_2}{r} = \frac{(1.6\times10^{-19}\ \text{C})(-1.6\times10^{-19}\ \text{C})}{(5\times10^{-10}\ \text{m})}\left(\frac{8.99\times10^9\ \text{N}\cdot\text{m}^2}{\text{C}^2}\right)$$

$$= -4.60\times10^{-19}\ \text{J} = -2.9\ \text{eV} \quad \Diamond$$

9. The HCl molecule is excited to its first rotational-energy level, corresponding to $J = 1$. If the distance between its nuclei is 0.1275 nm, what is the angular speed of the molecule around its center of mass?

Solution For the HCl molecule in the $J = 1$ rotational energy level, we are given $r_0 = 0.1275$ nm.

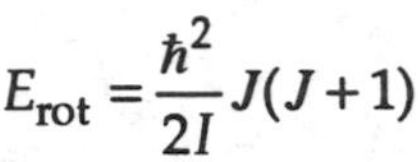

$$E_{\text{rot}} = \frac{\hbar^2}{2I} J(J+1)$$

Taking $J = 1$, $\quad E_{\text{rot}} = \frac{\hbar^2}{I} = \frac{1}{2} I\omega^2 \quad$ and $\quad \omega = \sqrt{\frac{2\hbar^2}{I^2}} = \sqrt{2}\frac{\hbar}{I}$

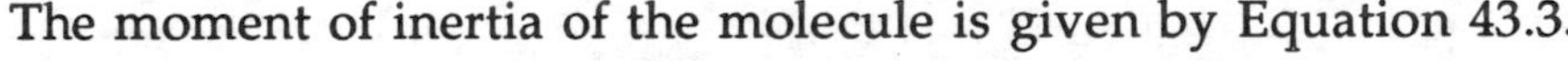

The moment of inertia of the molecule is given by Equation 43.3.

$$I = \mu r_0^2 = \left(\frac{m_1 m_2}{m_1 + m_2}\right) r_0^2 = \left[\frac{(1\text{ u})(35\text{ u})}{1\text{ u} + 35\text{ u}}\right] r_0^2 = (0.972\text{ u}) r_0^2$$

$$I = (0.972\text{ u})(1.66 \times 10^{-27}\text{ kg/u})(1.275 \times 10^{-10}\text{ m})^2 = 2.62 \times 10^{-47}\text{ kg} \cdot \text{m}^2$$

Therefore, $\quad \omega = \sqrt{2}\frac{\hbar}{I} = \sqrt{2}\left(\frac{1.055 \times 10^{-34}\text{ J} \cdot \text{s}}{2.62 \times 10^{-47}\text{ kg} \cdot \text{m}^2}\right) = 5.69 \times 10^{12}\text{ rad/s} \quad \Diamond$

---

11. If the effective force constant of a vibrating HCl molecule is $k = 480$ N/m, estimate the energy difference between the ground state and the first vibrational level.

Solution The reduced mass of the pair of atoms is

$$\mu = \frac{m_1 m_2}{m_1 + m_2} = \frac{(1.01\text{ u})(35.5\text{ u})}{1\text{ u} + 35.5\text{ u}}\left(1.66 \times 10^{-27}\text{ kg/u}\right) = 1.63 \times 10^{-27}\text{ kg}$$

The energy difference between adjacent vibration states is

$$\Delta E_{\text{vib}} = \frac{h}{2\pi}\sqrt{\frac{k}{\mu}} = \left(\frac{6.63 \times 10^{-34}\text{ J} \cdot \text{s}}{2\pi}\right)\sqrt{\frac{480\text{ N/m}}{1.63 \times 10^{-27}\text{ kg}}} = 5.72 \times 10^{-20}\text{ J} = 0.358\text{ eV} \quad \Diamond$$

---

16. Consider a one-dimensional chain of alternating positive and negative ions. Show that the potential energy of an ion in this hypothetical crystal is

$$U(r) = -k_e \alpha \frac{e^2}{r}$$

where α = 2 ln 2 (the Madelung constant) and r is the interionic spacing. [*Hint:* Use the series expansion for $\ln(1 + x)$.]

Solution

The total potential energy is obtained by summing over all pairs of interactions:

$$U = \sum_{i \neq j} k_e \frac{q_i q_j}{r_{ij}} = -k_e \left[\frac{e^2}{r} - \frac{e^2}{r} + \frac{e^2}{2r} + \frac{e^2}{2r} - \frac{e^2}{3r} - \frac{e^2}{3r} + \frac{e^2}{4r} + \frac{e^2}{4r} \right]$$

$$U = -2k_e \frac{e^2}{r} \left[1 - \frac{1}{2} + \frac{1}{3} - \frac{1}{4} + \ \ldots \right]$$

But
$$\ln(1+x) = x - \frac{x^2}{2} + \frac{x^3}{3} - \frac{x^4}{4} + \ \ldots$$

Therefore, x = 1 for our series, and U becomes

$$U = -2 \ \ln(2) k_e \frac{e^2}{r} = -\alpha k_e \frac{e^2}{r} \quad \Diamond$$

---

21. Calculate the energy of a conduction electron in silver at 800 K if the probability of finding the electron in that state is 0.95. The Fermi energy is 5.48 eV at this temperature.

Solution Taking E_F = 5.48 eV for sodium at 800 K, and given f = 0.95, we find

$$f = \frac{1}{e^{(E - E_F)/k_B T} + 1} = 0.95$$

$$e^{(E - E_F)/k_B T} = \frac{1}{0.95} - 1 = 0.05263$$

$$\frac{E-E_F}{k_BT} = \ln(0.05263) = -2.944$$

$$E - E_F = -2.944\ k_BT = -2.944(1.38\times10^{-23}\ \text{J/K})(800\ \text{K})$$

$$E - E_F = -3.25\times10^{-20}\ \text{J} = -0.203\ \text{eV}$$

$$E = 5.28\ \text{eV} \quad \Diamond$$

---

22. Show that the average kinetic energy of a conduction electron in a metal at 0 K is $E_{\text{av}} = \frac{3}{5}E_F$. (*Hint:* In general, the average kinetic energy is

$$E_{\text{av}} = \frac{1}{n}\int EN(E)\,dE$$

where n is the density of particles, and $N(E)\,dE$ is given by Equation 43.22.

Solution $$E_{\text{av}} = \frac{1}{n}\int_0^\infty EN(E)\,dE$$

where $$N(E) = \frac{CE^{1/2}}{e^{(E-E_F)/k_BT}+1} = Cf(E)E^{1/2}$$

But at $T = 0$, $f(E) = 0$ for $E > E_F$

Also, $f(E) = 1$ for $E < E_F$

So we can take $N(E) = CE^{1/2}$

$$E_{\text{av}} = \frac{1}{n}\int_0^{E_F} CE^{3/2}\,dE = \frac{2C}{5n}E_F^{5/2}$$

But from Equation 43.24 we have $$\frac{C}{n} = \frac{3}{2}E_F^{-3/2}$$

so $$E_{\text{av}} = \left(\frac{2}{5}\right)\left(\frac{3}{2}\right)\left(E_F^{-3/2}\right)E_F^{5/2} = \frac{3}{5}E_F \quad \Diamond$$

---

25. (a) Consider a system of electrons confined to a three-dimensional box. Calculate the ratio of the number of allowed energy levels at 8.5 eV to the number at 7.0 eV. (b) Copper has a Fermi energy of 7.0 eV at 300 K. Calculate the ratio of the number of occupied levels at an energy of 8.5 eV to the number at the Fermi energy. Compare your answer with that obtained in part (a).

Solution The density of states at the energy E is $g(E) = CE^{1/2}$. Hence, the required ratio is

(a)
$$\frac{g(8.5\text{ eV})}{g(7.0\text{ eV})} = \frac{C(8.5)^{1/2}}{C(7.0)^{1/2}} = 1.10 \quad \Diamond$$

(b) From Equation 43.22 we see that the number of occupied states having energy E is

$$N(E) = \frac{CE^{1/2}}{e^{(E-E_F)/k_BT}+1}$$

Hence, the required ratio is
$$\frac{N(8.5\text{ eV})}{N(7.0\text{ eV})} = \left(\frac{8.5}{7.0}\right)^{1/2}\left[\frac{e^{(7-7)/k_BT}+1}{e^{(8.5-7)/k_BT}+1}\right]$$

At T =300 K, $k_BT = 0.0259$ eV, so
$$\frac{N(8.5\text{ eV})}{N(7.0\text{ eV})} = (1.10)\left(\frac{2}{e^{1.5/0.0259}+1}\right) = 1.55\times10^{-25} \quad \Diamond$$

Comparing this result with (a), we conclude that very few states with $E > E_F$ are occupied.

---

29. Most solar radiation has a wavelength of 10^{-6} m or less. What energy gap should the material in a solar cell have in order to absorb this radiation? Is silicon appropriate (see Table 43.5)?

Solution To absorb the longest-wavelength photons, the energy gap should be no larger than the photon energy.

$$hf = \frac{hc}{\lambda} = \frac{(6.63\times10^{-34}\text{ J}\cdot\text{s})(3.00\times10^{8}\text{ m/s})}{10^{-6}\text{ m}}\left(\frac{1\text{ eV}}{1.6\times10^{-19}\text{ J}}\right) = 1.24\text{ eV} \quad \Diamond$$

So silicon, with gap of 1.14 eV < 1.24 eV, is appropriate.

---

35. Show that the ionic cohesive energy of an ionically bonded solid is given by Equation 43.17. (*Hint:* Start with Equation 43.16, and note that $dU/dr = 0$ at $r = r_0$.)

Solution The total potential energy is given by Equation 43.16:

$$U_{\text{total}} = -\alpha k_e \frac{e^2}{r} + \frac{B}{r^m}$$

The potential energy has its minimum value U_0 when $r = r_0$, where r_0 is the equilibrium spacing. At this point, the slope of the curve U versus r is *zero*. That is,

$$\left.\frac{dU}{dr}\right|_{r=r_0} = 0$$

$$\frac{dU}{dr} = \frac{d}{dr}\left(-\alpha k_e \frac{e^2}{r} + \frac{B}{r^m}\right) = \alpha k_e \frac{e^2}{r^2} - \frac{mB}{r^{m+1}}$$

Taking $r = r_0$ and setting this equal to zero, we find $\alpha k_e \dfrac{e^2}{{r_0}^2} - \dfrac{mB}{{r_0}^{m+1}} = 0$

Therefore, $$B = \alpha \frac{k_e e^2}{m} {r_0}^{m-1}$$

Substituting this value of B into U total gives

$$U_0 = -\alpha k_e \frac{e^2}{r_0} + \alpha \frac{k_e e^2}{m} {r_0}^{m-1}\left(\frac{1}{r^m}\right) = -\alpha \frac{k_e e^2}{r_0}\left(1 - \frac{1}{m}\right) \quad \Diamond$$

Chapter 44

Superconductivity

Chapter 44

SUPERCONDUCTIVITY

INTRODUCTION

The phenomenon of superconductivity has always been very exciting, both for fundamental scientific interest and because of its many applications.

Classical physics cannot explain the behavior and properties of superconductors. In fact, the superconducting state is now known to be a special quantum condensation of electrons. This quantum behavior has been verified through such observations as the quantization of magnetic flux produced by a superconducting ring.

In this chapter, we first give a brief historical review of superconductivity, beginning with its discovery in 1911 and ending with recent developments in high-temperature superconductivity. In describing some of the electromagnetic properties displayed by superconductors, we use simple physical arguments whenever possible. The essential features of the theory of superconductivity are reviewed with the realization that a detailed study is beyond the scope of this text. We then discuss many of the important applications of superconductivity and speculate on potential future applications.

NOTES FROM SELECTED CHAPTER SECTIONS

44.1 Brief Historical Review

The temperature at which a material becomes superconducting is called the *critical temperature.*

When superconductors in the presence of a magnetic field are cooled below their critical temperature, *the magnetic flux is expelled from the interior of the superconductor.* This behavior is known as the *Meissner effect.*

44.2 Some Properties of Type I Superconductors

The dc resistance of a type I superconductor is zero below their critical temperature. Type I superconductors are characterized by low critical field values and for this reason cannot by used to construct high field magnets—superconducting magnets.

The magnetic flux in the interior of a superconductor cannot change in time. This is because the electric field in its interior must be zero.

A type I superconductor is therefore not only a perfect conductor, but also a perfect diamagnet; this is an essential property of the superconducting state.

44.3 Type II Superconductors

Type II superconductors are characterized by two critical magnetic fields. Below the lower critical field, they behave as type I superconductors. Above the upper critical field, the superconducting state is destroyed just as for type I materials. For magnetic fields between the values of the upper and lower critical field values, the materials are in a *mixed state* referred to as a *vortex state.*

Most type II superconductors are compounds formed from elements of the transition and actinide series. These materials are well suited for constructing high-field *superconducting magnets* which can sustain large magnetic fields with *no power consumption.*

44.4 Other Properties of Superconductors

Once a current is set up in a superconducting material, *persistent currents* will continue in the superconductor for long time intervals (on the order of a year) *without any applied voltage* and with *no measurable losses.*

The *coherence length* is the distance over which the electrons in a *Cooper pair* remain together. A superconductor will be type I if the coherence length is greater than the penetration depth.

44.5 Electronic Specific Heat

The *electronic specific heat* is the heat energy absorbed by the conduction electrons per unit increase in the temperature of the system. The temperature variation of the specific heat of a superconductor below T_c can be used to measure the energy gap of the material.

44.6 The BCS Theory

The central feature of the BCS theory is that two electrons in the superconductor are able to form a bound state called a *Cooper pair* if they experience an *attractive interaction.* The interaction between the two electrons occurs via the crystal lattice as it 'distorts' in the 'wake' of one of the electrons. The quantized lattice vibrations are called *phonons* and the electron-electron attractive force is called a *phonon mediated mechanism.*

The superconducting state is a state in which the electrons act collectively, rather than independently. All Cooper pairs are in the same quantum state and the system exhibits quantum effects on a *macroscopic* scale. The *condensed state of the Cooper pairs is represented by a single coherent wave function.*

44.7 Energy Gap Measurements

The energy gap in superconductors can be measured by *single particle tunneling* experiments and by experiments based on *absorption of electromagnetic radiation.*

44.9 High-Temperature Superconductivity

Recently, superconductivity has been found to occur in several materials at temperatures well above the boiling point of liquid nitrogen (77 K). The interest in these materials is due to the following factors:

- The metallic oxides are relatively easy to fabricate and, hence, can be investigated at smaller laboratories and universities.
- They have very high T_c values and very high upper critical magnetic fields, estimated to be greater than 100 K in several materials.
- Their properties and the mechanisms responsible for their superconducting behavior represent a great challenge to theoreticians.
- They may be of considerable technological importance for both current applications and their potential use in nitrogen-temperature superconducting electronics and large-scale applications such as energy generation and transport, and magnetic levitation for high-speed transportation.

EQUATIONS AND CONCEPTS

The superconducting state is destroyed when the material is in the presence of a magnetic field greater than the *critical magnetic field* B_c. The temperature dependence of B_c is given by Equation 44.1.

$$B_c(T) \cong B_c(0)\left[1-\left(\frac{T}{T_c}\right)^2\right] \quad (44.1)$$

When a material is in a superconducting state, its magnetization opposes the external magnetic field, and the magnetic susceptibility has its maximum negative value.

$$\mathbf{M} = -\frac{\mathbf{B}_{\text{ext}}}{\mu_0} = \chi \mathbf{B}_{\text{ext}} \quad (44.3)$$

The superconducting state has a lower free energy per unit volume than the normal state. This is the defining equation for the critical field.

$$E_s + \frac{B_c^2}{2\mu_0} = E_n \quad (44.4)$$

The *energy gap* E_g in a superconductor represents the energy needed to break up one Cooper pair. At T = 0 K, the energy gap is proportional to the critical temperature.

$$E_g = 3.53 k_B T_c \quad (44.7)$$

The magnetic flux trapped in a superconductor by a persistent current is quantized in units of the magnetic flux quantum Φ_0.

$$\Phi = \frac{nh}{2e} = n\Phi_0 \quad (44.5)$$

$$\Phi_0 = \frac{h}{2e} \quad (44.6)$$

When a dc voltage is applied across a Josephson junction, an *alternating current* is generated.

$$I = I_{max} \sin(\delta - 2\pi f t) \quad (44.9)$$

The frequency of the Josephson current is proportional to the applied dc voltage.

$$f = \frac{2eV}{h} \quad (44.10)$$

REVIEW CHECKLIST

▷ Be familiar with the historical development of the phenomenon of superconductivity, and the properties of high temperature superconductors which give rise to research interest in those materials.

▷ Understand and be able to explain the fundamental properties and characteristics of type I superconductors in terms of critical temperature, critical magnetic field, and penetration depth. Explain how materials which are type II superconductors differ from type I superconductors.

▷ Describe the general features of the BCS theory of superconductivity, including the concept of a Cooper pair.

▷ Understand Josephson tunneling as it applies to the dc Josephson effect, the ac Josephson effect, and quantum interference.

SOLUTIONS TO SELECTED END-OF-CHAPTER PROBLEMS

3. Determine the current generated in a superconducting ring of niobium metal 2.0 cm in diameter if a 0.020-T magnetic field directed perpendicular to the ring is suddenly decreased to zero. The inductance of the ring is $L = 3.1 \times 10^{-8}$ H.

Solution

From Faraday's law (Eq. 32.1), we find

$$|\mathcal{E}| = \frac{\Delta\Phi}{\Delta t} = A\frac{\Delta B}{\Delta t} = L\frac{\Delta I}{\Delta t} \qquad \text{or} \qquad \Delta I = \frac{A\,\Delta B}{L} = \frac{\pi(0.01\text{ m})^2(0.02\text{ T})}{3.1\times10^{-8}\text{ H}} = 200\text{ A} \quad \Diamond$$

---

7. *Persistent Currents.* In an experiment carried out by S.C. Collins between 1955 and 1958, a current was maintained in a superconducting lead ring for 2.5 years with no observed loss. If the inductance of the ring was 3.14×10^{-8} H and the sensitivity of the experiment was 1 part in 10^9, determine the maximum resistance of the ring. (*Hint:* Treat this as a decaying current in an *RL* circuit, and recall that $e^{-x} \cong 1 - x$ for small x.)

Solution If a current is set up in an *RL* circuit, the current decays with time according to Equation 32.10:

$$I = I_0 e^{-(R/L)t}$$

Hence, $\quad \dfrac{I}{I_0} = e^{-(R/L)t} \cong 1 - \dfrac{R}{L}t \qquad\qquad \dfrac{R}{L}t = 1 - \dfrac{I}{I_0} = \dfrac{I_0 - I}{I_0} = 10^{-9}$

Since $t = 2.5\text{ y} = 7.88 \times 10^7$ s, and $L = 3.14 \times 10^{-8}$ H,

$$R_{\max} = 10^{-9}\frac{L}{t} = \frac{10^{-9}(3.14\times10^{-8}\text{ H})}{7.88\times10^{7}\text{ s}} = 3.98\times10^{-25}\ \Omega \quad \Diamond$$

---

11. Calculate the energy gap for each superconductor in Table 44.1 as predicted by the BCS theory. Compare your answers with the experimental values given in Table 44.3.

Solution For aluminum, Table 44.1 gives critical temperature $T_c = 1.196$ K. Then the BCS theory predicts energy gap

$$E_g = 3.53k_BT_c = 3.53(1.38\times10^{-23}\text{ J/K})(1.196\text{ K})(1\text{ eV}/1.6\times10^{-19}\text{ J}) = 0.364\text{ meV} \quad \lozenge$$

differing from the value in Table 44.3 by 7%.

We find the others similarly:

Superconductor	T_c (K)	E_g BCS, meV	E_g exptl, meV
Al	1.196	0.364	0.34
Ga	1.083	0.330	0.33
Hg	4.153	1.26	1.65
In	3.408	1.04	1.05
Nb	9.26	2.82	3.05
Pb	7.193	2.19	2.73
Sn	3.722	1.13	1.15
Ta	4.47	1.36	1.4
Ti	0.39	0.119	
V	5.30	1.61	
W	0.015	0.0046	
Zn	0.85	0.26	0.24

We have good agreement overall.

15. *Cooper Pairs.* A Cooper pair in a type I superconductor has an average particle separation of approximately 1.0 × 10^{-4} cm. If these two electrons can interact within a volume of this diameter, how many other Cooper pairs have their centers within the volume occupied by one pair? Use the appropriate data for lead, which has $n_s = 2.0 \times 10^{22}$ electrons/cm^3.

Solution Each pair will occupy volume of

$$V_{\text{pair}} = \tfrac{4}{3}\pi R^3 = \tfrac{4}{3}\pi(5\times10^{-5}\text{ cm})^3 \cong 5.2\times10^{-13}\text{ cm}^3$$

Since there are 2 × 10^{22} electrons per cm^3 of conduction electrons, each cm^3 will contain 1 × 10^{22} pairs of electrons, so

$$N = \left(1\times10^{22}\ \frac{\text{pairs}}{\text{cm}^3}\right)(5.2\times10^{-13}\text{ cm}^3) = 5.2\times10^{9}\text{ pairs} \quad \Diamond$$

That is, when the material is in the superconducting state, each Cooper pair effectively moves with 5 billion other pairs.

---

19. If a magnetic flux of 1.0 × 10^{-4} Φ_0 ($\frac{1}{10000}$ of the flux quantum) can be measured with a device called a SQUID (Fig. P44.19), what is the smallest magnetic field change ΔB that can be detected with this device for a ring having a radius of 2.0 mm?

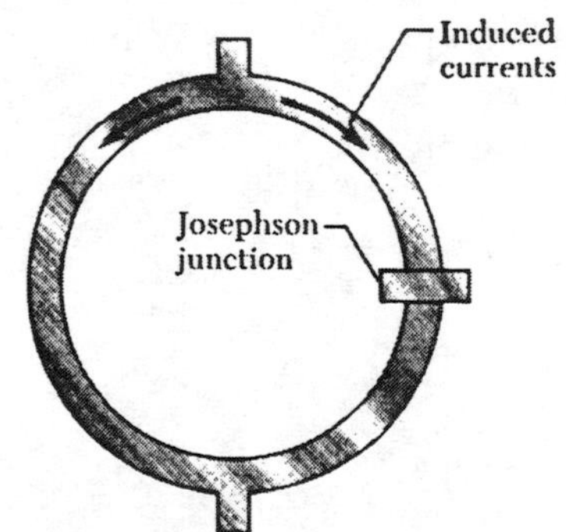

Figure P44.19

Solution The flux quantum is given by Equation 44.6:

$$\Phi_0 = \frac{h}{2e} = 2.07\times10^{-15}\text{ T}\cdot\text{m}^2$$

Since $\Phi = BA$, the smallest field change that can be detected is

$$\Delta B = \frac{\Delta\Phi}{A} = \frac{10^{-4}\ \Phi_0}{\text{A}} = \frac{2.07\times10^{-19}\text{ T}\cdot\text{m}^2}{\pi(2\times10^{-3}\text{ m})^2} = 1.6\times10^{-14}\text{ T} \quad \Diamond$$

---

25. *"Floating" a Wire.* Is it possible to "float" a superconducting lead wire of radius 1.0 mm in the magnetic field of the Earth? Assume the horizontal component of the Earth's magnetic field is 4.0×10^{-5} T.

Solution

If the wire can "float," the downward force of gravity must be balanced by the upward magnetic force ILB as shown. That is, $ILB = Mg$.

|F| = ILB
⊗B
I
I
mg

Since $M = \rho V = \rho(\pi r^2 L)$,

we find $I = \dfrac{\rho \pi r^2 g}{B}$

For lead, $\rho = 1.135 \times 10^4 \text{ kg/m}^3$;

Taking $r = 10^{-3}$ m and $B = 4 \times 10^{-5}$ T

gives $$I = \frac{(1.135 \times 10^4 \text{ kg/m}^3)(\pi)(10^{-3} \text{ m})^2(9.8 \text{ m/s}^2)}{4 \times 10^{-5} \text{ T}} = 8.74 \text{ kA}$$

The magnetic field that this current would produce at the surface of the wire (from Ampère's law) is

$$B = \frac{\mu_0 I}{2\pi r}$$

Substituting and solving for B, $$B = \frac{(4\pi \times 10^{-7} \text{ N/A}\cdot\text{m}^2)(8740 \text{ A})}{2\pi(10^{-3} \text{ m})} = 1.8 \text{ T}$$

Since this exceeds the critical field for lead, $B_c = 0.080$ T,

The lead wire cannot be suspended. ◊

26. *Magnetic Field Inside a Wire.* A type II superconducting wire of radius R carries current uniformly distributed through its cross section. If the total current carried by the wire is I, show that the magnetic energy per unit length inside the wire is $\mu_0 I^2/16\pi$.

Solution Ampère's law is given by

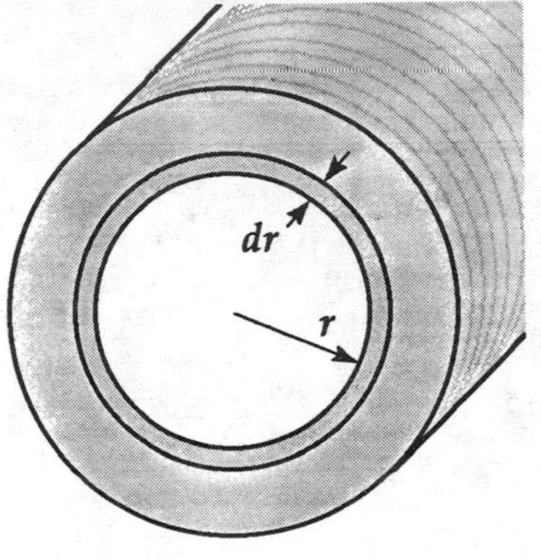

$$\oint \mathbf{B} \cdot d\mathbf{s} = \mu_0 I'$$

where I' is the current through the area surrounded by the path of integration. Taking the path to be a circle of radius $r < R$, we see that $I' < I$, where

$$I' = \left(\frac{\pi r^2}{\pi R^2}\right) I = \frac{r^2}{R^2} I \qquad (r < R)$$

Hence, $\oint \mathbf{B} \cdot d\mathbf{s} = B(2\pi r) = \mu_0 \dfrac{r^2}{R^2} I$ or $B = \left(\dfrac{\mu_0 I}{2\pi R^2}\right) r$

Since the energy per unit volume is $\dfrac{B^2}{2\mu_0}$, the total energy in a length L of the wire is

$$U = \int \frac{B^2}{2\mu_0} dV = \frac{1}{2\mu_0} \int_0^R \left(\frac{\mu_0 I r}{2\pi R^2}\right)^2 2\pi r L dr$$

Hence, the energy per unit length is

$$\frac{U}{L} = \frac{\mu_0 I^2}{4\pi R^4} \int_0^R r^3 dr = \frac{\mu_0 I^2}{16\pi} \qquad \Diamond$$

29. *Entropy Difference.* The entropy difference per unit volume between the normal and superconducting states is

$$\frac{\Delta S}{V} = -\frac{\partial}{\partial T}\left(\frac{B^2}{2\mu_0}\right)$$

where $B^2/2\mu_0$ is the magnetic energy per unit volume required to destroy superconductivity. Determine the entropy difference between the normal and superconducting states in 1.0 mol of lead at 4.0 K if the critical magnetic field $B_c(0) = 0.080$ T and $T_c = 7.2$ K.

Solution Since $B = B_c(0)\left[1-\left(\frac{T}{T_c}\right)^2\right]$, we get

$$\frac{\Delta S}{V} = -\frac{\partial}{\partial T}\left\{\frac{B_c^2(0)}{2\mu_0}\left[1-\left(\frac{T}{T_c}\right)^2\right]^2\right\} = -\frac{B_c^2(0)}{2\mu_0}\frac{\partial}{\partial T}\left[1-\left(\frac{T}{T_c}\right)^2\right]^2 = \frac{2B_c^2(0)}{\mu_0 T_c}\left[\frac{T}{T_c}-\left(\frac{T}{T_c}\right)^3\right]$$

At $T = 0$ K, $B_c(0) = 0.08$ T and $T_c = 7.2$ K for lead, so

$$\frac{\Delta S}{V} = \frac{2(0.08\text{ T})^2}{(4\pi\times10^{-7})(7.2)}\left[\frac{4\text{ K}}{7.2\text{ K}}-\left(\frac{4\text{ K}}{7.2\text{ K}}\right)^3\right] = 543\text{ J/m}^3\cdot\text{K}$$

Since lead has an atomic weight of 206 and a density of $\rho = 11.35$ g/cm^3, one mol occupies a volume of $\frac{206}{11.35} = 18.1\text{ cm}^3 = 18.1\times10^{-6}\text{ m}^3$. Therefore,

$$\Delta S = (543\text{ J/m}^3\cdot\text{K})(18.1\times10^{-6}\text{ m}^3\text{/mol}) = 9.8\times10^{-3}\text{ J/mol}\cdot\text{K} \quad \Diamond$$

Note that the superconducting state has lower entropy than the normal state.

Chapter 45

Nuclear Structure

Chapter 45

NUCLEAR STRUCTURE

INTRODUCTION

In this chapter we discuss the properties and structure of the atomic nucleus. We start by describing the basic properties of nuclei, and then discuss nuclear forces and binding energy, nuclear models, and the phenomenon of radioactivity. We also discuss nuclear reactions and the various processes by which nuclei decay.

NOTES FROM SELECTED CHAPTER SECTIONS

45.1 Some Properties of Nuclei

Important quantities in the description of nuclear properties are:

- The *atomic number*, Z, which equals the number of protons in the nucleus.
- The *neutron number*, N, which equals the number of neutrons in the nucleus.
- The *mass number*, A, which equals the number of nucleons (neutrons plus protons) in the nucleus.

The nuclei of all atoms of a particular element contain the same number of protons but often contain different numbers of neutrons. Nuclei that are related in this way are called isotopes. The isotopes of an element have the same Z value but different N and A values.

The *atomic mass unit*, *u*, is defined such that the mass of one atom of the isotope ^{12}C is exactly 12 u.

Experiments have shown that most nuclei are approximately spherical and all *have nearly the same density*. The stability of nuclei is due to the *nuclear force*. This is a *short range, attractive* force which acts between all nuclear particles. Nuclei have *intrinsic angular momentum* which is quantized by the *nuclear spin quantum number* which may be integer or half integer.

The *magnetic moment of the nucleus* is measured in terms of the *nuclear magneton*. When placed in an external magnetic field, nuclear magnetic moments precess with a frequency called the *Larmor precessional frequency*.

45.3 Binding Energy and Nuclear Forces

The total mass of a nucleus is always less than the sum of the masses of its individual nucleons. The *binding energy* of the nucleus is mass difference multiplied by c^2.

45.4 Nuclear Models

The *liquid-drop* model, proposed by Bohr in 1936, treats the nucleons as if they were molecules in a drop of liquid. The nucleons interact strongly with each other and undergo frequent collisions as they "jiggle around" within the nucleus. This is analogous to the thermally agitated motion of molecules in a liquid. The three major effects which influence the binding energy according to the liquid-drop model are the volume effect, the surface effect, and the Coulomb effect.

The *independent-particle model*, often called the *shell model*, is based upon the assumption that *each nucleon moves in a well-defined orbit within the nucleus and in an averaged field produced by the other nucleons.* In this model, the nucleons exist in quantized energy states and there are few collisions between nucleons.

A third model of nuclear structure, known as the *collective model*, combines some features of both the liquid-drop model and the independent-particle model. The nucleus is considered to have some "extra" nucleons moving in quantized orbits, in addition to the filled core of nucleons.

45.5 Radioactivity

There are three processes by which a radioactive substance can undergo decay:

- alpha (α) decay, where the emitted particles are ^{4}He nuclei,
- beta (β) decay, in which the emitted particles are either electrons or positrons,
- gamma (γ) decay, in which the emitted "rays" are high-energy photons.

A positron is a particle similar to the electron in all respects except that it has a charge of $+e$, and is the antimatter twin of the electron. In Beta decay, the electron is denoted by a β^-, while the positron is designated with a β^+.

The three types of radiation have quite different penetrating powers. Alpha particles barely penetrate a sheet of paper, beta particles can penetrate a few millimeters of aluminum, and gamma rays can penetrate several centimeters of lead.

45.6 The Decay Processes

Alpha decay can occur because, according to quantum mechanics, some nuclei have barriers that can be penetrated by the alpha particles (the tunneling process). Beta decay is energetically more favorable for those nuclei having a large excess of neutrons. A nucleus can undergo beta decay in two ways. It can emit either an electron (β^-) and an antineutrino ($\overline{\nu}$), or a positron (β^+) and a neutrino (ν). In the electron-capture process, the nucleus of an atom absorbs one of its own electrons (usually from the K shell) and emits a neutrino.

The neutrino has the following properties:

- It has zero electric charge.
- It has a rest mass smaller than that of the electron, and in fact its rest mass may be zero (although recent experiments suggest that this may not be true).
- It has a spin of 1/2, which satisfies the law of conservation of angular momentum.
- It interacts very weakly with matter and is therefore very difficult to detect.

In gamma decay, a nucleus in an excited state decays to its ground state and emits a gamma ray. The value (disintegration energy) is the energy released as a result of the decay process.

45.7 Natural Radioactivity

Radioactive nuclei are generally classified into two groups: (1) unstable nuclei found in nature, which give rise to what is called *natural radioactivity*, and (2) nuclei produced in the laboratory through nuclear reactions, which exhibit *artificial radioactivity.*

45.8 Nuclear Reactions

Nuclear reactions are events in which collisions change the identity or properties of nuclei. The total energy released as a result of a nuclear reaction is called the *reaction energy, Q*.

An *endothermic reaction* is one in which Q is negative, and the minimum energy for which the reaction will occur is called the *threshold energy.*

EQUATIONS AND CONCEPTS

Most nuclei are approximately spherical in shape and have an average radius which is proportional to the cube root of the mass number, or total number of nucleons. This means that the volume is proportional to A and that all nuclei have nearly the same density.

$$r = r_0 A^{1/3} \tag{45.1}$$

The nucleus has an angular momentum and a corresponding nuclear magnetic moment associated with it. The nuclear magnetic moment is measured in terms of a unit of moment called the nuclear magneton μ_n.

$$\mu_n \equiv \frac{e\hbar}{2m_p} = 5.05 \times 10^{-27}\ \text{J/T} \tag{45.3}$$

The binding energy of any nucleus can be calculated in terms of the mass of a neutral hydrogen atom, the mass of a neutron, and the atomic mass of the associated compound nucleus.

$$E_b(\text{MeV}) = \left(Zm_p + Nm_n - M_A\right) \times 931\ \text{MeV/u} \tag{45.4}$$

The semiempirical binding energy formula is based on the liquid-drop model of the nucleus.

$$E_b = C_1 A - C_2 A^{2/3} - C_3 \frac{Z(Z-1)}{A^{1/3}} - C_4 \frac{(N-Z)^2}{A} \tag{45.5}$$

The number of radioactive nuclei in a given sample which undergoes decay during a time interval Δt depends on the number of nuclei present. The number of decays depends also on the decay constant λ which is characteristic of a particular isotope.

$$\frac{dN}{dt} = -\lambda N \tag{45.6}$$

The number of nuclei in a radioactive sample decreases exponentially with time. The plot of number of nuclei N versus elapsed time t is called a decay curve.

$$N = N_0 e^{-\lambda t} \tag{45.7}$$

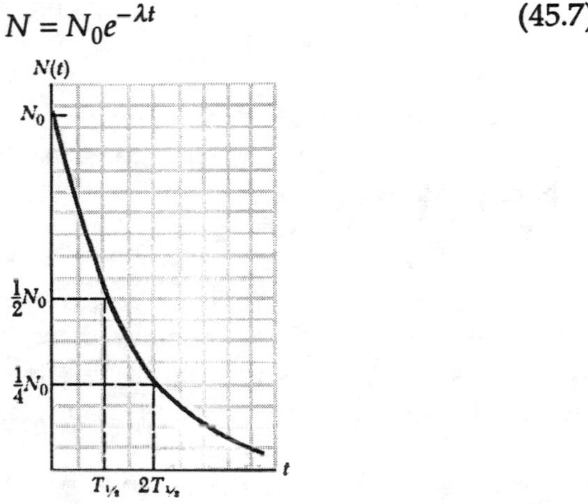

The decay rate R or activity of a sample of radioactive nuclei is defined as the number of decays per second.

$$R = \left|\frac{dN}{dt}\right| = R_0 e^{-\lambda t} \tag{45.8}$$

The half-life $T_{1/2}$ is the time required for half of a given number of radioactive nuclei to decay.

$$T_{1/2} = \frac{0.693}{\lambda} \tag{45.9}$$

When a nucleus decays by *alpha emission*, the parent nucleus loses two neutrons and two protons. In order for alpha emission to occur, the mass of the parent nucleus must be greater than the combined mass of the daughter nucleus and the emitted alpha particle. The mass difference is converted into energy and appears as kinetic energy shared (unequally) by the alpha particle and the daughter nucleus.

$$ {}^{A}_{Z}X \rightarrow {}^{A-4}_{Z-2}Y + {}^{4}_{2}He \qquad (45.10)$$

$$ {}^{238}_{92}U \rightarrow {}^{234}_{90}Th + {}^{4}_{2}He \qquad (45.11)$$

The disintegration energy Q can be calculated in MeV when the masses are expressed in u.

$$Q = (M_X - M_Y - M_\alpha)(931.494 \text{ MeV/u}) \qquad (45.14)$$

When a radioactive nucleus undergoes *beta decay*, the daughter nucleus has the same mass number as the parent nucleus, but the charge number (or atomic number) increases by one. The electron that is emitted is created within the parent nucleus by a process which can be represented by a neutron transformed into a proton and an electron. The total energy released in beta decay is greater than the combined kinetic energies of the electron and the daughter nucleus. This difference in energy is associated with a third particle called a neutrino.

$$ {}^{A}_{Z}X \rightarrow {}^{A}_{Z+1}Y + \beta^- \qquad (45.15)$$

$$ n \rightarrow p + \beta^- + \bar{\nu} \qquad (45.19)$$

Nuclei which undergo alpha or beta decay are often left in an excited energy state. The nucleus returns to the ground state by emission of one or more photons. Gamma-ray emission results in no change in mass number or atomic number.

$$ {}^{A}_{Z}X^* \rightarrow {}^{A}_{Z}X + \gamma \qquad (45.21)$$

Nuclear reactions can occur when target nuclei are bombarded with energetic particles. In these reactions the structure, identity, or properties of the target nuclei are changed.

$$a + X \rightarrow Y + b \quad (45.24)$$

The quantity of energy required to balance the equation representing a nuclear reaction (e.g. Eq. 44.24) is called the Q value of the reaction. The Q value can be calculated in terms of the total mass of the reactants minus the total mass of the products or as the kinetic energy of the reactants. Q is positive in the case of exothermic reactions and negative for endothermic reactions.

$$Q = (M_a + M_X - M_Y - M_b)c^2 \quad (45.25)$$

TABLE 44.3 Various Decay Pathways

Alpha Decay	${}^{A}_{Z}X \rightarrow {}^{A-4}_{Z-2}X + {}^{4}_{2}He$
Beta Decay (β^-)	${}^{A}_{Z}X \rightarrow {}_{Z+1}^{A}X + \beta^- + \overline{\nu}$
Beta Decay (β^+)	${}^{A}_{Z}X \rightarrow {}_{Z-1}^{A}X + \beta^+ + \nu$
Electron Capture	${}^{A}_{Z}X + {}^{0}_{-1}e \rightarrow {}_{Z-1}^{A}X + \nu$
Gamma Decay	${}^{A}_{Z}X^* \rightarrow {}^{A}_{Z}X + \gamma$

SUGGESTIONS, SKILLS, AND STRATEGIES

The rest energy of a particle is given by $E = mc^2$. It is therefore often convenient to express the unified mass unit in terms of its equivalent energy, 1 u = 1.660559×10^{-27} kg or 1 u = 931.50 MeV/c^2. When masses are expressed in units of u, energy values are then $E = m(931.50 \text{ MeV/u})$.

Equation 45.7 can be solved for the particular time t after which the number of remaining nuclei will be some specified fraction of the original number N_0. This can be done by taking the natural log of each side of Equation 45.7 to find

$$t = \frac{1}{\lambda}\ln\left(\frac{N_0}{N}\right)$$

REVIEW CHECKLIST

▷ Use the appropriate nomenclature in describing the static properties of nuclei.

▷ Discuss nuclear stability in terms of the strong nuclear force and a plot of N vs. Z.

▷ Account for nuclear binding energy in terms of the Einstein mass-energy relationship. Describe the basis for energy released by fission and fusion in terms of the shape of the curve of binding energy per nucleon vs. mass number.

▷ Identify each of the components of radiation that are emitted by the nucleus through natural radioactive decay and describe the basic properties of each. Write out typical equations to illustrate the processes of transmutation by alpha and beta decay and explain why the neutrino must be considered in the analysis of beta decay.

▷ State and apply to the solution of related problems, the formula which expresses decay rate as a function of the decay constant and the number of radioactive nuclei. Describe the process of carbon dating as a means of determining the age of ancient objects.

▷ Calculate the Q value of given nuclear reactions and determine the threshold energy of endothermic reactions.

SOLUTIONS TO SELECTED END-OF-CHAPTER PROBLEMS

9. Certain stars at the end of their lives are thought to collapse, combining their protons and electrons to form a neutron star. Such a star could be thought of as a gigantic atomic nucleus. If a star of mass equal to that of the Sun ($M = 1.99 \times 10^{30}$ kg) collapsed into neutrons ($m_n = 1.67 \times 10^{-27}$ kg), what would the radius be? (*Hint:* $r = r_0 A^{1/3}$.)

Solution The number of nucleons in a star of one solar mass is

$$A = \frac{1.99 \times 10^{30}\ \text{kg}}{1.67 \times 10^{-27}\ \text{kg}} = 1.20 \times 10^{57}$$

Therefore, $$r = r_0 A^{1/3} = (1.2 \times 10^{-15}\ \text{m})\sqrt[3]{1.20 \times 10^{57}} = 12.7\ \text{km} \quad \Diamond$$

13. (a) Use energy methods to calculate the distance of closest approach for a head-on collision between an alpha particle having an initial energy of 0.50 MeV and a gold nucleus (^{197}Au) at rest. (Assume the gold nucleus remains at rest during the collision.) (b) What minimum initial speed must the alpha particle have in order to get as close as 300 fm?

Solution

(a) The initial kinetic energy of the alpha particle must equal the electrostatic potential energy of the two-particle system at the distance of closest approach, r_{min}. That is,

$$K_\alpha = U = k_e \frac{qQ}{r_{min}}$$

or $$r_{min} = k_e \frac{qQ}{K_\alpha} = \frac{(9.0\times10^9\ \mathrm{N\cdot m^2/C^2})(2)(79)(1.6\times10^{-19}\ \mathrm{C})^2}{(0.50\ \mathrm{MeV})(1.6\times10^{-13}\ \mathrm{J/MeV})} = 4.55\times10^{-13}\ \mathrm{m} \quad \Diamond$$

(b) Since $K_\alpha = \frac{1}{2}mv^2 = k_e \frac{qQ}{r_{min}}$, we find

$$v = \sqrt{\frac{2k_e qQ}{m r_{min}}} = \sqrt{\frac{2(9.0\times10^9\ \mathrm{N\cdot m^2/C^2})(2)(79)(1.6\times10^{-19}\ \mathrm{C})^2}{4(1.67\times10^{-27}\ \mathrm{kg})(3.00\times10^{-13}\ \mathrm{m})}} = 6.0\times10^6\ \mathrm{m/s} \quad \Diamond$$

---

21. Calculate the minimum energy required to remove a neutron from the $^{43}_{20}\mathrm{Ca}$ nucleus.

Solution Removal of a neutron from $^{43}_{20}\mathrm{Ca}$ would result in the residual nucleus $^{42}_{20}\mathrm{Ca}$. If the required separation energy is S_n, the overall process can be described by

$$\mathrm{mass}\left(^{43}_{20}\mathrm{Ca}\right) + S_n = \mathrm{mass}\left(^{42}_{20}\mathrm{Ca}\right) + \mathrm{mass(n)}$$

or $$S_n = (41.95863 + 1.008665 - 42.958770)\ \mathrm{u}$$

$$S_n = (0.008525\ \mathrm{u})(931.5\ \mathrm{MeV/u}) = 7.94\ \mathrm{MeV} \quad \Diamond$$

---

22. A pair of nuclei for which $Z_1 = N_2$ and $Z_2 = N_1$ are called mirror isobars (the atomic and neutron numbers are interchangeable). Binding energy measurements on these nuclei can be used to obtain evidence of the charge independence of nuclear forces (that is, proton-proton, proton-neutron, and neutron-neutron forces are approximately equal). Calculate the difference in binding energy for the two mirror isobars $^{15}_{8}O$ and $^{15}_{7}N$.

Solution For $^{15}_{8}O$ we have (using Equation 45.4)

$$E_b = [8(1.007825)\text{ u} + 7(1.008665)\text{ u} - (15.003065)\text{ u}](931.5\text{ MeV/u}) = 111.96\text{ MeV}$$

For $^{15}_{7}N$ we have

$$E_b = [7(1.007825)\text{ u} + 8(1.008665)\text{ u} - (15.000109)\text{ u}](931.5\text{ MeV/u}) = 115.49\text{ MeV}$$

Therefore, the difference in the two binding energies is $\Delta E_b = 3.53\text{ MeV}$ ◊

---

23. The isotope $^{139}_{57}La$ is stable. A radioactive isobar (see Problem 19) of this lanthanum isotope, $^{139}_{59}Pr$, is located above the line of stable nuclei in Figure 45.3 and decays by β^+ emission. Another radioactive isobar of ^{139}La, $^{139}_{55}Cs$ decays by β^- emission and is located below the line of stable nuclei in Figure 45.3. (a) Which of these three isobars has the highest neutron-to-proton ratio? (b) Which has the greatest binding energy per nucleon? (c) Which do you expect to be heavier, ^{139}Pr or ^{139}Cs?

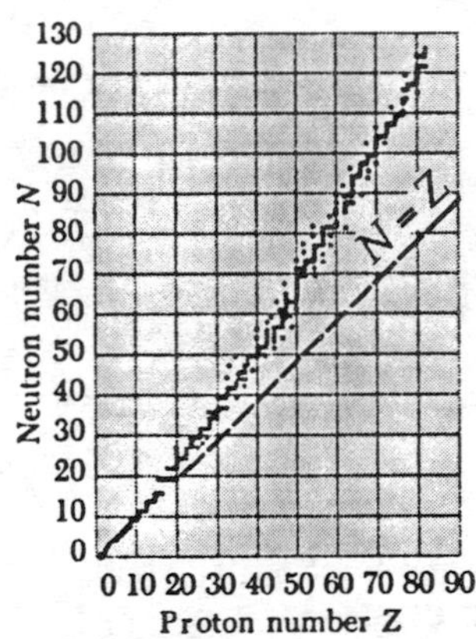

Figure 45.3

Solution

(a) For $^{139}_{59}Pr$ the neutron number is 139 – 59 = 80. For $^{139}_{55}Cs$ the neutron number is 84, so this cesium isotope has the greatest neutron-to-proton ratio. ◊

(b) Binding energy per nucleon measures stability so it is greatest for the stable nucleus, the lanthanum isotope. Note also that it has a magic number of neutrons, 82. ◊

(c) Cs–139 has 55 protons and 84 neutrons. Pr–139 has 59 protons and 80 neutrons. Since neutrons are heavier than protons (by about 1.3 MeV), we would expect that the more neutron-rich nucleus Cs–139 would be heavier. ◊

---

25. Using the graph in Figure 45.9, estimate how much energy is released when a nucleus of mass number 200 is split into two nuclei each of mass number 100.

Solution The curve of binding energy shows that a heavy nucleus of mass number $A = 200$ has total binding energy about

$$(7.4 \text{ MeV/nucleon})(200 \text{ nucleons}) = 1.5 \text{ GeV}.$$

Thus, it is less stable than its potential fission products, two middleweight nuclei of $A = 100$, having total binding energy

$$2(8.4 \text{ MeV/nucleon})\ 100 \text{ nucleons} = 1.7 \text{ GeV}.$$

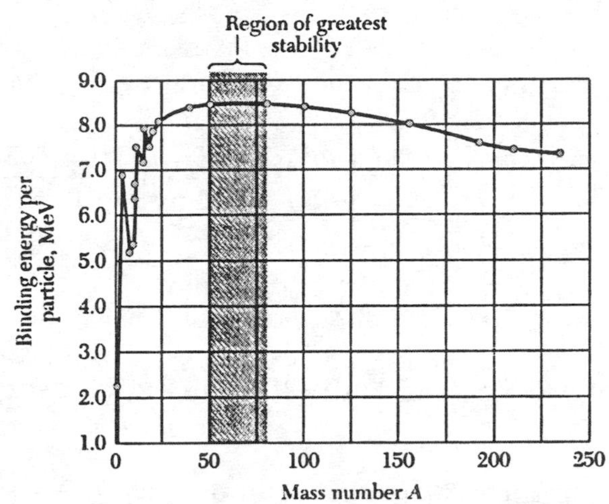

Figure 45.9

The energy released in the fission can be about 1.7 GeV − 1.5 GeV = 200 MeV. ◊

This is the energy source of uranium bombs and of nuclear electric-generating plants.

---

29. A sample of radioactive material contains 10^{15} atoms and has an activity of 6.00×10^{11} Bq. What is its half-life?

Solution $\dfrac{dN}{dt} = -\lambda N$

$$\lambda = \frac{1}{N}\left(-\frac{dN}{dt}\right) = 10^{-15}(6.00 \times 10^{11} \text{ s}^{-1}) = 6.00 \times 10^{-4} \text{ s}^{-1}$$

$$T_{1/2} = \frac{\ln 2}{\lambda} = 1160 \text{ s} \quad \Diamond \quad (19.3 \text{ min})$$

---

31. A freshly prepared sample of a certain radioactive isotope has an activity of 10 mCi. After 4.0 h, its activity is 8.0 mCi. (a) Find the decay constant and half-life. (b) How many atoms of the isotope were contained in the freshly prepared sample? (c) What is the sample's activity 30 h after it is prepared?

Solution From the rate equation, $R = R_0 e^{-\lambda t}$, we can solve for λ to get

(a) $$\lambda = \frac{1}{t}\ln\left(\frac{R_0}{R}\right) = \left(\frac{1}{4.0\text{ h}}\right)\ln\left(\frac{10\text{ mCi}}{8.0\text{ mCi}}\right) = 5.58\times10^{-2}\text{ h}^{-1} = 1.6\times10^{-5}\text{ s}^{-1} \quad \Diamond$$

$$T_{1/2} = \frac{\ln 2}{\lambda} = 12\text{ h} \quad \Diamond$$

(b) $R_0 = 10\text{ mCi} = 10\times10^{-3}\times(3.7\times10^{10}\text{ decays/s}) = 3.7\times10^{8}\text{ decays/s}$

Since $R_0 = \lambda N_0$, we find $$N_0 = \frac{R_0}{\lambda} = \frac{3.7\times10^{8}\text{ decays / s}}{1.55\times10^{-5}\text{ s}^{-1}} = 2.4\times10^{13}\text{ atoms} \quad \Diamond$$

(c) $$R = R_0 e^{-\lambda t} = (10\text{ mCi})e^{-(5.58\times10^{-2}\text{ h}^{-1})(30\text{ h})} = 1.9\text{ mCi} \quad \Diamond$$

---

37. The radioactive isotope ^{198}Au has a half-life of 64.8 h. A sample containing this isotope has an initial activity (t = 0) of 40.0 μCi. Calculate the number of nuclei that decay in the time interval between t_1 = 10.0 h and t_2 = 12.0 h.

Solution First, let us find λ and N_0 from the given information:

$$\lambda = \frac{\ln 2}{T_{1/2}} = \frac{0.693}{64.8\text{ h}} = 0.0107\text{ h}^{-1} = 2.97\times10^{-6}\text{ s}^{-1}$$

$$N_0 = \frac{R_0}{\lambda} = \frac{(40\ \mu\text{Ci})(3.7\times10^{4}\text{ decays/s/}\mu\text{Ci})}{2.97\times10^{-6}\text{ s}^{-1}} = 4.98\times10^{11}\text{ nuclei}$$

Since $N = N_0 e^{-\lambda t}$, the number of nuclei which decay between times t_1 and t_2 is

$$N_1 - N_2 = N_0(e^{-\lambda t_1} - e^{-\lambda t_2})$$

Substituting in the values for λ and N_0 gives

$$N_1 - N_2 = (4.98\times10^{11})\left[e^{-(0.0107\text{ h}^{-1})(10\text{ h})} - e^{-(0.0107\text{ h}^{-1})(12\text{ h})}\right] = 9.46\times10^{9}\text{ nuclei} \quad \Diamond$$

---

39. Find the energy released in the alpha decay

$$^{238}_{92}\text{U} \rightarrow {}^{234}_{90}\text{Th} + {}^{4}_{2}\text{He}$$

You will find the following mass values useful:

$$M\left(^{238}_{92}\text{U}\right) = 238.050\ 786\ \text{u}$$

$$M\left(^{234}_{90}\text{Th}\right) = 234.043\ 583\ \text{u}$$

$$M\left(^{4}_{2}\text{He}\right) = 4.002\ 603\ \text{u}$$

Solution $Q = (M_u - M_{Th} - M_{He})(931.5\ \text{MeV/u})$

$$Q = (238.050786 - 234.043583 - 4.002603)(931.5) = 4.28\ \text{MeV} \quad \lozenge$$

---

45. Enter the correct isotope symbol in each open square in Figure P45.45.

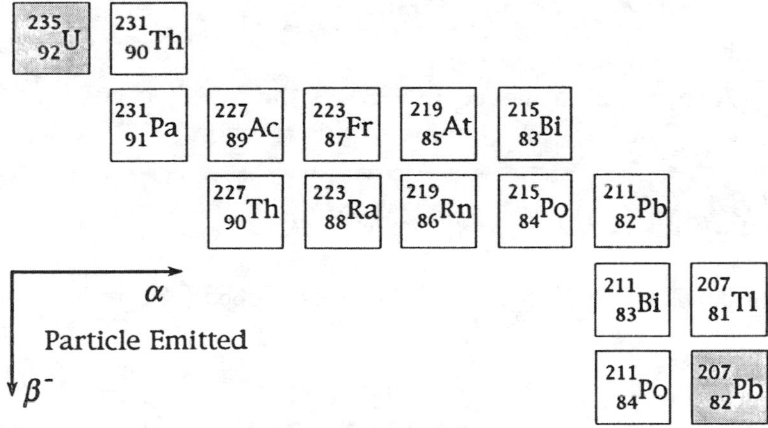

Figure P45.45 (modified)

Solution Whenever an $\alpha = {}^{4}_{2}\text{He}$ is emitted, Z drops by 2 and A by 4. Whenever a $\beta^- = {}^{0}_{-1}\beta^-$ is emitted, Z increases by 1 and A is unchanged. We find the chemical name by looking up Z in a periodic table. The values in the shaded boxes (^{235}U and ^{207}Pb) were given; all others have been filled in as part of the solution.

---

47. The nucleus ${}^{15}_{8}O$ decays by electron capture. Write (a) the basic nuclear process and (b) the decay process. (c) Determine the energy of the neutrino. Disregard the daughter's recoil.

Solution

(a) $e^- + p \to n + \nu$ ◊

(b) Add 7 protons, 7 neutrons, and 7 electrons to each side to give ${}^{15}O \to {}^{15}N + \nu$ ◊

(c) From Table A.3,

$$m({}^{15}O) = m({}^{15}N) + Q/c^2$$

$$\Delta m = 15.003065 - 15.000109 = 0.002956 \text{ u}$$

$$Q = (931.5 \text{ MeV/u})(0.002956 \text{ u}) = 2.75 \text{ MeV} \quad ◊$$

---

49. The reaction ${}^{27}_{13}Al(\alpha, n){}^{30}_{15}P$, achieved in 1934, is the first known in which the product nucleus is radioactive. Calculate the Q value of this reaction.

Solution $Q = \left[M_\alpha + M({}^{27}Al) - M({}^{30}P) - M_n\right](931.5)$

$$Q = (4.002603 + 26.981541 - 29.078310 - 1.008665)(931.5)$$

$$Q = -2.64 \text{ MeV} \quad ◊$$

The reaction is endoenergetic; you must put in 2.64 MeV to make it happen.

---

52. Natural gold has only one isotope, ${}^{197}_{79}\text{Au}$. If natural gold is irradiated by a flux of slow neutrons, β^- particles are emitted. (a) Write the appropriate reaction equations. (b) Calculate the maximum energy of the emitted beta particles. The mass of ${}^{198}_{80}\text{Hg}$ is 197.96675 u.

Solution

(a) The Au–197 will absorb a neutron to become ${}^{198}_{79}\text{Au}$ which emits a β^- to become ${}^{198}_{80}\text{Hg}$. The reaction for nuclei is ${}^{197}_{79}\text{Au}$ nucleus + ${}^{1}_{0}\text{n} \to {}^{198}_{80}\text{Hg}$ nucleus + ${}^{0}_{-1}\beta + \bar{\nu}$ Add 79 electrons to both sides to have the reaction for neutral atoms.

$$ {}^{197}_{79}\text{Au} + {}^{1}_{0}\text{n} \to {}^{198}_{80}\text{Hg} + \bar{\nu} \quad \Diamond $$

(b) From Table A.3,

$$ 196.96656 + 1.008665 = 197.96675 + 0 + Q/c^2 $$

$$ Q = \Delta mc^2 = (0.008475 \text{ u})(931.5 \text{ MeV/u}) $$

$$ Q = 7.89 \text{ MeV} \quad \Diamond $$

---

53. Using the appropriate reactions and Q values from Table 45.5, calculate the mass of ${}^{8}\text{Be}$ and ${}^{10}\text{Be}$ in atomic mass units to four decimal places.

Solution $\quad {}^{9}\text{Be} + \text{n} \to {}^{10}\text{Be} + 6.810 \text{ MeV}$

$$ m({}^{10}\text{Be}) = m({}^{9}\text{Be}) + m_n - \frac{6.810 \text{ MeV}}{931.5 \text{ MeV/u}} = 9.01218 + 1.008665 - 0.007311 = 10.0135 \text{ u} \quad \Diamond $$

$$ {}^{9}\text{Be} + 1.666 \text{ MeV} \to {}^{8}\text{Be} + \text{n} $$

$$ m({}^{8}\text{Be}) = m({}^{9}\text{Be}) - m_n + \frac{1.666 \text{ MeV}}{931.5 \text{ MeV/u}} = 9.01218 - 1.008665 + 0.001789 = 8.0053 \text{ u} \quad \Diamond $$

---

57. (a) One method of producing neutrons for experimental use bombards light nuclei with alpha particles. In one particular arrangement, alpha particles emitted by polonium are incident on beryllium nuclei:

$$^4_2\text{He} + {}^9_4\text{Be} \rightarrow {}^{12}_6\text{C} + {}^1_0\text{n}$$

What is the Q value? (b) Neutrons are also often produced by small-particle accelerators. In one design, deuterons accelerated in a Van de Graaf generator bombard other deuterium nuclei:

$$^2_1\text{H} + {}^2_1\text{H} \rightarrow {}^3_2\text{He} + {}^1_0\text{n}$$

Is this reaction exothermic or endothermic? Calculate its Q value.

Solution

(a) $Q = \left[m(^9\text{Be}) + m(^4\text{He}) - m(^{12}\text{C}) - m(^1\text{n})\right](931.5 \text{ MeV/u})$

$Q = [(9.01292 + 4.002603 - 12.0000 - 1.008665) \text{ u}](931.5 \text{ MeV / u}) = 5.70 \text{ MeV}$ ◊

(b) $Q = \left[2m(^2\text{H}) - m(^3\text{He}) - m(^1\text{n})\right](931.5 \text{ MeV/u})$

$Q = [2(2.014102) - 3.016029 - 1.008665 \text{ u}](931.5 \text{ MeV / u}) = 3.27 \text{ MeV}$ ◊

The reaction is exothermic since Q is positive.

---

59. A by-product of some fission reactors is the isotope $^{239}_{94}\text{Pu}$, an alpha emitter having a half-life of 24 000 years:

$$^{239}_{94}\text{Pu} \rightarrow {}^{235}_{92}\text{U} + \alpha$$

Consider a sample of 1 kg of pure $^{239}_{94}\text{Pu}$ at $t = 0$. Calculate (a) the number of $^{239}_{94}\text{Pu}$ nuclei present at $t = 0$ and (b) the initial activity in the sample. (c) How long does the sample have to be stored if a "safe" activity level is 0.1 Bq?

Solution

(a) $\text{Number of nuclei} = \dfrac{\text{mass parent}}{\text{mass of 1 nucleus}} = \dfrac{1\text{ kg}}{(239\text{ u})(1.66\times10^{-27}\text{ kg/u})} = 2.52\times10^{24}$ ◊

(b) $\lambda = \dfrac{0.693}{T_{1/2}} = \dfrac{0.693}{(2.4\times10^{4}\text{ y})(3.156\times10^{7}\text{ s/y})} = 9.149\times10^{-13}\text{ s}^{-1}$ ◊

$R_0 = \lambda N_0 = (9.149\times10^{-13}\text{ s}^{-1})(2.52\times10^{24}) = 2.306\times10^{12}\text{ decays/s}$ ◊ $= 2.306\times10^{12}\text{ Bq}$

(c) $R = R_0 e^{-\lambda t}$ or $e^{-\lambda t} = \dfrac{R}{R_0} = \dfrac{0.1\text{ Bq}}{2.306\times10^{12}\text{ Bq}} = 4.336\times10^{-14}$

$\lambda t = -\ln(4.336\times10^{-14}) = 30.77$ and

$$t = \frac{30.77}{\lambda} = \frac{30.77}{9.149\times10^{-13}\text{ s}^{-1}} = 3.363\times10^{13}\text{ s} = 1.07\times10^{6}\text{ y}$$

t = 1.07 million years ◊

---

61. A large nuclear power reactor produces about 3000 MW of thermal power in its core. Three months after a reactor is shut down, the core power from radioactive by-products is 10 MW. Assuming that each emission delivers 1.0 MeV of energy to the thermal power, estimate the activity in becquerels three months after the reactor is shut down.

Solution $P = 10\text{ MW} = 10^{7}\text{ J/s}$

If each decay delivers 1 MeV = 1.6×10^{-13} J, then

$$\text{Number of decays/s} = \frac{(10^{7}\text{ J/s})}{(1.6\times10^{-13}\text{ J/decay})} = 6.25\times10^{19}\text{ Bq}$$ ◊

---

65. (a) Find the radius of the ${}^{12}_{6}\text{C}$ nucleus. (b) Find the force of repulsion between a proton at the surface of a ${}^{12}_{6}\text{C}$ nucleus and the remaining five protons. (c) How much work (in MeV) has to be done to overcome this electrostatic repulsion in order to put the last proton into the nucleus? (d) Repeat parts (a), (b), and (c) for ${}^{238}_{92}\text{U}$.

Solution

(a) $R = R_0 A^{1/3} = 1.2 A^{1/3}$ fm

When A = 12,

$$R = 1.2(12)^{1/3} \text{ fm} = 2.75 \text{ fm} \quad \Diamond$$

(b) Since the proton interacts with Z – 1 other protons, where Z = 6 and R = 2.75 fm, we find

$$F = k_e \frac{(Z-1)e^2}{R^2} = \frac{(9\times10^9 \text{ N}\cdot\text{m}^2/\text{C}^2)(5)(1.6\times10^{-19}\text{ C})^2}{(2.75\times10^{-15}\text{ m})^2} = 152 \text{ N} \quad \Diamond$$

(c) $$U = k_e \frac{q_1 q_2}{R} = k_e \frac{(Z-1)e^2}{R}$$

Solving, $$U = \frac{(9\times10^9 \text{ N}\cdot\text{m}^2/\text{C}^2)(5)(1.6\times10^{-19}\text{ C})^2}{2.75\times10^{-15}\text{ m}}$$

thus $$U = 4.19\times10^{-13}\text{ J} = 2.62 \text{ MeV} \quad \Diamond$$

(d) For ${}^{238}_{92}\text{U}$, we take $A = 238$ and $Z = 92$ to find

$$R = 7.44 \text{ fm} \quad \Diamond,$$

$$F = 379 \text{ N} \quad \Diamond,$$

and $$U = 17.6 \text{ MeV} \quad \Diamond$$

69. Carbon detonations are powerful nuclear reactions that temporarily tear apart the cores of massive stars late in their lives. These blasts are produced by carbon fusion, which requires a temperature of about 6×10^8 K to overcome the strong Coulomb repulsion between carbon nuclei. (a) Estimate the repulsive energy barrier to fusion, using the required ignition temperature for carbon fusion. (In other words, what is the kinetic energy for a carbon nucleus at 6×10^8 K?) (b) Calculate the energy (in MeV) released in each of these "carbon-burning" reactions:

$$^{12}\text{C} + {}^{12}\text{C} \rightarrow {}^{20}\text{Ne} + {}^{4}\text{He}$$

$$^{12}\text{C} + {}^{12}\text{C} \rightarrow {}^{24}\text{Mg} + \gamma$$

(c) Calculate the energy (in kWh) given off when 2 kg of carbon completely fuses according to the first reaction.

Solution

(a) At 6×10^8 K, each carbon nucleus has thermal energy of

$$\tfrac{3}{2}k_B T = (1.5)(8.62 \times 10^{-5}\ \text{eV/K})(6 \times 10^8\ \text{K}) = 7.7 \times 10^4\ \text{eV} \quad \Diamond$$

(b) Energy released $= \left[2m(\text{C}^{12}) - m(\text{Ne}) - m(\text{He}^4)\right]c^2$

$$= (24.000000 - 19.992440 - 4.002603)(931.49)\ \text{MeV} = 4.62\ \text{MeV} \quad \Diamond$$

Energy released $= \left[2m(\text{C}^{12}) - m(\text{Mg}^{24})\right](931.49)\ \text{MeV/u}$

$$= (24.000000 - 23.985042)(931.49)\ \text{MeV} = 13.9\ \text{MeV} \quad \Diamond$$

(c) Energy released = the energy of reaction of the number of carbon nuclei in a 2-kg sample, which corresponds to

$$[(2 \times 10^3\ \text{g} \times 6.02 \times 10^{23}\ \text{atoms/mol}/12\ \text{g/mol})$$

$$(1\ \text{fusion event}/2\ \text{nuclei})(4.62\ \text{MeV})\ 1\ \text{kWh}/(2.25 \times 10^{19}\ \text{MeV})]$$

$$\Delta E = \frac{(1.0 \times 10^{26})(4.62)}{2(2.25 \times 10^{19})}\ \text{kWh} = 10.3 \times 10^6\ \text{kWh} \quad \Diamond$$

73. "Free neutrons" have a characteristic half-life of 12 min. What fraction of a group of free neutrons at thermal energy (0.04 eV) will decay before traveling a distance of 10 km?

Solution

The fraction that will remain is given by the ratio N/N_0, where $N/N_0 = e^{-\lambda t}$ and t is the time it takes the neutron to travel a distance of $d = 10$ km.

Since $K = \frac{1}{2}mv^2$, the time is given by

$$t = \frac{d}{v} = \frac{d}{\sqrt{\frac{2K}{m}}} = \frac{10\times10^3 \text{ m}}{\sqrt{\frac{2(0.04 \text{ eV})(1.6\times10^{-19} \text{ J/eV})}{1.675\times10^{-27} \text{ kg}}}} = 3.61 \text{ s}$$

The decay constant is then,

$$\lambda = \frac{0.693}{T_{1/2}} = \frac{0.693}{720 \text{ s}} = 9.63\times10^{-4} \text{ s}^{-1}$$

Therefore,

$$\lambda t = (9.63\times10^{-4} \text{ s}^{-1})(3.61 \text{ s}) = 3.47\times10^{-3}$$

so

$$\frac{N}{N_0} = e^{-\lambda t} = e^{-3.47\times10^{-3}} = 0.9965$$

Hence, the fraction that has decayed in this time is

$$1 - \frac{N}{N_0} = 0.0035 \quad \text{or} \quad 0.35\% \quad \Diamond$$

75. The decay of an unstable nucleus by alpha emission is represented by Equation 45.10. The disintegration energy Q given by Equation 45.13 must be shared by the alpha particle and the daughter nucleus in order to conserve both energy and momentum in the decay process. (a) Show that Q and K_α, the kinetic energy of the alpha particle, are related by the expression

$$Q = K_\alpha\left(1 + \frac{M_\alpha}{M}\right)$$

where M is the mass of the daughter nucleus. (b) Use the result of (a) to find the energy of the alpha particle emitted in the decay of ^{226}Ra. (See Example 45.8 for the calculation of Q.)

Solution

(a) Let us assume that the parent nucleus (mass M_p) is initially at rest, and let us denote the masses of the daughter nucleus and alpha particle by M_d and M_α, respectively. Applying the equations of conservation of momentum and energy for the alpha decay process gives

$$M_d v_d = M_\alpha v_\alpha \qquad (1)$$

$$M_p c^2 = M_d c^2 + M_\alpha c^2 + \tfrac{1}{2} M_\alpha v_\alpha{}^2 + \tfrac{1}{2} M_d v_d{}^2 \qquad (2)$$

The disintegration energy Q is given by

$$Q = (M_p - M_d - M_\alpha)c^2 = \tfrac{1}{2} M_\alpha v_\alpha{}^2 + \tfrac{1}{2} M_d v_d{}^2 \qquad (3)$$

Eliminating v_d from Equations (1) and (3) gives

$$Q = \tfrac{1}{2} M_d \left(\frac{M_\alpha}{M_d} v_\alpha\right)^2 + \tfrac{1}{2} M_\alpha v_\alpha{}^2 = \frac{1}{2}\frac{M_\alpha{}^2}{M_d} v_\alpha{}^2 + \frac{1}{2} M_\alpha v_\alpha{}^2$$

$$= \tfrac{1}{2} M_\alpha v_\alpha{}^2 \left(1 + \frac{M_\alpha}{M_d}\right) = K_\alpha \left(1 + \frac{M_\alpha}{M_d}\right)$$

(b) $$K_\alpha = \frac{Q}{1 + \frac{M_\alpha}{M_d}} = \frac{4.87 \text{ MeV}}{1 + \frac{4}{226}} = 4.79 \text{ MeV} \quad \Diamond$$

Chapter 46

Fission and Fusion

Chapter 46

FISSION AND FUSION

INTRODUCTION

This chapter is concerned primarily with the two means by which energy can be derived from nuclear reactions: fission, in which a large nucleus splits, or fissions, into two smaller nuclei, and fusion, in which two small nuclei fuse to form a larger one. In either case, there is a release of energy that can then be used either destructively (bombs) or constructively (production of electric power). We also examine the interaction of radiation with matter and several devices used to detect radiation. The chapter concludes with a discussion of some industrial and biological applications of radiation.

NOTES FROM SELECTED CHAPTER SECTIONS

46.1 Interactions Involving Neutrons

The probability that neutrons will be captured as they move through matter *generally increases with decreasing neutron energy.* A *thermal neutron* is one that has an energy of approximately k_BT.

46.2 Nuclear Fission

Nuclear fission occurs when a heavy nucleus, such as ^{235}U, splits, or fissions, into two smaller nuclei. In such a reaction, *the total rest mass of the products is less than the original rest mass.*

The sequence of events in the fission process is

- The ^{235}U nucleus captures a thermal (slow-moving) neutron.
- This capture results in the formation of ^{236}U*, and the excess energy of this nucleus causes it to undergo violent oscillations.
- The ^{236}U* nucleus becomes highly distorted, and the force of repulsion between protons in the two halves of the dumbbell shape tends to increase the distortion.
- The nucleus splits into two fragments, emitting several neutrons in the process.

46.3 Nuclear Reactors

A nuclear reactor is a system designed to maintain a *self-sustained chain reaction.* The *reproduction constant* K is defined as the average number of neutrons released from each fission event that will cause another event. In a power reactor, it is necessary to maintain a value of K close to 1.

46.4 Nuclear Fusion

Nuclear fusion is a process in which two light nuclei combine to form a heavier nucleus. A great deal of energy is released in such a process. The major obstacle in obtaining useful energy from fusion is the large Coulomb repulsive force between the charged nuclei at close separations. Sufficient energy must be supplied to the particles to overcome this Coulomb barrier and thereby enable the nuclear attractive force to take over.

The temperature at which the power generation exceeds the loss rate is called the *critical ignition temperature.* The confinement time is the time the interacting ions are maintained at a temperature equal to or greater than the ignition temperature.

46.6 Radiation Damage in Matter

In biological systems, it is common to separate radiation damage into two categories: *somatic damage* and *genetic damage.* Somatic damage is radiation damage associated with all the body cells except the reproductive cells. Genetic damage affects only the reproductive cells of the person exposed to radiation.

The roentgen (R) is defined as that amount of ionizing radiation that will produce $\frac{1}{3}\times 10^{-9}$ C of electric charge in 1 cm^3 of air under standard conditions.

One *rad* is that amount of radiation that deposits 10^{-2} J of energy into 1 kg of absorbing material.

The *RBE* (relative biological effectiveness) factor for a given type of radiation is defined as *the number of rad of x-radiation or gamma radiation that produces the same biological damage as 1 rad of the radiation being used.*

The *rem* (radiation equivalent in man) is defined as the product of the dose in rad and the RBE factor.

EQUATIONS AND CONCEPTS

The fission of a uranium nucleus by bombardment with a low energy neutron results in the production of fission fragments and typically two or three neutrons. The energy released in the fission event appears in the form of kinetic energy of the fission fragments and the neutrons.

$$ {}_0^1n + {}_{92}^{235}U \rightarrow {}_{92}^{236}U^* \rightarrow X + Y + 3{}_0^1n \quad (46.2) $$

These fusion reactions seem to be most likely to be used as the basis of the design and operation of a fusion power reactor. The stated Q values refer to the amount of energy released per reaction.

$$ {}^{2}_{1}H + {}^{2}_{1}H \rightarrow {}^{3}_{2}He + {}^{1}_{0}n \quad (Q = 3.27\ \text{MeV}) \tag{46.4} $$

$$ {}^{2}_{1}H + {}^{2}_{1}H \rightarrow {}^{3}_{1}H + {}^{1}_{1}H \quad (Q = 4.03\ \text{MeV}) $$

$$ {}^{2}_{1}H + {}^{3}_{1}H \rightarrow {}^{4}_{2}He + {}^{1}_{0}n \quad (Q = 17.59\ \text{MeV}) $$

Lawson's criterion states the conditions under which a net power output of a fusion reactor is possible. In these expressions, n is the plasma density (number of ions per cubic cm) and τ is the plasma confinement time (the time during which the interacting ions are maintained at a temperature equal to or greater than that required for the reaction to proceed).

$$ n\tau \geq 10^{14}\ \text{s/cm}^3 \quad \text{(D-T reaction)} \tag{46.5} $$

$$ n\tau \geq 10^{16}\ \text{s/cm}^3 \quad \text{(D-D reaction)} $$

The radiation dose in rem is the product of the dose in rad and the relative biological effectiveness factor.

$$ 1\ \text{rem} \equiv 1\ \text{rad} \times \text{RBE} \tag{46.6} $$

1 roentgen is the amount of ionizing radiation that deposits 0.0876 J of energy into 1 kg of air.

1 rad is the amount of radiation that deposits 0.01 J of energy into 1 kg of any absorbing material.

RBE is a factor defined as the number of rad of x-radiation or gamma-radiation that produces the same biological damage as 1 rad of the radiation being used.

REVIEW CHECKLIST

▷ Write an equation which represents a typical fission event and describe the sequence of events which occurs during the fission process. Use data obtained from the binding energy curve to estimate the disintegration energy of a typical fission event.

▷ Describe the basic design features and control mechanisms in a fission reactor including the functions of the moderator, control rods, and heat exchange system. Identify some major safety and environmental hazards in the operation of a fission reactor.

▷ Describe the basis of energy release in fusion and write out several nuclear reactions which might be used in a fusion powered reactor.

▷ Define the roentgen, rad, and rem as units of radiation exposure or dose.

▷ Describe the basic principle of operation of the Geiger counter, semiconductor diode detector, scintillation detector, photographic emulsion, cloud chamber, and bubble chamber.

SOLUTIONS TO SELECTED END-OF-CHAPTER PROBLEMS

3. List the nuclear reactions required to produce ^{233}U from ^{232}Th under fast neutron bombardment.

Solution First, ${}^{1}_{0}\text{n} + {}^{232}_{90}\text{Th} \rightarrow {}^{233}_{90}\text{Th}$

With an extra neutron, this isotope is unstable against beta decay:

$${}^{233}_{90}\text{Th} \rightarrow {}^{233}_{91}\text{Pa} + \beta^- + \overline{\nu}$$

Protactinium-233 has more neutrons than the more stable Protactinium-231, so it too decays by beta emission:

$${}^{233}_{91}\text{Pa} \rightarrow {}^{233}_{92}\text{U} + \beta^- + \overline{\nu}$$

7. Suppose enriched uranium containing 3.4 percent of the fissionable isotope ${}^{235}_{92}U$ is used as fuel for a ship. The water exerts an average frictional drag of 1.0×10^5 N on the ship. How far can the ship travel per kilogram of fuel? Assume that the energy released per fission event is 208 MeV and that the ship's engine has an efficiency of 20 percent.

Solution

One kilogram of enriched uranium contains this much of the uranium-235:

$$m_{235} = (1000\text{ g})(0.034) = 34\text{ g}$$

In terms of number of nuclei, this is equivalent to

$$N_{235} = 34\text{ g}\left(\frac{1\text{ mol}}{235\text{ g}}\right)\left(\frac{6.02\times 10^{23}\text{ atoms}}{\text{mol}}\right) = 8.71\times 10^{22}\text{ nuclei}$$

If it all fissions, it puts out thermal energy:

$$\left(8.71\times 10^{22}\text{ nuclei}\right)\left(\frac{208\text{ MeV}}{\text{fission}}\right)\left(\frac{1.6\times 10^{\ 19}\text{ J}}{1\text{ eV}}\right) = 2.90\times 10^{12}\text{ J}$$

Now, for the engine:

$$\text{efficiency} = \frac{\text{work output}}{\text{heat input}}$$

$$e = \frac{Fs\cos\theta}{Q_h}$$

$$s = \frac{e\,Q_h}{F\cos 0°} = \frac{0.20\left[2.90\times 10^{12}\text{ J}\right]}{1.0\times 10^{5}\text{ N}} = 5.8\times 10^{6}\text{ m} \quad \Diamond$$

9. It has been estimated that there are 10^9 tons of natural uranium available at concentrations exceeding 100 parts per million, of which 0.7 percent is ^{235}U. If all the world's energy needs (7×10^{12} J/s) were to be supplied by ^{235}U fission, how long would this supply last? (This estimate of uranium supply was taken from K. S. Deffeyes and I. D. MacGregor, *Scientific American*, January 1980, p. 66.)

Solution The mass of natural uranium reserves is $m = 10^9 \times 10^3 \text{ kg} = 1 \times 10^{12} \text{ kg}$

The reserves of fissionable uranium are $m_{235} = \left(10^{12} \text{ kg}\right)(0.0070) = 7.0 \times 10^9 \text{ kg}$

$$N_{235} = 7 \times 10^{12} \text{ g}\left(\frac{6.02 \times 10^{23} \text{ atoms}}{235 \text{ g}}\right) = 1.79 \times 10^{34} \text{ nuclei}$$

Following Example 46.2, we take the fission energy as 208 MeV/fission:

$$E = 1.79 \times 10^{34} \text{ nuclei}\left(\frac{208 \text{ MeV}}{1 \text{ fission}}\right)\left(\frac{1.6 \times 10^{-19} \text{ J}}{1 \text{ eV}}\right) = 5.97 \times 10^{23} \text{ J}$$

Now $P = E/t$: $$t = \frac{E}{P} = \frac{5.97 \times 10^{23} \text{ J}}{7 \times 10^{12} \text{ J/s}} = 8.53 \times 10^{10} \text{ s} = 2700 \text{ y} \quad \Diamond$$

---

13. To understand why plasma containment is necessary, consider the rate at which an uncontained plasma would be lost. (a) Estimate the rms speed of deuterons in a plasma at 4×10^8 K. (b) Estimate the time such a plasma would remain in a 10-cm cube if no steps were taken to contain it.

Solution The average kinetic energy per particle $\left(\frac{1}{2}m\overline{v^2}\right)$ must equal the thermal energy $\frac{3}{2}k_BT$.

That is, $\frac{1}{2}m\overline{v^2} = \frac{3}{2}k_BT$ or $v_{\text{rms}} = \sqrt{\frac{3k_BT}{m}}$

(a) $$v_{\text{rms}} = \sqrt{\frac{3(1.38 \times 10^{-23} \text{ J/K})(4 \times 10^8 \text{ K})}{2(1.67 \times 10^{-27} \text{ kg})}} = 2 \times 10^6 \text{ m/s} \quad \Diamond$$

(b) $$t = \frac{x}{v} = \frac{0.10 \text{ m}}{2 \times 10^6 \text{ m/s}} = 5 \times 10^{-8} \text{ s} \quad \Diamond$$

17. A building has become accidentally contaminated with radioactivity. The longest-lived material in the building is strontium-90 ($^{90}_{38}Sr$ has an atomic mass 89.907 7, and its half-life is 28.8 years). If the building initially contained 1.0 kg of this substance uniformly distributed throughout the building (a very unlikely situation) and the safe level is less than 10.0 counts/min, how long will the building be unsafe?

Solution The number of nuclei in the original sample is

$$N_0 = \frac{\text{mass present}}{\text{mass of nucleus}} = \frac{1.0 \text{ kg}}{89.9077 \text{ u } (1.66 \times 10^{-27} \text{ kg/u})}$$

$$N_0 = 6.70 \times 10^{24} \text{ nuclei}$$

$$\lambda = \frac{0.693}{T_{1/2}} = \frac{0.693}{28.8 \text{ y}} = 2.4063 \times 10^{-2} \text{ y}^{-1} = 4.575 \times 10^{-8} \text{ min}^{-1}$$

$$R_0 = \lambda N_0 = (4.575 \times 10^{-8} \text{ min}^{-1})(6.70 \times 10^{24})$$

$$R_0 = 3.07 \times 10^{17} \text{ counts/min}$$

$$\frac{R}{R_0} = e^{-\lambda t} = \frac{10 \text{ counts/min}}{3.07 \times 10^{17} \text{ counts/min}} = 3.26 \times 10^{-17}$$

and $\lambda t = -\ln(3.26 \times 10^{-17}) = 37.96$ giving $t = 1600$ y ◊

---

21. A "clever" technician decides to heat some water for his coffee with an x-ray machine. If the machine produces 10 rad/s, how long will it take to raise the temperature of a cup of water by 50°C?

Solution 1 rad = 10^{-2} J/kg $\qquad Q = mc\,\Delta T \qquad P \cdot t = mc\,\Delta T$

$$t = \frac{mc\,\Delta T}{P} = \frac{m(4186 \text{ J / kg} \cdot \text{C}°)(50°\text{C})}{(10 \text{ rad / s})(10^{-2} \text{ J / kg} \cdot \text{rad})(m)} = 2.09 \times 10^{6} \text{ s} \cong 24 \text{ days}$$ ◊

(Note that the power P is the product of dose rate and mass.)

---

25. In a Geiger tube, the voltage between the electrodes is typically 1.0 kV and the current pulse discharges a 5.0-pF capacitor. (a) What is the energy amplification of this device for a 0.50-MeV electron? (b) How many electrons are avalanched by the initial electron?

Solution

(a) $$\frac{E}{E_0} = \frac{\frac{1}{2}CV^2}{0.5 \text{ MeV}} = \frac{(0.5)(5\times10^{-12} \text{ F})(1\times10^3 \text{ V})^2}{(0.5 \text{ MeV})(1.6\times10^{-13} \text{ J/MeV})} = 3.1\times10^7 \quad \Diamond$$

(b) $$N = \frac{Q}{e} = \frac{CV}{e} = \frac{(5\times10^{-12} \text{ F})(1\times10^3 \text{ V})}{1.6\times10^{-19} \text{ C}} = 3.1\times10^{10} \text{ electrons} \quad \Diamond$$

---

29. The half-life of tritium is 12 years. If the TFTR fusion reactor contains 50 m^3 of tritium at a density equal to 2.0×10^{14} particles/cm^3, how many curies of tritium are in the plasma? Compare this value with a fission inventory (the estimated supply of fissionable material) of 4×10^{10} Ci.

Solution The number of hydrogen-3 nuclei is

$$N = 50 \text{ m}^3 (2\times10^{14} /\text{cm}^3)(10^2 \text{ cm/m})^3 = 10^{22}$$

$$\lambda = \frac{\ln 2}{T_{1/2}} = \frac{0.693}{12 \text{ y}}\left(\frac{1 \text{ y}}{3.16\times10^7 \text{ s}}\right) = 1.83\times10^{-9} /\text{s}$$

The activity is

$$R = \lambda N = (1.83\times10^{-9} / \text{s})10^{22} = 1.83\times10^{13} \text{ Bq} = 1.83\times10^{13} \text{ Bq}\left(\frac{1 \text{ Ci}}{3.7\times10^{10} \text{ Bq}}\right) = 490 \text{ Ci} \quad \Diamond$$

This is smaller than 4×10^{10} Ci by an order-of-magnitude of a hundred million.

---

37. Assuming that a deuteron and a triton are at rest when they fuse according to

$$^{2}\mathrm{H} + {}^{3}\mathrm{H} \rightarrow {}^{4}\mathrm{He} + \mathrm{n} + 17.6\ \mathrm{MeV},$$

determine the kinetic energy acquired by the neutron.

Solution

The vector momentum of the alpha particle and that of the neutron must add to zero:

$$(1.0087\ \mathrm{u})v_n = (4.0026\ \mathrm{u})v_\alpha$$

At the same time, their kinetic energies must add to Q:

$$E = \tfrac{1}{2}(1.0087\ \mathrm{u})\ {v_n}^2 + \tfrac{1}{2}(4.0026\ \mathrm{u})\ {v_\alpha}^2 = 17.6\ \mathrm{MeV}$$

Substitute $v_\alpha = 0.2520 v_n$ to obtain

$$(0.50435\ \mathrm{u}){v_n}^2 + (0.12710\ \mathrm{u}){v_n}^2 = 17.6\ \mathrm{MeV}\left(\frac{1\ \mathrm{u}\ c^2}{931.5\ \mathrm{MeV}}\right)$$

$$v_n = \sqrt{\frac{0.0189c^2}{0.63145}} = 0.173c = 51.9 \times 10^6\ \mathrm{m/s}$$

Then the kinetic energy of the neutron is

$$\tfrac{1}{2}(1.0087\ \mathrm{u})(0.173c)^2\left(\frac{951.5\ \mathrm{MeV}}{1\ \mathrm{u}\ c^2}\right) = 14.1\ \mathrm{MeV} \quad \lozenge$$

39. (a) Calculate the energy (in kilowatt hours) released if 1.00 kg of ^{239}Pu undergoes complete fission and the energy released per fission event is 200 MeV. (b) Calculate the energy (in electron volts) released in the D-T fusion:

$$^{2}_{1}H + ^{3}_{1}H \rightarrow ^{4}_{2}He + ^{1}_{0}n$$

(c) Calculate the energy (in kilowatt hours) released if 1.00 kg of deuterium undergoes fusion. (d) Calculate the energy (in kilowatt hours) released by the combustion of 1.00 kg of coal if each $C + O_2 \rightarrow CO_2$ reaction yields 4.2 eV. (e) List the advantages and disadvantages of each of these methods of energy generation.

Solution

(a) $1000\text{ g}\left(\frac{6.02\times10^{23}\text{ nuclei}}{239\text{ g}}\right)\left(\frac{200\times10^{6}\text{ eV}}{1\text{ nucleus}}\right)\left(\frac{1.6\times10^{-19}\text{ J}}{1\text{ eV}}\right)\left(\frac{1\text{ kWh}}{3.6\times10^{6}\text{ J}}\right) = 2.24\times10^{7}\text{ kWh}$ ◊

(b) From Appendix A.3,

$$\text{mass before} = 2.014102\text{ u} + 3.016049\text{ u} = 5.030151\text{ u}$$

$$\text{mass after} = 4.002603\text{ u} + 1.008665\text{ u} = 5.011268\text{ u}$$

$$\Delta m = 0.018883\text{ u}$$

$$E = \Delta mc^2 = 0.018883\text{ u}\left(\frac{931.5\text{ MeV}}{\text{u}}\right) = 17.59\text{ MeV} \quad ◊$$

(c) $1000\text{ g deuterium}\left(\frac{6.02\times10^{23}\text{ deuterons}}{2.014\text{ g}}\right)\left(\frac{17.59\times1.6\times10^{-13}\text{ J}}{\text{deuteron fusion}}\right)\left(\frac{1\text{ kWh}}{3.6\times10^{6}\text{ J}}\right)$

$= 2.34\times10^{8}\text{ kWh}$ ◊

(d) Coal is essentially pure carbon:

$$(1000\text{ g C})\left(\frac{6.02\times10^{23}\text{ C atoms}}{12\text{ g}}\right)\left(\frac{4.2\times1.6\times10^{-19}\text{ J}}{\text{reaction}}\right)\left(\frac{1\text{ kWh}}{3.6\times10^{6}\text{ J}}\right) = 9.36\text{ kWh} \quad ◊$$

(e) You pay the electric company to burn coal. Coal is cheap at this moment in human history. We hope that fusion reactions can be controlled to make energy much cheaper.

43. Consider the two nuclear reactions

(I) A + B → C + E and (II) C + D → F + G

(a) Show that the net disintegration energy for these two reactions ($Q_{net} = Q_I + Q_{II}$) is identical to the disintegration energy for the reaction A + B + D → E + F + G.

(b) One chain of reactions in the proton-proton cycle in the Sun's interior is

$$ {}_1^1H + {}_1^1H \to {}_1^2H + \beta^+ + \nu \qquad {}_1^1H + {}_1^2H \to {}_2^3H + \gamma \qquad {}_1^1H + {}_2^3He \to {}_2^4He + \beta^+ + \nu $$

Based on part (a), what is Q_{net} for this sequence?

Solution $Q_I = m_A + m_B - (m_C + m_E)$

$$Q_{II} = m_C + m_D - (m_F + m_G)$$

$$Q_{net} = (m_A + m_B + m_D) - (m_E + m_F + m_G)$$

(a) This is identical to the Q for the reaction A + B + D → E + F + G. Thus, any product (e.g., "C") that is a reactant in a subsequent reaction disappears from the energy balance. ◊

(b) Eliminating ${}_1^2H$ and ${}_2^3He$ because each is used up in a subsequent reaction, the net process is

$$(4){}_1^1H \to {}_2^4He + (2){}_1^0e + 2\nu + \gamma$$

The Q for any reaction is defined as the change in the *rest* masses of the reactants and products. Since neutrinos and gamma rays both have zero rest mass,

$$Q = 4\left({}_1^1H\right) - \left({}_2^4He + 2{}_1^0e\right)$$

$$Q = [4(1.007825\text{ u}) - 4.002603\text{ u}]\left(931.5\ \frac{\text{MeV}}{\text{u}}\right) - 2(0.511\text{ MeV})$$

$$Q = 25.7\text{ MeV} \quad ◊$$

44. The carbon cycle, first proposed by Hans Bethe in 1939, is another cycle by which energy is released in stars and hydrogen is converted to helium. The carbon cycle requires higher temperatures than the proton-proton cycle. The series of reactions is

$$^{12}\text{C} + {}^{1}\text{H} \rightarrow {}^{13}\text{N} + \gamma$$

$$^{12}\text{N} \rightarrow {}^{13}\text{C} + \beta^{+} + \nu$$

$$^{13}\text{C} + {}^{1}\text{H} \rightarrow {}^{14}\text{N} + \gamma$$

$$^{14}\text{N} + {}^{1}\text{H} \rightarrow {}^{15}\text{O} + \gamma$$

$$^{15}\text{O} \rightarrow {}^{15}\text{N} + \beta^{+} + \nu$$

$$^{15}\text{N} + {}^{1}\text{H} \rightarrow {}^{12}\text{C} + {}^{4}\text{He}$$

(a) If the proton-proton cycle requires a temperature of 1.5×10^{7} K, estimate the temperature required for the first step in the carbon cycle. (b) Calculate the Q value for each step in the carbon cycle and the overall energy released. (c) Do you think the energy carried off by the neutrinos is deposited in the star? Explain.

Solution

(a) Roughly, $\frac{7}{2}(15 \times 10^{6})$ K or 52×10^{6} K ◊ since the highest coulombic barrier that must be surmounted looks like $7e(k_e e)/r$ instead of $2e(k_e e)/r$.

(b) For the first step,

$$Q = \Delta mc^{2} = (12.00000 + 1.007825 - 13.005738 \text{ u})(931.5 \text{ MeV / u}) = 1.943 \text{ MeV}$$

Overall, $Q = 1.943 + 1.709 + 7.551 + 7.297 + 2.242 + 4.966 = 25.75$ MeV ◊

The net effect is $^{12}_{6}\text{C} + 4\text{p} \rightarrow {}^{12}_{6}\text{C} + {}^{4}_{2}\text{He}$

(c) Most of the energy is lost since neutrinos have such low cross section (no charge, little mass, etc.).

Chapter 47

Particle Physics and Cosmology

Chapter 47

PARTICLE PHYSICS AND COSMOLOGY

INTRODUCTION

In this concluding chapter, we examine the properties and classifications of the various known subatomic particles and the fundamental interactions that govern their behavior. We also discuss the current theory of elementary particles, in which all matter is believed to be constructed from only two families of particles, quarks and leptons. Finally, we discuss how clarifications of such models might help scientists understand cosmology, which deals with the evolution of the Universe.

NOTES FROM SELECTED CHAPTER SECTIONS

47.1 The Fundamental Forces in Nature

There are four fundamental forces in nature: *strong* (hadronic), *electromagnetic*, *weak*, and *gravitational*. The strong force is the force between nucleons that keeps the nucleus together. The weak force is responsible for beta decay. The electromagnetic and weak forces are now considered to be manifestations of a single force called the *electroweak* force.

The fundamental forces are described in terms of particle or quanta exchanges which *mediate* the forces. The electromagnetic force is mediated by photons, which are the quanta of the electromagnetic field. Likewise, the strong force is mediated by field particles called *gluons*, the weak force is mediated by particles called the W and Z *bosons*, and the gravitational force is mediated by quanta of the gravitational field called *gravitons*.

47.2 Positrons and Other Antiparticles

An antiparticle and a particle have the same mass, but opposite charge. Furthermore, other properties may have opposite values such as lepton number and baryon number. It is possible to produce particle-antiparticle pairs in nuclear reactions if the available energy is greater than $2mc^2$, where m is the rest mass of the particle.

Pair production is a process in which a gamma ray with an energy of at least 1.02 MeV interacts with a nucleus, and an electron-positron pair is created.

Pair annihilation is an event in which an electron and a positron can annihilate to produce two gamma rays, each with an energy of at least 0.511 MeV.

47.3 Mesons and the Beginning of Particle Physics

The interaction between two particles can be represented in a diagram called a *Feynman diagram.*

47.4 Classification of Particles

All particles (other than photons and gravitons) can be classified into two categories: *hadrons* and *leptons*.

There are two classes of hadrons: *mesons* and *baryons* which are grouped according to their masses and spins. It is believed that hadrons are composed of units called *quarks* which are more fundamental in nature.

Leptons have no structure or size and are therefore considered to be truly elementary particles.

47.5 Conservation Laws

In all reactions and decays, quantities such as energy, linear momentum, angular momentum, electric charge, baryon number, and lepton number are strictly conserved. Certain particles have properties called *strangeness* and *charm*. These unusual properties are conserved only in those reactions and decays that occur via the strong force.

Whenever a nuclear reaction or decay occurs, the sum of the baryon numbers before the process must equal the sum of the baryon numbers after the process.

The sum of the electron-lepton numbers before a reaction or decay must equal the sum of the electron-lepton numbers after the reaction or decay.

47.6 Strange Particles and Strangeness

Whenever the strong force or the electromagnetic force causes a reaction or decay to occur, the sum of the strangeness numbers before the process must equal the sum of the strangeness numbers after the process.

47.8 Quarks — Finally

Recent theories in elementary particle physics have postulated that all hadrons are composed of smaller units known as *quarks*. Quarks have a fractional electric charge and a baryon number of 1/3. There are six flavors of quarks, up (u), down (d), strange (s), charmed (c), top (t), and bottom (b). All baryons contain three quarks, while all mesons contain one quark and one antiquark.

According to the theory of *quantum chromodynamics*, quarks have a property called *color*, and the strong force between quarks is referred to as the *color force.*

EQUATIONS AND CONCEPTS

Pions and muons are very unstable particles. A decay sequence is shown in Equation 47.1.

$$\pi^- \to \mu^- + \bar{\nu} \qquad (47.1)$$

$$\mu^- \to e^- + \nu + \bar{\nu}$$

Hubble's law states a linear relationship between the velocity of a galaxy and its distance R from the Earth. The constant H is called the *Hubble parameter.*

$$v = HR \qquad (47.4)$$

$$H = 17 \times 10^{-3} \text{ m/(s·lightyear)}$$

The *critical mass density* of the Universe can be estimated based on energy considerations.

$$\rho_c = \frac{3H^2}{8\pi G}$$

Chapter 47

REVIEW CHECKLIST

▷ Be aware of the four fundamental forces in nature and the corresponding field particles or quanta via which these forces are mediated.

▷ Understand the concepts of the antiparticle, pair production, and pair annihilation.

▷ Know the broad classification of particles and the characteristic properties of the several classes (relative mass value, spin, decay mode).

▷ Determine whether or not a suggested decay can occur based on the conservation of baryon number and the conservation of lepton number. Determine whether or not a predicted reaction/decay will occur based on the conservation of strangeness for the strong and electromagnetic interactions.

SOLUTIONS TO SELECTED END-OF-CHAPTER PROBLEMS

3. One of the mediators of the weak interaction is the Z^0 boson, mass 96 GeV/c^2. Use this information to find an approximate value for the range of the weak interaction.

Solution The rest energy of the Z^0 boson is E_0 = 96 GeV. The maximum time a virtual Z^0 boson can exist is found from $\Delta E \Delta t \geq \hbar/2$, or

$$\Delta t \approx \frac{\hbar}{2\Delta E} = \frac{1.055\times10^{-34}\ \text{J}\cdot\text{s}}{2(96\ \text{GeV})(1.6\times10^{-10}\ \text{J/GeV})} = 3.44\times10^{-27}\ \text{s}$$

The maximum distance it can travel in this time is

$$d = c(\Delta t) = (3\times10^{8}\ \text{m/s})(3.44\times10^{-27}\ \text{s}) = 1.1\times10^{-18}\ \text{m} \quad \Diamond$$

The distance d is an approximate value for the range of the weak interaction.

6. A neutral pion at rest decays into two photons according to

$$\pi^0 \rightarrow \gamma + \gamma$$

Find the energy, momentum, and frequency of each photon.

Solution Since the pion is at rest and momentum is conserved, the two gamma-rays must have equal momenta in opposite directions. So, they must share equally in the energy of the pion.

$$m_{\pi}{}^0 = 135 \text{ MeV}/c^2 \qquad \text{(Table 47.2)}$$

Therefore,

$$E_\gamma = 67.5 \text{ MeV} \quad \Diamond$$

$$p = \frac{E}{c} = 67.5 \text{ MeV}/c \quad \Diamond$$

$$f = \frac{E}{h} = 1.63 \times 10^{22} \text{ Hz} \quad \Diamond$$

---

7. Name one possible decay mode (see Table 47.2) for Ω^+, $\overline{\text{K}^0}$, $\overline{\Lambda^0}$, and $\overline{\text{n}}$.

Solution These decay into the antiparticles of the listed particles into which their antiparticles decay:

$$\Omega^+ \rightarrow \overline{\Lambda^0} + \text{K}^+$$

$$\overline{\text{K}^0} \rightarrow \pi^+ + \pi^- \quad \text{or} \quad \pi^0 + \pi^0$$

$$\overline{\Lambda^0} \rightarrow \overline{\text{p}} + \pi^+$$

$$\overline{\text{n}} \rightarrow \overline{\text{p}} + \text{e}^+ + \nu_e$$

---

10. The following reactions or decays involve one or more neutrinos. Supply the missing neutrinos (ν_e, ν_μ, or ν_τ).

(a) $\pi^- \to \mu^- + ?$

(b) $K^+ \to \mu^+ + ?$

(c) $? + p^+ \to n + e^+$

(d) $? + n \to p^+ + e^-$

(e) $? + n \to p^+ + \mu^-$

(f) $\mu^- \to e^- + ? + ?$

Solution

(a) $\pi^- \to \mu^- + \bar{\nu}_\mu$ $\qquad L_\mu: \ 0 \to 1 - 1$

(b) $K^+ \to \mu^+ + \nu_\mu$ $\qquad L_\mu: \ 0 \to -1 + 1$

(c) $\bar{\nu}_e + p^+ \to n + e^+$ $\qquad L_e: \ -1 + 0 \to 0 - 1$

(d) $\nu_e + n \to p^+ + e^-$ $\qquad L_e: \ 1 + 0 \to 0 + 1$

(e) $\nu_\mu + n \to p^+ + \mu^-$ $\qquad L_\mu: \ 1 + 0 \to 0 + 1$

(f) $\mu^- \to e^- + \bar{\nu}_e + \nu_\mu$ $\qquad L_\mu: \ 1 \to 0 + 0 + 1$ and $L_e: \ 0 \to 1 - 1 + 0$

---

12. Determine which of the reactions below can occur. For those that cannot occur, determine the conservation law (or laws) violated.

(a) $p^+ \to \pi^+ + \pi^0$

(b) $p^+ + p^+ \to p^+ + p^+ + \pi^0$

(c) $p^+ + p^+ \to p^+ + \pi^+$

(d) $\pi^+ \to \mu^+ + \nu_\mu$

(e) $n^0 \to p^+ + e^- + \bar{\nu}_e$

(f) $\pi^+ \to \mu^+ + n$

Solution

(a) $p^+ \to \pi^+ + \pi^0$ — Baryon number is violated: $1 \to 0 + 0$

(b) $p^+ + p^+ \to p^+ + p^+ + \pi^0$ — This reaction can occur.

(c) $p^+ + p^+ \to p^+ + \pi^+$ — Baryon number is violated: $1 + 1 \to 1 + 0$

(d) $\pi^+ \to \mu^+ + \nu_\mu$ — This reaction can occur.

(e) $n^0 \to p^+ + e^- + \bar{\nu}_e$ — This reaction can occur.

(f) $\pi^+ \to \mu^+ + n$ — Violates baryon number: $0 \to 0 + 1$ and violates muon-lepton number: $0 \to -1 + 0$

---

13. Determine whether or not strangeness is conserved in the following decays and reactions:

(a) $\Lambda^0 \rightarrow p^+ + \pi^-$

(b) $\pi^- + p^+ \rightarrow \Lambda^0 + K^0$

(c) $p^- + p^+ \rightarrow \overline{\Lambda^0} + \Lambda^0$

(d) $\pi^- + p^+ \rightarrow \pi^- + \Sigma^+$

(e) $\Xi^- \rightarrow \Lambda^0 + \pi^-$

(f) $\Xi^- \rightarrow p^+ + \pi^-$

Solution

We look up the strangeness quantum numbers in Table 47.2.

(a) $\Lambda^0 \rightarrow p^+ + \pi^-$ Strangeness: $-1 \rightarrow 0 + 0$
(strangeness is not conserved)

(b) $\pi^- + p^+ \rightarrow \Lambda^0 + K^0$ Strangeness: $0 + 0 \rightarrow -1 + 1$
(0 = 0 and strangeness is conserved)

(c) $p^- + p^+ \rightarrow \overline{\Lambda^0} + \Lambda^0$ Strangeness: $0 + 0 \rightarrow +1 - 1$
(0 = 0 and strangeness is conserved)

(d) $\pi^- + p^+ \rightarrow \pi^- + \Sigma^+$ Strangeness: $0 + 0 \rightarrow 0 - 1$
(0 does not equal –1 so strangeness is not conserved)

(e) $\Xi^- \rightarrow \Lambda^0 + \pi^-$ Strangeness: $-2 \rightarrow -1 + 0$
(–2 does not equal –1 so strangeness is not conserved)

(f) $\Xi^- \rightarrow p^+ + \pi^-$ Strangeness: $-2 \rightarrow 0 + 0$
(–2 does not equal 0 so strangeness is not conserved)

19. Analyze each reaction in terms of constituent quarks:

(a) $\pi^- + p^+ \rightarrow K^0 + \Lambda^0$

(b) $\pi^+ + p^+ \rightarrow K^+ + \Sigma^+$

(c) $K^- + p^+ \rightarrow K^+ + K^0 + \Omega^-$

(d) $p^+ + p^+ \rightarrow K^0 + p^+ + \pi^+ + ?$

In the last reaction, identify the mystery particle.

Solution We look up the quark constituents of the particles in Table 47.4.

(a) $d\bar{u}$ + uud $\rightarrow$ $d\bar{s}$ + uds

(b) $\bar{d}u$ + uud $\rightarrow$ $u\bar{s}$ + uus

(c) $\bar{u}s$ + uud $\rightarrow$ $u\bar{s}$ + $d\bar{s}$ + sss

(d) uud + uud $\rightarrow$ $d\bar{s}$ + uud + $u\bar{d}$ + uds   A uds is a Λ^0

---

21. A distant quasar is moving away from Earth at such high speed that the blue 434-nm hydrogen line is observed at 650 nm, in the red portion of the spectrum. (a) How fast is the quasar receding? (See the hint in Problem 20.) (b) Using Hubble's law, determine the distance from Earth to this quasar.

Solution

(a) $$\frac{\lambda'}{\lambda} = \frac{650 \text{ nm}}{434 \text{ nm}} = 1.498 = \sqrt{\frac{1+\frac{v}{c}}{1-\frac{v}{c}}} \qquad \frac{1+\frac{v}{c}}{1-\frac{v}{c}} = 2.243$$

$v = 0.383c$ or 38.3% the speed of light ◊

(b) Using Equation 47.4, $v = HR$, we find

$$R = \frac{v}{H} = \frac{(0.383)(3.0\times 10^8 \text{ m/s})}{1.7\times 10^{-2} \text{ m/s}\cdot\text{lightyear}} = 6.7\times 10^9 \text{ lightyear} \quad ◊$$

---

24. What are the kinetic energies of the proton and pion resulting from the decay of a Λ^0 at rest:

$$\Lambda^0 \rightarrow p^+ + \pi^-$$

Solution

Given that $\Lambda^0 \rightarrow p + \pi^-$, we look up the energy of each particle:

$$m_\Lambda c^2 = 1115.6 \text{ MeV}$$

$$m_p c^2 = 938.3 \text{ MeV}$$

$$m_\pi c^2 = 139.6 \text{ MeV}$$

The difference between starting mass-energy and final mass-energy is the kinetic energy of the products:

$$(K_p + K_\pi) = (1115.6 - 938.3 - 139.6) \text{ MeV} = 37.7 \text{ MeV}$$

In addition, since momentum is conserved,

$$|p_p| = |p_\pi| = p$$

Applying conservation of relativistic energy:

$$\left(\sqrt{(938.3)^2 + p^2c^2} - 938.3\right) + \sqrt{(139.6)^2 + p^2c^2} - 139.6 = 37.7 \text{ MeV}$$

Solving the algebra yields $p_\pi c = p_p c = 100.4$ MeV.

Thus $\qquad K_p = \sqrt{(m_p c^2)^2 + (100.4)^2} - m_p c^2 = 5.4 \text{ MeV}$ ◊

and $\qquad K_\pi = \sqrt{(139.6)^2 + (100.4)^2} - 139.6 = 32.3 \text{ MeV}$ ◊

25. The energy flux of neutrinos from the Sun is estimated to be on the order of 0.4 W/m^2 at Earth's surface. Estimate the fractional mass loss of the Sun over 10^9 years due to the radiation of neutrinos. (The mass of the Sun is 2×10^{30} kg. The Earth-Sun distance is 1.5×10^{11} m.)

Solution Since the neutrino flux from the Sun reaching the Earth is 0.4 W/m^2, the total energy emitted per second by the Sun in neutrinos is

$$(0.4\ \text{W/m}^2)(4\pi r^2) = (0.4\ \text{W/m}^2)(4\pi)(1.5\times10^{11}\ \text{m})^2 = 1.13\times10^{23}\ \text{W}$$

In a period of 10^9 y, the Sun emits a total energy of

$$(1.13\times10^{23}\ \text{J/s})(10^9\ \text{y})(3.156\times10^7\ \text{s/y}) = 3.56\times10^{39}\ \text{J}$$

in the form of neutrinos. This energy corresponds to an annihilated mass of

$$mc^2 = 3.56\times10^{39}\ \text{J} \qquad m = 3.96\times10^{22}\ \text{kg}$$

Since the Sun has a mass of about 2×10^{30} kg, this corresponds to a loss of only about 1 part in 50,000 000 of the Sun's mass over 10^9 y in the form of neutrinos. ◊

---

27. If a K^0 meson at rest decays in 0.90×10^{-10} s, how far will a K^0 meson travel if it is moving at 0.96*c* through a bubble chamber?

Solution The K^0 is relativistic, with time-dilated lifetime

$$T = \gamma T_0 = \frac{0.9\times10^{-10}\ \text{s}}{\sqrt{1-v^2/c^2}} = \frac{0.9\times10^{-10}\ \text{s}}{\sqrt{1-(0.96)^2}} = 3.214\times10^{-10}\ \text{s}$$

$$\text{distance} = (0.96)(3\times10^8\ \text{m/s})(3.214\times10^{-10}\ \text{s}) = 9.3\ \text{cm} \quad ◊$$